D0983573

FORM 125 M
Cop. 1
SCIENCE &
TECHNOLOGY DEPARTMENT
The Chicago Public Library
Received Oct. 14, 1970

Chicago Public Library
C
P
L
REFERENCE
Form 178 rev. 11-00

Structures Technology
for Large Radio and Radar
Telescope Systems

Structures Technology for Large Radio and Radar Telescope Systems

James W. Mar and Harold Liebowitz, Editors

INTERNATIONAL SYMPOSIUM ON STRUCTURES TECHNOLOGY FOR LARGE RADIO AND RADAR TELESCOPE SYSTEMS.

THE MIT PRESS
Massachusetts Institute of Technology
Cambridge, Massachusetts, and London, England

Set in IBM Selectric Press Roman. Printed and bound in the United States of America by The Maple Press Company.

SBN 262 13046 7

Library of Congress catalog card number: 68-25377

PREFACE

The Office of Naval Research and the Massachusetts Institute of Technology co-sponsored an International Symposium on Structures Technology for Large Radio and Radar Telescope Systems held at M.I.T. from 18 to 20 October 1967. Topics which were covered included stress analysis of complex structures, radome design, passive and active control of antenna shape, structural features of existing large radio telescopes, design studies of proposed telescopes, as well as background papers on radio and radar astronomy. The primary emphasis was on the problem areas associated with very large radio and radar astronomy systems, and with the tools used for their structural design.

Sessions were held in the Little Theatre of Kresge Auditorium in the morning and afternoon. On Friday afternoon there was a trip to the Lincoln Laboratory Haystack Facility located in Westford, Massachusetts, wherein is located the 120-foot diameter, steerable antenna inside of a 150-foot diameter radome.

There was a banquet at the M.I.T. Faculty Club on Thursday evening. After dinner there was a panel discussion of the future needs and requirements of radio and radar astronomy.

It was evident from the papers that a very large radio telescope is feasible and furthermore that the structures' community sees in the design and construction of such a system many extremely interesting and challenging problems. The requirement that the parabolic shape of the antenna remain within very close tolerances, the trade-offs involved in a radome enclosure for the antenna, the desire not to be limited operationally by moderate winds and temperature excursions and, above all, the need to minimize costs are system parameters which will require engineering of the highest order. It became evident, also, that the design will require close collaboration between the hardware designers and the astronomers who will be the users.

The complete program follows:

Wednesday morning session: October 18, 1967, 9:30 a.m.

Session Chairman: E. M. Purcell, Harvard University

Jerome B. Wiesner: Welcome

E. G. Bowen, "The Design and Application of Large Steerable Telescopes"

A. B. Youmans, "Design of a 300-foot Research Antenna"

O. Hachenberg, "Design of the Bonn University 100-meter Telescope"

H. G. Weiss, "Design Studies for a 440-foot-diameter Radio/Radar Telescope"

Wednesday afternoon session: October 18, 1967, 1:45 p.m.

Session Chairman: J. W. Mar, M.I.T.

C. Scruton, "Some Considerations of Wind Effects on Large Structures"

M. S. Katow, "Techniques Used to Evaluate the Performance of the NASA/JPL 210-foot Reflector Structure Under Environmental Loads"

T. G. Butler, "Analytical Determination of the Structural Transfer Functions of the ROSMAN I Tracking Antenna"

M. H. Jeffery, "Construction and Performance of the 150-foot NRC Antenna at Algonquin Radio Observatory, Ontario, Canada"

I. K. Shah, H. Simpson, and H. D. Smith, "On the Prediction of Antenna Deformations"

Thursday morning session: October 19, 1967, 9:00 a.m.

Session Chairman: W. H. Gayman, Jet Propulsion Laboratory

H. C. Minnett, D. E. Yabsley, and M. J. Puttock, "Structural Performance of the Parkes 210-foot Paraboloid"

H. A. Cress and S. G. Talbert, "Determination of Approximate Structural Parameters of a Large Steerable Antenna"

W. Weaver, Jr., "Computer-Aided Design of a Large Steerable Antenna Structure"

H. Rothman and F. K. Chang, "Maintaining Surface Accuracy of Large Radio Telescopes by Active Compensation"

P. Weidlinger, "Control of RMS Surface Error in Large Antenna Structures"

Thursday afternoon session: October 19, 1967, 1:45 p.m.

Session Chairman: N. Harper, Skidmore, Owings, Merrill

A. D. Kuzmin and P. D. Kalachev,* "Structures of Large Radio Telescopes"

*Unfortunately, Dr. Kuzmin and Dr. Kalachev could not be present. Mr. Fanning presented a summary.

S. von Hoerner, "Homologous Deformations of Tiltable Telescopes"

D. R. Strome and B. E. Greene, "ASTRA–Boeing's Advanced Structural Analyzer"

C. W. McCormick, "Application of the NASA General Purpose Structural Analysis Program to Large Radio Telescopes"

R. D. Logcher, "ICES STRUDL–An Integrated Approach to a Structural Computer System"

Friday morning session: October 20, 1967, 9:00 a.m.

Session Chairman: M. J. Holley, M.I.T.

J. Ruze, "Electromagnetic Loss of Metal Space Frames"

R. D'Amato, "Metal Space Frame Radome Design"

D. T. Wright, "Instability in Reticulated Spheroids: Experimental Results and the Effects of Nodal Imperfections"

L. Berke and R. H. Mallett, "Automated Large Deflection and Stability Analysis of Three-Dimensional Bar Structures"

J. J. Connor, "Non-Linear Analysis of Elastic Framed Structures"

Friday afternoon session: October 20, 1967, 1:45 p.m.

Haystack Visit. Buses leave from Kresge Auditorium parking lot.

G. Pettingill, "The Scientific Program at the Haystack Facility"

W. R. Fanning, F. A. Folino, R. A. Muldoon, and H. G. Weiss, "Design of the Haystack Facility"

Papers which could not be presented but which are part of this volume are as follows:

M. Ginat, "The Nançay Radio Telescope"

Lewis V. Smith, Jr., "Pointing and Tracking Accuracy. Recommended Standards"

S. Dean Lewis and E. A. Witmer, "Buckling Tests on Space Frame Radome"

Finally, it is a pleasure to express thanks to some of the many persons who contributed valuably to the Symposium: to the program committee, J. M. Crowley, M. J. Holley, Jr. and H. G. Weiss, and to Miss Ann Gorrasi who handled all the voluminous details with great effectiveness and good humor.

J. W. Mar
H. Liebowitz
Co-Chairman of the Symposium

CONTENTS

Structures Technology
for Large Radio and Radar
Telescope Systems

E. G. Bowen
Commonwealth Scientific and Industrial Research Organization
Australia

THE ROLE OF THE GIANT STEERABLE TELESCOPE IN RADIO ASTRONOMY

Some Milestones in the History of Radio Astronomy

Optical astronomy has been practiced for some two thousand years and, by comparison, radio astronomy has an exceedingly short history.

Radio astronomy began in the 1930's with the discovery by Jansky of the Bell Telephone Laboratories that residual noise signals were coming from the direction of the Milky Way. This was followed up by Grote Reber, who in a remarkable individual effort at his home in Illinois studied the radiation at considerably higher frequencies. There was much speculation as to where this radiation came from, and at the time there was reason to suppose that it came not from the stars but from the very tenuous matter in interstellar space. To Reber belongs the credit of building the first of the steerable paraboloids, an instrument 31 ft in diameter, which is preserved to this day at the National Radio Astronomy Observatory, Green Bank, W. Va.

The next important steps were taken soon after the end of World War II in Europe and Australia, when the two-element interferometer was introduced to give improved resolution at the rather long wavelengths then employed. Perhaps the most significant finding of this era was the discovery by Bolton that much of the noise from outer space came from discrete sources, a small number of which he was able to identify with comparatively well-known optical objects like the Crab Nebula.

This indicated that signals were coming not, as previously supposed, from matter in interstellar space but from discrete objects, many of which turned out to be of a nebular character, and many of which were outside the local galactic system.

The next exciting development came soon afterward from the optical astronomers. They found that the second strongest radio source in the sky, namely that in the constellation Cygnus, came from an exceedingly faint nebulosity with a red shift of one-third the velocity of light—at that time the most distant object known to mankind. This indicated the possibility that many of the radio sources much weaker than that in Cygnus, and so far not identified with optical objects, might well be beyond the range of the largest optical telescopes. Thus, far from being in a fledgling state, radio astronomy gave promise of penetrating to parts of the universe that had never previously been sounded.

The next basic discovery was made here in Massachusetts by Ewen and Purcell, who, in 1951, detected radiation from hydrogen in the ground state, on a frequency of 1420.5 Mc/sec. This discovery marked the opening of still another epoch in radio astronomy; it provided a tool for recognizing the presence of hydrogen, the most abundant element in the universe, and for measuring the velocity of this hydrogen relative to the earth by means of the Doppler shift.

To this discovery must be added the detection some 10 years later of radiation from the OH radical by Barrett and his co-workers here at MIT, followed soon afterward by the discovery of the recombination lines of helium and hydrogen and another as yet unidentified element.

The study of line radiation is now one of the most important and productive branches of radio astronomy. Along with the long-standing spectral observations of optical astronomy it has become the most powerful method of studying the constitution and abundance of elements in the universe. The great bulk of observational work on line radiation has been done by means of the steerable telescope, which is likely to remain the basic instrument for studies of this type.

Finally, I want to refer to the latest milestone, the discovery of quasars, the so-called quasi-stellar objects: that is, radio signals from objects that appear to be genuine starlike objects, as distinct from the radio galaxies that have become so familiar to us in the past 20 years. The excitement of this discovery is based on:

1. The red shifts found in many quasars now range up to 0.8 of the velocity of light. If one assumes that red shift is proportional to distance, it places them at a distance in excess of 10,000 million light years.

2. If the red-shift assumption is correct, the quasars are by far the brightest and most energetic bodies known to man. Their source of energy is so great that it cannot readily be accounted for by nuclear processes, and a great wave of speculation is in progress in an attempt to explain it.

3. If the red shift is not proportional to distance, most alternative explanations of the quasar's behavior run into serious difficulties and we have to admit that there is no satisfactory physical description of the behavior of quasars at this moment.

Quasars are a vast enigma, therefore, and the problems are certainly not clarified by recent observations. Their discovery has created great excitement in the astronomical world and the mystery, when resolved, is certain to revolutionize our concepts of how the universe is constituted.

The first accurate determination of the position and structure of a quasar came from the 210-ft telescope at Parkes, followed soon afterward by an optical identification and a determination of red shift from Mount Palomar. The number of properly identified quasars is now several hundred, mostly in the northern sky. Several hundred more exist in the southern sky, but their identification awaits the construction of optical telescopes of adequate power in the southern hemisphere.

Instrumentation for Radio Astronomy

Radio telescopes exist in a bewildering array of types. The layman is invariably confused by the different varieties and we have to admit that the radio astronomers are often confused too.

Radio telescopes may be classified into two broad types:

1. The steerable telescope, usually parabolic in shape.
2. The spaced arrays or unfilled aperture arrays, derived orginally from the two-element interferometer.

This is not the place to consider the relative merits of the two types.[1] It is sufficient to say that they are not competitive with one another—both have an important role to play in the future development of radio astronomy.

The choice between them is dictated almost entirely by one's requirements. Exaggerated claims are made for both types from time to time, but when the special pleading is stripped away, it will always be true to say that the steerable paraboloid is the general-purpose instrument, adaptable to any form of radio or radar astronomy, while arrays are special-purpose instruments, in the design of which great emphasis is placed on one objective, like the highest possible resolution. This is usually obtained at the sacrifice of some other characteristics, like ability to observe over a broad spectral range, or to receive signals in real time.

It follows that the unexpected discoveries of radio astronomy are more likely to come from the general-purpose instrument, which can be quickly adapted to new techniques and new discoveries as the science develops. There is less chance of a new discovery with an instrument that is specially designed with one preconceived purpose in mind.

If our purpose, therefore, is the widest possible range of astronomical studies covering source surveys, the spectra of sources over a wide frequency range, line radiation, polarization, and radar astronomy, there is only one choice, and that is the steerable telescope. If the purpose were different—for example a sky survey with the highest possible resolution or rapid-scan spectroscopy of the sun—the choice would also be different.

Fifteen years ago, the design of large radio telescopes was at a rudimentary level and few engineers had been introduced to the subject. Today the first generation of giant instruments exists and comprehensive measurements of their engineering performance have been collected. The remarkable agreement that exists between the calculated and the measured performance of some of these telescopes, notably the Haystack, the Parkes, and the Goldstone dishes, makes it clear that the design criteria are well understood and that larger instruments are entirely feasible.

My personal guess is that a diameter of 400 to 450 ft is attainable in the near future. However, such telescopes will only be built by employing the best engineering talent and the utmost sophistication in design. In 1952, the number of talented engineers involved in radio telescope design could be counted on the fingers of one hand. It is enormously encouraging that today the number actively engaged amounts to several hundred and most of them are represented in this Symposium.

Reference

1. J. P. Wild, "Instrumentation for Radio Astronomy," *Physics Today* **19** (No. 7), 28 (July 1966).

A. B. Youmans
U. S. Naval Research Laboratory
Sugar Grove, West Virginia

THE DESIGN OF A 300-FT RESEARCH ANTENNA

The performance of a large steerable antenna is governed by a complex of parameters most of which in themselves are neither unique nor dependent in either cause or effect. Furthermore, the design of such devices encompasses the theory and disciplines of numerous fields of the physical sciences and their related technologies. Although today with modern computers we strive to examine problems as a whole, we still find it necessary to separate them into logical parts to establish the behavior and variance of its important facets. In carrying out this work we must not forget that these parts, and the antenna as a whole, form only one element of a complete system to acquire or transfer information and, ultimately, it must be judged in terms of the performance of the over-all system.

Initially we can assume that certain requisites will be established which in addition to technical goals will include time, manpower, and money limitations, each of which must be appropriately considered with regard to the others. In most instances we can expect the nontechnical limitations to be more severe during the preliminary studies when we are determining feasibility and establishing the confines for the final design. The more restrictive these nontechnical limitations become, the more we must rely on experience, judgment, and satisfactory rather than optimized results. Stringent time requirements will generally demand the use of standard materials and established technologies. Although stringent monetary limitations will have a similar effect on the main support structures, it will not necessarily impose these same restraints in superstructures employing a large number of members with numerous connections or in areas where high precision is required. Both of these conditions result in a high cost per unit of material, thereby allowing the designer, with an adroit selection of materials, configuration, and techniques to be competitive on an over-all cost basis. Technical requirements including gain or effective area, pointing accuracy, tracking velocity, and environment also establish definite demands on the mode, scope, and extent of the work. Their relationship within themselves and with time and money are dependent upon the current "state of the art" for both theory and practice. Furthermore and foremost, we must consistently remember that "design" inherently implies "the ability to be realized" and consequently, we must ultimately wed, not divorce, the requirements and the constraints and form realistic, not idealistic solutions.

In the fall of 1962, preliminary studies were undertaken at the Naval Research Laboratory on a military-oriented research program that required a fully steerable, receiving antenna operable over an extended portion of the spectrum, specifically including the frequencies from 50 to 2000 Mc. The initial technical goals for the antenna included the tracking of near sidereal rate objects with aiming accuracies under operating environment commensurate to one-third its half-power beamwidth and an aperture efficiency exceeding 50 per cent over the prescribed frequency range.

Although a specific site was not designated, the environmental and geological conditions governing the design were assumed to be those expected in the National Radio Quiet Zone. Its terrain is rough and characterized by distinct mountains and narrow valleys typical of those found throughout the central portion of the Appalachian mountain range. Appropriate areas within the zone have a relatively mild climate where an antenna would not be subjected to particularly high winds or excessive ice and snow loads. Furthermore, this region is geologically stable and in certain areas its climatological conditions are well documented. Geographically, it is essentially a contiguous area of sedimentary rock principally in the form of shale or limestone. In the places where shale is found, it is either directly exposed or lightly covered by clay and mud. Such conditions are adaptable to a continuous ring footing. However, until a specific site is selected the foundation design cannot be appropriately considered.

The available contract funds for design was $100,000 with additional funds of about $10,000 for preliminary studies, experimentation, and consulting services. If authorized, the antenna cost, including the structure, drive mechanisms, control systems, electrical distribution, operating compartment and certain ancillary facilities, was estimated to be less than six million dollars.

With two possible exceptions, a review of the existing antennas, tentative designs, and design studies, which were sufficiently complete in 1962 to permit evaluation, revealed that a design based directly upon the then prac-

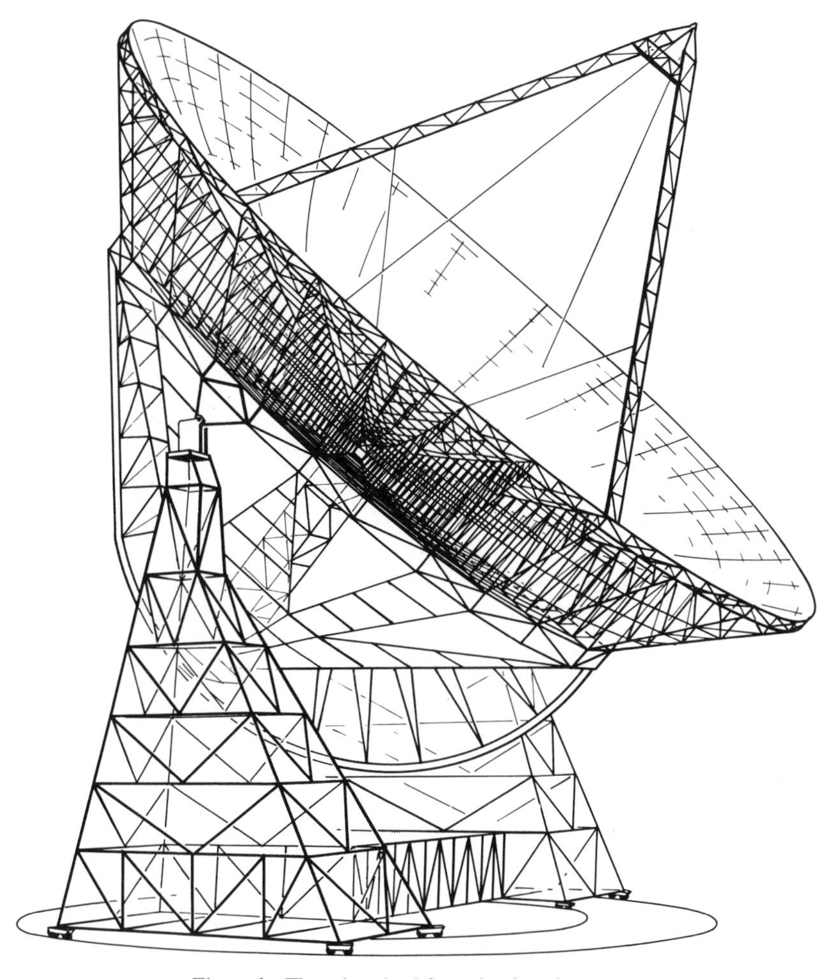

Figure 1. Three-hundred-foot altazimuth antenna.

ticed "state of the art" would either fail to meet our performance requirements or be too expensive. The two exceptions included the 210-ft Australian instrument and a design study by the National Radio Astronomy Observatory for an improved version of their 300-ft transit telescope. Both had similar sky coverage limitations. Government contract regulations and estimated costs for an antenna constructed in this country favored continuing the NRAO work. Consequently, Mr. E. R. Faelten, Consulting Engineer, was contracted to make a rudimentary engineering design and estimate approximate costs. Except for increased survival loads and hemispherical coverage, the design conditions were essentially identical to his previous work for NRAO.

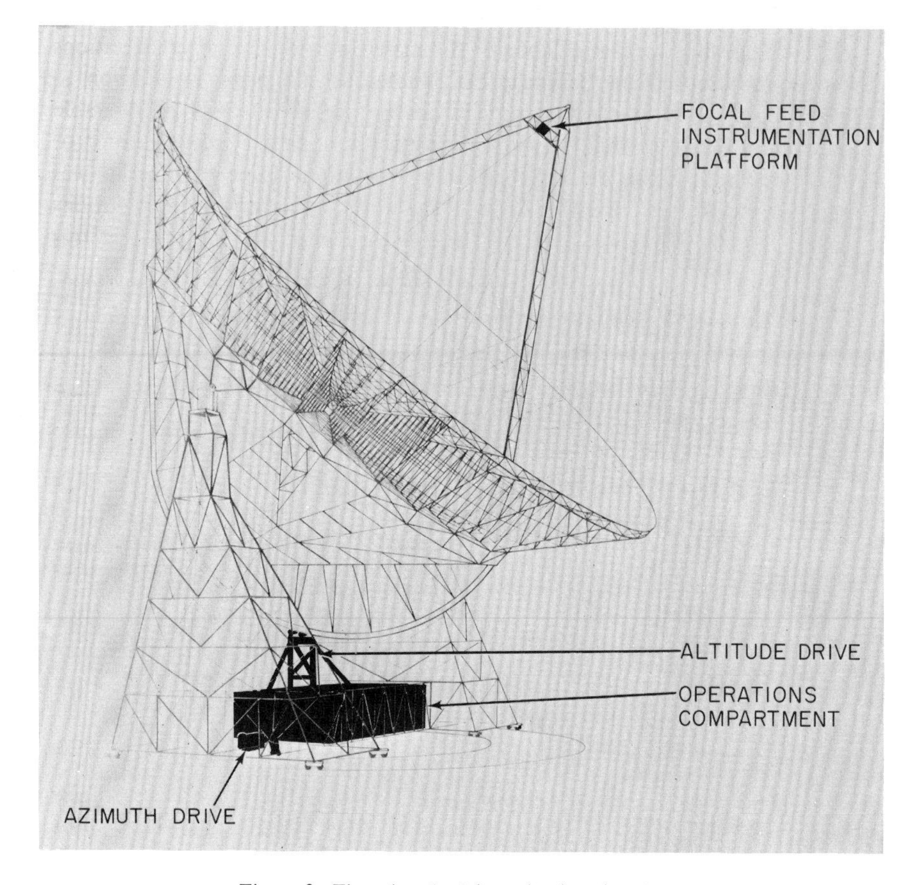

Figure 2. Three-hundred-foot altazimuth antenna.

Concurrently, the NRL staff formulated procedures for the conduct of the work and undertook the problem inherent in transforming the required antenna characteristics into applicable design criteria. Studies were made in special areas including the mesh surface, its attachment, and adjustment. Satisfactory solutions were achieved. The governing factors for operating and survival conditions were established, examined, and ultimately expressed in the form of standard codes. The AISC Manual of Steel Construction was selected as the basic document to be used throughout design, fabrication, and erection of the structure. Similar codes and industrial standards were adopted for the mechanical and electrical systems. This phase of the work was completed in June 1963 with the formalization of this information into technical specifications for the design contract. The contract for the design was awarded to Mr. Faelten in August 1963.

Before continuing with the specifics of this particular design, I would like to acknowledge the work and labor of the many individuals and organizations who contributed in the design effort. Numerous industries, in addition to furnishing detailed manufacturing and cost information, spent considerable time and effort to provide us with needed technical data. The cooperation of Dr. John Findlay and the National Radio Astronomy Observatory staff was especially helpful in the early stages of the work. A part of the contributions of the Battelle Memorial Institute is the subject of another paper at this

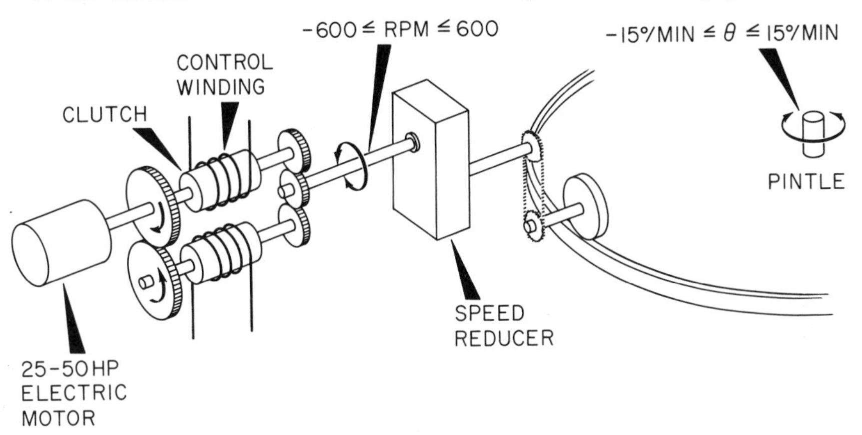

Figure 3. Azimuth drive (two required).

Symposium. The Supply Division at NRL and the ONR procurement office were extremely helpful. But principally the design herein described is the results of work by the staff of the Naval Research Laboratory and by Mr. E. R. Faelten, Consulting Engineer, Cummaquid, Massachusetts, under ONR contract Nonr-4286(00)(X). The contributions of Mr. C. E. Emerson of the NRL are especially noteworthy.

The geometry of the final design is illustrated in Figure 1. Pictorially, it is represented by a structural cone which, with its center tower, supports the antenna superstructure and feed towers. The lip of the cone is formed and reinformed by the rim, trunnion, and altitude cage girders. In addition to its structural functions, a significant portion of the cone serves as counterweight. The altitude wheel, although its specific configuration is primarily determined from drive considerations, is also an important structural element. All of these components rotate as an integral structure on trunnion bearings atop two support towers. The base section of the towers are connected by the turntable. Eight three-wheeled main trucks augmented by two drive wheels and two two-wheeled inner trucks support the entire weight of the antenna and permit its rotation in azimuth. A pintle bearing is used to confine lateral motions. Five rails are used of which the outer four are paired in a typical

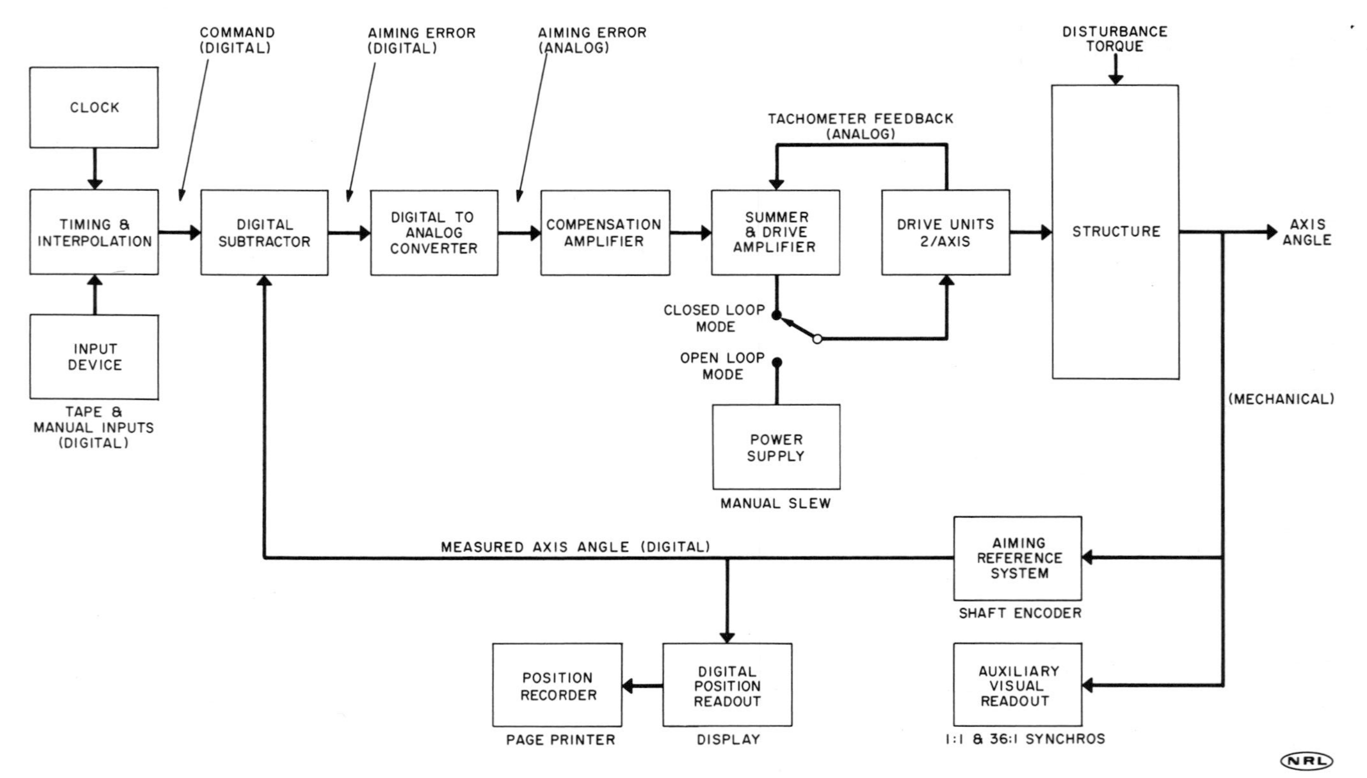

Figure 4. Block diagram of a drive system.

track arrangement to accommodate the three wheels of the main trucks. The top of each rail is flat and the wheels are contoured to reduce edge loading due to misalignments.

The structural design is definitive. For the most part, it utilized standard rolled shapes of A-36 steel with bolted connections. Where weldments are employed, a steel known to exhibit appropriate brittle fracture characteristics is specified. (An NDT of -30° was selected.) Standard ASTM tests supplemented by Federal Specifications for material quality control are used throughout.

Figure 2 shows the location of the principal operating space on the antenna. It is essentially a large shielded compartment containing 2700 square feet of floor space. Except for focal point installations, it will house all of the electronic instrumentation required by the receiving system. It is a self-contained unit with its own electrical substation and air-conditioning plant.

Figure 2 also depicts two of the three locations for the drive machinery. (The azimuth drive located under the left tower is hidden in the view.) The functional operation of an azimuth drive is illustrated in Figure 3.
The altitude drive is similar except that the output shafts of the two units are mechanically connected and the chain drives the altitude wheel. In either case,

Table 1. Design Characteristics of the 300-ft Antenna.

Physical characteristics	
Diameter	300 ft
Focal length	128.5 ft
F/D	0.428
Depth of paraboloid	43.7 ft
Surface area	78 x 10^3 sq ft
Surface material	5/8″ alum. square mesh
Over-all height	330 ft
Turntable dimensions	100 x 259.4 ft
Weight movable in altitude	1.35 x 10^3 ton
Moment of inertia about altitude axis	1.3 x 10^{10} lb ft^2
Locked rotor frequency–altitude axis	1.4 cps
Total weight movable in azimuth	3.45 x 10^3 tons
Moment of inertia about azimuth axis	5.0 x 10^{10} lb ft^2
Locked rotor frequency–azimuth axis	1.7 cps
Altitude rotation–manual modes	0 to 90°
tracking modes	0 to 80°
Azimuth rotation–all modes	1.5 revolutions (max)
Altitude and azimuth drives–manual modes	100 hp (available)
tracking modes	50 hp (available)
Performance parameters	
Aiming error (max operating environment)	$<$1.2 min of arc
Aperture efficiency–2000 Mc	~50%
Max velocity altitude axis manual	7.5°/min
Max velocity altitude axis tracking	112°/h
Max velocity azimuth axis manual	15°/min
Max velocity azimuth axis tracking	225°/h

the speed and direction of motion can be changed by connecting the clutch control windings in a typical push-pull circuit. Bias and tachometer feedback are used to achieve linearity and increase torsional stiffness. Figure 4 is a block diagram depicting a complete drive system for controlling the motion of one axis. The other axis is similar; however, the clock and input device will be common to both. The closed loop modes are used for normal operation and the open loop slew mode in conjunction with the synchro readout provides a simple method to change the position of the antenna for maintenance purposes.

Table 1 gives certain physical parameters and summarized expected performance. In essence, this information and the design as a whole represented our solution for a specific problem. However, during the course of the work, the time limitations originally imposed were relaxed. While this permitted more detailed evaluation of certain interesting facets, undoubtedly, if and when the antenna is constructed, it will be more costly for the same reasons that a home built today costs more than one built three years ago. This we would expect: since most of our parameters are time and position dependent, our conclusion must also be so.

Supplementary Readings

1. American Institute of Steel Construction, *Manual of Steel Construction* (1963), 6th ed.
2. Ammann and Whitney (Consulting Engineers), "Engineering Reports No. 2 and 3, Contract NBy-337788" (1962).
3. Battelle Memorial Institute, 24th Phase Report, "Study of Basic Drive Elements" (1958).
4. Battelle Memorial Institute, Final Report on "Determination of Approximate Structural Parameters of a Large Steerable Antenna and Their Relationship to the Antenna Performance."
5. C. F. White and J. W. Titus, "Elevation-Azimuth Servo System Specifications for Star Tracking," NRL Report 4732 (1956).
6. "Proceedings on Symposium on Communication Theory and Antenna Design," Electronics Research Directorate, Air Force Cambridge Research Center, Air Research and Development Command (1957).
7. A. S. Locke, "Guidance" (1955).
8. J. W. Titus, "Wind Induced Torques on a Large Antenna," NRL Report 5549 (1960).
9. "Lift and Drag Coefficients for Various Types of Radar Antenna Screens," Report R-293, Aero 697, Navy Department, David Taylor Model Basin (1951).
10. J. E. Alexander and J. M. Bailey, "Systems Engineering Mathematics" (1962).

O. Hachenberg
Max-Planck-Institut für Radioastronomie
Germany

THE 100-M RADIO TELESCOPE OF THE MAX PLANCK INSTITUTE FOR RADIO ASTRONOMY IN BONN

The single telescope with large filled aperture continues to be important to radio astronomy. The large telescope offers definite advantages for line-spectroscopic research in our galactic system. The apparent distribution of hydrogen with respect to the celestial sphere changes only slightly between points at a distance of less than a quarter of a degree from each other. Thus an extremely great angular resolution is not necessary for research work on interstellar hydrogen. Also, for investigations dealing with the structure of the spiral arms of the system and with star clusters and H II regions, an angular resolution of a few minutes of arc is sufficient and can still be realized by a filled aperture. On the other hand, the large telescope offers the possibility that, if one simultaneously measures in different frequency channels, a temperature resolution of 0.1°K or even less can be obtained in one channel relative to the others. The telescope combined with a multichannel spectrometer permits detection of fine structures in the lines which are hardly accessible to the array.

Further striking advantages of the large telescope become evident at measurements in the lower centimeter region, where the variations in the refractive index and in the transmission of the terrestrial atmosphere soon limit the use of arrays. The single telescope is able to visualize very feeble intensity differences.

The importance of the single antenna for absolute radiation measurements as well as for polarization measurements in the galactic continuum should also be mentioned, but will not be considered further here.

From the preceding it is evident that the large telescope has a wide range of application even for future radio-astronomical research. This range preferably lies in the 1 to 20 GHz frequency range..

In planning a new large telescope, it is necessary to strive for its highest possible adaptation to these applications. The telescope must therefore:

1. Be usable for as high frequencies as possible.
2. Have as low an antenna noise as possible in the centimeter-wave range.
3. Have a sufficiently high pointing and steering accuracy.

If special attention is given to the centimeter-wave range during the planning stage, there is a chance of obtaining an angular resolution of 1 min of arc or even less by a parabolic antenna. This might be realized, e.g., if it were possible to build a reflector of 100 m in diameter which could still be used for a wavelength of 2 cm. On the other hand, this leads to the difficulty of very high demands on the accuracy of the surface as well as on the accuracy of the pointing.

The Allowable Surface Deviations

The first problem facing the designer of a large telescope is to determine allowable surface deviations. A good estimate of the maximum allowable deviations can be derived with the aid of the theoretical work of J. Ruze. If σ is the rms error of the dish profile, the efficiency of the surface η_s is given by

$$\eta_s = \exp[-(2\pi \frac{\sigma}{\lambda})^2],$$

If, for instance, η_s is $\geqslant 0.67$, σ must remain $\leqslant \frac{1}{20}\lambda_0$, where λ_0 is the smallest usable wavelength. If we adopt $\lambda_0 = 2$ cm, σ is 1 mm.

In practice, the surface deviations of the reflector consist of

1. Errors resulting from the production of the single panels.
2. Deviations resulting from the elastic deformations of the single panels.
3. Adjusting errors of the panels.
4. Deviations resulting from the elastic deformation of the steel construction.

Let us suppose that the errors resulting from the production of the panels and the adjusting errors can be kept small. There remain only the deviations resulting from the elastic deformation of the carrying system, which have to be dealt with specially. In general, deformations of the carrying system are not distributed at random over the reflector surface; they occur over most of the surface and seem to have rather a systematic character; we likewise demand that the rms value σ_d of the distances of the real surface from the ideal paraboloid, which we measure at many equally distributed points, be $\sigma_d = \frac{1}{20}\lambda_0$

Ways of Reducing the Elastic Deformations of the Reflector Carrying System

If we postulate a value of σ=1.5 mm, which should be realized by the telescope, it is clear from the very beginning that it will not be easy to build a reflector of 100 m in diameter with such precision. The experience gathered from reflectors built in the past shows that the elastic deformations of the carrying system are 20 to 40 times higher than the demanded value.

Once again it becomes necessary to study all the possibilities available to keep the elastic deformations small. The following methods have already been discussed in the literature:

1. An automatically adjustable reflector surface was suggested for the first time in planning the telescope at Sugar Grove.
2. A dish construction with several independent supporting points has been proposed by Kalachev.
3. As the deformation of a construction can be exactly calculated, it is possible to provide the construction with additional forces, e.g., hydraulic links operated by a computer which makes adjustments that depend upon the elevation angle of the telescope.

Further possibilities exist.

The idea that a certain deformation should be allowed was first followed in the feasibility studies for the 100-m telescope at the Max Planck Institute. However, the deformation should be such that the parabolic surface, once adjusted at zenith position, is converted into a series of new paraboloids when the telescope is tilted about its elevation axis. The new paraboloids may have other axial directions and different focal lengths. The resulting displacement of the focal point can be corrected by an adequate displacement of the feed.

In other words, we must look for a construction method with minimum deflections of the real deformed surface from a best-fit paraboloid:

$$\phi \longrightarrow \Sigma(\triangle Z \cdot \cos\varphi)^2 = \Sigma d^2 \longrightarrow \text{minimum.}$$

How is it possible to derive a practical construction showing such qualities? We used approximation methods in our feasibility studies.

By means of electronic computers the displacement of each single joint can be calculated in three coordinates, even for complicated space frameworks. On the other hand, by the free choice of beam cross sections and material qualities, the steel framework has enough free parameters to influence to a certain degree the displacements of groups of joints—especially joints lying in a surface. If some joints of the framework are kept fixed, there is a certain chance of complying with the conditions postulated for the joints in the reflector surface.

Figure 1. Dish construction with four-point support.

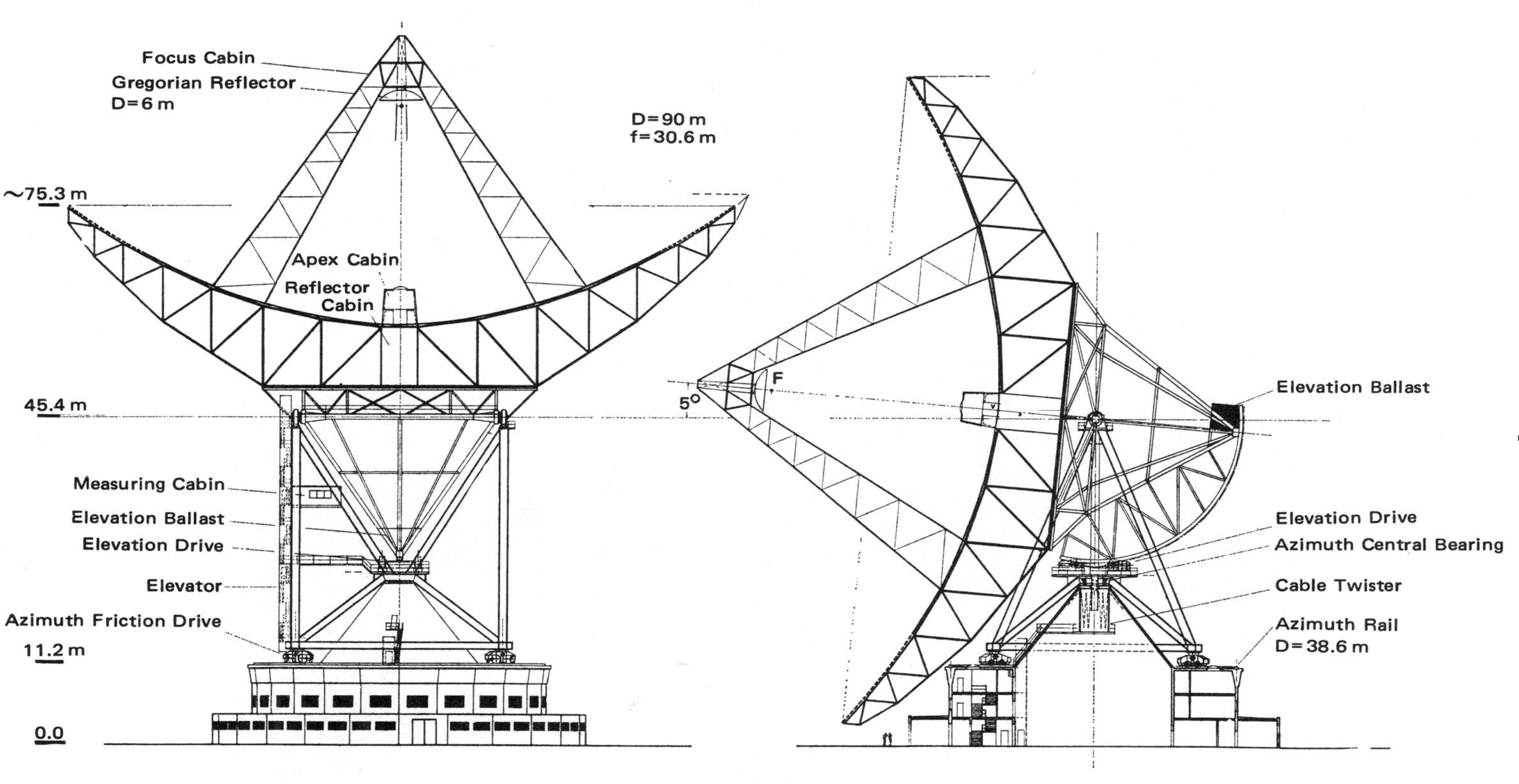

Figure 2. Dish construction with eight-point support.

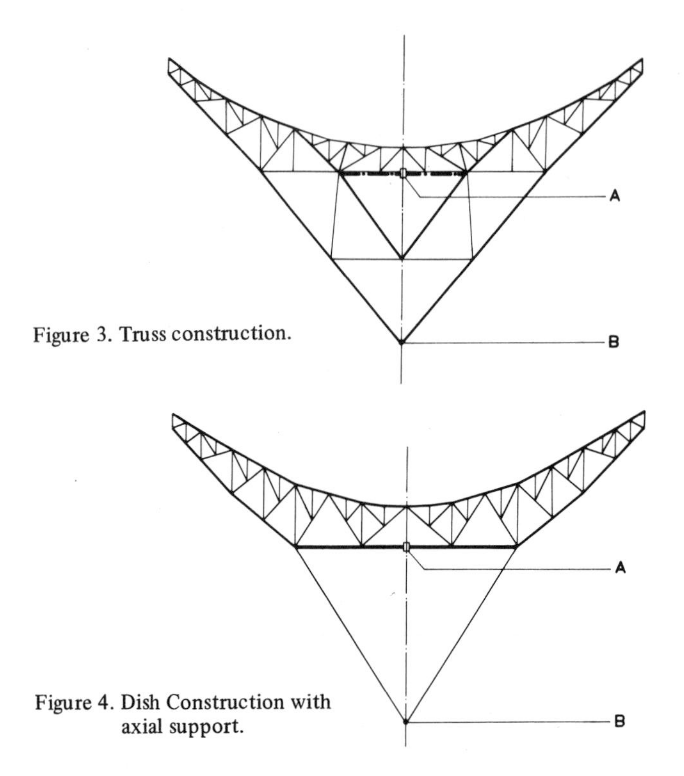

Figure 3. Truss construction.

Figure 4. Dish Construction with axial support.

Investigations of Special Models

This method of optimizing a reflector construction had not yet been tried, and no experience was yet available about the type of previous constructions most suitable for such a study. Thus four different types were used for trials.

In the first attempt we started with a dish construction for the reflector consisting of radial contilevers, ring girders, and upper and lower cross connections. The dish was supported at four independent points. Figure 1 shows the construction with the four-point support. A second test was based on a similar dish construction supported, however, at eight points formed by an eight-bar pyramid at the rear of the dish. This type is shown in a skeleton sketch in Figure 2.

Two further constructions are characterized by an axial support of the reflector. In one construction of this type the radial cantilevers were extended behind the reflector to a top on the paraboloid axis (Figure 3). The bars converging at this top form a cone. The radial cantilevers form trusses and, therefore, the name "truss construction" was used. In the middle–near the reflector apex–the trusses are kept at equal distance by means of a spoke wheel. On the other hand, they are held together by a ring beam at the reflector rim. If adequate tensions are introduced, the effect of the construction may be compared to an opened umbrella. Like an umbrella, the reflector thus "opened" is supported at two points A and B lying on the axis, one of them being the cone apex, the second the wheel hub near the paraboloid apex.

Finally, a modified type somewhat resembling the dish construction was examined. Figure 4 shows it in a skeleton sketch. The spoke wheel and the axial support have been maintained, but the dish resembles the normal dish construction. We call this type a "dish construction with axial support."

In all cases, the investigations of the elastic deformation of the systems were carried on in a similar way. The four or eight supporting points of the dish models or the two axial carrying points of the truss models were considered as fixed ones. The displacement of each single joint of the reflector construction was then calculated in three coordinates by the electronic computer, and afterward the displacement of those joints that lie in the reflector surface was examined. Through these joints a best-fit paraboloid was laid, and the distance of the joints from the latter was determined.

In various cycles of calculation the cross sections of single beams and groups of beams were changed step by step to bring the distance of the reflector surface joints from the best-fit paraboloid to a minimum and also to obtain the lowest possible total weight. The procedure was discontinued as soon as no further improvements could apparently be realized. In most cases four to six cycles were calculated for the various models.

For each construction the procedure proved successful to a certain degree. The result is perhaps most impressively characterized by comparing the

Table 1. Comparison of the Maximum Deformation with the Maximum Deflection of the Surface from the Best-Fit Paraboloid.

Model	A Max deformation upper rim	A Max deformation lower rim	B Max deflection from best-fit parab.	B : A
1. Dish construction with four-point support	+ 23 mm	− 19 mm	+ 4 mm − 3 mm	1 : 6.5
2. Dish construction with eight-point support	+ 60 mm	− 22 mm	±3 mm	1 : 15
3. Truss construction	+ 33 mm	− 33 mm	± 1.2 mm	1 : 25
4. Dish construction with axial support	+ 82 mm	− 53 mm	± 1.8 mm	1 : 36

maximum absolute deformation of the construction (which occurs when the dish is tilted from zenith position to an elevation of 5°) with the maximum distance of the surface from the best-fit paraboloid. Table 1 gives the results for the four different models.

For critical examination of the results it is also important to know by what amounts the best-fit paraboloid has been shifted in comparison with the original one.

The displacement is first described by the dislocation of the apex. The apex shifts by Z_S, in the direction of the paraboloid axis, i.e., the direction of the Z axis, and by Y_S in the direction of the Y axis, which is normal to the elevation axis and the Z axis. Displacements in the direction of the elevation axis are not to be expected for reasons of symmetry of forces. Second the changes of the axis direction β of the new paraboloid and the alteration of the focal length f must be known. All these data are compiled in Table 2 for the dish model with eight-point support and in Table 3 for the dish model with axial support. The tables also contain rms values of the deflections of the

Table 2. Characteristic Values of the Best-Fit Paraboloid (Dish Construction with Eight-Point Support).

D	Load	Elevation	rms	y_S	z_S	f	β
(m)		(deg)	(mm)	(mm)	(mm)	(mm)	(sec of arc)
	Weight	5	1.9	0.10	+ 1.51	30600.9	171
90	of	45	1.6	0.07	+ 12.3	30607.6	121
	framework	90	1.1	0.0	- 17.3	30610.7	0

Table 3. Characteristic Values of the Best-Fit Paraboloid (Dish Construction with Axial Mounting).

D	Load	Elevation	rms	y_S	z_S	f	β
(m)		(deg)	(mm)	(mm)	(mm)	(mm)	(sec of arc)
	Weight	0	0.28	337.6	+ 1.06	29361.7	1140
90	of	45	0.46	240.5	- 1.62	29379.2	846
	framework	90	0.50	0.0	- 3.05	29386.7	0
100		90	0.83	0.0	- 3.76	29383.3	0

surface. These values are derived in the following way: The paraboloid was assumed to have been adjusted in the unloaded (gravity-free) case in each elevation position; then the system was subjected to gravity and the subsequent deformations were calculated. This kind of calculation allows a particularly good insight into the process of deformation under dead load. In the practical operation of the telescope, only the relative deformations are important.

The Results of the Investigations

The results of the investigations of the various constructions clearly show that the chosen method seems especially adequate to the problem of building reflectors of highest accuracy.

In detail, the following statements can be made:

For dish constructions with four-point support an rms error of 1.3 mm was obtained when the reflector had been adjusted at zenith and then tilted by 90° in elevation. In the deformation diagram the influence of the four points of support can still be clearly seen. Though an improvement could be

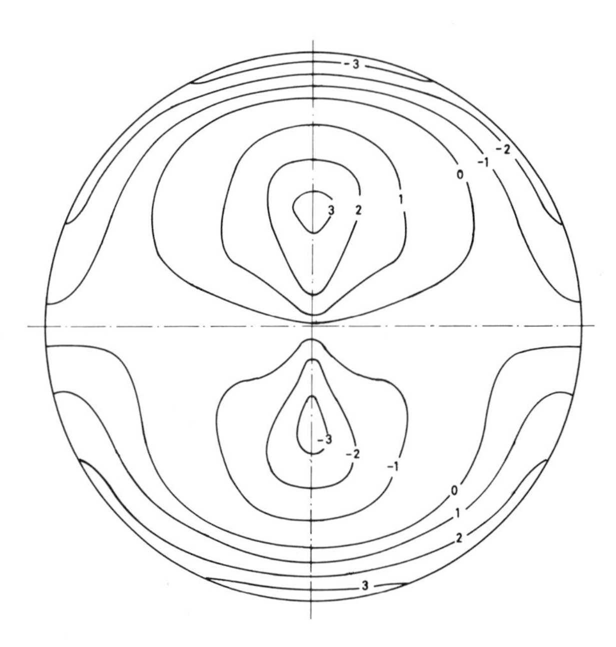

Figure 5. Lines of constant deviation from the best-fit paraboloid for the dish construction with eight-point support.

obtained in this type of construction, the dish construction with four-point support does not seem especially favorable for the chosen procedure.

For dish construction with eight-point support the influence of the supporting points becomes effaced. The surface accuracy is described by an rms error of 1.2 mm. The distribution of the deviation of the surface is represented in Figure 5. In this case, the effect of the quadripod with the focus cabin erected on the dish becomes noticeable. A further compensation of its influence could not be obtained. For both models it was advantageous to keep the alterations of focal length and axis direction of the best-fit paraboloid small. A correction would have been necessary only for wavelengths below 10 cm. With regard to the deformations, very good results could be realized on dish models with axial support. The rms error for the 90-m surface amounted only to 0.56 mm, that for 100-m to 0.8 mm. The distribution of the deviations of the surface from the best-fit paraboloid is shown in Figure 6.

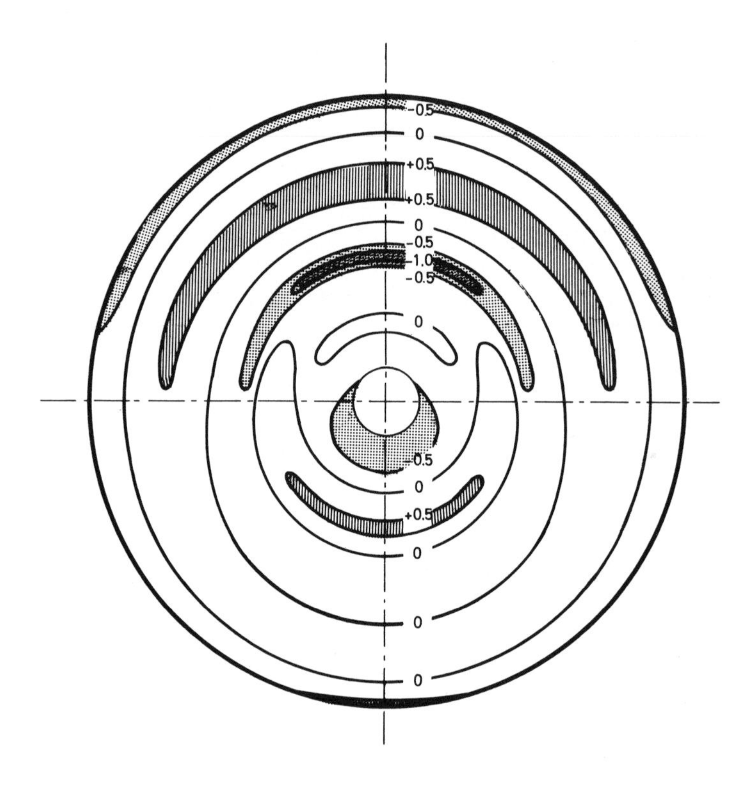

Figure 6. Lines of constant deviation from the best-fit paraboloid for the dish construction with axial support.

In this case, the deviations have an almost symmetrical form. In both constructions with axial support, however, the displacements of the best-fit paraboloids in relation to the original surface are greater. The correction of the feed horn in the focus amounts to about 12 cm in the Y direction, but only to a few millimeters in the axis direction. Feed horn and Gregorian reflector must be adjusted according to the elevation angle in this case.

The calculation method used is of an experimental nature; an estimation of remaining errors is impossible. Thus no final statement can be made about which construction is preferable. Only some general directions are possible which may prove useful in the successful application of the procedure.

1. Having one or some of the main points of support of the reflector in its very surface should be avoided. Otherwise, it will be very difficult to make the other joints of the surface lie on the best-fit paraboloid.
2. Singular loads in the surface should be eliminated as far as possible; the quadripod carrying the focus cabin should not be erected in the surface itself.
3. According to the available results, an axially symmetrical construction has certain advantages over asymmetrical ones. This decision, however, can still be influenced by the results gained in examining other systems.

The evaluation of the results of the design studies suggested basing the final construction of the 100-m telescope on the "dish construction with axial support." The determining factors for this decision were the very small remaining deformations and the preference given to the axial support. A photograph of a model of this type of construction is shown in Figure 7.

Some Technical Details

The support of the reflector at the two points of the axis requires an independent carrier system, which is clearly shown in Figure 8. It consists of an octahedron, one beam of which bridges the two elevation bearings as elevation axis. This beam penetrates the reflector construction without contact with the framework and is fastened to the hub of the spoke wheel by a pin. Two further beams of the octahedron go from the elevation bearings to the cone apex; they carry the reflector, especially in zenith position. The other beams serve to give additional support to the reflector at the hub when it is tilted. The upper apex of the octahedron is formed by the quadripod with the focus cabin. In this way, the quadripod is quite independent from the reflector construction. There is no influence of this weight on the deformation of the surface.

Up to a diameter of 80 m the surface will be covered with aluminum sheets. In the region from 80 to 90 m perforated aluminum sheets will be used in because of the wind load, while for the outer ring from 90 to 100 m in diam-

Figure 7. Model of the dish construction with axial support seen from the rear.

Figure 8. Model of the dish construction with axial support.

eter wire mesh will be used for covering. For the surface up to 90 m in diameter the following claims on its precision are made:

1. Elastic deformation	rms	$\leqslant$	0.6 mm
2. Manufacture of the panels	rms	$\leqslant$	0.4 mm
3. Adjustment of the panels	rms	$\leqslant$	0.5 mm
4. Bending of the panels and the reflector at wind load up to 18 m/sec	rms	$\leqslant$	0.6 mm
Total factor of the surface	rms	$\leqslant$	1.2 mm

Thus the reflector with a surface efficiency of $\eta_s = 0.7$ will be usable for $\lambda \approx 2.6$ cm up to a diameter of 90 m. With its whole surface it will be usable for $\lambda \geqslant 5$ cm.

The half-width of the antenna beam at $\lambda = 3$ cm will be about 1.2 min of arc. Therefore, a pointing accuracy of 8 sec of arc is needed. The drive and the control of the telescope will be regulated by a computer.

The telescope has been designed to be used in the primary focus from a focus cabin as well as in the secondary focus from an apex cabin. This apex cabin is for also used for measuring the contour surface. For the secondary focus a Gregorian system will be applied. This system has the advantage of a real primary focus. The Gregorian reflector can, therefore, remain firmly mounted under the floor of the focus cabin. The primary focus is then accessible by an opening in the Gregorian reflector. Furthermore it offers a possibility of fixing a shield collar at its rim; the latter screens the feed horn in the primary focus from disturbing radiation that might hit it laterally over the rim of the main reflector. A shielding at the main reflector of 1.8 m in height helps to reduce the antenna noise temperature; the dependence of the antenna noise temperature on the elevation angle turns out especially favorably in this way.

Further technical details have been compiled in Table 4.

Table 4. Technical Details of the Telescope.

Reflector diameter	100 m
Focal length	30 m
Gregorian reflector (diameter)	6.5 m
Focal length in Gregory focus	~64 m
Cone ring (diameter)	38 m
Elevation gear rim (radius)	28 m
Height of elevation axis above ground level	50 m
Azimuth rail (diameter)	64 m
Reflector weight including reflector surface and apex cabin	753 tons
Weight of intermediate carrier system including quadripod and counterweight	760 tons
Weight over azimuth rail	2800 tons
Motion range about elevation axis	$5^\circ - 94^\circ$
Motion about azimuth axis	$\pm 360^\circ$
Steering accuracy in the motion range of 0° to 2°/min	± 8 sec of arc

In spite of the high accuracy of the surface, the reflector weight of 753 tons is rather low. The intermediate carrier system (octahedron) and the quadripod with focus cabin have a weight of 760 tons.

In comparison with the reflector, the weight of the two azimuth towers is high. But the claimed high steering accuracy can only be reached if the azimuth towers are very stiff, and this leads to the high weight over the azimuth rail ring.

On 1 July 1967 a working group of the firms Krupp and MAN was assigned to build the telescope. The construction will take about 30 months, and the telescope is likely to be finished at the end of 1969.

Acknowledgments

The calculations for dish constructions were made by a group of MAN engineers led by Mr. Schneider and Dr. Schonbach. The calculations and the final design of the dish model with axial support were made by a group of Krupp engineers under guidance of Geldmacher, Hartmann, and Rusel. The Institute was assisted by Mr. Hooghoudt as consulting engineer.

H. G. Weiss
Massachusetts Institute of Technology
Lincoln Laboratory
Lexington, Massachusetts

DESIGN STUDIES FOR A 440-FT-DIAMETER RADIO AND RADAR TELESCOPE

The Cambridge Radio Observatory Committee (CAMROC) was established in 1965 by Harvard University, M.I.T., and the Smithsonian Astrophysical Observatory to develop plans for an advanced regional radio and radar astronomy research facility. The formation of CAMROC was in response to the very evident need for a large telescope within a few hours' drive of the greater Boston area to enable both the staff and students of the many New England universities to participate more actively in radio and radar astronomy. As the plans for this facility evolved, representatives of 14 universities in New England formed the Northeast Radio Observatory Corporation (NEROC) to cooperatively seek support to construct and operate this proposed research installation.

Performance Requirements

Detailed studies* were initiated at the onset of the program to define the principal research objectives[1] in radio and radar astronomy of direct interest to the scientists in New England. These objectives, which included spectral-line and continuum analysis of galactic and extragalactic sources, investigations of the interstellar medium, radiometric and radar studies of the solar system, high-resolution studies of low-brightness sources, and multiple-site interferometric research, were then used to establish the performance requirements[2] for a proposed telescope.

* With support provided by NSF Grant 5832 and by the participating institutions.

A variety of filled-aperture and array antennas was studied and it was concluded that the diverse needs of the New England scientific community could be fulfilled most effectively by a large, very precise, steerable, filled-aperture telescope providing it could function efficiently over a wide spectral range. Figure 1 summarizes the principal characteristics desired in the proposed instrument. While it was obvious that the largest and most precise steerable aperture would have the greatest scientific potential, it was also recognized that considerations of engineering feasibility and cost would have a major influence on the attainment of the necessary financial support for the project. Because several steerable antennas in the 200- to 300-ft size class are now in operation and because NRAO is currently investigating telescopes 600 ft and larger in size, the NEROC group decided to focus its attention on very high precision antennas in the 300- to 500-ft size range.

After studying the performance capabilities and limitations of all existing and planned telescopes, it was concluded that significant new opportunities for scientific exploration would become available if a 300- to 500-ft-diameter telescope could be made with the very high precision needed for efficient operation at wavelengths as short as 5 cm. However, it was recognized that it would be extremely difficult and costly to construct a large steerable precision antenna by just linearly scaling an existing design (see Figure 2). Studies were initially based on a 400-ft-diameter antenna and were eventually changed to a 440-ft-diameter to compensate for radome effects. If at all feasible at a reason-

BROAD SPECTRAL RANGE
SIMULTANEOUS SPECTRAL COVERAGE
HIGH RESOLUTION
LOW SIDELOBES
LARGE ABSOLUTE APERTURE
POLARIZATION MEASUREMENT CAPABILITY
LONG INTEGRATION TIMES
EFFECTIVE USE OF LOW-NOISE RADIOMETERS
MAXIMUM SKY COVERAGE
COMPATIBLE FOR RADAR USE
VERSATILE TRACKING CAPABILITY
OBSERVATION OF SHORT-TIME-PERIOD PHENOMENA
EASE OF EXPERIMENTAL CHANGEOVER

Figure 1. CAMROC antenna requirements.

able cost, it was desired that the telescope be usable at high efficiency at all wavelengths between 5 and 300 cm, and some performance degradation at shorter and longer wavelengths would be acceptable.

This requirement for 5-cm operation dictated that the contour errors of the reflector system should not exceed 0.1 in. (rms).[3] The ratio of the diameter to tolerance required to meet this performance criteria with a 440-ft telescope would have to be approximately 48,000 (see Table 1). The only known steerable telescope with this high a diameter-to-tolerance ratio is the 120-ft-diameter Haystack system.[4]

Selection of a Configuration

After the nominal size and tolerance of the telescope were established (see Figure 3), many different concepts for filled-aperture antennas were examined to determine the most economical configuration. A variety of conventional and nonconventional antennas (folded horns, tilting plates, floating spheres, off-axis paraboloids, etc.) were investigated[5,6] to ascertain which general configuration would be most suitable. These studies indicated that the cost of a telescope is determined not only by the area and precision of the reflecting surface, but also by: (1) the total weight of the structure, (2) the environmental operating and survival requirements, and (3) the required pointing precision. Configurations that required multiple surfaces (tilting plates and horns) and systems in which the reflector surface constituted only a small percentage of the over-all moving weight had greater over-all cost. A series of five parallel studies[7] carried out over a nine-month time span by five design

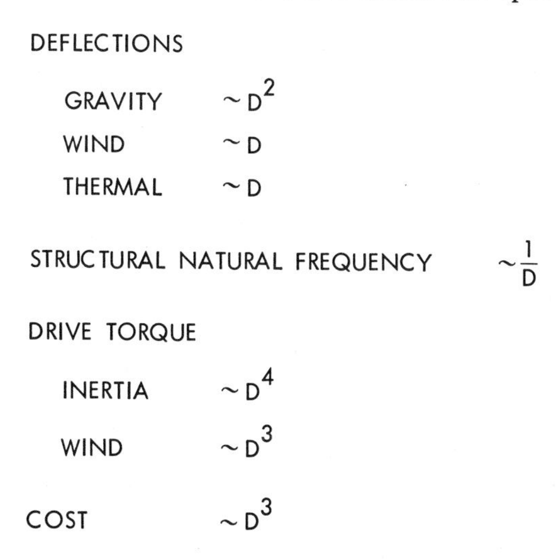

Figure 2. The effects of linearly scaling all dimensions of an exposed antenna.

Table 1. Typical Radio Telescopes.

	Diameter (D) (ft)	Tolerance (ϵ) rms (in.)	λ_m (cm)	Precision $D/\epsilon \times 10^{-3}$ (000)	Beamwidth (min)	Remarks
Lincoln Laboratory	28	0.008	0.26	42	1.3	
Lebedev	72	0.022	0.70	39	1.3	
Michigan	85	0.030	1.0	34	1.5	
Haystack	120	0.029	1.0	50	1.1	in radome
Green Bank	140	0.040	1.3	42	1.3	
Algonquin	150	0.045	1.4	36	1.5	
Parkes	210	0.14	4.5	18	2.8	elevation min = + 30°
Jodrell Bank	250	0.40	13	7.5	7.1	
Green Bank	300	0.47	15	7.6	7.0	transit instrument
Arecibo	1000	1.2	40	10	5.3	spherical reflector (± 20°scan)
Proposed						
Bonn	328	0.15	3	26	2.0	
Univ. of Manchester	400	0.50	16	9.6	5.5	
NEROC	440	0.10	3.2	53	1.0	in radome

Data from various sources; sometimes inconsistent (nighttime - no-wind conditions).
λ_m, where gain in maximum ($\lambda = 4\pi\epsilon$); reduction in gain due to tolerance errors at λ_m = 4.3 dB.
Half-power beamwidth at λ_m.

teams (Ammann & Whitney; Simpson, Gumpertz & Heger; The Rohr Corporation; P. Weidlinger; and Lincoln Laboratory) provided useful comparative data on a variety of antenna configurations. From these studies it was concluded that the conventional parabolic configuration offered the most promise for fulfilling the performance goals at the lowest over-all cost.

Pointing Considerations

It also became apparent from these studies that *the requirement to point the antenna with very high percision would be the most difficult design requirement to fulfill.* At the shortest operating wavelength of 3.2 cm, a 440-ft-diameter aperture will have a half-power antenna beamwidth (3 dB response points) of only 1 min of arc (0.017°). Efficient operation necessitated the control of the beam position with an uncertainty of less than about one-sixth of a beamwidth or approximately 0.003°. This pointing precision has not been achieved to date on any large exposed steerable reflector except during very

Reflector diameter	440 ft (178 m)
Surface accuracy	0.1 in. (0.25 cm) rms
Operating range	
(high efficiency)	5 cm (6 GHz) to 100 cm (300 MHz)
(reduced gain)	3.2 cm (9.6 GHz) to 300 cm (100 MHz)
Equivalent area	$\sim 1.3 \times 10^5$ ft^2
Beamwidth at 3.2 cm	60-sec of arc (0.02°)
Pointing accuracy	10-sec of arc (0.003°)
Sky coverage	azimuth 360°; elevation 0 to +85°

Figure 3. Proposed characteristics.

low wind conditions and at nighttime when the thermal gradients are low. The radome-enclosed Haystack telescope is, again, a unique example of a large steerable antenna capable of fulfilling this pointing requirement. The desire to operate at the shorter wavelengths requires that the reflecting surface be solid or have a very fine mesh.

Preliminary design studies clearly indicated that it would not be realistic to expect any 400-ft-diameter exposed antenna with a comparatively solid reflecting surface to fulfill this very rigorous pointing requirement due to the enormous stiffness required in the reflector mount, control, and alidade structure.

Environmental Considerations

The environment in which the telescope must survive and function has, of course, a very important influence on the design.[8] Two factors are of principal concern—the winds and the thermal environment. Figure 4 indicates the peak wind velocities averaged over a 5-min interval which have a 1-per cent likelihood of occurrence in a 25-year time span at a height 50 ft above the ground.[9,10] It will be noted that in New England, at locations at least 75 miles from the sea coast, the peak wind velocities are substantially lower than throughout a large fraction of the United States. While the actual winds in any area will be dependent on the local terrain features, the selection of telescope sites and the establishment of survival and performance criteria are very strongly affected by winds.

Figure 5 is a representative wind escalation curve relating the frequency of occurrence winds as a function of height above the ground.[11] These curves

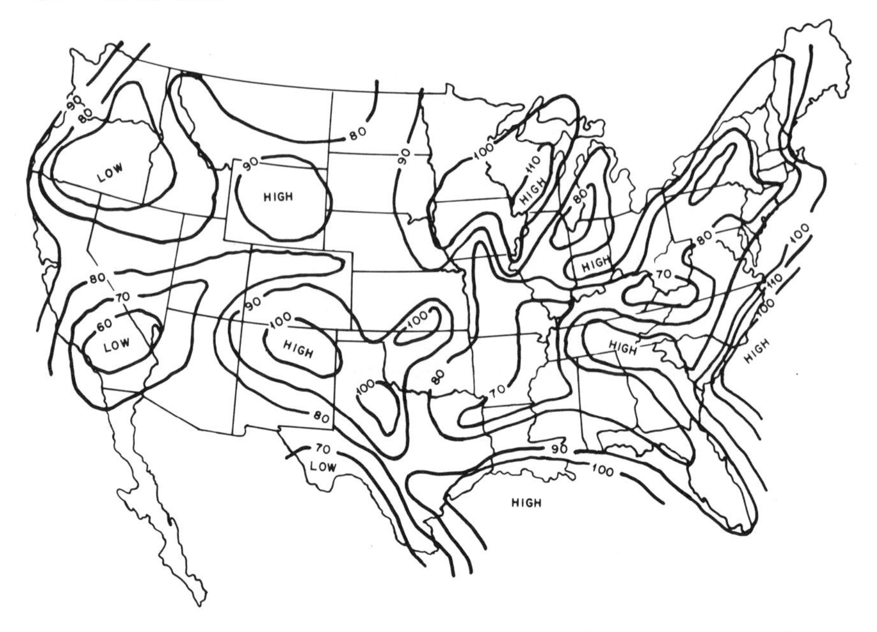

Figure 4. 5-min peak winds in the United States. Isopleths of 1 per cent of 25-year maximum 5-min wind in mph at height of 50 ft above ground.

indicate that, as the antenna size increases, the design wind speed must also be increased significantly if operation is to be maintained for a large fraction of the total time. For example, an antenna 50 ft above the ground designed to function in a 30-mph wind could be operable except for about 3 per cent of the time, whereas winds at a height of 400 ft will exceed 30 mph about 30 per cent of the time.

The Need for a Radome

Many of the smaller and less precise telescopes now in use are not capable of operating at full efficiency when in direct sunlight or when the winds exceed 20 to 30 mph. It was recognized that the severity of the wind and thermal problem would increase as a function of the antenna area and precision and would be further aggravated by the requirement for a fine mesh or solid reflector surface. These considerations, as well as the very consistent performance of the Haystack antenna and the clear indication that there will continue to be a copious supply of scientific problems and investigations to keep the proposed telescope in use essentially 100 per cent of the time, motivated a study to ascertain the influence of a radome upon telescope performance and cost.

Because of their extensive experience on the JPL 210-ft Goldstone antenna and other radio telescopes, the Rohr Corporation was asked to study the

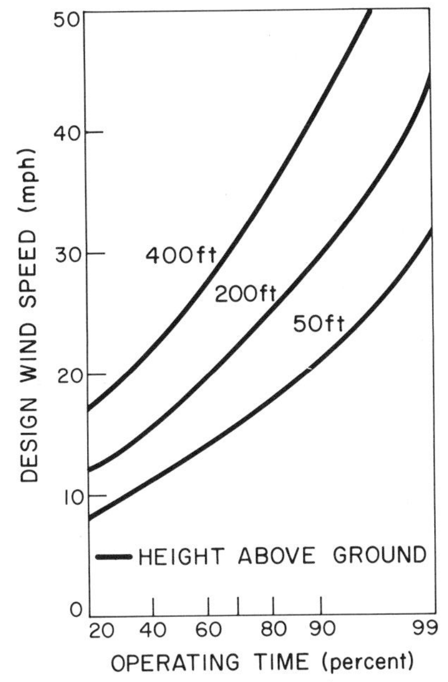

Figure 5. Wind escalation versus height.

comparative cost and feasibility of exposed and radome-protected telescopes 210, 328, and 400 ft in diameter. Their studies[12] indicated that it would not be feasible at any reasonable cost to fulfill the ±0.003° angular pointing requirement in an exposed, large, steerable microwave antenna. Within the radome, however, the desired performance objectives seemed completely feasible.

The attractiveness of the radome concept stems from the ability to deal independently with those design considerations that influence the precision of the telescope and those that influence its survival and performance. The price for this design freedom is not just the need to increase the diameter of the aperture by about 10 per cent to overcome the blockage and noise contributed by the radome, but also the cost of the radome must now be included in the over-all cost of the telescope.

How a Radome Simplifies the Design and Reduces the Antenna Cost

The underlying philosophy for the design of a radome-protected antenna differs substantially from that for an antenna that must operate in an exposed environment. Figure 6 lists some of the ways in which a radome can simplify

The reflecting surface and its supporting structure can be light in weight.

Simple techniques can be used to compensate for the only variable loads on the structure (gravity).

The antenna can have a low natural frequency (~0.2 cps) (low mass – low stiffness).

Small diameter bearings can be used since the overturning moments and weight are reduced.

The power required to drive and control the antenna system will be low (~20 hp).

Figure 6. Why the cost of an antenna can be reduced when it is protected by a radome.

and reduce the cost of the antenna system. In contrast to an outdoor antenna, the reflecting surface and many of the other components can be made exceedingly light in weight, limited primarily by fabrication, shipping, and erection requirement. For example, detailed studies show that the over-all weight for a reflector surface and the structure needed to attach it to the main backup support members can be as low as 0.25 lb/ft^2, only about 15 per cent of the weight of the equivalent structural elements in a typical exposed antenna. This low panel weight simplifies the supporting structure and control system.

With the advent of advanced computer techniques, which make it possible to calculate precisely the elastic deformations that will occur as the antenna is rotated about its elevation axis, it becomes possible in the controlled environment of a radome to introduce inexpensive preprogrammed compensation techniques. A variety of methods for correcting the gravity-induced elastic deformations have been studied[13] including (see Figure 7):

1. The adjustment of the reflector panels, each about 400 sq.ft in area, by controlled jacks at the corners of each panel;
2. The use of a small number of active motor-driven jacks or cam mechanisms in the reflector backup structure (displacement compensation);
3. The introduction of a variable prestress by means of control cables (force compensation);
4. The use of homology,[14] i.e., the maintenance of a parabolic contour by appropriately selecting the geometry and the size and stiffness of the structur-

- DYNAMIC CONTOUR CONTROL BY INDEPENDENTLY ADJUSTABLE PANELS
- RADIALLY SYMMETRIC REFLECTORS
  - HOMOLOGY
  - PASSIVE COMPENSATION
- NONSYMMETRIC REFLECTORS (VERTICAL TRUSS)
  - DISPLACEMENT COMPENSATION
  - FORCE COMPENSATION

Figure 7. Configurations studied for correcting deflections.

al members;

5. The use of appropriately coupled counterweights and linkages in the backup structure.

In the radome environment, because there are no rapidly changing wind loads and because of the light weight of the structure, the compensation mechanisms can be small and inexpensive. It was further found that the option was available to achieve the desired precision while maintaining an "invariant" parabolic contour, a capability not normally achieved in existing antennas. This eliminates the need to move the focal position as a function of the elevation angle.

One of the studies to investigate the improvement possible in the contour of a simple cantilevered truss as a function of the number of compensation elements is summarized in Figure 8. The deflections of a 200-ft-long cantilevered truss of the type that might be used in a radial rib of a 400-ft-diameter telescope were computed with the truss horizontal and vertical. The deflection at the tip was about 10 in., but a new parabolic contour with an rms error of about 2.2 could be "best fitted" to the changed contour. It was shown that by placing a single controllable actuator in the appropriate location in the truss the residual rms contour error could be reduced to about 0.08 in., an improvement of about 25 to 1. The use of additional compensators resulted in the further improvements as listed.

The absence of wind forces and the comparatively slow acceleration and tracking requirements for observing distant stars permits the design of a telescope with a comparatively low stiffness and structural natural frequency. In contrast to exposed antennas, which must be very stiff and have a lowest resonant frequency of at least one cycle if they are to point accurately in the presence of winds, the radome-enclosed antenna can have a natural frequency as low as 0.2 cps. This allows a major reduction in the material required in the structure and is a significant factor in permitting a low-cost design.

Equally important cost savings result because the lack of wind forces permits the telescope to be supported on a relatively small bearing and alidade structure. This, in turn, lowers the total weight and inertia of the telescope, reduces the bearing "stiction" problem and the cost. All of these factors, plus the absence of wind loads makes it feasible to control the antenna with a low-power drive motor. In contrast, one of the alternative designs for use with an exposed 400-ft antenna required in excess of 2000 hp on each axis of the telescope.

A Description of a Specific Telescope Configuration

Figure 9 shows the type of vertical-truss telescope now under study. Before selecting this concept we studied a variety of structural configurations for large parabolic reflectors. The basic types studied may be classified broadly into three groups: those with radial symmetry (radial ribs and circumferential trusses), those that are symmetrical about a single axis of rotation (vertical trusses), and those based upon shell configurations. While each of these configurations has specific advantages and limitations, when a radome is used and the structure only rotates about a single axis, there appears to be an advantage in employing a support structure based upon a series of deep vertical trusses. The rationale for this approach lies in the realization that an efficient use of the material in the backup structure involves the selection of a geometry with a high bending moment perpendicular to the axis of rotation; i.e., a deep truss oriented perpendicular to the elevation axis. In addition, the vertical-truss geometry is compatible with a variety of deflection compensation techniques

NUMBER OF COMPENSATION POINTS PER TRUSS	RMS DEFLECTION (in.)	RMS GAIN
0	2.17	1
1	0.077	28
2	0.033	66
3	0.013	167
4	0.0087	250
.	.	.
.	.	.
.	.	.
9	0.0011	1970

Figure 8. Deflection improvement versus number of compensation points (Weidlinger study).

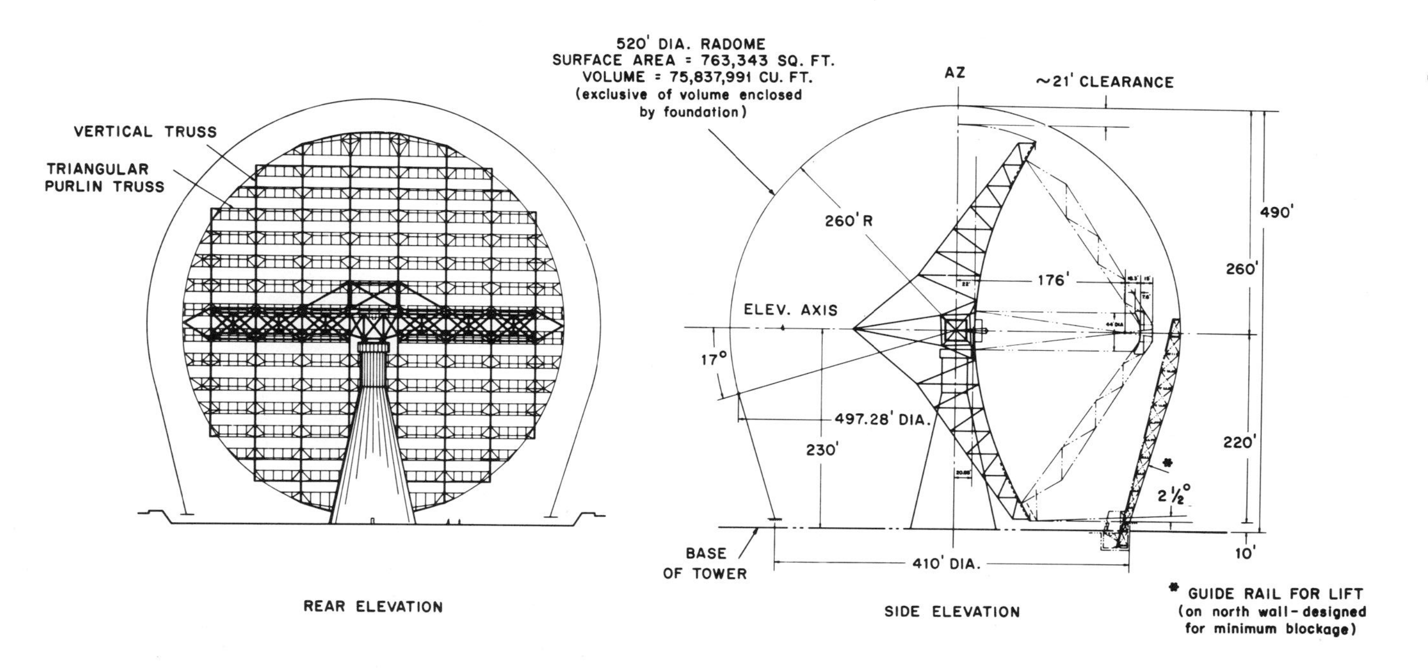

Figure 9. NEROC vertical truss telescope.

and can be analyzed with only moderate complexity. It also provides a geometry that permits a small azimuth bearing to be used and allows a large instrumentation laboratory to be located directly behind the apex of the reflector. Its disadvantages are that it is a novel configuration and, therefore, must be very intensively studied; being symmetric about only one axis, it may be more susceptible to temperature effects.

Preliminary studies showed that because of the lightweight design the weight on the azimuth axis would not exceed about 1000 tons, about 40 per cent of the moving weight of the 210-ft Goldstone antenna and of the 140-ft Green Bank antenna. The weight could easily be carried on an azimuth bearing only about 30 ft in diameter. In addition, a small azimuth support system would allow the bearing to be located between the center set of vertical trusses. With this geometry the inertia and cost of the azimuth support system would be very low compared to alternative configurations that used large-diameter bearing systems. The use of a tall, central tower to support the antenna provided a suitable enclosure for housing the control computer radar and other instrumentation. The general nature of this central tower is shown in Figure 10.

The pointing precision requirement indicated that it would be advantageous to "float" the antenna on an oil-film hydrostatic bearing. The 14-ft-diameter Haystack hydrostatic bearing, which supports a 120-ft radome-enclosed antenna (Figure 11), has been very satisfactory and has contributed to the attainment of a very high pointing precision. The geometry of the antenna led to the use of only four hydrostatic pads each with an area of about 20 sq. ft. A sketch of the bearing system now under study for NEROC by the Franklin Research Institute, Philadelphia, Pa., is shown in Figure 12.

Supported by the azimuth bearing system is a large, framed steel truss analogous in many ways to that used on the "Hammerhead" crane. This truss contains eight elevation bearings, one for each vertical truss. Surrounding the horizontal truss is a rotating torque tube that interconnects the eight vertical trusses.

The loads as seen by the Hammerhead truss are the same for different elevation angles of the antenna. This allows the elevation bearing axis to be adjusted only once during the final assembly stage. It has been possible to obtain 85° of elevation rotation without resorting to any design complications, and 90 deg of elevation travel could be achieved if it were essential.

The vertical trusses are spaced approximately 56 ft apart and are spanned by a series of lightweight purlin trusses, which in turn support the reflector panels. The base members of the purlin trusses are 14 ft apart and the trusses are also 14 ft apart. The reflector surface consists of a thin (about 0.011 in.) perforated aluminum sheet, about 62 per cent open, attached to a 14 x 14-ft lightweight but stiff doubly contoured frame. A grid of smaller framing

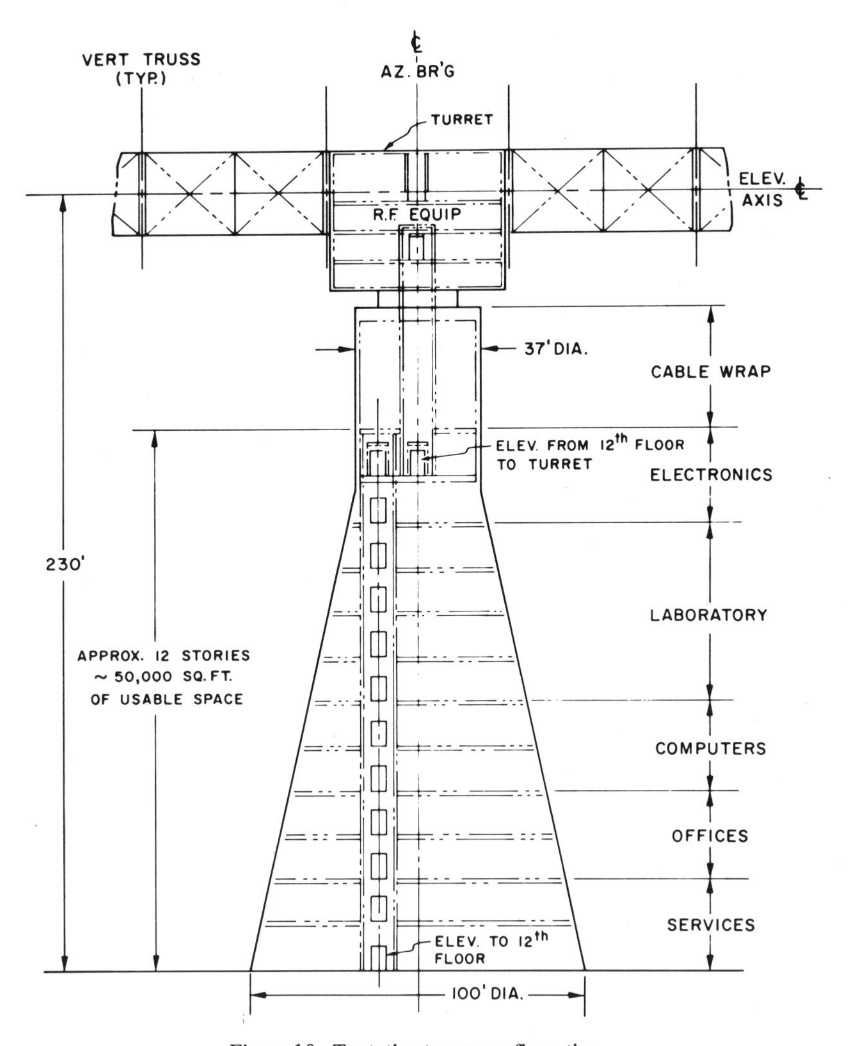

Figure 10. Tentative tower configuration.

members subdivide the large spans. To minimize construction, fabrication,and alignment problems, each of the 14 x 14-ft reflector panels, when viewed from the face of the aperture, will have the same projected rectangular shape. The reflector surface will deform if stepped on but will support the weight of a person, By the careful selection of materials, it has been possible to keep the weight of a 14 x 14-ft section of the panel and its supporting framework to approximately 0.25 lb/ft^2.

The vertical trusses have been extensively analyzed and optimized in a computer. By proper selection of geometry and member areas (homology) and

Table 2. Tolerance Budget (in.).

Primary reflector	Peak	rms	rms^2
Gravity			
Panels and purlin trusses	0.10	0.033	0.001089
Vertical truss	0.12	0.04	0.0016
Thermal			
Panels and purlin trusses	0.07	0.023	0.000529
Vertical trusses	0.08	0.027	0.00073
Geometry and manufacture panels	0.117	0.039	0.001525
Measurement and rigging panels	0.085	0.028	0.0008
Dynamic			
Vertical truss	0.03		
Torque tube wrapup	0.06		
Subtotal	0.09	0.03	0.0009
Jacks	0.09	0.03	0.0009
Hammerhead	0.072	0.024	0.000569
Total primary reflector			0.008642

rms error of primary reflector
$\sqrt{0.008642} = 0.093$ in.
Total rms error of secondary reflector
$\sqrt{0.00094} = 0.031$ in.
Total primary and secondary rms error
$\sqrt{0.008642 + 0.00094} = 0.0971$ in.

with a single compensator in each of the eight upper and eight lower trusses, it has been possible to maintain an "invariant" parabolic contour to within the tolerance budget shown in Table 2. The compensation will most likely be achieved by appropriate coupling of the counterweight to the truss structure.

Electromagnetic Considerations

Because of the desire to operate the antenna over a wide spectral range (5–100 cm) with high efficiency, it has been decided to employ both prime focus and Gregorian feed systems. The prime focus feed system would operate at wavelengths between 100 and 21 cm and the Gregorian system would be used between 21 and 3.2 cm. At the prime focus position an equipment room approximately 20 x 20 x 20 ft with a 4000-lb instrumentation capacity will be provided directly behind the Gregorian subreflector. The prime focus feed will be withdrawn within the instrumentation room when it is not in use. Access to this equipment room would be achieved when the antenna is pointed toward the horizon by means of an elevated platform support from the radome.

For Gregorian operation, feeds would be mounted on the face of an equip-

Figure 11. Haystack hydrostatic bearing.

ment room which will extend approximately 10 ft forward of the vertex of the prime reflector. Figure 13 is one view of a typical feed configuration at the vertex of the antenna. A number of feeds could be located in this area and be operated either concurrently or in sequence with appropriate corrections to the beam pointing instructions. Figure 14 is a photograph of a model of the telescope which shows the feed elements.

The achievement of an over-all 0.1-in. rms tolerance in an antenna with over 150,000 sq. ft of surface area is, indeed, a rigorous requirement. The care which must be utilized in the design, manufacture, and erection of this large precision antenna will become very evident by a brief study of the allowable error budget in Table 2. Nevertheless, based upon the experience of the Haystack system, these goals appear achievable at low risk by utilizing a very simple but carefully evaluated design concept and by paying great attention to the quality control, the assembly of the system and the environment within the radome.

Since contour errors due to tilting a reflector can now be made negligibly small, it is becoming increasingly evident that *the precision in a large reflector will, in the limit, be determined by thermal environment.* This is particularly true for an exposed antenna, but careful control of temperature changes

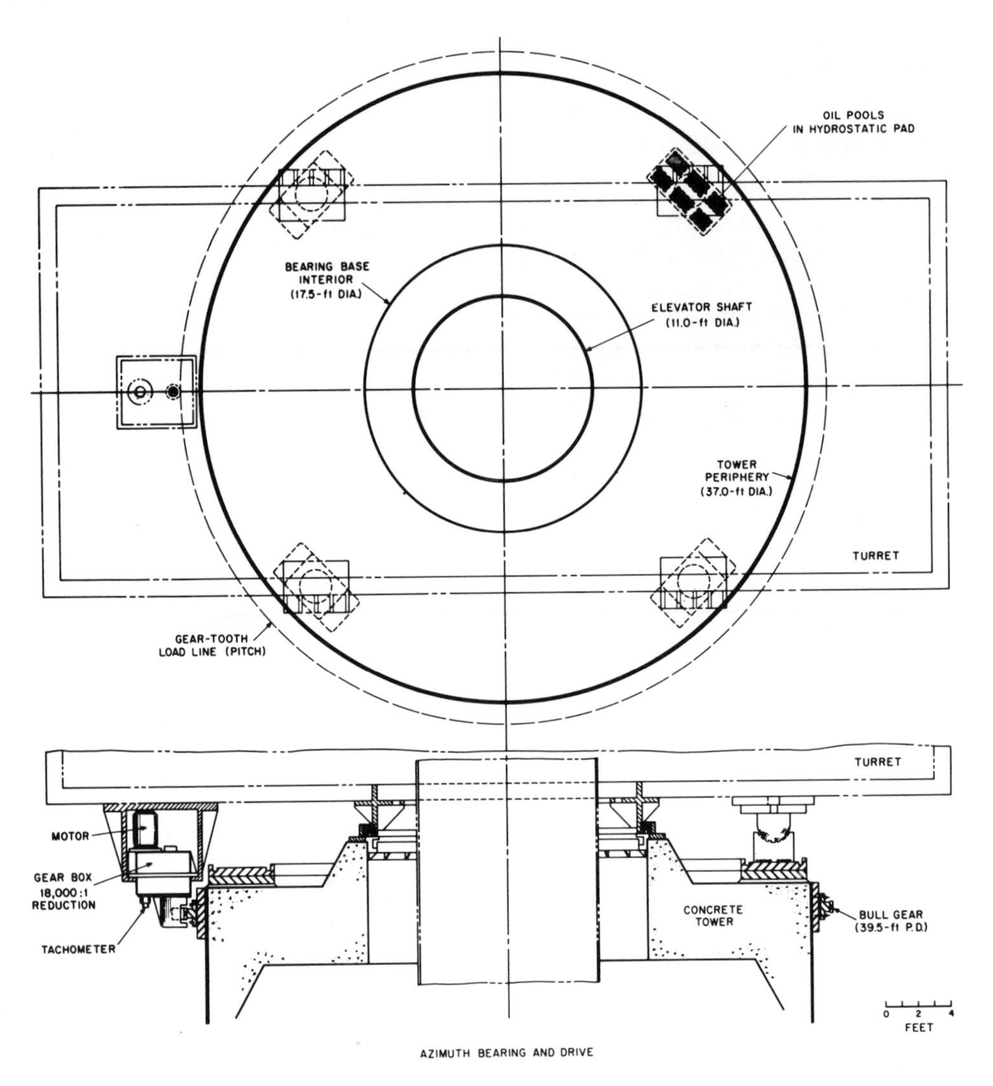

Figure 12. Four-pad bearing system.

and temperature gradients will also be needed if a radome is used. It is just not feasible to build very high precision large antennas at low cost unless materials such as steel, aluminum, or fiberglass are used; and since these materials have thermal coefficient of expansion of about 1 part in 100,000 per °F precisions higher than about 1 part in 30,000 are difficult to maintain unless the structure is in a very stable thermal environment. Antennas with closed-loop feedback systems or which use the circulation of liquids throughout the structure to minimize the thermal effects become complex and costly.

Selection of the 440-ft Antenna Size

After completion of a series of studies to evaluate the feasibility and cost of fully steerable reflectors in the 300–500-ft size range, there appeared to be sound, scientific arguments for choosing a size with an equivalent effective aperture of not less than 400 ft. Studies of the electromagnetic performance that might be achieved in large radomes (Figure 15) indicated that the aperture blockage could be in the vicinity of 0.7 dB and the noise temperature contribution of the radome would be about 5°K. To overcome these system losses it appeared appropriate to increase the antenna diameter by approximately 10 per cent to achieve a gain/temperature ratio equal to that of an ideal 400-ft telescope (see Figure 16).

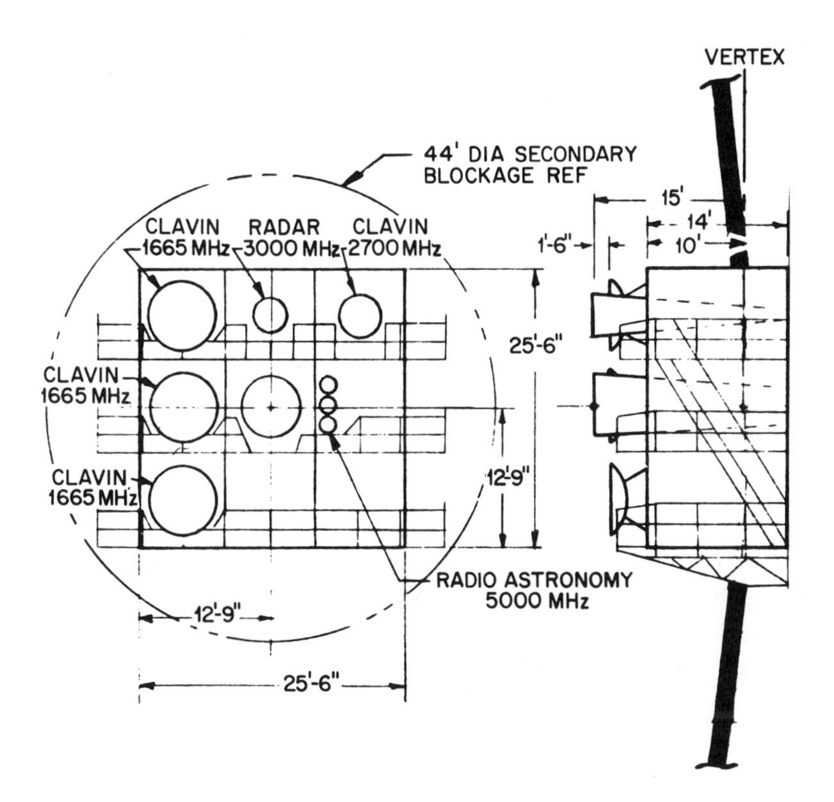

Figure 13. Typical vertex feed configuration.

Radome Studies[4,5,7]

The over-all geometry of the radome to protect the 440-ft antenna is shown in Figure 17. While the exact design details are still under investigation, the tentative characteristics of the space frame and the membrane of the radome are shown in Figure 18. The challenge in the design of a very large radome is to provide a structure that will survive under the environmental loads and at the same time have a minimum influence on the electromagnetic properties of the antenna. This requires a detailed understanding about the environment and the structural properties of these reticulated shells.

A number of three-dimensional linear and nonlinear mathematical studies have been undertaken to establish buckling criteria for large space-frame

Figure 14. Model of telescope showing feed elements.

radomes. To check the validity of the calculations, a structural model of a 14-ft-diameter radome space frame has been constructed (Figure 19) and its performance measured under controlled load conditions. In addition, wind tunnel tests have been conducted to obtain pressure profile distributions around the radome under a variety of conditions.

One concept under investigation is to provide a pressurization system within the radome to minimize the likelihood of buckling under severe wind conditions. A pressure of approximately 0.25 $lb/in.^2$ can be achieved within the radome with only about 50-hp hours of energy. Studies are also under way to determine the most appropriate means for maintaining the proper thermal gradients within the radome as well as for preventing snow from forming on the top of the structure. While the exact design for this air circulation and heating system is still under study, it is expected that the eventual system will, in many respects, resemble that in use at Haystack.

Based upon tests on a variety of possible dielectric panel materials, it is currently planned to employ pre-pregnated fiberglass panels approximately 0.040 in. thick. These panels, which will be probably shipped to the site in rolls, will have metal edging strips bonded to the panels; these edging strips in turn will be fastened into recessed grooves in the metal space frame. To obtain a structure with the greatest electromagnetic transparency, the fiberglass panels will be used to stabilize in-plane buckling of the metal beams. Accelerated life tests on a variety of plastic materials have been started and, to date, the structural properties seem suitable for at least a 30-year life. Many other tests to evaluate the dielectric properties, the absorption of moisture, and optical characteristics of radome materials have been completed.

A number of different techniques are under investigation for erecting this large radome. It is evident that between 25 and 30 per cent of the total cost of the structure will result from the on-site erection costs. In addition to the more obvious erection methods, one possible technique under investigation involves the assembly of the radome from the top down with successive tiers being raised as the assembly proceeds. This technique is attractive because all

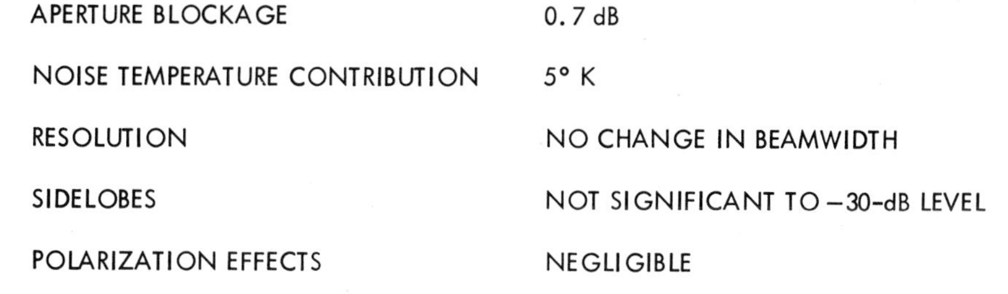

APERTURE BLOCKAGE	0.7 dB
NOISE TEMPERATURE CONTRIBUTION	5° K
RESOLUTION	NO CHANGE IN BEAMWIDTH
SIDELOBES	NOT SIGNIFICANT TO –30-dB LEVEL
POLARIZATION EFFECTS	NEGLIGIBLE
POINTING ERRORS	LESS THAN 1/100 OF A BEAMWIDTH

Figure 15. Design goals for radome.

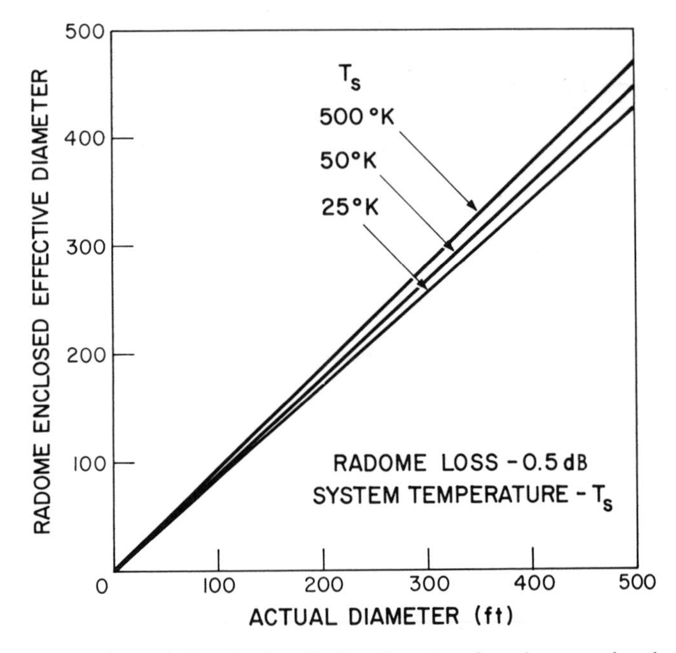

Figure 16. Ratio of actual diameter to effective diameter of a radome-enclosed antenna.

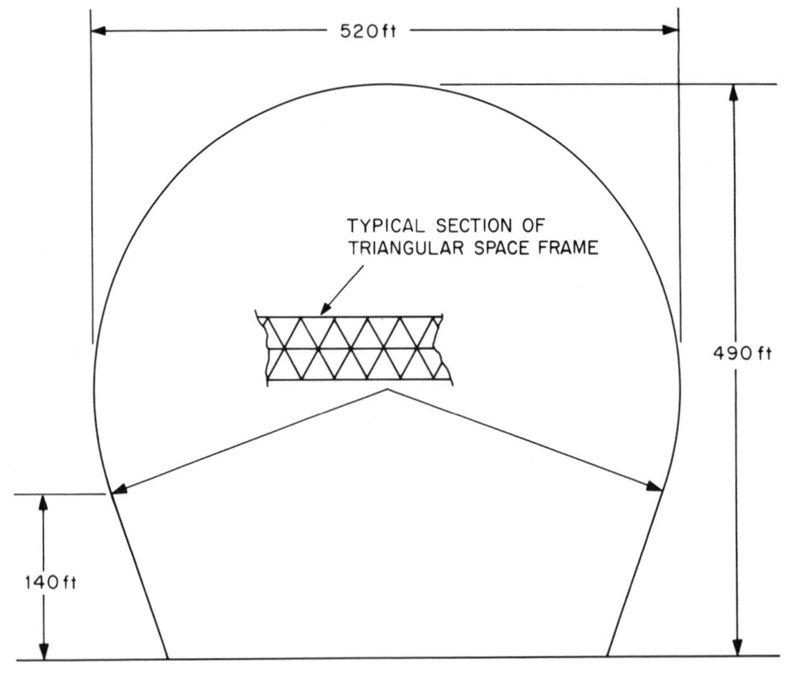

Figure 17. External dimensions of a radome.

SPACE FRAME

BEAM LENGTH (average)	30 ft
BEAM CROSS SECTION	3 × 14 in.
MATERIAL	STEEL (60,000 psi, yield)
TOTAL LENGTH OF BEAMS	9×10^4 ft
TOTAL WEIGHT OF STEEL	2400 TONS

MEMBRANE

MATERIAL	FIBERGLASS
THICKNESS	0.040 in.

Figure 18. Space frame and membrane characteristics.

of the assembly work is accomplished within 30 ft of the ground since the dielectric panels would also be installed to stiffen the space frame as the assembly proceeds.

Cost Studies

The attainment of a realistic cost estimate for the CAMROC antenna and radome has been a pivotal part of the design studies. To obtain meaningful data, several analyses were conducted on a preliminary design by independent groups both within industry and Lincoln Laboratory.[7,15] These cost studies show that because of a large number of factors all resulting more or less because a radome is used, the cost of the antenna alone is substantially below that which would be anticipated for a large, exposed antenna. Figures 20 and 21 indicate that the actual construction cost for the antenna should be about \$5.5 million and that the total assignable cost including engineering and contingencies are estimated to total \$9.6 million. The radome structure and the air heating and circulation system are also estimated to cost about \$5.5 million, with the total over-all radome cost of \$8.1 million.

It is difficult to place these costs in perspective because 400-ft-diameter antennas with the desired surface tolerance and pointing precision do not now exist. However, an attempt has been made to compare these costs with the known cost of two relatively large and precise antennas that have been placed into operation during the past two years (see Figure 22). One of these antennas is the 150-ft National Research Council antenna at Algonquin Park,

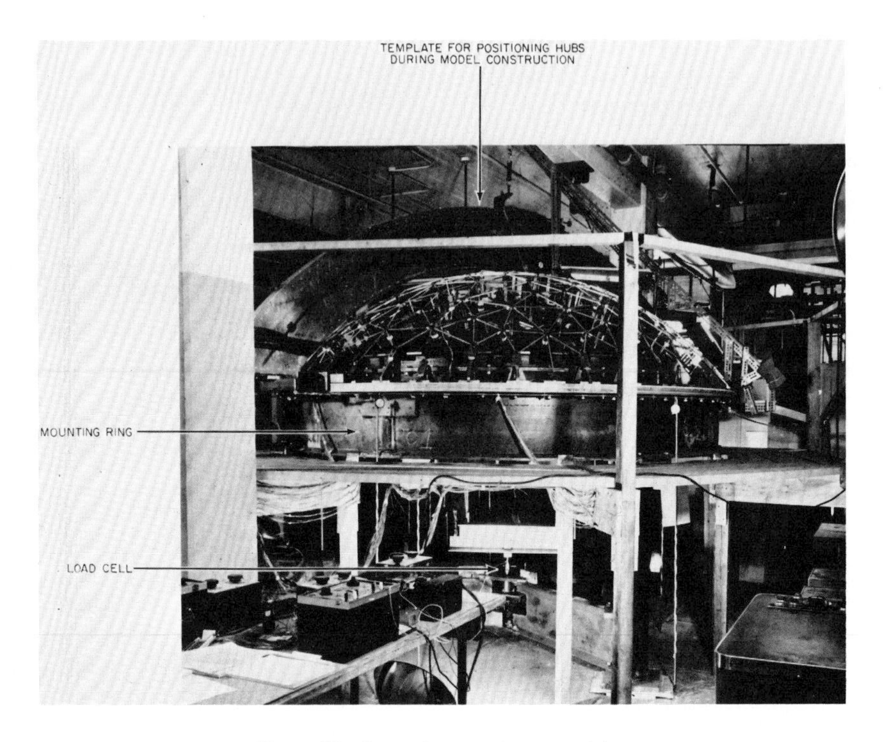

Figure 19. Space-frame radome model.

Ontario, Canada. Its cost of \$3.5 million (which could probably not be duplicated at a competitive cost in the United States because much of the design and engineering was completed in England and Germany), when scaled to 400 ft in accordance with a cost versus size relationship of $\$D^{2.5}$ is estimated to cost about \$40 million. The other comparative example is scaled from the 210-ft JPL "Goldstone" antenna, which probably provides a more appropriate basis for construction estimates in the United States. This same scaling law indicates that a 400-ft-diameter exposed telescope would cost about \$65 million. While the validity of these cost scaling relations is certainly open to considerable question, it is believed that there is still major significance in the approximate 3 to 1 cost spread between the radome-enclosed and the equivalent exposed telescopes. In addition, *cost considerations notwithstanding there appears little likelihood that it would be possible to achieve the desired surface tolerance and pointing precision with an antenna of any known design if it were exposed to typical environmental conditions.*

An attempt has been made to compare the cost of the proposed design on the basis of cost per pound and cost per square foot using the NRC and JPL systems for comparison. This information is shown in Figure 23 and 24 and seems to indicate that the budgetary estimates generated by our studies are

Antenna (440 ft – 0. 1-in. rms)	DOLLARS (in millions)
Primary reflector	$2.35
Secondary dish	0.13
Reflector support structures	0.17
Lab. space behind reflector	0.09
Tower foundation	0.43
Azimuth bearing	0.28
Azimuth drive	0.10
Elevation bearing	0.48
Elevation drive	0.14
Miscellaneous items	1.26
Construction total	5.43
Engineering	1.40
Engineering supervision	0.85
Contingency 25%	1.92
Total Antenna Cost	$9.60

Figure 20. Antenna cost summary.

realistic in that we are allowing substantially more dollars for each pound of material in the structure. At the same time, the over-all cost per square foot of useful antenna aperture is significantly lower than has been achieved to date in large, steerable antenna systems.

Conclusions

The studies that have been carried out during the past 18 months have outlined an attractive concept for a very high performance, steerable telescope. In addition, there now is a reasonable basis for predicting the performance and cost for this type of instrument. It is anticipated that one additional year of design and engineering effort will complete the planning and establish a firm design and more detailed cost information. The results to date also suggest that substantially larger steerable antennas can be constructed at a

Radome (520-ft dia.)	DOLLARS (in millions)
Steel space frame	$2.20
Fiberglass	1.46
Foundation	0.30
Miscellaneous	0.50
Structure total	4.46
Air circulation and heating	1.00
Construction total	5.46
Engineering design	0.70
Engineering supervision	0.30
Contingency 25%	1.64
Total Radome Cost	$8.10

Figure 21. Radome cost summary.

comparatively low cost per square foot provided radomes can be increased in size without degrading their electromagnetic performance.

Acknowledgments

In this presentation, I am privileged to be the spokesman for a very large and competent team of scientists, engineers, industrial consultants, and design and manufacturing organizations who have, and are, continuing to participate in these studies. While it is not possible to give proper credit to the approximately 40 members of the Harvard and M.I.T. family who are active in this program, the white-covered progress reports being distributed during this conference list in detail the many individuals and organizations participating in these studies. It would be remiss, however, if specific mention were not made of the major contributions of P. Stetson on the antenna configuration and R. D'Amato on the radome.

The Lincoln Laboratory is operated with support from the U.S. Air Force.

References

1. "Objectives and Study Programs for a Regional Radio and Radar Astrono-

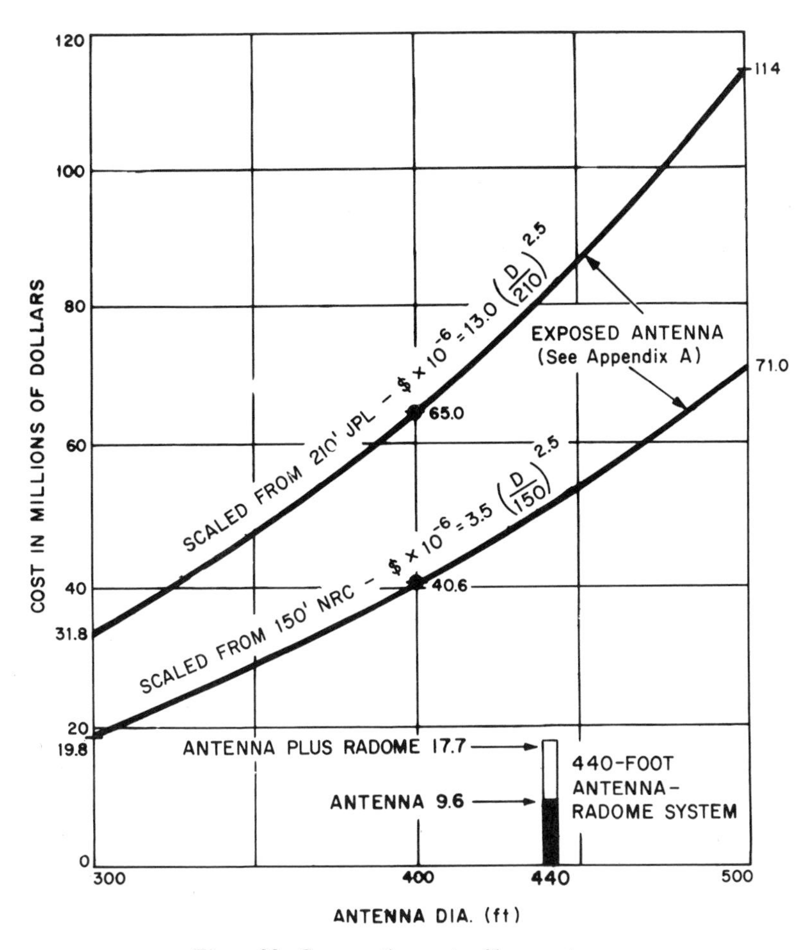

Figure 22. Comparative costs of large antennas.

my Research Facility," Lincoln Laboratory Internal Report (February 1966).

2. "Engineering Design Objectives for a Large Radio Telescope" (October 1965); reissued in two volumes as Report, Defense Documentation Center, Washington, D.C. (January 1966).
3. J. Ruze, "Antenna Tolerance Theory—A Review," *Proc. IEEE* **54**, 633 (1966).
4. H. G. Weiss, "The Haystack Microwave Research Facility," *IEEE Spectrum* **2**, 50 (1965).
5. J. Ruze, "Radomes and Large Steerable Antennas: Conference Proceedings" (17–18 June 1966), Paper 10.

Antenna	Rotating Weight (tons)	Cost	Cost per Pound
150-ft NRC (1966)	900	$ 3,500,000	$1.95
210-ft JPL (1966)	2500	13,000,000	2.60
440-ft NEROC	1050	9,600,000	4.60 (in budget)

Figure 23. Cost per pound.

Antenna	Aperture Area (Sq Ft)	Cost	Cost per Sq Ft of Antenna Aperture
150-ft NRC (1966)	17,650	$ 3,500,000	$200
210-ft JPL (1966)	34,600	13,000,000	375
440-ft NEROC	152,000	17,700,000*	115 (in budget)

* Radome cost of $8,100,000 included.

Figure 24. Cost per square foot.

6. "Large Antenna Configurations," prepared for CAMROC Conference (17–18 June 1966).
7. "A Large Radio/Radar Telescope–CAMROC Design Concepts," Lincoln Laboratory Internal Report (15 January 1967), Vols. I and II.
8. "Large Steerable Radio Antennas–Climatological and Aerodynamic Considerations," *Ann. N.Y. Acad. Sci.* **116**, 1–355 (26 June 1964).
9. N. Sissenwine and I. Gringorten, "Radomes and Large Steerable Antennas: Conference Proceedings" (17–18 June 1966), Paper 2.
10. H.C.S. Thorn, "Distribution of Extreme Winds in the United States" (U.S. Weather Bureau, Washington, D.C., 1959), 23 pp.
11. "Winds," in *Handbook of Geophysics and Space Environments* (McGraw-Hill Book Company, Inc., New York, 1966), Chap. 4, pp. 4-1 - 4-54.
12. "A Large Radio/Radar Telescope–CAMROC Design Concepts," Appendix A–"A Study to Evaluate the Effects of a Radome Environment upon the Performance and Cost of a Large - Diameter Radio Telescope," Rohr Corporation, Chula Vista, California (21 October 1966).
13. "A Large Radio/Radar Telescope–CAMROC Design Concepts," Lincoln Laboratory Internal Report (15 January 1967), Appendixes C, D, E, F, G.
14. S. von Hoerner, "Design of Large Steerable Antennas," *Astron. J.* **72**, 35 (1967).
15. "A Large Radio/Radar Telescope–Proposal for a Research Facility," Lincoln Laboratory Internal Report (June 1967).

M. Ginat
Obseratoire de Paris
Section D'Astrophysique
Paris, France

THE NANÇAY RADIO TELESCOPE

The Nançay radio telescope is a meridian instrument similar to the Ohio State University radio telescope (J. D. Kraus). It has a plane reflector 200 x 40 m (600 x 120 ft) located 460 m (1400 ft) away from the second reflector 300 x 35 m (900 x 100 ft), which is a portion of a sphere whose radius is 560 m (1700 ft). Both reflectors are symmetrical with respect to the meridian plane. (See Figures 1 - 3.)

Designed for 1420 MHz (21 cm), it is used also at 2700 and 5000 MHz (11 and 6 cm) (see Figures 4 and 5). The aperture efficiency in the axis is less than for a parabolic reflector of the same aperture, but the observational field is larger (± 7.5° about the axis). The position of the plane reflector with respect to the center of the sphere results from a compromise between the loss in aperture efficiency and the geometrical aberrations (coma, field curvature, etc.) Sphericity aberration is corrected according to Schmidt's procedure: slight deformations of the east and west edges of the plane reflector. Phase correction introduced varies with declination of observation between 0.3λ and 0.6λ at 5000 MHz.

The main focus is located 2 m above the ground. To allow the tracking of radio sources before and after meridian transit, the feeds are made movable on the focal sphere, horizontally on a rail track and vertically with a lift.

The reflecting surfaces are made of welded mesh of 1.5-mm-diameter steel wires, at 12.5-mm centers, galvanized after welding. The rms errors of the surfaces as built with respect to theoretical ones are:

Figure 1. Radio telescope, aerial photograph (by courtesy French War Minister).

± 3 mm (0.12 in.) for the plane reflector and for the central and lower part of the spherical reflector (5/6 of the surface);
± 5 mm (0.2 in.) for the higher part of the spherical reflector.

Spherical Reflector

The backup structure of the spherical reflector is made of 31 vertical pylons 35 m high, set radially 10 m apart on a circle 560 m in radius. Horizontal ribs each 5.50 m brace the pylons and support secondary vertical beams spaced 2.50 m apart. On this structure are stretched horizontal steel cables (9 mm diameter) that support the vertical mesh strips. The spacing of the cables is 0.80 m. The surface is such that:

1. Its section by a diametrical vertical plane is an arc of a circle 560 m in radius.
2. Its section by a horizontal plane is a polygonal line of 120 elements 2.50 m long.

Errors due to this design, with respect to the theoretical sphere, amount to 1.5 mm. Each pylon was set in its theoretical place by means of a precision

Figure 2. Spherical reflector: backup structure.

geodetic survey. Adjusting devices are located at each intersection of the cables and vertical secondary beams and pylons. Each of them was set independently on the theoretical sphere by means of theodolite measurements. A survey was made by microtriangulation on targets distributed along the

Figure 3. Plane reflector: backup structure.

main pylons in the central part of the reflector. The arithmetic mean of absolute values of deviations from the sphere is 2.4 mm.

Plane Reflector

The plane reflector is made of ten independent identical panels 40 m high and 20 m wide, facing south. Each panel rotates 90° about a horizontal east-west shaft that lies 21 m high on the top of steel towers. The reflector is used between 30° and 90°, i.e., between +76° and −44° declination. The horizontal position is used only for storage. The backup structure of each panel takes the shape of a large half-wheel bearing on the drive system. In order to obtain a sufficient stiffness in the vertical plane, the shaft is made of a bos-girder. It is connected by means of radial struts to a semicircular rim (radius 11 m) bearing on the driving drum. The rim is fixed to a double cantilevered polygonal structure, symmetrical with respect to the shaft, which supports the reflecting plane made of horizontal H beams.

The cables that bear the mesh, spaced 0.75 m apart, are stretched vertically by adjustment bolts and rest on the horizontal beams. The surface of the panels was made plane, by direct survey in the horizontal position, to ± 2 mm accuracy before it was covered by the mesh.

The shaft of each panel ends by two trunnions rotating on top of steel towers. A gangway at 9 m above ground, extending between towers, acts as bracing and bears the driving mechanisms. Each element weighs 40 tons; the ten shafts were aligned (or set parallel according to Schmidt's correction) in the east-west direction by means of precision triangulations to ± 2 mm accuracy.

The deflections of the surface during rotation were obtained by two different methods:

1. Direct measurement of deflection targets distributed along one meridian section of the panels;
2. Determination of the equation of the mean plane by geodetical survey, giving coordinates of targets distributed on the whole surface of two elements. During rotation, the maximum deflections between reflecting surface and mean plane are ± 6 mm. If the mean inclination angle of the panel is measured by the inclination of the slope line joining regions 12 m apart from shaft, the rms error is ± 6 sec of arc.

A mechanical cam corrector, driven by the trunnion, takes into account the differential rotation between shaft and mean reflecting plane. The corrector is adjusted and periodically tested with a pentagonal prism telescope, giving the reference line joining targets fixed to the structure 12 m from the shaft. The output shaft of the corrector drives directly a Ferranti digitizer (moire

Figure 4. 1420 MHz carriage and equipment.

Figure 5. "Continuum" carriage, feeds for 1405,2700,and 5000 MHz.

fringes system). As the digitizer undergoes any movement of the supporting tower, the inclination of the angular reference (zero of digitizer) is measured with an accuracy of ± 1 sec of arc by an Electrolevel device (British Aircraft Corporation). This system is used for compensation of the thermal deformations of the tower supporting the shaft and its digitizer. The semiautomatic control system for pointing the panel to a given inclination angle was also designed by Ferranti.

Four panels only have been given an inclination measuring system; they control six others by two chains of servos:

1. One for coarse adjustment, which controls the high-speed electric motors of the drive (6° of rotation per time minute), using Honeywell magnetic proximity switches with sensitivity threshold of ± 3 min of arc.

2. One for final adjustment, which controls the low-speed electric motors (6′ of rotation per time minute), using proximity detectors with air-coil differential transformers and sensitivity threshold of ± 6 sec of arc. The rotation of the panels causes complicated oscillatory modes in the structure (frequencies around 1.6 Hz, amplitude 160 sec of arc peak to peak); it was assumed that the plane reflector would always be used in fixed inclination during observations and that the durations of final pointing would be sufficient to allow a complete damping of the oscillations caused by the high-speed rotation.

The parallelism of the ten panels is checked from the control console by means of air-coil differential transformers described above. Every parameter included, we may assume that the rms error in pointing the plane reflector rises to ± 20 sec of arc in the best environmental conditions (middle of night, or cloudy windless weather). Wind speeds of 10 m/sec are the utmost limit for observations.

Focal Systems

Radio sources are observed from half an hour before to half an hour after local meridian transit time. Antennas are carried by two carriages (mean weight 9 tons). The first carriage bears the hydrogen-line equipment: hog horn feed[1] and correlation receiver with cooled parametric amplifier.[2] The second carriage bears "continuum" equipment: hog horns for 1405, 2700, 5000 MHz and correlation receivers for both polarizations with parametric amplifiers.[2] On each carriage, vertical movement of the feed to take into account the change of zenith distance of the image of the source during tracking is made with a hydraulic lift system. Horizontal tracking is made along a 100-m-long rail track, developing concentrically to the spherical reflector. This rail track includes four supporting rails and one guiding rail. One vertical face of the guiding rail has been grounded and bears a rack for the drive. Position along the rail track is measured on each carriage with an Invar wheel rolling along the guiding rail and driving an angular digitizer. The accuracy of the radius of curvature of the guiding rail is ± 10 mm.

Control of the carriages and of the lifts carrying the feeds is made through Ferranti numerical control equipment: each 30 sideral seconds the carriage under control receives from a punched tape:

1. A position order along the rail track;
2. A speed order for horizontal movement (adjustable between speeds 0 and 20 cm/sec);
3. A position order for vertical movement.

Any kind of tracking can be realized (accelerated, delayed, drifts, etc.). The accuracy of positioning along the rail track is ± 2 mm.

For the frequency 5000 MHz, the ground temperature is 70° K and the aperture efficiency 16 per cent. Aperture efficiencies for the other frequencies are:

45 per cent for 1420 MHz;
22 per cent for 2700 MHz.

The instrument was completed in November 1966.

References

1. F.Biraud, "Sources primaires pour le grand radioteléscope de NANÇAY," *Onde Electr. Fr.* No. 479 (February 1967).
2. J. Delannoy and J. C. Ribes, "Amplificateurs parametriques pour la radioastronomie," *Onde Electr. Fr.* No. 479 (February 1967).

Supplementary Readings

J. Arsac, "Calcul du grand radioteléscope de Nançay," internal report.

M. Ginat, "Le radiotelescope a deux miroirs de NANÇAY: étude de la structure (App P H D CNRS A O 1044 PARIS 1966)," *J. Observateurs Fr.* (in press).

M. Ginat and J. L. Steinberg, "Le radiotelescope à deux miroirs de NANÇAY: étude du pointage," *Rev. Phys. Appl. (J. Rhys. Radium Fr;)* **II**, 2 (1967).

C. Scruton
Aerodynamics Division
National Physical Laboratory

SOME CONSIDERATIONS OF WIND EFFECTS ON LARGE STRUCTURES

1. Introduction

Since the international conferences in 1963 on Wind Effects on Buildings and Structures[1] and on Large Steerable Radio Antennas–Climatological and Aerodynamic Considerations,[2] considerable progress has been made in the understanding and the treatment of the various aspects of loading due to wind. Except for some *ad hoc* measurements of wind loads on models in wind tunnels (e.g., References 3–5) little of this work is directly applicable to large radio telescopes or their radomes. Nevertheless, the researches on the aerodynamics of bluff structural shapes and on the response of flexible structures to atmospheric winds will be of interest to the designer of radio telescopes, and the improvements in wind tunnel techniques, particularly with regard to the simulation of natural winds, are equally applicable to radio telescopes and radomes as to other structures.

For major constructions wind effects are conveniently considered under three categories. Briefly these are:

1. The maximum time-averaged wind loads to which the structure will be subject during its lifetime.

2. The dynamic response to the direct fluctuating wind loads arising from the turbulence of atmospheric winds.

3. The oscillatory and divergent instabilities due to wind.

In the experience of the writer, the structural designs of most large structures (including radio telescopes) continue to be based, at least in the preliminary stages, on the time-averaged wind loads, but there is a growing tendency to check the design for dynamic response employing the statistical concepts advocated by Davenport.[6] A difficulty of the latter approach is that the aerodynamic transfer functions for converting from the power spectra of wind speed to those of wind loads are rarely available. Tests of aeroelastic models in wind tunnels with proper simulation of atmospheric winds are valuable but are expensive and not always practicable.

The problems of the estimation of the time-averaged and the fluctuating wind loads on large structures will now be discussed, together with some of the more interesting illustrative results of recent investigations.

Notation

A		Area
C_D	=	Drag/$\frac{1}{2}\rho V^2 A$, drag coefficient
C_L	=	Lift/$\frac{1}{2}\rho V^2 A$, lift coefficient
C_m	=	Pitching moment/$\frac{1}{2}pV^2 AR$, pitching moment coefficient
C_{pb}	=	Pressure/$\frac{1}{2}\rho V^2$, base pressure coefficient
D		Diameter or width of structure
E		Young's modulus
$F(n)$		Power spectral density function
H_a, H_s		Aerodynamic and structural stiffnesses per unit length
K_a, K_s		Aerodynamic and structural damping coefficients per unit length
K		Surface drag coefficient
h_a	=	$H_a/\rho N^2 D^2$
k_a	=	$K_a/\rho ND^2$
L		Length
L_x		Longitudinal scale of turbulence
M		Mass per unit length
N		Structural frequency
n		Frequency of turbulence
q		Dynamic pressure of wind
S		Area of working section
s		Reference height for wind speed
t		Shell thickness
u		Velocity fluctuation due to longitudinal component of turbulence

V	Wind speed
$\bar{V}$	Mean wind speed
$\bar{V}_s$, $\bar{V}_Z$	Mean wind speed at reference and current height, respectively
V_r	= V/ND, reduced velocity
Z	Height
z	Displacement
α	Wind incidence *or* Exponent of power-law profile of wind speed with height
ρ	Air density
σ	Structural density
η_0	= z_0/D, nondimensional amplitude

2. The Time-Averaged Forces

For many decades the designer has made his estimates of wind loading using a design wind speed based on the maximum wind or gust speed to which the structure will be subjected during its lifetime,* together with wind load coefficients found from tests carried out in the smooth airflows of conventional wind tunnels. These airflows failed to represent the conditions in the atmosphere in at least the following respects:

1. The variation of wind speed with height (wind shear).
2. The variation of the direction of the wind with height.
3. The turbulence properties (intensity and scale).
4. The Reynolds number of the flow.
5. The walls and roofs of enclosed working section wind tunnels unnaturally constrain the airflow (blockage effects).

With the exception of the variation of wind direction with height, which according to Sutton[7] could amount to between 30 to 45° in 1000 ft, attempts have been made in a number of aerodynamics laboratories both to obtain airflows in wind tunnels more closely representative of natural wind and to derive corrections (e.g., blockage corrections) to account for the lack of similitude.

The influence of the Reynolds number is not very marked for exposed reflector bowls but may be important for an enclosing radome. For both structures wind shear and turbulence can be expected to have a pronounced effect on the wind loads. The drag of flat plates normal to a turbulent airstream has been measured by Vickery[9] and by Bearman at the NPL. The mean base pressure decreased (leading to higher drag forces) as both the scale and the intensity of the turbulence were increased. The results for plates of various

* The problems of assessing a design wind speed are discussed in Reference 8.

shapes and sizes in relation to the scale of the turbulence were found to correlate if plotted against $(\sqrt{\bar{u}^2}/\bar{V})(L_x/D)^2$, indicating that the mean base pressure is more sensitive to changes of scale than of intensity of turbulence. Vickery[10] found that the base pressure of square section prisms increases (and hence the drag decreases) when turbulence is present in the airstream (Figure 1). The

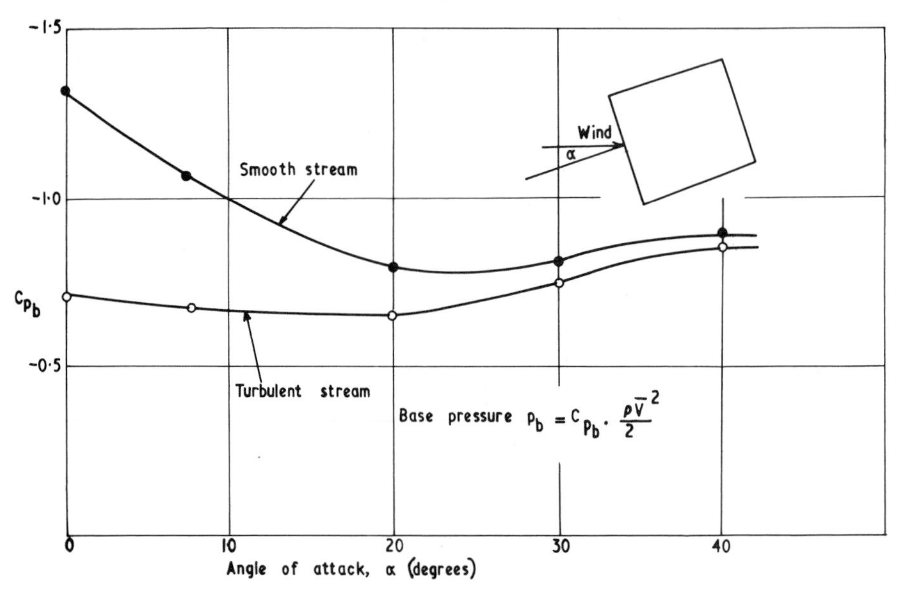

Figure 1. Variation of base pressure coefficient with angle of attack on an "infinite" square section prism in smooth and turbulent airstreams.

mean drag of cylinders and spheres (a probable shape for a radome) are influenced by the turbulence properties to an extent dependent on the Reynolds number. The results for a cylinder by Bearman[11] (Figure 2) suggest that the turbulence had little effect for low values of Reynolds number, but the effect on base pressure and drag ($-C_D \approx C_{pb}$) was very large for the range $2 \times 10^4 < Re < 4 \times 10^5$. Above this range of Re, increases of drag with turbulence can be expected. For spheres Dryden[12] *et al.* plotted critical values of Reynolds number as a function of the parameter $(\sqrt{\bar{u}^2}/\bar{V})(D/L_x)^{1/5}$. This plot is reproduced in Figure 3 together with a similar plot for circular cylinders. In the region below the curve the wind forces are sensibly independent of the Reynolds number and turbulence, but considerable differences in the wind load coefficients can be expected between those found for conditions below and above the curve.

On tall structures wind shear produces considerable changes in the pressure and drag distributions with height. In general, wind forces obtained in wind tunnel tests in uniform wind will be greater than those found with wind shear

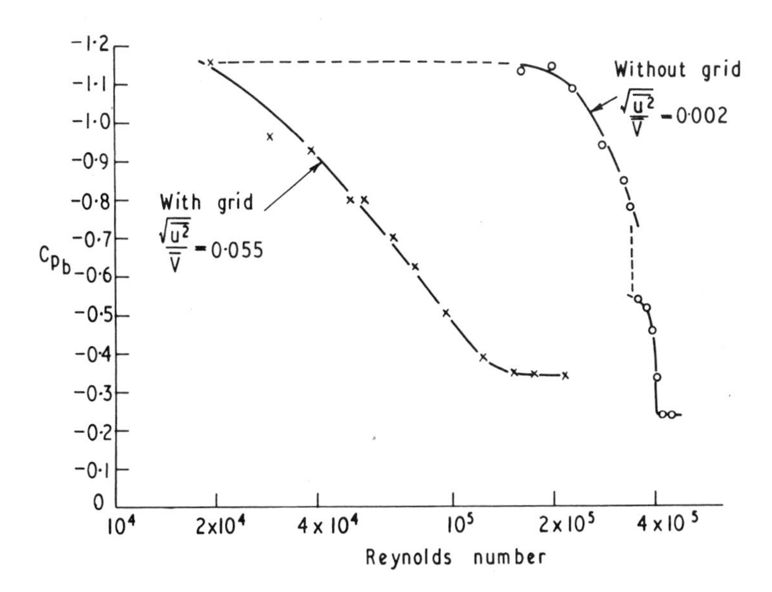

Figure 2. Circular cylinder - variation of base pressure coefficient with Reynolds number; with and without a turbulence grid.

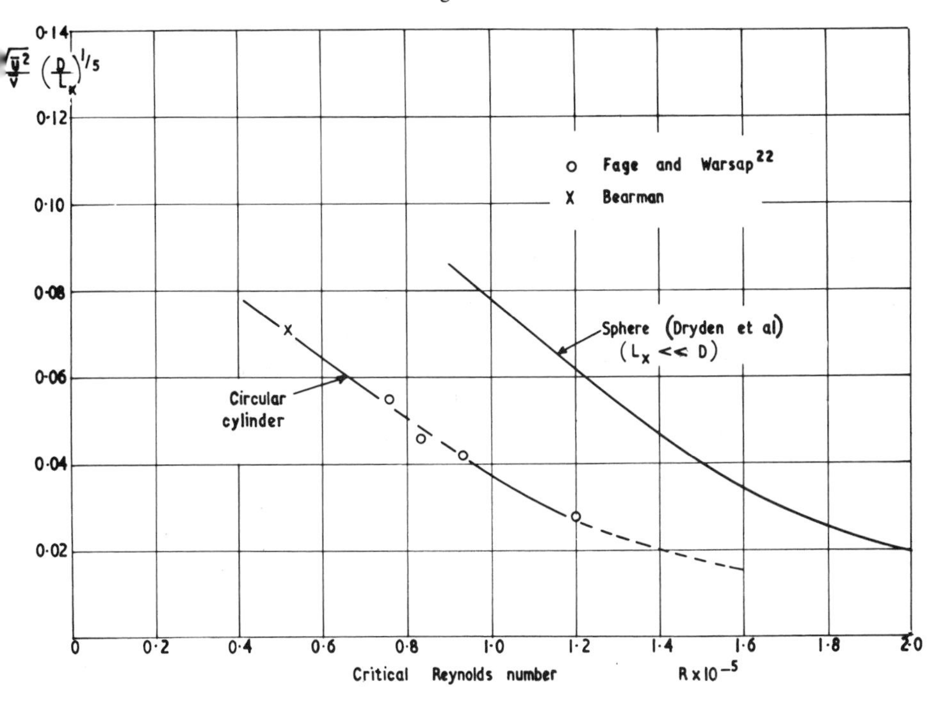

Figure 3. Critical Reynolds number as a function of $(\sqrt{\bar{u}^2}/\bar{v})\ (D/L_X)^{1/5}$.

present, when the wind speeds are referred to a height near the top of the structure. This is due to the reduction of dynamic pressure near the ground. However, as might be expected from the asymmetry of the wind speed with respect to the pitching axis, the pitching moment due to wind on a radio telescope might be adversely affected by the wind gradient. Figure 4 shows

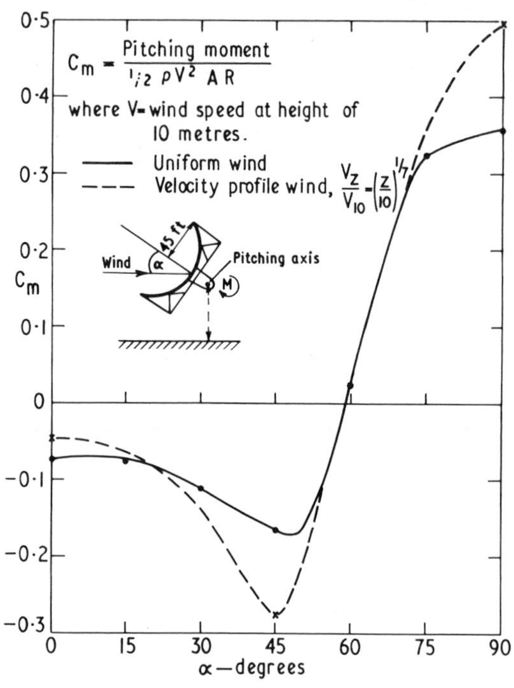

Figure 4. Influence of velocity profile on the pitching moment of a 90-ft-diameter Satellite tracker.

the effect of a power-law profile with an exponent of 1/7 on the pitching moment of a model satellite tracker shown in Figure 5. A grid of rods was used to produce the profile in the wind tunnel, and this grid would also have introduced some turbulence in the airstream.

Aerodynamic data obtained in wind tunnel tests for a number of paraboloidal reflectors used for radio telescopes, satellite trackers, etc., have been reported[3,4,5,13]; usually the three forces and three moments are measured with respect to a set of axes aligned with the wind direction and are then converted to refer to a system of body axes with the origin at the intersection of the altitude and azimuth axes. The wind forces and moments are sensitive to changes in structural detail, and especially to those of the rim and backup structure of the reflector bowl. A specific wind-tunnel investigation is therefore recommended for each major project.

Figure 5. 1/45th scale model of a Marconi 90-ft-diameter fully steerable satellite tracker–mounted in the NPL 7 ft x 7 ft wind-tunnel fitted with velocity profile grid. (Reproduced by permission of the Director of the National Physical Laboratory.)

3. The Dynamic Response to Turbulent Winds

The dynamic response of structures to the direct forcing action of the randomly fluctuating wind forces resulting from the velocity fluctuations of turbulent winds was first treated by Davenport[6] using statistical methods.

Useful contributions have been made by Harris[14] and by Vickery.[15] Structures of small frontal area, low natural frequency, and low structural damping are the most responsive to turbulent winds. The response also depends significantly on the aerodynamic characteristics of the structure and on the roughness of the surrounding terrain; because of the higher turbulence intensities it is greater in city centers than in the more open country where radio telescopes are likely to be sited. Calculations have shown that for some building and structures the rms and maximum deflections in turbulent winds may be larger than the deflection calculated on the assumption of steady wind loading. The comparison, however, depends very much on the averaging period used to determine the design wind speed. Only exceptionally are the differences very large provided that the design wind speed for the steady wind loading is based on the maximum mean speed for a short duration gust. The mean minute speed used in the U.K. or the "mile of wind" speed used in the U.S.A. represent too long an averaging period. Scruton and Newberry[8] recommend averaging periods of 10 and 3 sec, respectively, for the computation of the overturning moment and of the pressures on the cladding of a structure.

Although the basic theory for the calculation of the dynamic response of structures has been developed, the aerodynamic data required (e.g., the "aerodynamic admittance," the transfer function relating the power spectrum of wind speed to that of the wind loading) are not usually known nor, except perhaps for lattice structures,[9] can they be derived theoretically. For this reason the assessment of the dynamic response of a specific structure is most reliably made by tests of aeroelastic models in wind tunnels with flow characteristics appropriately modeled with respect to wind shear, and intensity and scale of turbulence.

4. The Oscillatory and Divergent Instabilities due to Wind

Wind forces in smooth airflow can be written in terms of the components in-phase and out-of-phase with the motion

$$F = H_a z + K_a \dot{z},$$

where H_a and K_a coefficients dependent on the wind speed, amplitude, Reynolds number, and frequency. When acting on a simple mass-damped-elastic system

$$M\ddot{z} + (K_s + K_a)\dot{z} + (H_s + H_a)z = 0.$$

If K_a, the aerodynamic damping coefficient, is negative and numerically exceeds K_s, the (positive) structural damping coefficient, oscillations will result. If H_a, the aerodynamic stiffness, is negative and numerically exceeds H_s, the (positive) structural stiffness, nonoscillatory divergence will result.

4.1. Wind-Excited Oscillations

There are a number of aerodynamic mechanisms that can cause flexible structures or their structural members to oscillate in wind. Of these the following may be of concern in the design of radio telescopes:

1. Vortex-excitation.
2. Galloping-excitation.
3. Stalling flutter.
4. Classical flutter.

A frequently experienced manifestation of vortex excitations is to be found in the wind-excited oscillations of tall chimney stacks.[16,17] Failures due to vortex-excited vibrations of circular-section members of the supporting frame of radar antenna have been reported.[18] Simple explanations of the phenomenon are based on the oscillatory nature of the vortex-wake behind bluff bodies. The oscillations usually occur in the cross-wind direction because of the oscillating "lift" forces induced by the vortex-shedding but the vortex-shedding also produces oscillatory "drag" forces and occasionally oscillations in line with the wind direction are reported. The in-line excitation is much weaker than that in the cross-wind direction, and only a very small amount of structural damping is sufficient to suppress the oscillations (see Figure 6). The cross-wind oscillations may also be suppressed either by added damping (for small members the Stockbridge damper appears to be very successful in this respect), or by fitting various aerodynamic devices. The devices developed by Scruton and Walshe[19] and by Weaver,[18] in which projections are wound round the surface as helices, have proved successful in application. Perforating of the cylindrical shell is also very effective (see Figure 7), and the perforated shroud device suggested by Price[20] (Figure 8) has proved successful in model tests, and might in some circumstances be preferred to the other methods. Figure 9 shows the very marked influence of aspect ratio on the amplitude of oscillation for a quadrafoil section prism.[21]

Of the common structural sections in use the circular section has the most favorable aerodynamic stability characteristics. Such a section has less vortex excitation than most other sections and it does not suffer from the galloping-type excitation.

Galloping-type excitation arises from the destabilizing characteristics of the variation of the steady wind forces with the incidence of the wind and, as with vortex excitation, tends to promote oscillations in the cross-wind direction. The simple criterion for galloping-type instability has been given by Den Hartog[22] for small amplitude oscillations as

$$dC_L/d\alpha + C_D < 0.$$

Graphical and numerical methods for the computation of galloping characteristics have been suggested respectively by Scruton[23] and by Parkinson.[24]

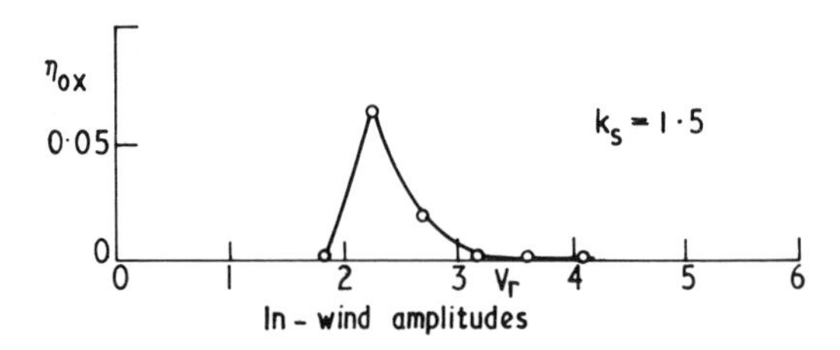

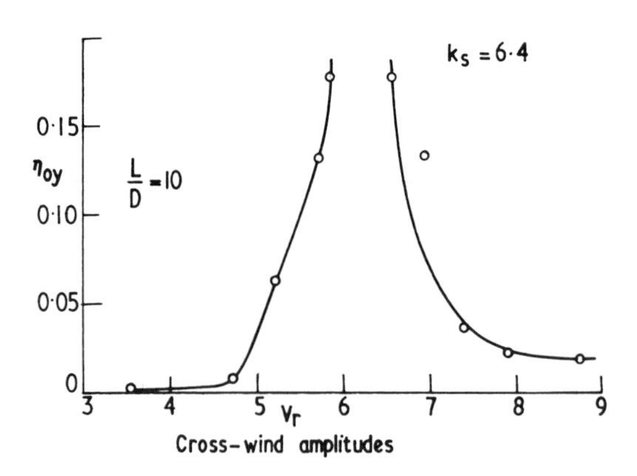

Figure 6. Tip amplitudes of a finite cylinder oscillating in a linear mode about its base.

Galloping excitation is found on rectangular and on some other polygonal sections, and also on structural sections distorted by ice accretion.

Turbulence in the airstream may have a marked influence on the vortex excitation and hence on the response amplitudes. The results of some tests on a circular stack model[25] (the model used was that of Figure 8 but without the perforated shroud) are given in Figure 10. The reduced velocity for maximum excitation was reduced from 6.4 in smooth airflow to 4.9 in a turbulent airstream and, except at high amplitudes, the value of the maximum excitation was increased. These tests were carried out at subcritical Reynolds number for smooth airflow and the change in the reduced velocity for maximum excitation was in the direction that would have been expected if the introduction of turbulence had induced supercritical conditions. It should be noted that the scale of turbulence $L_X/D = 0.54$ was too small to be representative of atmospheric turbulence.

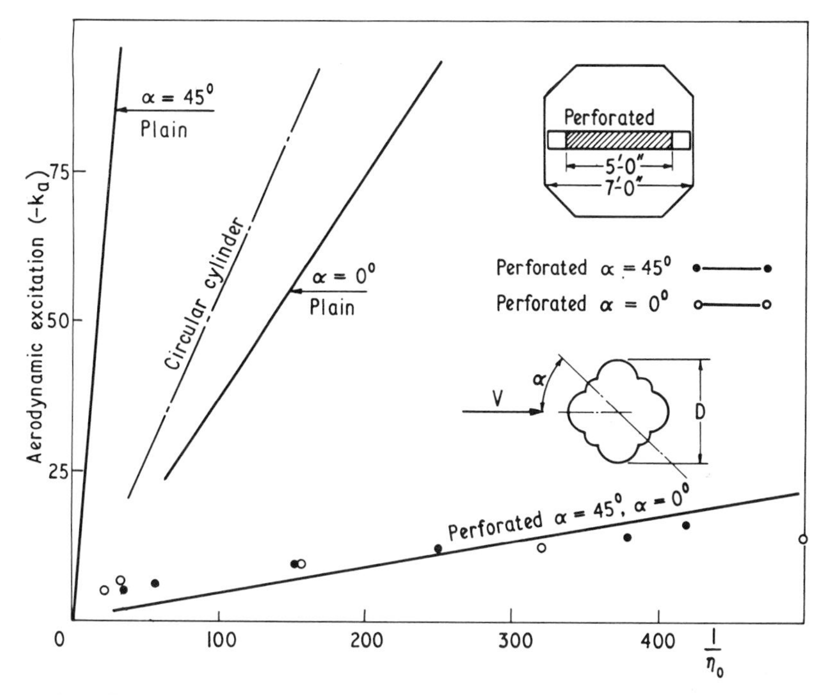

Figure 7. The effect of perforations on the excitation of a quadrafoil section prism.

The influence of turbulence was much more marked on a model of a square section tower block building. In smooth airflow the results given in Figure 11 show the peaked amplitude response typical of a narrow-band frequency excitation. In turbulent airflow the amplitude increased steadily with wind speed. In both the smooth and turbulent flows the oscillations were in the cross-wind direction; they were of fairly constant amplitude in smooth airflow but fluctuated randomly in amplitude in the turbulent flow.

The most probable aerodynamic causes of instability of reflector bowls are stalling and classical flutter. The following explanation of stalling flutter is adapted from that given by Halfman[26] *et al.* for the stalling flutter of aircraft wings. Referring to Figures 4 and 12 it can be inferred that a reflector bowl will be "stalled" with almost complete detachment of the flow from the concave surface for all incidences below 45°. Above incidences of 45° the flow reattaches, at least partly, to the surface. Near the stalling angle angular movements of the bowl modify the incidence at which attachment and reattachment of the flow take place and hence the incidences at which the bowl stalls and recovers from the stall. The hysteresis loop in the pitching moment is shown at A in Figure 12. Its clockwise direction indicates that energy is being extracted from the stream to yield negative aerodynamic

Figure 8. Linear-mode dynamic model of a proposed 850-ft-high multiflue chimney stack fitted with a perforated shroud – mounted in a 9 ft × 7 ft wind-tunnel fitted with a turbulence grid. (Reproduced by permission of the Director of the National Physical Laboratory.)

damping. For incidence away from the stalling angle, hysteresis loops B occur but in the anticlockwise direction, and positive aerodynamic damping results. Some rudimentary wind tunnel tests[3] indicated the possibility of stalling flutter oscillations occurring in pitching motion about the trunnion axes on

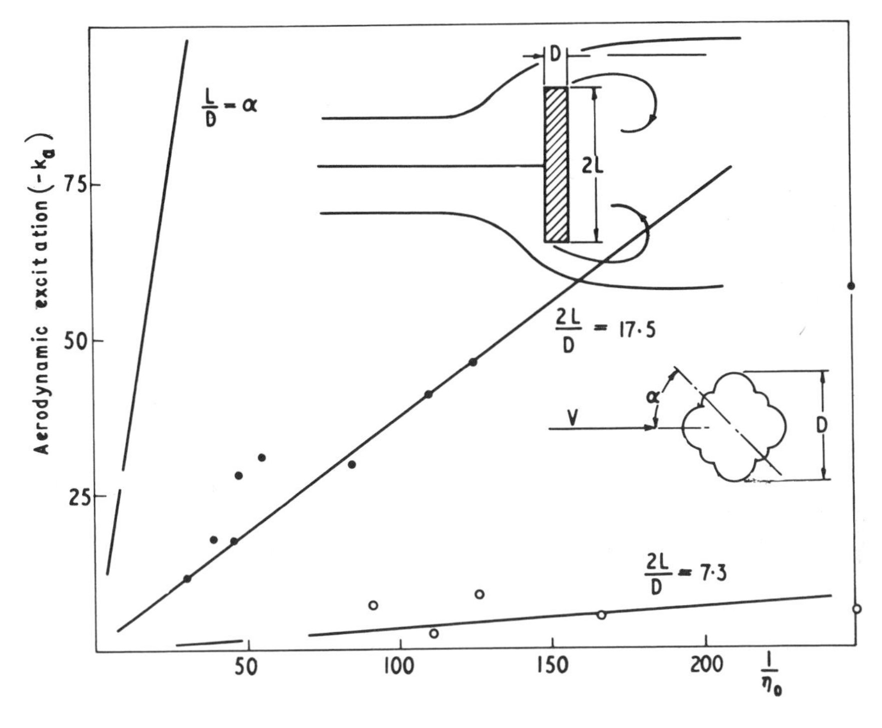

Figure 9. The influence of aspect ratio on the excitation of a quadrafoil section prism.

the 250-ft-diameter radio telescope at Jodrell Bank, and it was to ensure freedom from such oscillations that the "bicycle wheel" damper was fitted. Oscillations of flexibly mounted "rigid" paraboloidal reflectors due to classical flutter have been found in model tests. They occur because of the couplings, which may be elastic, inertial, or aerodynamic, between the various degrees of freedom provided by the mounting. In the model tests the flutter involved fore-and-aft movements and twisting motions about the vertical axis. Hull[27] has suggested a theoretical treatment of the flutter problem for a polar-mounted reflector; he concludes that flutter instability may become a problem, and recommends tests of appropriately scaled aeroelastic models.

The classical flutter problem has been studied extensively in aeronautical engineering, especially with regard to flexure-torsion flutter of aircraft wings. It may become a problem with long-span suspension bridges. After any tendency for aerodynamic instability in one degree of freedom had been eliminated by suitable aerodynamic design of the suspended roadway, classical

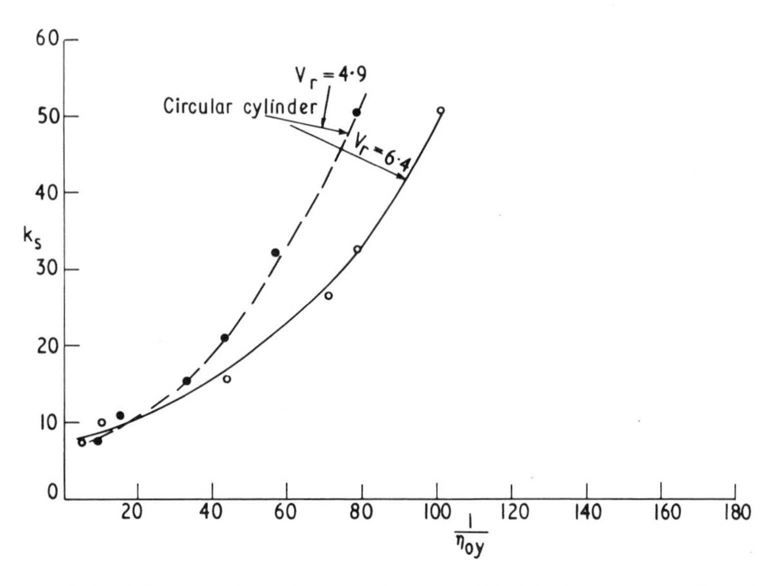

Figure 10. Variation with damping of maximum amplitude in the cross-wind direction.

flutter became the crucial type of instability of the recently completed suspension bridge over the River Severn in England, and it became especially critical during the erection stage of the bridge before the final stiffness and frequencies of the bridge were realized.

4.2. Divergent Instabilities

Some divergences of a simple type have been found in tests with elastically mounted models of bridges in wind tunnels. More complicated are the buckling instabilities of thin cylindrical shells under wind loading.

Buckling collapses under wind loading have occurred on steel cylindrical tanks. In a recent occurrence in England the tanks were under erection and were without the stiffening effects of the top cover or of liquid contents. Such instabilities have been studied both experimentally and theoretically by Holownia.[28] The buckling occurred as a curved plate (not as a cantilever) at a critical dynamic pressure of the wind given by

$$q_{crit} = \frac{2.16\, E(t/D)^{2 \cdot 5}}{(H/D)^{1 \cdot 3}}\,.$$

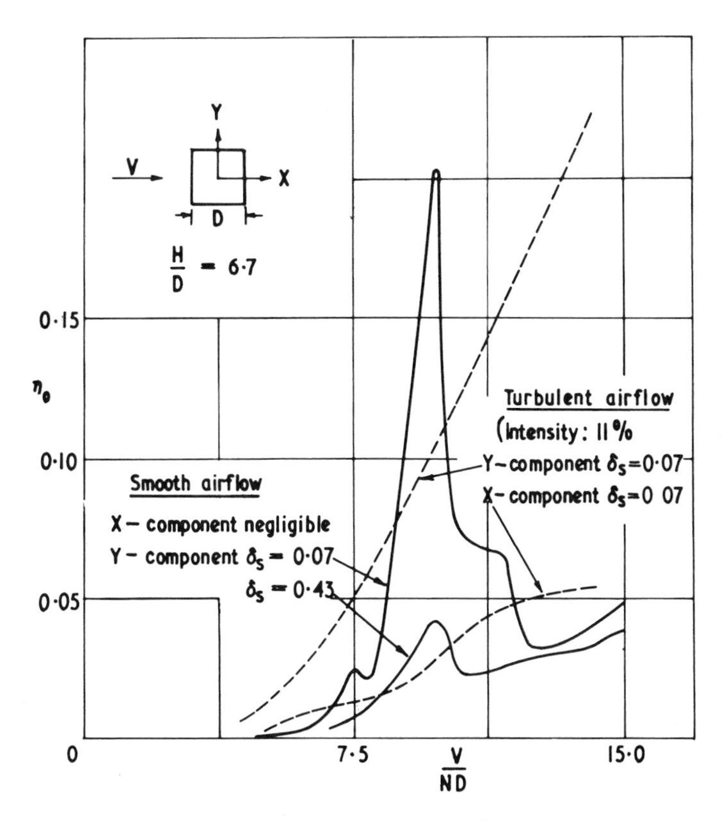

Figure 11. Amplitude response curves for a cantilever of square cross section in smooth and turbulent airflows.

Tests on the buckling of cooling tower models have been made in the NPL compressed air tunnel.[29] The models (see Figure 13) were made by electrodepositing copper on a perspex former and also from plastic sheet. Tests were made with different shapes of cooling tower and with different shell thicknesses in order to establish the relationship between the critical dynamic pressure and the shell thickness. The results gave a somewhat similar dependence to that found by Holownia for cylindrical tanks

$$q_{crit} \propto E(t/D)^{2.3}.$$

When applied to some very large thin-walled full-scale cooling towers in reinforced concrete, these results yielded a critical wind speed for buckling of over 200 mph, and it would seem that buckling collapse is not likely to prove a problem for cooling towers. The writer is not aware of any similar examination of the buckling collapse due to wind of spherical radomes; buckling under

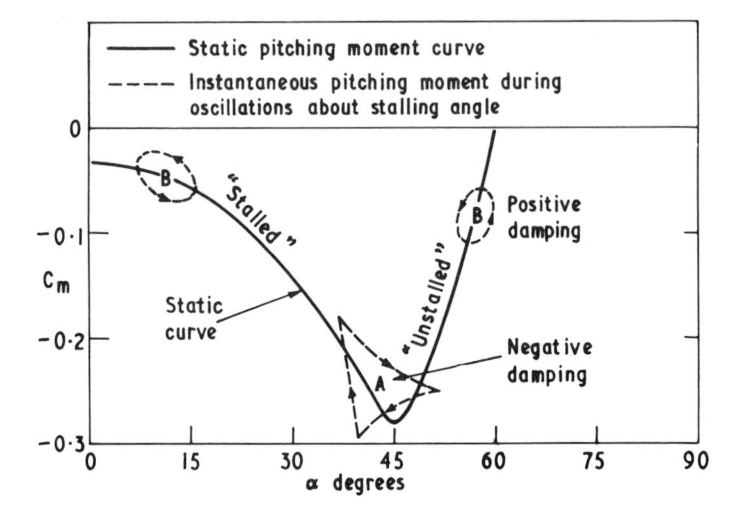

Figure 12. Stalling flutter of reflector bowl.

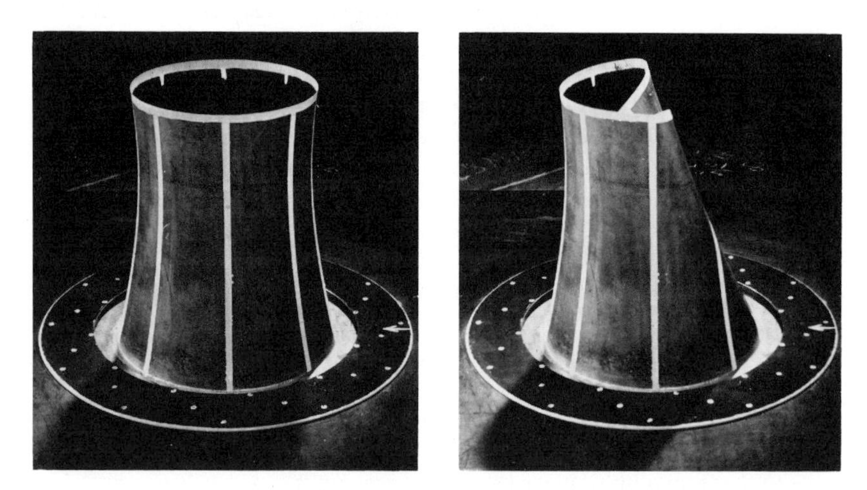

Figure 13. Aeroelastic models of cooling towers before and after buckling under wind pressure. (Reproduced by permission of the Director of the National Physical Laboratory.)

wind pressure may become a design consideration as the size of the radome increases. The buckling will be influenced by the internal pressure in the radome, and the most adverse conditions might arise during a stage of erection with the cladding only partially complete.

5. The Prediction of Wind Effects

Despite a number of limiting reservations which must be made with regard to the reliability of wind tunnel investigations, wind loading and instability information for design purposes is most reliably and rapidly acquired by wind tunnel tests on models of proposed structures. Careful consideration must be given to the conditions of test, particularly with regard to the simulation of the natural wind. The similarity requirements for steady force and pressure measurements applicable to structures with insignificant dynamic response to fluctuating wind loads can be stated as similarity of external shapes of model and prototype, together with similarity of the following airflow characteristics:

1. Wind shear.
2. Reynolds number.
3. Intensities of turbulence $\sqrt{\bar{u}^2}/\bar{V}$, etc.
4. Scales of turbulence L_X/D, etc.

For structures with significant dynamic response to the unsteady wind loads, aeroelastic models are required for which the similarity requirements, in addition to those of the airflow given earlier, include similarity of the mass, stiffness, and structural damping distributions with equal values of δs and the following structure-to-air ratios: σ/ρ, $E/\rho V^2$, gl/V^2. These strict requirements are very severe: Some relaxations and compromises are necessary, and are often permissible, to make the design and construction of aeroelastic models feasible. Methods have now been developed for correcting measured values of wind loads in wind tunnels with enclosed working sections to freestream values.[30,31]

A number of methods of reproducing wind shear and turbulence in wind tunnels representative of natural winds have been suggested, but none is entirely satisfactory. The two main methods are:

1. To use the boundary layer developed by representative surface roughness on the floor for a sufficient length ahead of the model.
2. To develop the desired flow characteristic by grids or other turbulence generators placed upstream of the model.

Because of the similarity of the mechanism producing the wind shear and turbulence, the former method would be expected to produce the more representative wind structure, and to model the wind structure automatically if the site topography is reproduced for a sufficient area surrounding the site. A very long working section is evidently required for such tests, but experience with such modeling has not been entirely satisfactory, perhaps because of the constraints due to the wind tunnel walls and roof.

The grid method for providing the required profile and turbulence characteristics is more suitable for tunnels with short working sections—one disadvantage is that assumptions must be made with regard to the wind structure at the site before attempts can be made to reproduce the appropriate conditions in the wind tunnel. Meteorological surveys at the site are not always practicable, but typical power-law profiles for different types of terrain may be assumed ($V_Z \propto Z^{\alpha}$), and be associated with "universal" turbulence spectra for atmospheric winds suggested by

$$\text{Davenport}^{32} \qquad \frac{n\,F(n)}{K\bar{V}_S^{\,2}} = \frac{4x^2}{(1+x^2)^{4/3}}$$

or

$$\text{Harris}^{33} \qquad \frac{n\,F(n)}{K\bar{V}_S^{\,2}} = \frac{4x}{(2+x^2)^{5/6}},$$

where $x = 4000\,n/\bar{V}_S$ in ft-sec units, K is the surface drag coefficient, the value of which depends on the terrain ($K \approx 0.005$ for open country), and $\bar{V}_S$ is the mean speed at the reference height.

Davenport[32] gives an expression for variation of turbulence intensity with height

$$\sqrt{\overline{u^2}}/\bar{V}_Z = 2.45\,K^{1/2}(Z/s)^{-\alpha},$$

where α is the exponent of the power-law profile of mean hourly wind speed.

Harris[33] has suggested turbulence scale profiles given by

$$L_X = 367(Z/s)^{\alpha}\ \text{ft}.$$

If these universal expressions are used the turbulence intensity is found to decrease from a value of about 0.17 near the ground to 0.10 at 500 ft, while the longitudinal scale increases from 370 to 550 ft. Turbulence could be expected to have an important influence on the wind loads on a structure of comparable dimension as for example, a 500-ft-diameter spherical radome. It is desirable therefore to simulate correctly the shear and turbulence of atmospheric winds in wind tunnel tests. The longitudinal scale of turbulence produced by a coarse mesh-type grid (see Figure 8) is about one-half the mesh size, and the maximum scale that can be produced in a wind tunnel by this method and still retain uniform flow distribution is approximately $\sqrt{S}/10$, where S is the cross-sectional area of the working section. Tests of a 500-ft radome in a turbulent airstream would require a large-section wind tunnel to enable a model of adequate size to be built, and preferably one with some degree of pressurization to attain a satisfactory value of the Reynolds number.

A wind profile is readily attained in a wind tunnel by the use of variously spaced horizontal rod grids. Such a grid is shown in Figure 5. A simple method for designing such a grid for specified power-law profiles has been given by

Cowdrey.[34] The grids also introduce turbulence in the airstream but at present there are no criteria for the quantitative prediction of the characteristics.

The wind shear and turbulence screen shown in Figure 14 was used for the tests of the dynamic behavior in wind of the 1350-ft-high twin tower blocks proposed for the World Trade Center in New York.[32] The grid, together with

Figure 14. 1/400th scale linear mode dynamic model of the 1350-ft-high twin tower blocks proposed for the World Trade Center, New York – mounted in a wind-tunnel fitted with a grid for providing wind shear and turbulence. (Reproduced by permission of the Director of the National Physical Laboratory.)

the forked ground fence, was chosen to give a representative wind velocity profile. No deliberate attempt was made to design the grid to obtain specific turbulence properties of the airstream. Similar tests were also carried out in the wind tunnel of Colorado State University but with the use of a model of the New York topography to determine the appropriate flow characteristics. The normalized spectra obtained in the two wind tunnels together with that obtained from observations in New York are plotted in Figure 15, and scales

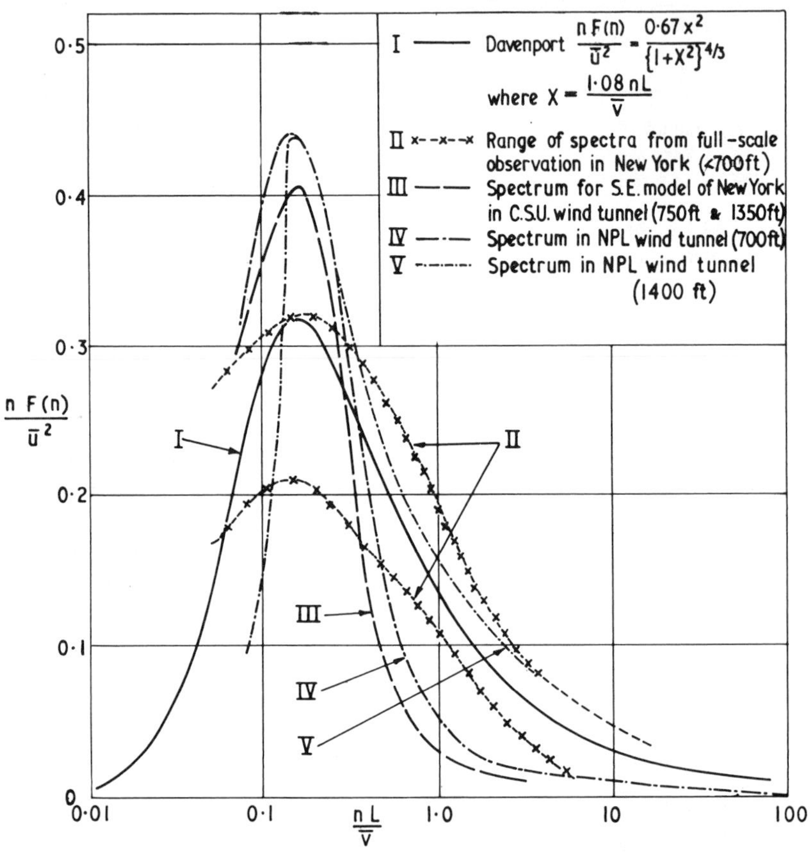

Figure 15. Comparison of the turbulence spectra obtained at NPL (grid induced) and the C.S.U. (roughness induced) with measured spectra for New York.

and intensities of turbulence are quoted in Table 1. These results show differences between the turbulence characteristics obtained in the two wind tunnel experiments and also between those observed in New York and found from wind tunnel tests using the topographic model of New York; in particular, the longer wavelengths found on full scale were not reproduced in the wind tunnel airflows. It is important, therefore, to establish how sensitive the

Table 1. Comparison of the Scales and Intensities of Turbulence Induced (1) by the Modelling of New York in the CSU Wind Tunnel, (2) by Grids in the NPL Wind Tunnel and (3) Full-Scale Observations in New York.

		Height z	Intensity		Scale[a]	
		(ft)	$\frac{\sqrt{\overline{u^2}}}{\bar{V}_z} \times 100$	$\frac{\sqrt{\overline{v^2}}}{\bar{V}_z} \times 100$	L_x (ft)	L_y (ft)
II	Full-scale observation in New York	500 920			2280	
III	S.E. model of New York in CSU wind tunnel	1350 750	6.5 14.7		320 320	
IV	Grid induced in NPL wind tunnel	1400	3.8	3.4	850	
V		700	24.3	18.0	910	425

[a] Scale of turbulence L is defined as $\lambda/2\pi$ where λ is the wavelength corresponding to the peak value of the normalized spectrum $nF(n)/\bar{V}^2$.

dynamic response of flexible structures is to the scale and the intensity of turbulence, and some current research at NPL is directed toward this end. The agreement in the results obtained for amplitude response of the tower blocks between the tests at NPL and CSU was reasonably satisfactory, especially when adjustments suggested by Whitbread[35,36] for differences in the model and the test conditions had been applied. These differences included those of the roughness of the model surfaces, the density ratio σ/ρ, the spectral density of turbulence, and the period for observation of the maximum amplitude.

6. Conclusions

1. Wind shear and turbulence of the airstream exert a marked influence on the steady and unsteady wind loading of structures.

2. In addition to the effects of mean wind on the stability of structures and on the pressures of their cladding, for many structures attention must be given to the dynamic response to the randomly fluctuating forces due to turbulence, and to the oscillatory and divergent instabilities due to wind.

3. Erection stages of construction are often more critical to wind action than that of the completed structure.

4. The most direct and reliable approach to the prediction of wind effects is by model tests in wind tunnels with appropriate airflow characteristics.

The precise simulation of the characteristics of atmospheric winds in wind tunnels is difficult to achieve. Further development of techniques for reproducing wind shear and turbulence in wind tunnels is required, together with an assessment of the importance of departures from precise simulation.

References

1. *Proceedings of the Conference on Wind Effects on Buildings and Structures held at NPL, Teddington, June 1963* (Her Majesty's Stationary Office, London, 1965).
2. "Large Steerable Radio Antenna–Climatological and Aerodynamic Considerations." *Ann. New York Acad. Sci.* **116** (June 1964).
3. W. G. Raymer, H. L. Nixon, and L. Woodgate, "Aeordynamic Tests of a Model Radio Telescope," unpublished report National Physical Laboratory/Aero/275 (January 1955).
4. R. E. Whitbread, "An Investigation for the Static Wind Forces on a Model of a Radio Telescope for the National Research Council of Canada," unpublished National Physical Laboratory Aero Report 1023 (June 1962).
5. R. B. Blaylock, B. Dayman, Jr., and H. L. Fox, "Wind Tunnel Testing of Antenna Models," *Ann. N. Y. Acad. Sci.* **116,** 239 (June 1964).
6. A. G. Davenport, "Application of Statistical Concepts to the Wind Loading of Structures," *Proc. Inst. Civil Engrs.* **19,** 449 (August 1961), Paper 6480.
7. O. G. Sutton, *Atmospheric Turbulence* (Methuen & Co., Ltd., London, 1949).
8. C. Scruton and C. W. Newberry, "On the Estimation of Wind Loads for Building and Structural Design," *Proc. Inst. Civil Engrs.* **25,** 97 (June 1963), Paper 6654.
9. B. J. Vickery, "On the Flow behind a Coarse Grid and its Use as a Model of Atmospheric Turbulence in Studies Related to Wind Loads on Buildings," unpublished National Physical Laboratory Aero Report 1143 (March 1965).
10. B. J. Vickery, "Fluctuating Lift and Drag on a Long Cylinder of Square Cross-Section in a Smooth and Turbulent Stream," unpublished National Physical Laboratory Aero Report 1146 (April 1965).
11. P. Bearman, "The Flow around a Circular Cylinder in the Critical Reynolds Number Regime in a Smooth and a Turbulent Airstream," National Physical Laboratory Aero Special Report (in preparation).
12. H. L. Dryden, G. B. Schubauer, W. C. Mock, and H. R. Skramstead, "Measurement of Intensity and Scale of Wind-Tunnel Turbulence and their Relation of the Cricital Reynolds Number of Spheres," NACA Report No. 581 (1937).
13. C. F. Cowdrey and M. Churchlow, "Wind Tunnel Tests on a 1/45th

Scale Model of a Marconi 90 ft. Diameter Fully Steerable Satellite Tracker," unpublished National Physical Laboratory Aero Report 1222 (February 1967).

14. R. I. Harris, "The Random Vibration of Distributed Parameter Mechanical Systems with Reference to Wind Loading Problems," Electrical Research Association Report No. 5110 (1965).
15. B. J. Vickery, "On the Assessment of Wind Effects on Elastic Structures," *Civil Eng. Trans. Inst. Engrs., Australia* (October 1966), Paper 2116.
16. C. Scruton, "On the Wind-Excited Oscillations of Stacks, Towers and Masts," in *Proceedings of the Conference on Wind Effects on Buildings and Structures held at NPL, Teddington June 1963* (Her Majesty's Stationary Office, London, 1965), Paper 16, pp. 798-832.
17. C. Scruton and A. R. Flint, "Wind-Excited Oscillations of Structures," *Proc. Inst. Civil Engrs.* **27**, *673* (April 1964).
18. W. Weaver, "Wind-Induced Vibrations in Antenna Members," *Proc. Am. Soc. Civil Engrs., J. Eng. Mech. Div.* 87 **(1961).**
19. C. Scruton and D. E. Walshe, "A Means for Avoiding Wind-Excited Oscillations of Structures with Circular or Nearly Circular Cross-Sections," Report National Physical Laboratory/Aero/335, Teddington (1959).
20. P. Price, "Suppression of the Fluid-Induced Vibration of Circular Cylinders," *J. Eng. Proc. Am. Soc. Civil Engrs.* **82** (No. EM 3) (1956), Paper 1030.
21. B. J. Vickery and D. E. J. Walshe, "An Aerodynamic Investigation for a Proposed Multi-Flue Smoke Stack at Fawley Power Station," unpublished National Physical Laboratory Aero Report 1132 (December 1964).
22. J. P. Den Hartog, *Mechanical Vibrations, 4th ed.* (McGraw-Hill, New York, 1956).
23. C. Scruton, "The Use of Wind Tunnels in Industrial Aerodynamics Research," AGARD Rep. 309 (October 1960).
24. G. V. Parkinson and N. P. H. Brooks, "On the Aeroelastic Instability of Bluff Cylinders," *J. Appl. Mech. 28 Trans. Am. Soc. Civil Engrs.* E 83, (1961).
25. D. E. J. Walshe, "The Aerodynamic Investigation for the proposed 850 ft. high Chimney Stack for Drax Power Station," unpublished National Physical Laboratory Aero Report 1227 (August 1967).
26. R. L. Halfman *et al.*, "Evaluation of High-Angle-of-Attack Aerodynamic Derivative Data and Stall-Flutter Prediction," NACA TN 2533 (1951).
27. F. H. Hull, "Dynamic Stability and Aeroelastic Considerations in the Design of Large Steerable Antennas," *Ann. N. Y. Acad. Sci.* **116**, 311 (June 1964).
28. B. P. Holownia, "Buckling of Cylindrical Shells under Wind Loading," Australian Dept. of Supply, Note ARL/SM292.
29. P. W. Bailey and R. Fidler, "Model Tests on the Buckling of Hyperbolic

Shells," CERL Laboratory Note No. RD/L/N113/65. Central Electricity Research Laboratories, Leatherhead, England.

30. F. C. Maskell, "A Theory of Blockage Effects on Bluff Bodies and Stalled Wings in a Closed Wind Tunnel," A. R. C. R & M 3400 (Her Majesty's Stationery Office, London, 1965).
31. C. F. Cowdrey, "The Application of Maskell's Theory of Wind-Tunnel Blockage to very Large Solid Models," National Physical Laboratory Aero Report 1247 (1967).
32. A. G. Davenport, "The Relationship of Wind Structure to Wind Loading," in *Proceedings of the Conference on Wind Effects on Buildings and Structures held at NPL, Teddington, June 1963* (Her Majesty's Stationery Office, London, 1965), Paper 2.
33. R. I. Harris, "On the Spectrum and Auto-Correlation Function of Gustiness in High Winds," Electrical Research Association Report SP1/T14.
34. C. F. Cowdrey, "A Simple Method for the Design of Wind-Tunnel Velocity-Profile Grids, unpublished National Physical Laboratory Aero Note 1055 (May 1967).
35. R. E. Whitbread and C. Scruton, "An Investigation of the Aerodynamic Stability of a Model of the Proposed Tower Blocks for the World Trade Center, New York." Part I. "An Investigation of the Amplitudes of Oscillation in the Fundamental Mode for a Range of Wind Conditions," National Physical Laboratory Aero Report 1156 (July 1965).(Report with restricted circulation.)
36. R. E. Whitbread, "An Investigation of the Aerodynamic Stability of a Model of the Proposed Tower Blocks for the World Trade Center New York." Part II. "Further Wind-Tunnel Studies Relating Mainly to the Response of the Tower Blocks to the Turbulent Wind," National Physical Laboratory Aero Special Report 001 (September 1967). (Report with restricted circulation.)

Lewis V. Smith, Jr.
Communications Satellite Corporation
Washington, D. C.

POINTING AND TRACKING ACCURACY. RECOMMENDED STANDARDS

1. Introduction

Pointing and tracking accuracy has been specified many times for large and small antenna systems. Each specification writer has his own way of stating the requirement based on what he hopes to achieve and his particular engineering background. In the same way, the engineer who interprets the specification in order to build equipment that conforms to the requirement will interpret the specification according to his own experience and sometimes his own convenience. Two basic questions can be asked: first, what does the specification writer really want? Second, how can a design be made to conform to a given specification without over- or underdesigning due to improper interpretation of the specification?

The first question can only be answered by the individual requirement to be specified. However, the method of specifying the accuracy should be related to the method of interpreting it. The specifier may very well ask for something quite different if he knows how the designer will interpret his requirement. It is very difficult to prove that a specification has been met when it can be interpreted in more than one way. The fact that antenna design is a relatively new field that has begun to mature in the last five to ten years means that very few standards exist, and each design organization works to its own interpretation of a given specification. There is no long history of equipment that has met or failed to meet the specifications. Too often the equipment must be used regardless of how effective it is. On the other hand, there is equipment in use which is better than it needs to be for the function that it performs, because the designer had to be conservative in order to meet a specification which was not clearly defined or understood. Several standard approaches have been developing in the specification of pointing accuracy and tracking accuracy. These, however, need to be reconciled with each other and published as a standard by one of the national engineering societies or interested technical centers.

Although there is bound to be some disagreement in principle, it is the intention of this paper to suggest a standard for specifying pointing and tracking accuracy. By the establishment of such a standard, all buyers and sellers of antenna equipment should understand the requirement and how it will be satisfied. In this way, no one will be able to take advantage of a loose interpretation of a specification in order to obtain a price advantage or, on the other hand, be penalized for an overconservative interpretation. With time, a history of actual performance versus that specified can be obtained so that we will know how to improve the specifications to provide the most economical designs.

Along with the proper specification and equipment implementation, it is necessary to obtain meaningful information about the actual performance of the equipment. In the past, there has been a limited amount of performance feedback to the designer and this has resulted in uneconomical designs in some instances.

The errors to be included in the pointing and tracking error budgets result from structural, mechanical, servo, rf and alignment considerations. One or several of these errors are frequently overlooked because so many engineering techniques are covered, and the tendency is to emphasize those techniques with which the designer is most familiar. By standardizing the error budget, a checklist is established, and valid comparisons may be made between several alternate designs.

The following sections will define the errors, the factors that contribute errors, and a method for combining them to obtain a specified value.

2. Definitions

2.1. Pointing Accuracy

Pointing accuracy is specified in order to establish limits within which the rf axis of an antenna must lie when position information is provided either manually or in a program mode. *The pointing error can be defined as the space angle difference between the command vector* and the actual position of the*

* The command vector can be represented by the azimuth and elevation angles necessary to point the antenna at a desired target.

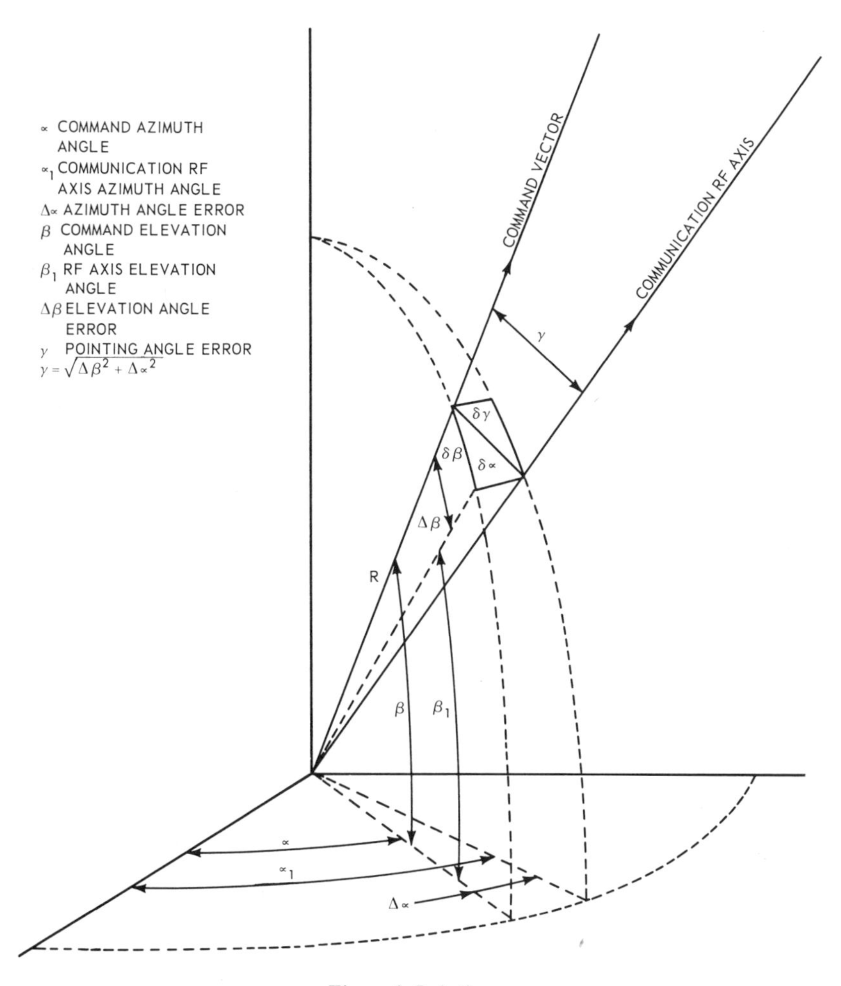

Figure 1. Pointing error.

antenna communication rf axis. (See Figure 1.) Pointing and readout errors are often thought to be the same. The readout error, however, includes all of the errors of pointing plus those errors associated with a tracking-type feed where used. These additional errors are inherent in the tracking feed null axis alignment and the tracking receiver. Since pointing, readout, and tracking accuracy are dependent upon different combinations of the same independent errors, the specifications of a particular value for all three would be redundant. The primary task of a satellite communications earth station is to communicate through a satellite. It must, therefore, look at the satellite continuously. This is accomplished either by programming the satellite position into the antenna, based on calculated orbits, or by tracking a beacon at the satellite. These two basic techniques require that the pointing and tracking accuracy be specified clearly. The predicted orbital information that can be derived from tracking, telemetry, and command stations can be obtained to sufficient accuracy, if the antenna pointing and tracking errors are limited, and a smoothing technique is used to minimize the final readout errors. Bias errors in the observations can be detected and compensated at this time. It is sufficient to specify pointing and tracking accuracy since the resultant readout accuracy will be a function of these two.

2.2. Tracking Accuracy

Tracking accuracy is specified to provide limits within which the rf axis of an antenna must lie with respect to the satellite or target while automatically tracking. Position information readout is not usually required while tracking so that many of the independent structural and alignment errors of the pointing error are excluded from the tracking error. *The tracking error can be defined as the space angle difference between the communication rf axis of the antenna and the vector to the rf source (see Figure 2) being tracked.* It is possible to provide a very accurate tracking antenna that is quite inaccurate for pointing because the tracking accuracy is primarily a function of the tracking feed and related tracking receiver and servo controls. The pedestal and structure can be relatively flexible while tracking as long as it does not introduce any vibrations which affect the stability of the drive system.

2.3. Implications of Specification

The method of specifying pointing and tracking accuracy is very important if it is to be understood correctly. It has been specified in many different ways, including a simple statement as one extreme and a detailed breakdown of errors as the other. A standard method would form a basis for understanding between the buyer and seller of antenna equipment.

The accuracy of pointing or tracking within given limits during a given period of operation can be considered to be a function of joint probabilities. Each independent error that contributes to the total error will vary considerably with time and will depend on the antenna attitude, the environment, and the speed of the target. The way these errors combine to form the total error depends on their distribution and frequency of occurrence.

The magnitude of the allowable error is some function of the antenna beamwidth which relates it to a permissible gain degradation of the received or transmitted signal. It is not the intention here to recommend the relationship

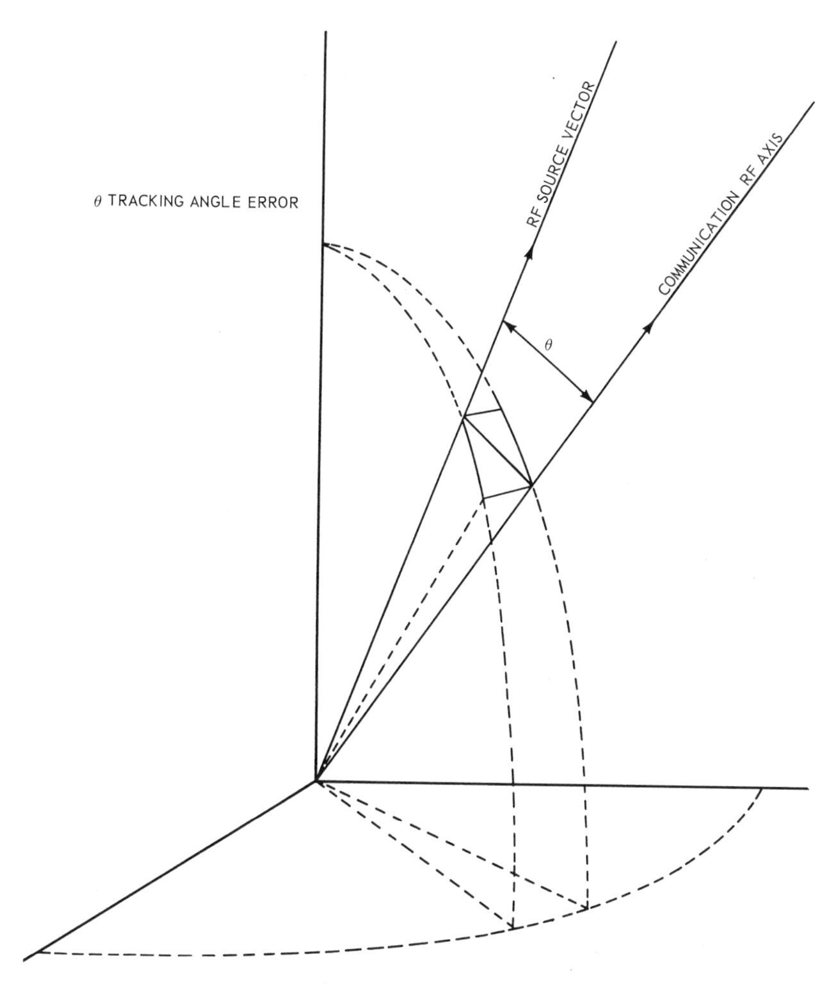

Figure 2. Tracking error.

of the allowable error to the antenna beamwidth, but rather to indicate what can be expected from a given method for specifying the accuracy.

It must be recognized that all independent factors affecting antenna error do not have a known distribution nor are they all systematic or constant. Constant errors are not considered in the analysis since we cannot predict the as-built condition, but the peak possible constant error can be included. If all the errors were systematic or constant, it would be possible to calibrate the antenna and reduce the error to a very low value. This process is very difficult and time consuming, since all combinations of elevation and azimuth angle would require calibration. The actual calibration would require accurately known target locations throughout the sky area of interest.

When long periods of operation over the entire field of view of the antenna are considered, it is possible to assume that each independent error has a distribution approaching a normal or Gaussian distribution even though they may vary systematically or in some other way. If such a distribution is allowed, a simple standard procedure for specifying and calculating the error may be established based on reasonable theory. The use of such a standard will assist in comparing the relative advantages of different antenna systems. Eventually a performance history will be obtained for each antenna so that changes can be made in the specification to improve the design.

On the basis of the assumption that all independent errors contributing to the antenna error follow a normal distribution, the peak of each independent error may be used as the basic measure of the total error. The peak error, for practical considerations, is equivalent to three times the standard deviation σ under this assumption.

$$\text{Peak error} = 3\sigma. \tag{1}$$

The standard deviation σ, sometimes called the root-mean-square (rms) deviation,* is the square root of the average value of the squares of the deviations from the arithmetic mean:

$$\sigma_a = \sqrt{(a_1^2 + a_2^2 + a_3^2 \cdots a_x^2)/x}\ , \tag{2}$$

where $a_1, a_2, a_3, \ldots, a_x$ are the errors or deviations from the arithmetic mean of set A, which consists of observations as a function of time of a component contributing to the antenna error; σ_a is the standard deviation of set A.

The square of the standard deviation is known as the variance (Var) of a random variable:

$$\sigma_a^2 = \text{Var}\,(A). \tag{3}$$

The normal distribution of errors implies that 68.27 per cent of the observations for each independent component contributing error will be within the 1σ value, 95.45 per cent will be within the 2σ value, and 99.73 per cent will be within the 3σ value. Since the variance of the sum of two or more independent random variables is equal to the sum of their variances, the error contributed by each independent component of an antenna system may be combined by adding their variances:

* This assumes the mean to be a zero value and the errors to be deviations from this value.

$$\text{Var (antenna)} = \text{Var (A)} + \text{Var (B)} + \text{Var (C)} \ldots . \tag{4}$$

It follows that the standard deviation of the antenna system error σ_{ant} is the square root of the sum of the squares of the standard deviations of the errors of each independent component of an antenna system or the root sum square (rss) of the independent standard deviations.

$$\sigma_{ant} = \sqrt{\sigma_a^2 + \sigma_b^2 + \sigma_c^2 \ldots} \; . \tag{5}$$

This also represents the rms error in a system of independent errors. The probability of exceeding the 1σ or rms value of the error is 0.3173 if a Gaussian distribution is assumed for each independent error; however, since all the independent errors have signs, some error compensation will result in combining the independent errors. The actual probability of exceeding the 1σ value will be about one-half that indicated by the normal distribution. While there are several assumptions and uncertainties in this method of specifying and interpreting and rms pointing and tracking error, the advantages obtained from an agreed-upon standard of this sort should more than compensate. In choosing the degradation allowed by a specified error, the specification writer must keep in mind the fact that the value specified will be exceeded part of the time. Since the relationship of gain degradation to pointing angle is a smooth function for small angles, effective antenna operation can still be obtained when the errors are in the 2σ and 3σ ranges.

It seems logical to specify an rms value for pointing and tracking errors. This should be some fraction of the antenna beamwidth which is a function of gain degradation. The specification would be satisfied analytically by considering the peak value of each independent component error for each axis. Since by agreement

$$\text{Peak error} = 3\sigma_a, \tag{6}$$

each independent component rms error = $1\sigma_a$. Therefore, the total rms error for azimuth σ_{az} and for elevation σ_{el} is

$$\sigma_{az} = \tfrac{1}{3}\sqrt{(3\sigma_a)^2 + (3\sigma_b)^2 + (3\sigma_c)^2 \cdots} \; , \tag{7}$$

$$\sigma_{el} = \tfrac{1}{3}\sqrt{(3\sigma_a)^2 + (3\sigma_b)^2 + (3\sigma_c)^2 \cdots} \; . \tag{8}$$

The azimuth and elevation errors are then combined vectorially to obtain the total antenna error σ_t.

$$\sigma_t = \sqrt{(\sigma_{az})^2 + (\sigma_{el})^2} \tag{9}$$

The field measurement of pointing and tracking error is difficult since many measurements under different conditions are required to duplicate the rms value calculated. Some peak values can be measured practically, and by combination with other calculated values a measure can be obtained of how the antenna conforms to the specification. This will not be covered here since the subject of measuring the accuracy of an antenna should be fully explored in a separate discussion.

3. Factors Contributing to Pointing Errors

The factors that contribute to the over-all pointing error may be classified into four categories. These categories are servo, structural, mechanical, and alignment. Many of the factors in these categories represent independent error contributions, but some of them combine with others under certain conditions and, therefore, must be considered dependent.

The following errors should be considered as a minimum for each category in determining the error of an az - el antenna mount. Others may exist for special equipment. Such special equipment should be individually evaluated using the suggested factors as a basis.

The servo errors to be considered are those due to (1) velocity lag, (2) acceleration lag, (3) wind gust, (4) encoder, (5) secant potentiometer, (6) tachometer, (7) motor cogging, (8) amplifier drift, and (9) servo noise. All of the errors may not be significant. For instance, a type II servo will have a zero velocity lag error. Nevertheless, such factors should be included in any checklist to avoid possible omission.

The factors to be considered in the mechanical area are (1) friction, (2) backlash, (3) encoder coupling, (4) azimuth axis wobble, and (5) elevation axis wobble. Some of these must be evaluated based on their effect on the servo so that they are frequently included with the servo error. Structural errors result from ice load structural distortion, dead load structural distortion, wind load structural distortion, thermal structural distortion, and acceleration structural distortion.

These may combine for some antenna orientations so that the errors may be directly additive. The dead load and ice load are additive for all orientations. The independent error to be included in the rms summation should be the arithmetic sum of the individual peak errors where errors are additive. If a larger error is obtained for another antenna orientation where the errors act independently, then that case should be used. Errors based on antenna orientation with respect to wind should be consistent between the azimuth axis and the elevation axis. Several conditions may have to be evaluated to determine the worst case orientation. The effect of the foundation should not be overlooked.

Alignment errors result from inaccuracies in manufacture and field installation of the equipment. These will generally be constant or systematic errors after the equipment has been installed. A peak limitation will establish the precision of manufacture and installation. For the sake of error budgeting and analysis, the peak error is considered the 3σ value since it can lie between zero and the peak value in the final installation. Alignment errors include (1) level alignment, (2) north alignment, (3) az - el axis orthogonality, and (4) rf axis - elev. axis orthogonality.

The nature of the error sources given and the method for calculating their peak values are known generally to designers of large antennas, so a detailed definition will not be presented. The purpose here is to identify the sources of error and how they are combined.

4. Factors Contributing to Tracking Error

Some of the independent errors that contribute to the tracking error are the same as those that contribute to the pointing error. Since tracking is accomplished with a tracking-type feed, nominally located at the focus of the reflector, the error signal supplied by the tracking feed to the control system to drive the antenna will bypass all of the structural-type errors and most of the alignment errors. However, additional errors are contributed by the tracking feed and the tracking receiver.

The servo-type errors that are present in the tracking error are (1) velocity lag (2) acceleration lag, (3) wind gust, (4) tachometer, (5) motor cogging, (6) amplifier drift, and (7) servo noise.

Mechanical errors consist of (1) friction, and (2) backlash.

The only significant alignment error is null axis - beam axis alignment.

The tracking receiver contributes an error that is not present in the pointing error.

It is generally much easier to obtain a good degree of accuracy through tracking than by pointing an antenna, because the errors contributed by the structural distortions and the axis alignments are excluded from the tracking loop. The tracking acceleration under which the accuracy is to be met should be based on the type of satellites or targets to be tracked, since the acceleration lag error becomes quite large with high acceleration rates, and the servo design becomes more difficult.

5. Recommendations

The following suggestions should be considered as a first step toward establishing a standard for specifying and interpreting the pointing and tracking accuracy for antennas.

Suggested pointing and tracking accuracy specification.

The pointing and tracking accuracy for an antenna shall be specified in the following way.

1. The rms pointing accuracy shall be within x beamwidth at y frequency while subject to the specified environment,
2. the rms tracking accuracy shall be within z beamwidth at y frequency while subject to the specified environment,

where x and z are some fraction of beamwidth which depends on the allowable gain degradation at the operating frequency y.

Table 1. Pointing Errors

		Azimuth	Elevation
a.	Velocity lag	3σ value	3σ value
b.	Acceleration lag	3σ value	3σ value
c.	Wind gust (servo)	3σ value	3σ value
d.	Friction (breakaway)	3σ value	3σ value
e.	Encoder (servo)	3σ value	3σ value
f.	Secant potentiometer	3σ value	3σ value
g.	Tacho meter	3σ value	3σ value
h.	Motor cogging	3σ value	3σ value
i.	Amplifier drift	3σ value	3σ value
j.	Backlash	3σ value	3σ value
k.	Servo noise	3σ value	3σ value
l.	Encoder coupling	3σ value	3σ value
m.	Level	3σ value	3σ value
n.	North alignment	3σ value	3σ value
o.	Az-el axis orthogonality	3σ value	3σ value
p.	rf axis – elev. axis orthogonality	3σ value	3σ value
q.[a]	Dead load and ice load struct. distortion	3σ value	3σ value
r.[a]	Wind load – struct. distortion	3σ value	3σ value
s.[a]	Thermal struct. distortion	3σ value	3σ value
t.[a]	Acceleration struct. distortion	3σ value	3σ value
u.	Az axis wobble	3σ value	3σ value
v.	El axis wobble		

$$\text{Az error} = \sqrt{a_{az}^2 + b_{az}^2 + c_{az}^2 \text{ etc.}}$$

$$\text{El error} = \sqrt{a_{el}^2 + b_{el}^2 + c_{el}^2 \quad \text{etc.}}$$

$$\text{Total pointing error} = \tfrac{1}{3}\sqrt{\text{Az error}^2 + \text{Elev error}^2}$$

[a] Where combinations of dead load, ice load, wind load, thermal load, and acceleration load affect the structure so that the errors are directly additive, the independent error to be included in the rms summation shall be the arithmetic sum of the 3σ values for structural distortion. All structures affected such as reflector, pedestal, and foundation shall be included in the error analysis.

Table 2. Tracking Errors

		Azimuth	Elevation
a.	Velocity lag	3σ value	3σ value
b.	Acceleration lag	3σ value	3σ value
c.	Wind gust (servo)	3σ value	3σ value
d.	Friction (breakaway)	3σ value	3σ value
e.	Tachometer	3σ value	3σ value
f.	Motor cogging	3σ value	3σ value
g.	Amplifier drift	3σ value	3σ value
h.	Backlash	3σ value	3σ value
i.	Servo noise	3σ value	3σ value
j.	Tracking receiver	3σ value	3σ value
k.	Null axis – beam axis alignment	3σ value	3σ value

Az error = $\sqrt{a_{az}^2 + b_{az}^2 + c_{az}^2 \text{ etc.}}$

El error = $\sqrt{a_{el}^2 + b_{el}^2 + c_{el}^2 \text{ etc.}}$

Total tracking error = $\frac{1}{3}\sqrt{\text{Az error}^2 + \text{Elev error}^2}$

Suggested antenna error budget.

When specifying accuracies according to the suggested method, the independent errors contributing to the total rms error shall be combined according to the suggested antenna error budget (Tables 1 and 2).

In the table errors should be considered as rms errors. Root-mean-square pointing errors those given in Table 1 and any additional independent errors peculiar to a given system. Each (3σ) independent error shall be determined for the specified environment. The 3σ value of the error shall be used for tabulating and combining the errors as shown in Table 1.

Root-mean-square tracking errors shall include those given in Table 2 and any additional independent errors peculiar to a given system. Each (3σ) independent error shall be determined for the specified environment. The 3σ value shall be used for tabulating and combining the errors.

John Ruze
MIT Lincoln Laboratory
Lexington, Massachusetts

LOSS CALCULATION ON METAL SPACE FRAME RADOMES

1. Introduction

The purpose of this note is to present a method of determining the loss due to a metal space frame. Fundamentally, the method does not differ from that introduced by Kennedy[1] and extended by Kay.[2] After certain simplifying assumptions, these methods consist of subtracting the forward-scattered field of each individual member from the unobstructed aperture axial field. It is hoped that the procedure suggested in this note will prove convenient to electrical and structural designers.

It must be realized that a rigorous formulation of the radome space frame as an electromagnetic scattering problem represents a tremendously difficult endeavor. Even such simplified problems as the transmission through a square grid have only approximate solutions and these only in the long- or short-wavelength limit. If a rigorous formulation were available, suitable for use with modern computing machines, one would question its engineering usefulness, since it would require as one of its inputs accurate knowledge of the scattering coefficients of various structural shapes and at all aspect angles.

It has therefore been customary to make certain simplifying assumptions to make the problem amenable to calculation. Principal among these is that the elements scatter as independent, infinite cylinders placed in front of an aperture with uniform phase. This implies the neglect of end effects, of circulating currents at member junctions, mutual scattering, and near field effects. In addition, simplifying the calculations to average various effects implies that such an average has meaning and the number of such elements is large.

To evaluate the success of this suggested procedure, recourse must be made to experimental verifications on actual radome geometries. Unfortunately, there is a dearth of precise radome loss measurements over a range of frequencies. The procedure, if applied to a number of structurally acceptable designs, should perhaps yield those preferable from an electromagnetic standpoint.

2. Theoretical Discussion

Flat Sheet Radome

We first consider the spherical radome as a flat sheet and determine the fraction of the aperture area blocked (Figure 1) as

$$\eta = \frac{2\sqrt{3}\,WL}{(L+2\pi)^2} + \frac{2\pi\, r^2}{\sqrt{3}\,(L+2\pi)^2} = \eta_m + \eta_h. \tag{1}$$

The first term is the member blockage and the second the contribution of the hubs. The latter is generally 10–30 per cent of the member blockage. Equation 1 is derived on the basis of equilateral triangles. For a random geometry the individual member lengths vary. It can be shown that for a variation of ± 10 per cent in length the error made by using the mean length in Equation 1 is 2 per cent (too small).

At very high frequency (optical limit) the reduction in axial field is then simply

$$e/e_0 = 1 - \eta_m - \eta_h \,. \tag{2}$$

At lower frequencies the blockage must be modified by a factor that accounts for the relative scattering efficiency of the members. This factor,[1,2] called the induced current or field ratio, is simply the ratio of the forward-scattered field of the member to the forward-radiated field of an incident plane wave of the same width. It is a complex number that depends on the member cross section, their shape, and on their orientation, θ, relative to the polarization vector. In the optical limit it approaches minus one. Equation 2 then becomes

$$e/e_0 = 1 + \eta_m[\overline{ICR}\bot\cos^2\Theta + \overline{ICR}\|\sin^2\Theta] - \eta_h \,, \tag{3}$$

where we have not modified the smaller hub contribution and where we have resolved the polarization vector along and perpendicular to the member axis.

If the members have a random or equiangular arrangement, then their summation in Equation 3 yields

$$e/e_0 = 1 + \eta_m\,\overline{ICR} - \eta_h, \tag{4}$$

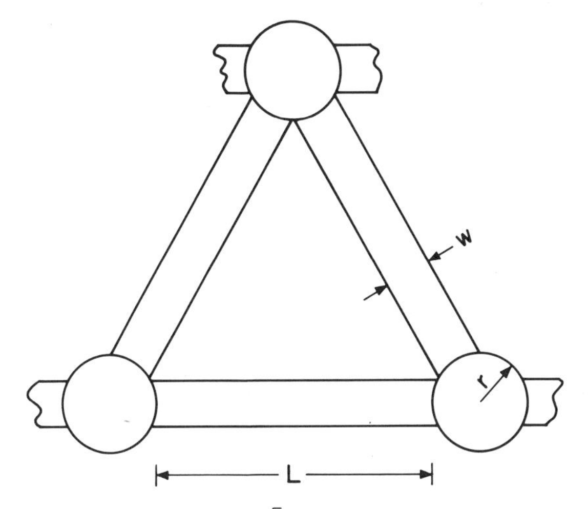

Figure 1. Blockage $= \dfrac{2\sqrt{3}wL}{(L+2r)^2} + \dfrac{2\pi}{\sqrt{3}}\left(\dfrac{r}{L+2r}\right)^2 = \eta_m + \eta_h$.

where $\overline{ICR}$ is the average induced current ratio for the two orthogonal polarizations.

The loss in axial power is then

$$p/p_o = |1 + \eta_m \overline{ICR} - \eta_h|^2 \approx 1 + 2\,\eta_m \mathcal{R}e\overline{ICR} - 2\eta_h, \tag{5}$$

where the approximation is valid as the blockage is hopefully small. (See Appendix.) Now, by the "shadow theorem, [3] the real part of the forward-scattered field is proportional to the total scattering cross section or

$$\mathcal{R}e\overline{ICR} = -\,\sigma_{sc}/2w = -\,g(w), \tag{6}$$

where 2w is the optical cross section and σ_{sc} the average total scattering cross section. We have alternately for the loss of axial power

$$p/p_o = 1 - 2\eta_h - 2\eta_m y(w). \tag{7}$$

The advantage in using the total scattering cross section instead of the more correct complex induced current ratio is that many references calculate the former. Figure 2 plots the total scattering cross sections for round cylinders. Figure 3 gives the average total cross section for rectangular cylinders.[4–9]

The flat radome loss is then calculated by Equation 7 with the use of Equation 1 and Figure 3. This loss is independent of the aperture illumination taper.

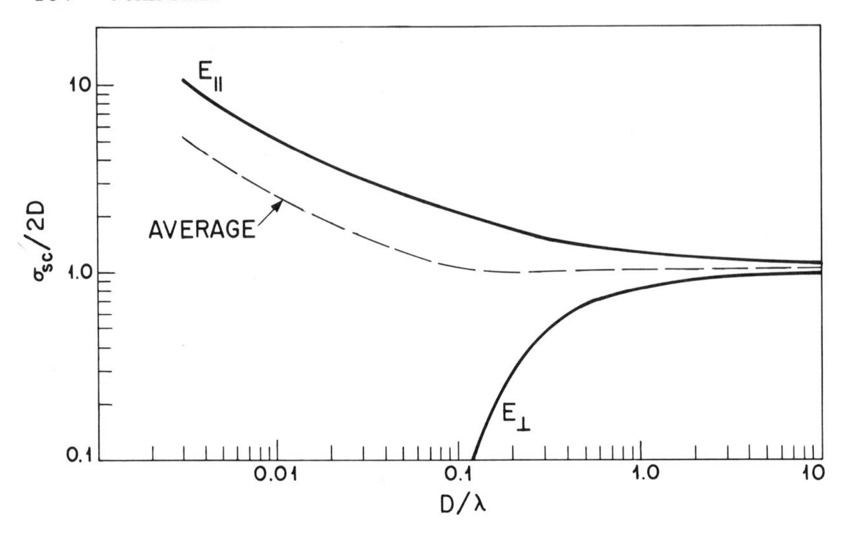

Figure 2. Total scattering cross section. Round metal cylinders.

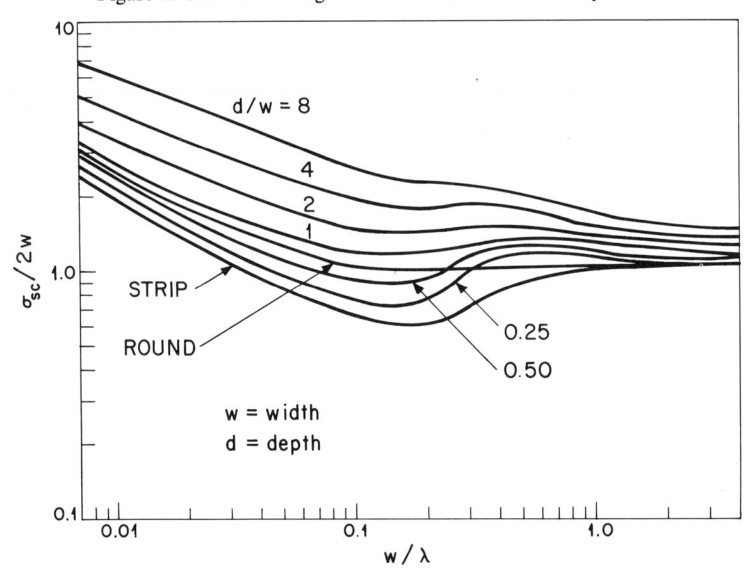

Figure 3. Total average scattering cross section. Rectangular cylinders.

Spherical Radomes

In a spherical radome the aspect of the members change with aperture position. It is therefore necessary to include this change in a modified scattering cross section or ICR and to weigh this changing contribution with the aperture illumination taper. The aspect of the members is bounded by two limiting cases; namely, (1) those members directed along great circle paths through the

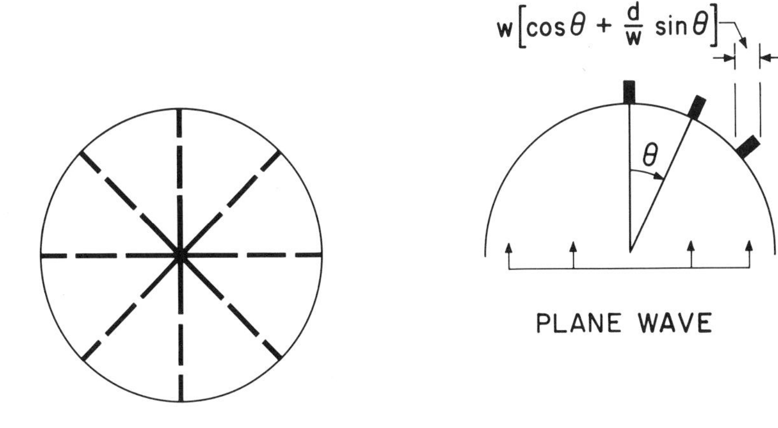

Figure 4. Limiting disposition of members.

beam axis, and (2) those members orthogonal to these great circles (Figure 4).

For the great circle members the scattering is reduced by the cosine of the angle of incidence since these members are foreshortened by the radome curvature. For the orthogonal members, the scattering is increased due to the greater projection of deep members. In addition, the apparent concentration of the orthogonal members increase as the cosine of the incident angle. We would therefore expect the scattering of the orthogonal system of rectangular members to vary in the optical limit (high frequency) as

$$(1/\cos\Theta)[\cos\Theta + (d/w)\sin\Theta]. \tag{8}$$

In the low-frequency limit, the rotation of the member is immaterial since the total scattering cross section here depends only on the perimeter in wavelength. Generally, we can represent the scattering of the orthogonal members as (Figure 5)

$$(1/\cos\Theta)[\cos\Theta + (d/w)\sin\Theta]^{n(\lambda)}. \tag{9}$$

The function $n(\lambda)$ varies from unity at the optical limit to zero at low frequencies. This function was determined, for rectangular members, from References 5 and 6 and is plotted in Figure 6. It was found that when it was plotted as a function of the perimeter in wavelengths that it was essentially independent of the d/w ratio. We note from Figure 6 that when the perimeter is less than half a wavelength the total scattering cross section is independent of element

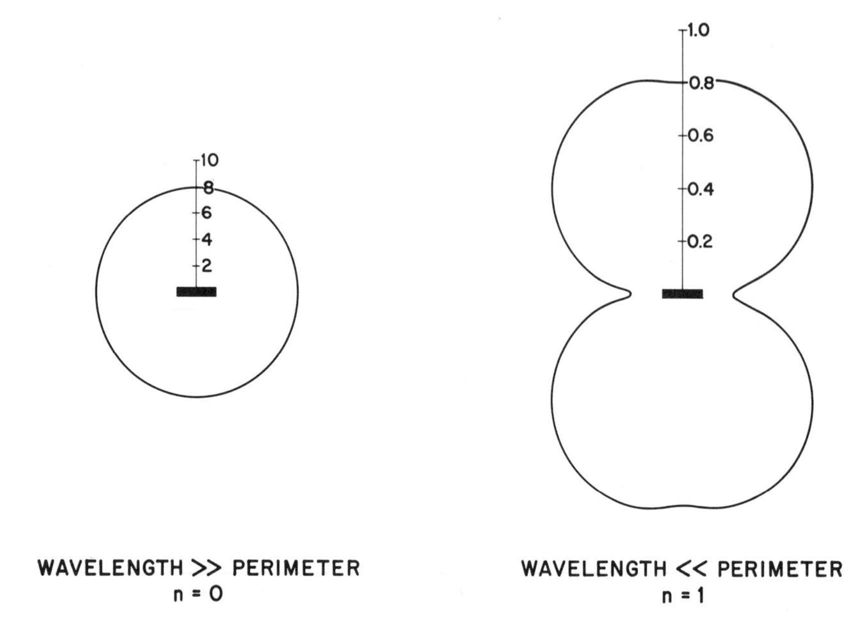

Figure 5. Relative total scattering cross section, generally $\bar{\sigma}/2w = g(w)[\cos\Theta + (d/w)\sin\Theta]^{h(\lambda)}$.

rotation; whereas when the perimeter is over ten wavelengths the scattering cross section is essentially the projected optical cross section. For round members there are no rotational effects and η may be taken as zero.

As we must weigh these effects by the illumination taper and normalize by the same, we have for the loss of axial gain

$$\frac{p}{p_0} = 1 - 2\eta_h - 2\,\eta_m g(w)\,\frac{\int_0^1 f(r)\cos\Theta r\,dr + \int_0^1 \frac{[\cos\Theta + (d/w)\sin\Theta]^{n(\lambda)} f(r) r\,dr}{\cos\Theta}}{2\int_0^1 f(r) r\,dr} \qquad (10)$$

where we have summed or integrated over circular rings and assumed that the extremes of member aspect behavior can be approximated by their mean.

3. Application

The result of our analysis is then given by Equation 10 or Figure 7. Here

1. The hub blockage is calculated from Equation 1 and is not modified by the curvature factor since the hubs are thin compared to their diameter. As they are uniformly distributed, their effect is independent of the aperture taper.

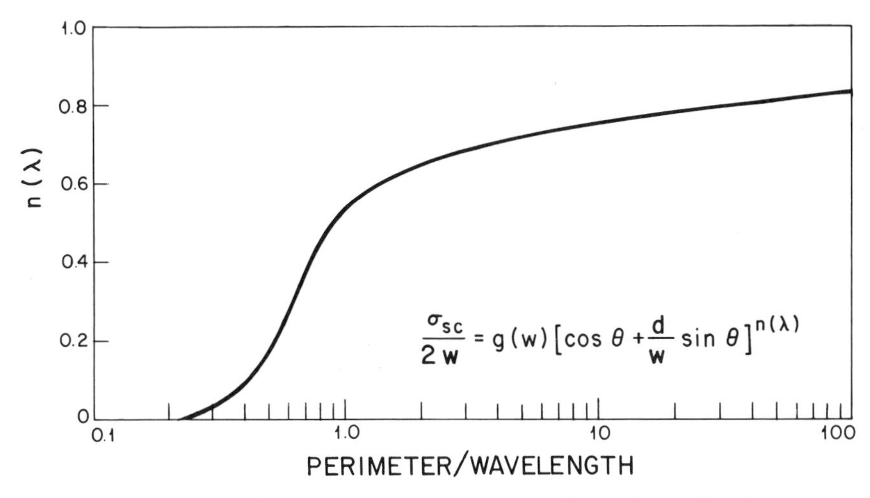

Figure 6. n(λ) plotted as a function of the perimeter in wavelengths.

2. The member blockage is also calculated by Equation 7. However, it must be modified by the $\overline{ICR}$ or relative average total scattering cross section, g(w) (Figure 3), and by a radome curvature factor.

3. The curvature factor depends naturally on the ratio of the radome to antenna diameter, on the aperture taper, and on the member depth-to-width ratio. The dependence is expressed in the bracketed term of Figure 7. For given parameters it can be evaluated and is given in Figures 8, 9, and 10. For $n(\lambda) = 0$ the curvature factor is essentially unity.

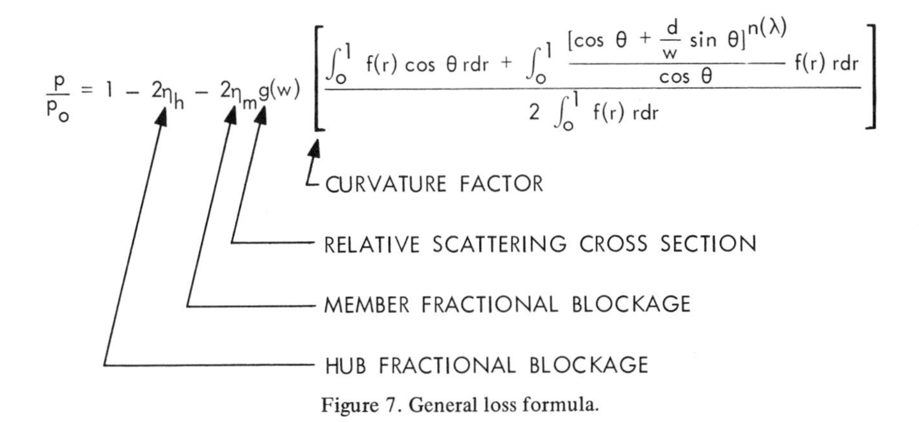

Figure 7. General loss formula.

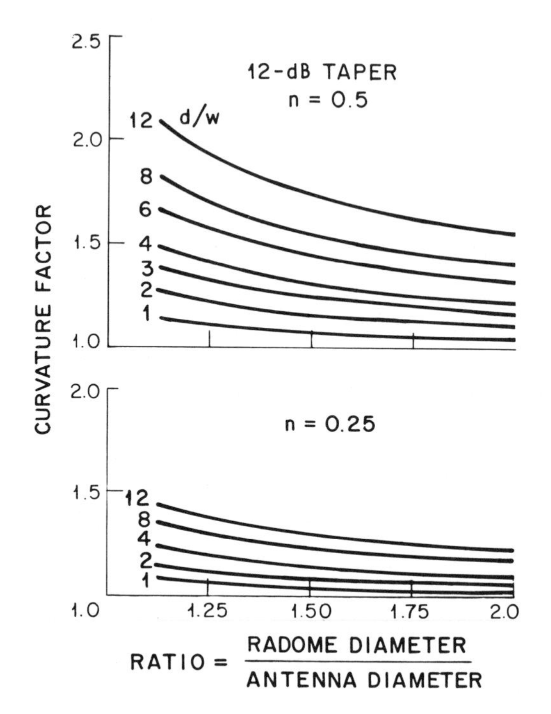

Figure 8. The curvature factor as a function of the ratio of radome to antenna diameter.

An example best illustrates the procedure. A proposed space frame has the following characteristics:

Radome diameter	500 ft
Antenna diameter	400 ft
Frequency	2700 Mc
Average member length, L	37.5 ft (35 ft to 40 ft var.)
Member width, W	2.5 in.
Member depth, σ	20 in.
Hub diameter, 2π	42 in.
Member perimeter, p	45 ft

From Equation 1 optical blockage

$$\eta = 1.02\,\frac{2\sqrt{3}(2.5)(450)}{(492)^2} + \frac{2\pi}{\sqrt{3}}\left(\frac{42}{492}\right)^2.$$

Note: 1.02 inserted due to member length variation.

$$\eta = \eta_m + \eta_h = 0.0164 + 0.0264.$$

At 2700 Mc, $\lambda = 4.5$in. and $p/\lambda = 10$; and $\eta = 0.75$ from Figure 6, as $w/\lambda = 2.5/4.5 = 0.555$; and $d/w = 8$. We have from Equation 10

$$p/p_0 = 1 - 2(0.0264) - 2(0.0164)(2.3)(1.9)$$

or

$$p/p_0 = 1 - 0.0528 - 0.1850 = 0.7622\ (-1.16\ \text{dB}).$$

4. Experimental Confirmation

To establish the validity of the foregoing procedure recourse must be made to experimental measurements on actual radome structures. Unfortunately, there is a dearth of precise measurements over a sufficiently wide frequency range and with diverse element shapes.

Figure 11 shows the correlation between the calculated and the experimentally measured data on the ESSCO model M-160 metal-space frame. The measured data were obtained by periodically rotating a half-radome scale model in front of the parabolic antenna. The vertical bars indicate the signal variability while the radome section was in front of the aperture. Even with no rotation there is a signal variability of about 0.02 dB.

5. Shaped Members

The present report presents data for loss calculations of radomes consisting of round or rectangular members. At times suggestions have been made to reduce the space frame loss by streamlining the members. This suggestion, no doubt, has its origin in the success obtained in reducing the radar cross section or backscattering coefficient of various missile shapes. Unfortunately, we are dealing in the space frame problem with the *total* and not the *back*scattering cross section.

It can be readily shown, at least in the high- and low-frequency limits, that any member shaping or streamlining should *not* offer any advantage.

In the high-frequency limit, the scattering cross section is the optical cross section or merely twice the projected area

$$\bar{\sigma}_{SC} = 2w.$$

For a flat radome the streamlining will therefore have no effect. For a spherical radome the streamlining will be deleterious due to the curvature factor of deep members.

In the low-frequency limit, the total *average* scattering cross section is given by

$$\bar{\sigma}_{SC} = \frac{\pi^2}{2k[\ln p/\lambda]^2} \tag{11}$$

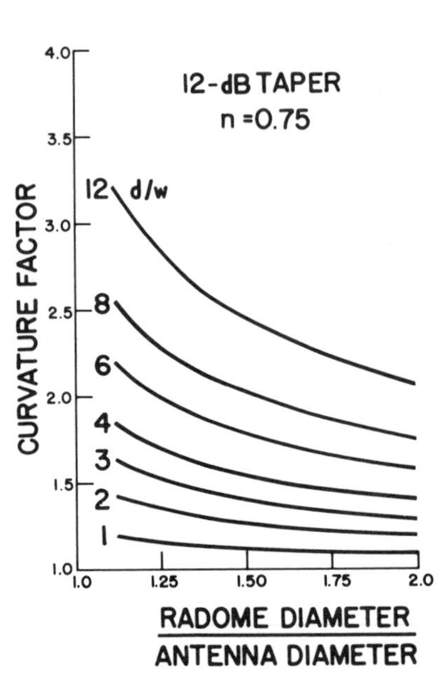

Figure 9. The curvature factor as a function of the ratio of radome to antenna diameter.

12-dB TAPER
n=1.0

12 d/w, 8, 6, 4, 3, 2, 1

CURVATURE FACTOR

RADOME DIAMETER / ANTENNA DIAMETER

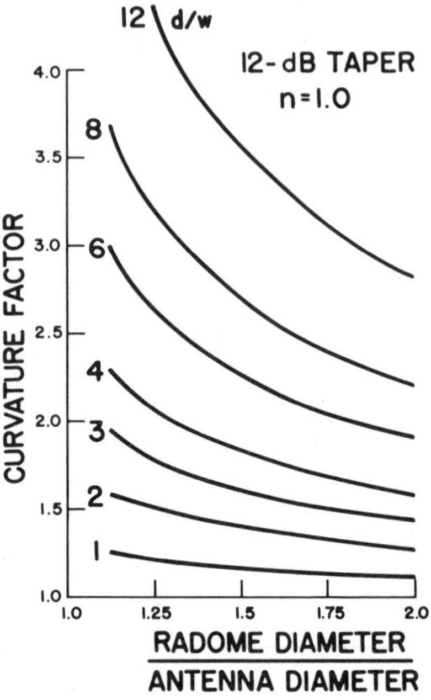

Figure 10. The curvature factor as a function of the ratio of radome to antenna diameter.

where $k = 2\pi/\lambda$ and p is the cross-section perimeter.

As for a fixed member, width streamlining will increase the perimeter and since the perimeter-to-wavelength ratio is less than unity, the streamlining will increase the scattering cross section and hence the space frame loss.

Acknowledgments

I am indebted to Albert Cohen and Adam Smolski of the Electronic Space Structure Corporation for providing the experimental data of Figure 11.

Also, to Joseph Morriello for the numerical integration of Equation 10.

Appendix

The approximation made in Equation 5 requires justification. To present the problem we note

$$p/p_0 = |1 + \eta\overline{ICR}|^2 = 1 + 2\eta|ICR|\cos\Theta + \eta^2|ICR|^2 \qquad (A.1)$$

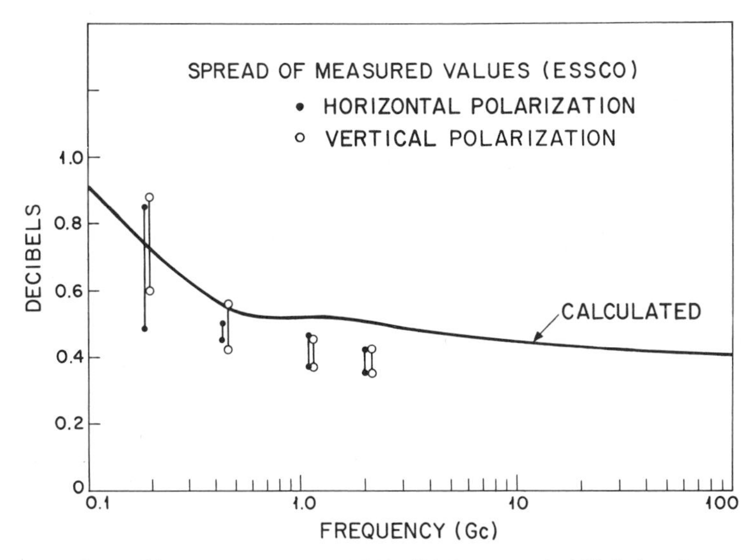

Figure 11. Space frame loss (M-160) (80-ft antenna in 160-ft dome).

in the optical limit the magnitude $|ICR| \rightarrow 1$ and its phase Θ approaches 180° or

$$p/p_0 = 1 - 2\eta + \eta^2 = 1 - 2\eta \qquad \text{(A.2)}$$

and the approximation is valid. However, in the low-frequency limit, the phase angle Θ approaches 90° and the approximation is not evident.

To investigate the behavior in the low-frequency limit, we insert the low-frequency asymptotic form into Equation A.1. As in the low-frequency limit, the induced current ratio or relative total scattering cross section does not depend on the member shape; we can use the values derived by Burke and Twersky[4] for elliptic cylinders. Using the dominant terms of their equation 41, we have

$$\overline{ICR} = -(D/kw)[1 - j(2L/\pi)], \qquad \text{(A.3)}$$

where 2w is the optical cross section,

$$D = \pi^2/(\pi^2 + 4L^2)$$
$$p = d/w$$
$$L = \ln \gamma x (1 + p)/2 .$$

References

1. P. D. Kennedy, "An Analysis of the Electrical Characteristics of Structurally Supported Radomes," Report No. 722-8 on Contract AF 80 (602)-1620, Ohio State University (November 1958).
2. A. F. Kay, "Electrical Design of Metal Space Frame Radomes," *IEEE Trans. Antennas Prop.* **AP-13**, 188 (March 1965).
3. J. E. Burke and V. Twersky, "On Scattering of Waves by an Elliptic Cylinder and by a Semielliptic Protuberance on a Ground Plane," *J. Opt. Soc. Am.* **54**, 732 (June 1964).
4. J. Van Bladel, *Electromagnetic Fields* (McGraw-Hill Book Company, New York, 1964), p. 375.
5. A. F. Kay and D. Paterson, "Design of Metal Space Frame Radomes," RADC-TDR-64-334 Air Force Rome Air Development, Rome, New York (1964); ASTIA AD-610037.
6. K. K. Mei and J. Van Bladel, "Scattering by Perfectly Conducting Rectangular Cylinders," *IEEE Trans. Antennas Prop.* **AP-11**, 185 (March 1963).
7. B. J. Morse, "Diffraction by Polygonal Cylinders," *J. Math. Phys.* **5**, 199 (1964).
8. R. W. P. King and T. T. Wu, *The Scattering and Diffraction of Waves* (Harvard University Press, Cambridge, Massachusetts, 1959).
9. J. R. Mentzer, *Scattering and Diffraction of Radio Waves* (Pergamon Press, New York, 1955), p. 56.

P. D. Kalachev and A. D. Kuzmin
P. N. Lebedev Physical Institute
Moscow, U. S. S. R.

THE CONSTRUCTIVE SCHEMES OF MULTISUPPORT STEERABLE PARABOLIC ANTENNAS

The achievement of high rigidity is the key problem in the construction of large steerable microwave radio telescopes.

There are three causes of distortion of the parabolic reflecting surface: weight loading, wind loading, and inhomogeneous heating. In the present paper some constructional ideas of reducing weight loading distortion are discussed.

Let us consider a two-support girder (Figure 1). It is known that there is the optimum position of supports corresponding to minimum sag. The value of

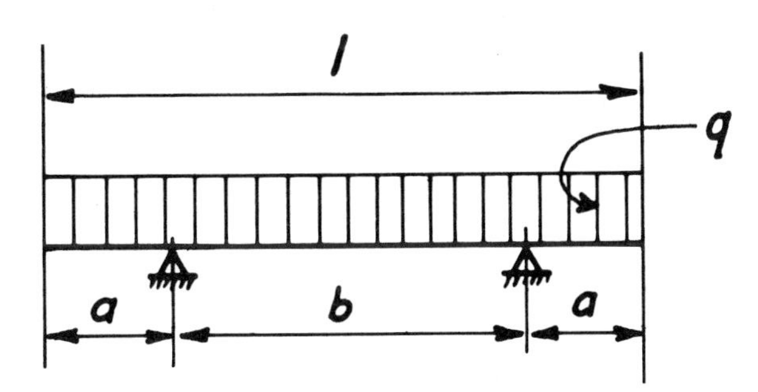

Figure 1. Two-support girder.

this sag is equal to

$$\delta = 0.119 \times 10^{-10}\, l^2 [(l^2/h^2)+24.3]. \qquad (1)$$

Here l is the girder length and h is its constructional height.[1]

The modification of this equation

$$\delta = 0.119 \times 10^{-10}\, \mathcal{K} l^2 [(l^2/h^2) + 24.3] \qquad (2)$$

can be applied to the evaluation of the reflector rigidity. We introduce here the loading factor

$$\mathcal{K} = G_{rfl}/G_{bear},$$

where G_{rfl} is the total weight of the reflector and its bearing elements and G_{bear} is the weight of diametral bearing elements.

The constructional height h is limited by engineering consideration. But sag $\delta = f(l^2)$ even if $h = \infty$. Therefore the only radical way of reducing the distortion is the shortening of the girder length. It can be accomplished by multi-support suspension of the reflector on radial symmetrical supports.

The scheme of mutisupport suspension of the reflector is shown in Figure 2.

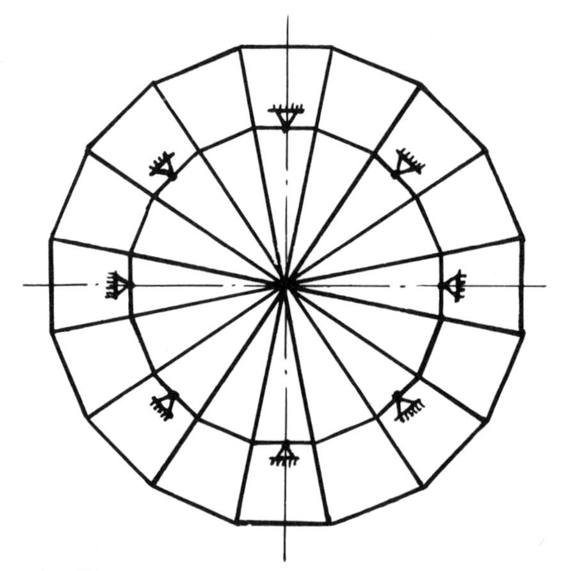

Figure 2. Multisupport suspension of reflector.

For calculation let us cut the chord elements of the frame (Figure 3). Then we have identical diametral bearing elements similar to the two-support

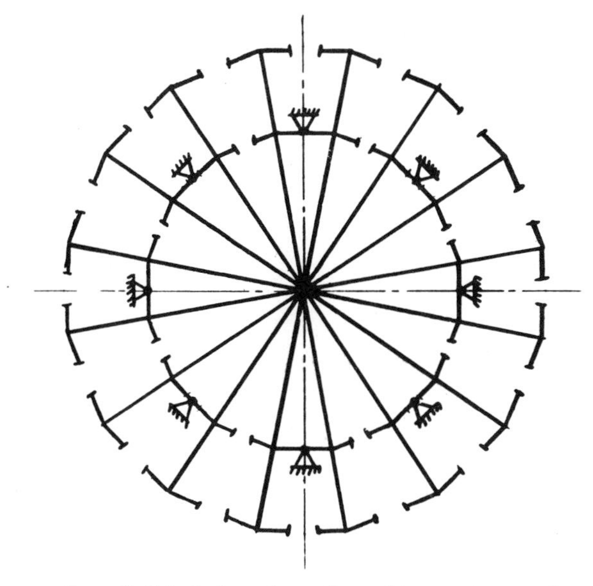

Figure 3. Calculation scheme for multisupport suspension.

girder. Maximum sag of each diametral bearing element is determined by Equation 2. The decrease of δ can be achieved by reduction of $\mathcal{K}$. But the latter leads to the increase of G_{rfl}. Nevertheless, it is profitable on a limited scale. This idea is illustrated by Figure 4, which shows the decrease of $\mathcal{K}$ by

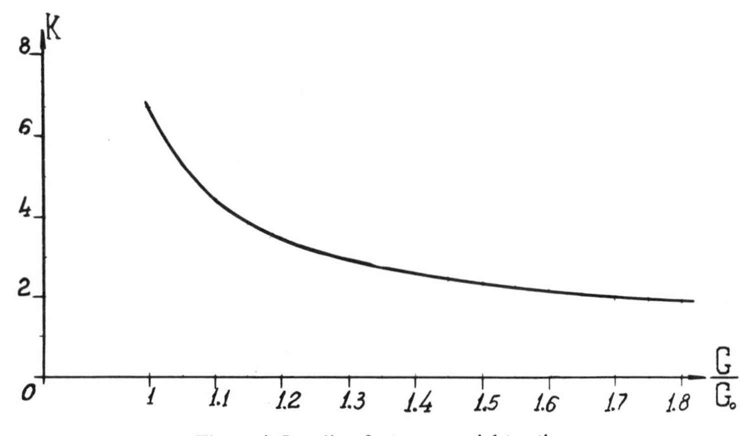

Figure 4. Loading factor vs. weight ratio.

the increase of weight G of the reflector, having initial weight G_0 and $\mathcal{K}_0 = 6.5$, typical of the bulk of the existing large antennas.

The scheme of multisupport suspension should satisfy the following require-

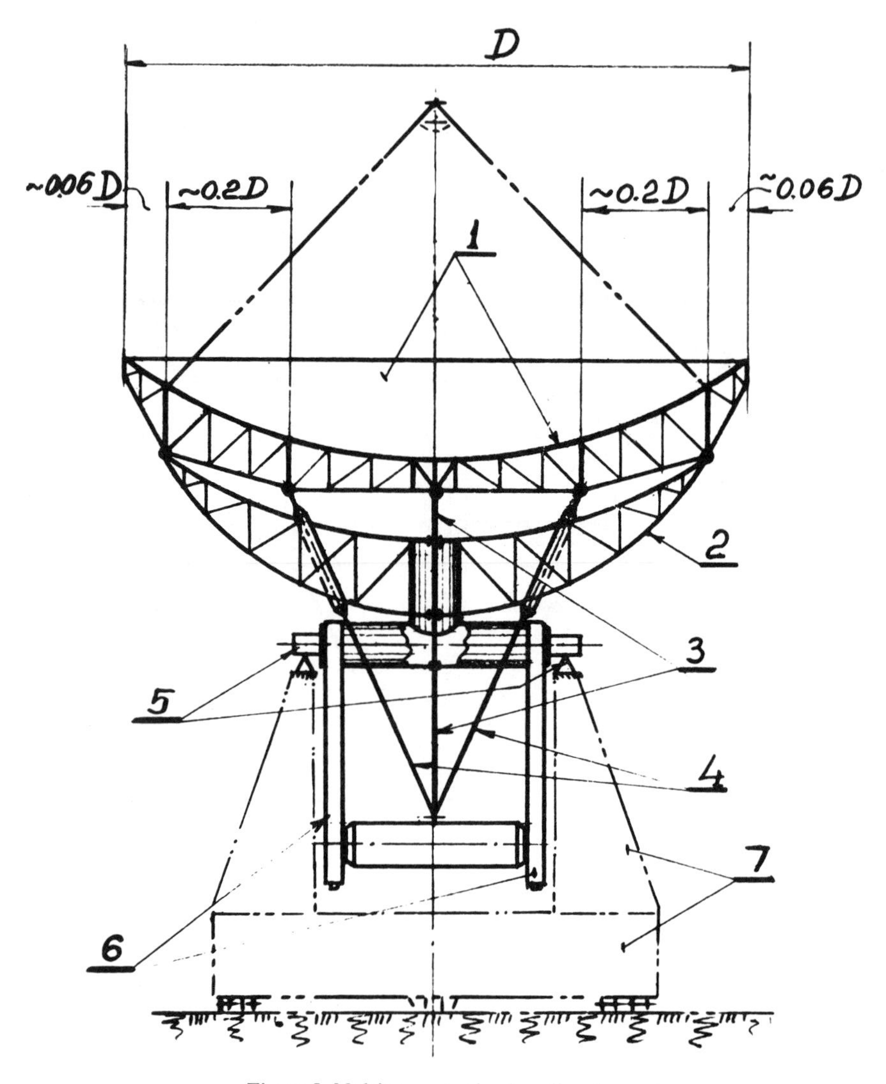

Figure 5. Multisupport mixed cyclic scheme I.

ments:

1. Rigidity of all supports must be identical.
2. All supports must be coplanar when the reflector is tilted.

The achievement of these requirements involves serious constructional difficulties.

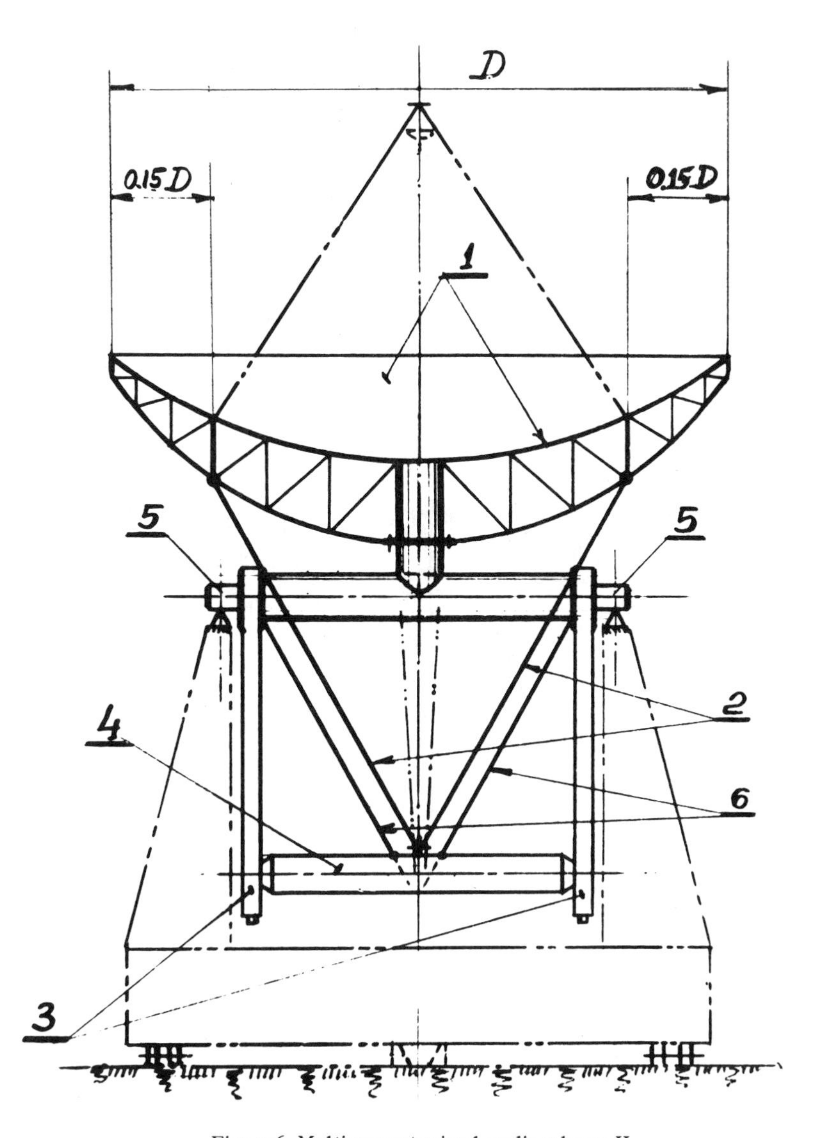

Figure 6. Multisupport mixed cyclic scheme II.

There are various ways of fulfilling the multisupport suspension of the reflector. One of them is the scheme of four-support suspension of the 22-m radio telescope of the Lebedev Physical Institute.[2] This scheme gives us an opportunity of reducing the weight distortion by half an order of magnitude.

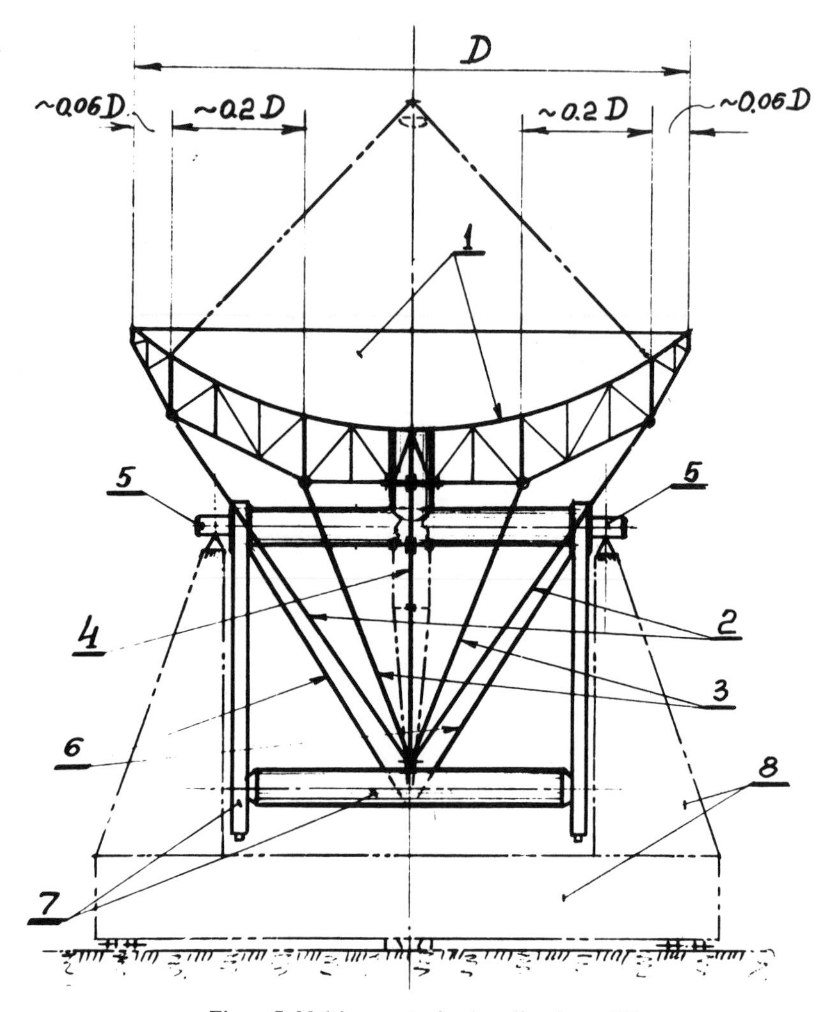

Figure 7. Multisupport mixed cyclic scheme III.

In the present paper we shall consider new possible schemes of multisupport suspension.

Figure 5 gives a multisupport mixed cyclic scheme. Here the following designation is used: 1–reflector; 2–intermediate multisupport frame; 3–central supporting pivot; 4–pivot pyramid; 5–bearing journals; 6–gears; 7–base of the antenna.

The intermediate multisupport frame has a cyclic construction. But it is loaded by the reflector. Therefore, for coplanar position of its supports some

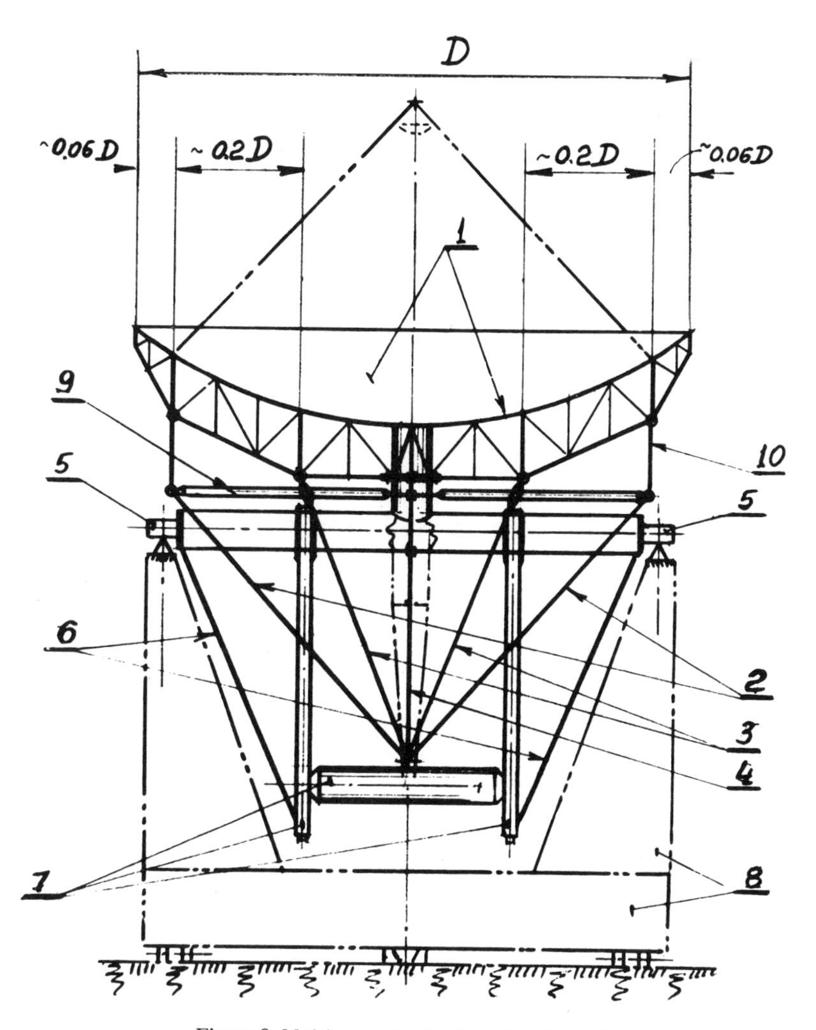

Figure 8. Multisupport mixed cyclic scheme IV.

additional requirements should be met.[3]

The pivot pyramid is the cyclic system and the coplanar position of the supports which are formed by its base takes place in case of any tilt of the reflector.

Figure 6 gives a multisupport cyclic scheme. The following designation is used: 1–reflector; 2–pivot pyramid; 3–gears; 4–counterweight beam; 5–bearing journals; 6–slants.

In the horizontal position of the reflector its weight is taken generally by the pivot pyramid and partly by the horizontal axis.

In the vertical position of the reflector its weight is taken entirely by the frame located in the plane coincident with the horizontal axis and is parallel to the reflector aperture. The moment of the reflector is taken by this frame and the pivot pyramid. The supports of the reflector are the base of this pyramid. These supports are coplanar.

Figure 7 gives a multisupport cyclic scheme with a double pivot pyramid and the central pivot. The designation 1, 2, 5, and 6 is the same as in Figure 6. Other designations are the following: 3–internal pivot pyramid; 4–central pivot; 7–gears. In this scheme the supports are the bases of pivot pyramids. However, they are coplanar for each pyramid separately. Mutual parallelism of these two planes is achieved by a special beam located at the top of one of the pyramids.[3]

Figure 8 gives a multisupport cyclic scheme with two pivot pyramids. But here the horizontal forces from the outer pyramid are taken by additional radial elements–9.

We consider that cyclic multisupport scheme of the parabolic reflector is the most promising and that constructional work in this direction should be continued.

References

1. P. D. Kalachev, "The Evaluation of Rigidity of a Parabolic Reflector Loaded by its own Weight," *Proc. Lebedev Phys. Inst.* **38**, 72 (1967).
2. P. D. Kalachev, "Elastic Deformation of the Reflector of the 22-Meter Radio Telescope of the Lebedev Phys. Inst. Loaded by its own Weight," *Proc. Lebedev Phys. Inst.* **28**, 183 (1964).
3. P. D. Kalachev, "The Problem of Designing a High-Resolution Parabolic Antenna," *Proc. Labedev Phys. Inst.* **28**, 52 (1964).

H. A. Cress and S. G. Talbert
Battelle Memorial Institute
Columbus, Ohio

DETERMINATION OF THE APPROXIMATE STRUCTURAL PARAMETERS OF A LARGE STEERABLE ANTENNA

Introduction

The performance of large parabolic antennas, such as the 300-ft reflector designed for the Naval Research Laboratory, is largely dependent upon the quality of the structural design to which they are built. Structural parameters, such as deflections, masses, and spring rates, are crucial in determining the uses to which the antenna can be put. Most of these parameters are not commonly computed when designing a structure because they do not determine its strength.

For antennas, however, knowledge of these other structural parameters is quite important before the antenna is built. Such knowledge permits definition of the operating limits – the performance envelope of the instrument – and identifies the critical areas of the structure that affect performance. The advent of the large high-speed computers makes calculation of these structure-dependent performance characteristics practical. A wealth of useful information can be gained in a reasonably short time.

Antenna designers have been leaders in the application of computer-based analysis to structural design. In our efforts in this area, several problems arose in using the computer as a design tool. First, and probably most important, the cost of the computer analysis became quite high as the structures became more complex. So long as the bulk of the computation could be done within the computer core memory, the machine time and, thus, the cost was reasonable. However, as the structures became more complex, it was necessary to use tape storage, which made the computations quite expensive because of the greater amount of machine time required. Also, because individual runs were expensive, the number of design variations that could be examined was restricted. Second, and more subjective, as the designer put larger and larger structures into machines, he lost contact with the important details – one might say that he lost his "feel" for the design.

To overcome these problems, an approximate technique of computer analysis was conceived and applied.

In a recent program conducted at Battelle for the Office of Naval Research, reasonable approximate results were obtained relative to the 300-ft-antenna design through iterative use of the computer, employing the stiffness-matrix method of analysis. The over-all objective of the research was to compute certain structural parameters of the design and to define their relationship to the anticipated over-all antenna performance. The major analysis conducted was structural, and was primarily intended to obtain surface deflections at various altitude angles and wind conditions.

In all cases, it is believed that approximate values sufficiently accurate for performance estimates were obtained economically.

Logic of Iterative Approach

In any large and complex structure there are too many members to permit a rapid and economical analysis of the structure as a whole. In even the largest computers the inherent machine limitations preclude simulation of the entire structure in a single step. To overcome this problem, small portions of the structure, such as a single rib or group of ribs, can be simulated by the substitution of a more simple structural unit.

For example, a complex truss might be analyzed and its characteristics determined. The complex truss could then be approximated with a simplified truss that duplicates the characteristics of the former at a few selected points. By successively applying this technique to the various portions of the structure, a simplified mathematical simulation of the actual structure can be built, piece by piece. When complete, this simplified mathematical model can, in turn, be analyzed to determine the structural characteristics of interest.

The size of the matrix is the determining factor in the size of the structure that can be accommodated. A 200 x 200 matrix, such as the one used in this program, will accept about 35 connecting points or nodes while permitting six degrees of freedom to be examined. If restrictions can be placed on the degrees of freedom in the system, significant increases can be made in the number of nodes. In many instances in an antenna structure, less than the maximum six degrees of freedom are needed. Until a detailed examination of the structure has been made, it is not possible to fix the number of degrees of freedom that must be assumed for each node.

The individual members that make up the simplified truss need not be physically realizable. As long as the computer will accept them, whether or not a steel mill will roll them is of no consequence. This freedom is of great value in matching the behavior of simple and complex units.

Analysis Procedure

The structural analysis of the NRL 300-ft-diameter antenna was divided into two major phases. In the first, the reflector itself was structurally modeled in as much detail as possible within the limitations of the computer core memory. The output of this part of the program was a simple structure that approximated the behavior of the complex reflector under normal loads. In the second phase of the analysis, the structure of the antenna as a whole was considered with the reflector and its backup structure combined to compute deflection patterns that were of primary importance in the radio-optical performance calculations. In addition, the towers and turntable were analyzed in a similar way to obtain trunnion tower deflections and drive-mount deflections. The latter results were the basis for calculations of the natural vibration frequencies of the lower structure.

Structural Analysis Details

The structural analysis was conducted to determine surface distortion at three altitude angles: 90° (zenith position), 45°, and 0°. The final number of beams and joints analyzed by the computer nearly eliminated the need for tape storage on the digital computer and substantially reduced computer costs. The mathematical basis for the computer program is described later.

Figure 1 is a sketch showing the structural design of the 300-ft antenna. The reflector portion consisted of 40 radial rib trusses, each interconnected with 15 circumferential hoops. These trusses were supported with a cone-shaped backup structure with the counterweight at its apex. The cone structure is 200 ft in diameter where it attaches to the rib trusses.

Wind loads for the following wind directions were applied:

0° azimuth angle, 0° altitude angle,

0° azimuth angle, 45° altitude angle.

The added deflections for an operating wind velocity (28 mph) were found to be small compared with the gravity deflections.

While no stress calculations were made for the models, attention was given to any unusually large deflections that might imply dangerous stress levels in the structure.

The approach used for obtaining simplified structural models was as follows. The actual structure was subdivided into small separate units, each of which could be completely evaluated with the computer program, such as rib trusses,

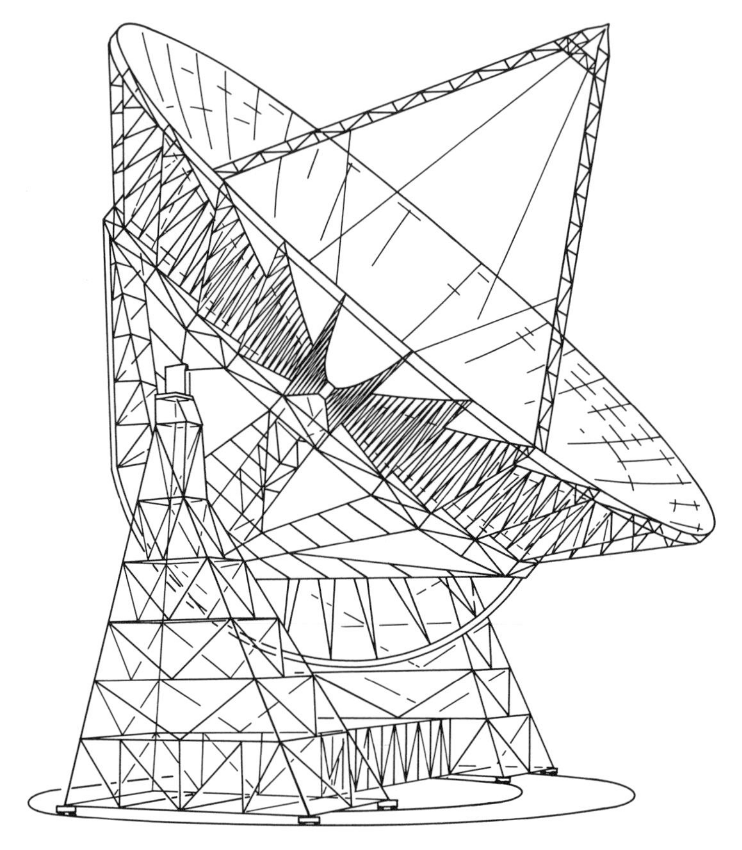

Figure 1. Three-hundred-foot antenna.

hoops, trunnion girder, rim-cage girder, etc. These small units were then replaced with a simplified structural model that accurately duplicated deflections at specific locations but which had fewer joints and beams. By combining these simplified units, a complete structural model was obtained of all the superstructure parts affecting surface deflections. This complete model consisted of 80 joints and over 300 beams, and indicated surface deflections at 34 points of a symmetrical half-section for dead loads and operating wind-pressure forces. These surface deflections were then fed into a radio-optical-analysis computer program to determine directive gain and pointing error at the various altitude angles.

Each structure analyzed on the computer was limited to about 35 joints, or nodes, if all six degrees of freedom were desired at each node. However, this number could be increased if, for example, all node rotations were eliminated and only translations in three directions were permitted. Naturally, one or more nodes must always be fixed to make the structure stable. Therefore, the reason for the modeling approach was to eventually arrive at an equivalent structure representing a symmetrical half-section of the superstructure, yet containing a manageable number of nodes.

The underlying assumption behind this structural modeling approach was that properly designed simplified units could all be combined to give an over-all model that would produce fairly accurate displacements at selected locations when equivalent external loads were applied. The computer program inherently assumed that each joint connection was a "fixed" or rigid connection. Loads and moments could be applied only at node locations.

Mathematical Analysis

The approach used in the analysis is based on the stiffness-matrix method. This method was designed to permit the analysis of complicated, highly redundant structures.

The stiffness-matrix method is applied to calculating the deflection and stresses in the structure in the following way. The structure is an assembly of straight beams. Now, from simple beam theory, it is possible to write a set of 12 equations relating the forces and moments on each end of the beam. In the matrix equation

$$\mathbf{F}_i = K_i \mathbf{u}_i ;$$

F and **u** are the 12-fold force and displacement vectors, respectively, and K is the 12 x 12 "stiffness matrix" for the beam segment. In the inverse equation

$$\mathbf{u}_i = K_i^* \mathbf{F}_i ;$$

the matrix K_i^* is called the "flexibility matrix." It is apparent that K_i^* is the inverse of K_i.

The stiffness-matrix approach consists of appropriately combining the stiffness matrices of each beam element to obtain a stiffness matrix for the entire structure. Then

$$\mathbf{F} = K\mathbf{u},$$

where **F** is the set of loads and moments applied to the structure at the points at which the segments are joined (hereafter these points will be called "nodes"), **u** is the vector of displacements of the node points, and K is the over-all stiffness matrix of the structure. In general K will be of the order of six times the number of node points (less the number of physical constraints put on the system).

Reflector Deflections

The entire superstructure, consisting of the dish and cone, will contribute to the deflections at the reflector (mesh) surface. The two basic parts – the dish and the cone – were modeled separately, and then combined to give the gross deflection pattern. It was assumed that the mesh surface had no rigidity. This and other assumptions necessary during the development of the models always favored the conservative approach, a more flexible structure.

Dish Analysis

Figure 2 shows views of the final simplified model. The actual dish portion itself consisted of 40 rib trusses, 15 hoops, and over 6000 connecting points. The final model of a symmetrical half-section of the dish had 10 ribs (actually 11 ribs, but two were half-ribs on the vertical centerline), 3 hoops, 46 nodes, and 148 beams.

The method of arriving at the design of the simplified rib trusses, hoops, and other assemblies, and of determining the concentrated load patterns is described below.

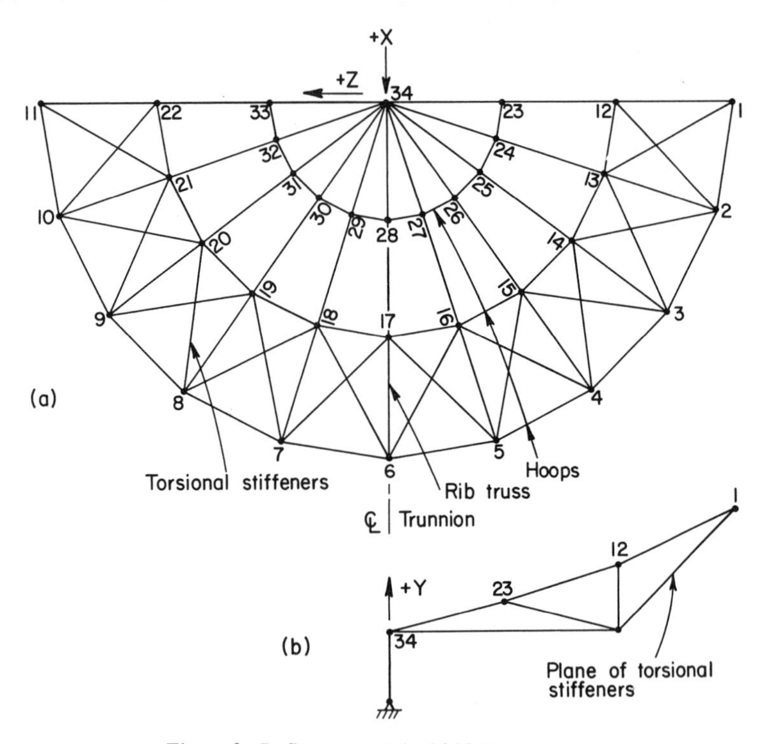

Figure 2. Reflector model of 300-ft antenna.

Rib Trusses. An actual rib truss had 32 nodes and 60 beams. Therefore, it could be handled nicely by the computer program. Usually, only the total dead-load distribution was examined, but one run was made at 90° with a 20-psf snow load included to check design deflections that had been previously calculated at the 150-ft radius. The design calculations indicated an outward deflection of 0.96 in. and a downward deflection of −1.28 in., while the computer run gave deflections of 1.01 and −1.36 in., respectively.

Because the computer program could handle only about 35 nodes with all degrees of freedom, it was decided that the simplified dish model could have about 10 ribs with nodes at 50-, 100-, and 150-ft radii. Therefore, each simplified rib represented two actual rib trusses. Their corresponding deflections at the 50-,100-, and 150-ft locations were matched closely for all three altitude angles and equivalent dead loads. Various types of simplified rib designs were tried that consisted of cantilever beams or simple tension and compression members. It was finally determined that the best model was also a truss, as similar as possible to the one being studied. Simple cantilever beams could not duplicate the two-dimensional deflection behavior of a rib truss. The rib model shown in Figure 2 was found to be the best over-all design. It consists of five nodes and seven beams. The beams act only in tension or compression.

Another difficulty that occurred when simplifying structural components was that loads could be applied only at nodes. Therefore, it was necessary to calculate new force vectors (load patterns) that would maintain similar total loads as well as moment balances.

The final choice of beam areas and force vectors for the simplified rib truss duplicated the corresponding deflections of the actual rib truss within about ± 9 per cent for altitude angles of 90°, 45°, and 0°. The magnitudes of the worst errors were less than 1/32 in. at 90° and 45°, and about 1/16 inch at 0°. One must also keep in mind that the rib models along the symmetrical centerline represent only one actual rib truss, so the areas of each beam are just half those of the other nine rib models. Likewise, the loads on these centerline nodes are half the magnitude of all the others at similar radii.

Hoops. The simplified dish model required only three hoop connections, at 50-, 100-, and 150-ft radii. Therefore, the 15 hoops in the actual dish had to be reduced to three equivalent hoops for the model. Actually, 14 hoops were reduced to 3, and the hoop at the 100-ft radius was left "as is" to maintain the rigidity of the dish where it fastened to the cone.

The following procedure was used to simplify the number and complexity of the hoops. Each actual hoop was analyzed with a separate computer run, with the hoop fixed on one side and loaded so as to determine its resistance to elongation and shear. Each hoop was then modeled as a single cantilever beam consisting of an area to resist elongation and a bending stiffness to resist shear. These areas and stiffnesses were calculated from the assumed loads and computer deflections for each hoop. Combining the simplified hoops required the following assumptions. Considering two actual rib trusses with 14 simplified hoop connectors, it was assumed that the center ends of the ribs were rigidly fastened to the spool, that the pin connections at the cone were essentially rigid points, and that the 100-ft hoop prevented any relative movement at its location. With the spool and pin connections acting at pivot points for the two ribs, it was further assumed that external separating and shearing forces were applied only at the 50- and 150-ft points and that the rib trusses acted as infinitely rigid members between the pivot points and the load locations. The deflections resulting from this set of arbitrary loads when all 14 hoops were in place should remain the same when the 14 hoops are combined to give two equivalent hoops at the 50- and 150-ft locations.

The 100-ft hoop was also investigated by a computer run, and then replaced with diagonal beams in the model at a 100-ft radius to give equal displacements for identical shearing forces.

Torsional Stiffeners. The model study showed that the size of the torsional stiffeners markedly affects the deflection pattern of the outer edge of the reflector surface at an altitude angle of 0°. The alignment of these torsional stiffeners in the bottom chord of the dish approximately parallels the location of the diagonals from the 100 to 150-ft radius in the dish model. Thus, each model diagonal was sized to represent two of these torsional stiffeners.

Dead Loads. The forces applied to the nodes of the dish model for each of the three altitude angles investigated had the same magnitude as the individual rib loads described above in the section on rib trusses. The loads, or forces, are actually the weight of the members concentrated at the nodes. The equivalent model weights were determined for altitude angles of 90°, 45°, and 0°, respectively. These node weights were then assumed to be identical for all other rib locations from the top around to the bottom.

The simulation of altitude angles of 90°, 45°, and 0° was accomplished by orienting the direction of gravity forces at those various angles. For example, at 90°, all of the deadweight was applied in the $-Y$ direction; whereas, at 0°, the deadweight was acting in the $+Z$ direction at each node.

Wind-Pressure Forces. The average pressure acting over the effective area for each node was estimated. The direction and magnitude of survival wind forces for pressure of 16.5 psf were scaled down by the ratio of the average operating wind pressure (2.0 psf) at that node to 16.5 psf. This total wind-force vector was then added to the dead-load-force vector to obtain the added displacements due to an operating wind velocity of approximately 28 mph.

Cone Analysis

The actual cone members could be modeled more directly than could the dish because of fewer beams and connecting points. The major simplification required was in the rim-cage girders. Also, because 11 rib locations were used for the dish model, the cone model had to have 11 connecting points at the same positions. Figure 3 shows the final simplified model configuration of the cone.

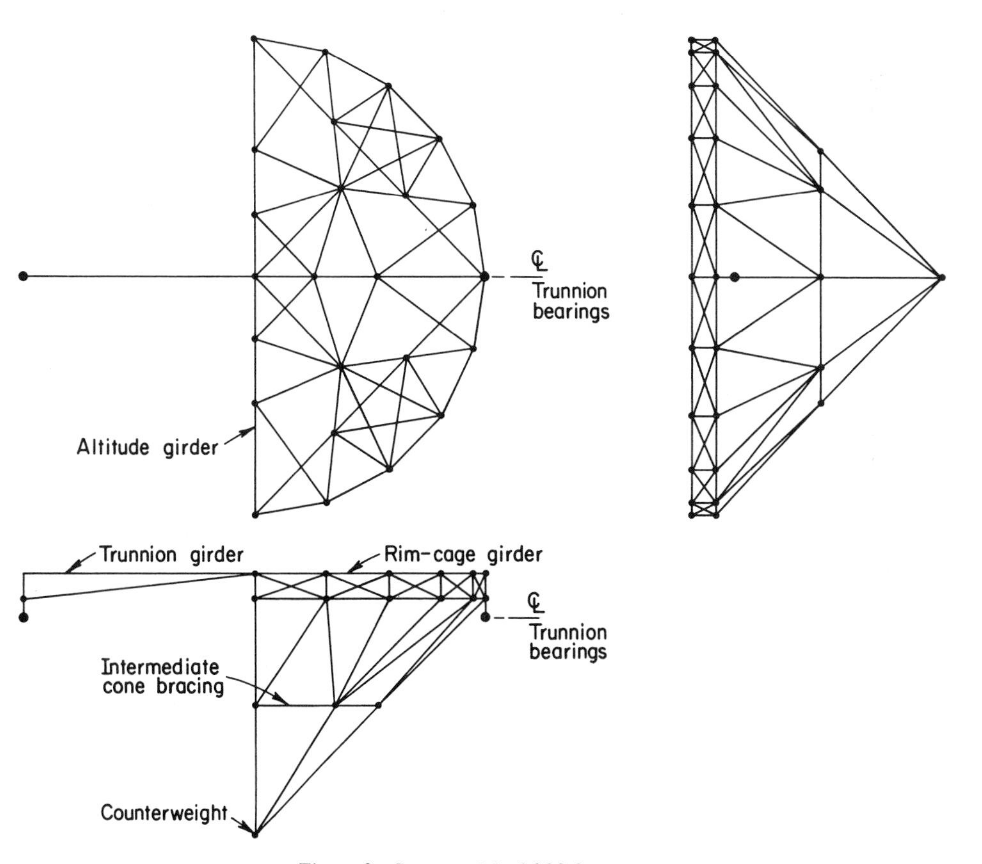

Figure 3. Cone model of 300-ft antenna.

Rim-Cage Girders. The actual rim-cage girders could be analyzed directly with the computer program, since they had 38 nodes and 141 beams. Again, a simple arrangement of cantilever beams was tried for a model, but since a force in one direction actually causes appreciable deflections in all three directions on the actual girder, a similar truss arrangement was needed. The final model chosen had only 16 nodes and 62 beams. The rim model had to have only four rib attachment points along its outer edge between the altitude and trunnion girders, while the actual rims had eight attachment points. Therefore, the eight actual rim displacements produced by the calculated rib reaction forces at the three altitude angles were interpolated to arrive at the four displacements required at the rim-model rib connections. The forces also had to be carefully rearranged on the model because of fewer nodes at different spacings.

Because this model was three dimensional and had so many beams, a trial-and-error method was used to find the required beam areas to match deflections for given forces vectors. One set of beam areas was finally found that satisfied all three altitude angles. The deflection patterns of the simplified model deviated no more than 0.03 in. in the X direction, 0.09 in. in the Y direction, and 0.06 in. in the Z direction for all three altitude-angle force vectors.

Trunnion Girder. A single equivalent cantilever beam was substituted for the actual trunnion girder, since the primary loading is at the center of its span. One-half the actual trunnion girder was put into the computer program with loads for bending in both directions, twisting about centerline, and stretching lengthwise. The properties of a simple cantilever were then calculated from the deflections caused by the corresponding forces. The single-beam properties were then divided in half so two beams could be used between the spool location and the trunnion ends to give proper attachment for the rim girders over the trunnion bearing.

The ends of the trunnion girder consist of large wide-flange beams to provide rigidity between the trunnion bearings and girder. This structure was also replaced by two cantilever beams after calculating the shear deflections that would arise along the Z direction.

Altitude Girder and Center Tower. Both of these structures were analyzed by the same procedure described for the trunnion girder, and were replaced with simple cantilever beams having the proper areas and bending stiffnesses.

Cone Members. All the other cone members could usually be included directly in the model with their actual areas and geometries. However, the effect of the altitude wheel's stiffening the beams between the counterweight and the ends of the altitude girder was included by substantially increasing the bending stiffness of the cone model beams in that direction. Also, the effect of the trusses stiffening the hanger beam was similarly included. The cone-bracing beams between the center tower and the hanger beam were run on the computer to determine actual bending stiffness, and then a simple cantilever beam substituted.

Weight Distribution. The total weight of the cone was maintained in the model, but since there are many fewer nodes in the model and since forces could be applied only at nodes, it was necessary to distribute the beam weights carefully to maintain the proper moment balance about the trunnion axis.

All weights occurring along the symmetrical centerline were reduced to half of their calculated values because only a half-section was considered. The beam areas and other beam properties were also divided in half for those beams lying along this boundary.

No forces were assumed in the altitude drive chain, since it was assumed that a balanced superstructure existed. However, feed-support forces were added.

Combined Dish and Cone Analysis

Spring-Supported Dish. The first attempt to analyze reflector surface deflections with the cone deflection effects included employed an iterative approach, using equivalent springs at the connection points. With the dish model containing 46 nodes and the cone model having 45 nodes, it was not actually possible to combine the models and retain all the degrees of freedom necessary for a complete deflection analysis without resorting to massive tape storage. A combined model would have a minimum of 78 nodes (the rib connecting points are common to both models.) Therefore, it was decided to mount cantilever "springs" beneath the dish model to simulate the flexibility of the cone at each point in all three axial directions. By making successive trial runs, alternating between obtaining dish support forces in the springs, and impressing these onto the cone model to obtain deflections, and thus "spring-rates," it was expected that deflection patterns at the two interfaces would eventually satisfactorily match one another. Then the surface deflection pattern obtained on the dish model (supported by springs) would include the effect of the cone flexibility. This method was tried with a fair degree of success. However, the iterations were tedious, and it appeared that convergence of the interface deflection patterns was slow.

Dish-on-Cone. The second attempt consisted of combining the dish and cone models into one model consisting of 80 nodes and 322 beams. Then the deflections could be obtained directly with one computer run. However, because of the matrix and memory size limitations of the digital computer, it was necessary to eliminate all node rotation in the combined model. Only translation in three directions was permitted, except at the symmetrical boundary where all nodes were fixed in the X direction and at the trunnion bearings where no motion was permitted. This restriction probably had little effect at nodes where all connected beams have negligible bending stiffness, compared to their tension or compression resistance. But many beams have sizable bending stiffness, especially at the trunnion ends, so fixing node rotations had the effect of restricting gross rotation of the model. It is not believed that the actual distortion of the dish surface was seriously affected, because comparisons between dish distortions for the "spring supported" dish and the dish-on-cone model showed similar patterns.

Deflections. Figure 4 shows the surface deflections along the vertical centerline. These were plotted from the computer data. The reference position used throughout the structural analysis is the nominal position of the beam connections or node points shown on the structural drawings. It is obvious from an examination of the results that an improved surface accuracy can be attained

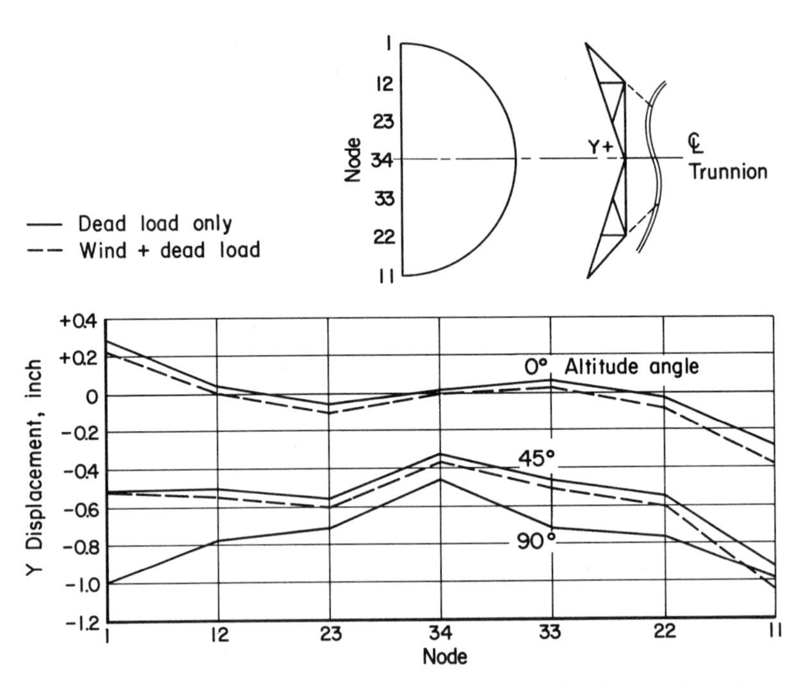

Figure 4. Plot of y displacements along dish vertical centerline for several conditions.

Table 1. Computer Deflections of 300-ft-Diameter Reflector Surface for Setting Surface at 45° Altitude Angle. Deflections Normalized to 45°, i.e., 45° Altitude Deflection = 0.

Node	0°	90°	Node	0°	90°	Node	0°	90°	Node	0°	90°
1			10			19			27		
x	0	0	x	−.004	−.013	x	+.059	−.035	x	+.012	−.011
y	+.808	−.458	y	+.609	−.065	y	+.349	−.155	y	+.425	−.143
z	+.339	−.422	z	−.034	−.300	z	+.006	−.159	z	+.100	−.190
2			11			20			28		
x	+.015	+.009	x	0	0	x	+.042	−.036	x	+.039	−.015
y	+.771	−.241	y	+.640	−.074	y	+.450	−.199	y	+.411	−.161
z	+.327	−.409	z	−.037	−.310	z	+.016	−.192	z	+.090	−.223
3			12			21			29		
x	+.013	−.008	x	0	0	x	+.020	−.020	x	+.003	−.020
y	+.660	−.380	y	+.539	−.271	y	+.500	−.214	y	+.369	+.076
z	+.286	−.381	z	+.184	−.315	z	+.026	−.217	z	+.076	−.214
4			13			22			30		
x	+.017	−.012	x	+.004	+.020	x	0	0	x	−.018	+.027
y	+.464	−.277	y	+.525	−.269	y	+.516	−.222	y	+.373	−.228
z	+.227	−.376	z	+.185	−.302	z	+.037	−.236	z	+.061	−.205
5			14			23			31		
x	+.068	−.042	x	+.015	+.029	x	0	0	x	−.015	+.018
y	+.127	−.138	y	+.476	−.255	y	+.497	−.155	y	+.389	−.243
z	+.170	−.209	z	+.174	−.274	z	+.126	−.236	z	+.152	−.202
6			15			24			32		
x	+.107	−.062	x	+.035	−.019	x	−.004	−.004	x	−.010	+.010
y	+.074	−.005	y	+.373	−.211	y	+.494	−.144	y	+.400	−.248
z	+.127	−.305	z	+.155	−.234	z	+.129	−.233	z	+.046	−.200
7			16			25			33		
x	+.069	−.045	x	+.064	−.005	x	−.005	−.006	x	0	0
y	+.148	+.083	y	+.236	−.149	y	+.482	−.150	y	+.406	−.246
z	−.037	−.296	z	+.119	−.194	z	+.123	−.233	z	+.053	−.209
8			17			26			34		
x	+.025	−.034	x	+.095	−.027	x	−.003	−.013	x	0	0
y	+.348	−.003	y	+.161	−.101	y	+.457	−.144	y	+.333	−.131
z	+.035	−.289	z	+.066	−.161	z	+.114	−.228	z	+.083	−.209
9			18								
x	+.001	−.020	x	+.073	−.026						
y	+.571	−.045	y	+.220	−.109						
z	+.006	−.293	z	+.019	−.146						

if the surface is set at some altitude angle other than 90°. It was decided that with the information available, the best assumption would be to make the dish shape perfect at an altitude angle of 45° and to use as errors the surface deflections from this 45° location.

Actually, the results indicate that an altitude angle of something less than 45°, say 30°, might be somewhat more desirable. However, deflection data are not available for 30° and the accuracy of the information available did not warrant such a minor change. Table 1 shows the calculated errors if the dish is perfectly set at 45°. These are the deflection values used in the radio-optical analysis.

Tower and Turntable Analysis

The structural-deflection analysis indicated the deflection of the trunnion-bearing centerline for the various load conditions as follows:

	Load	Deflection
1	Deadweight of cone and dish, −1300 kips.	Y = −0.4 in.
2	Survival wind on dish at 15° altitude, $F_x = -206$ kips and $F_y = -270$ kips.	X = 4.95 in. Y = −0.08 in.
3	Survival lateral wind on dish at 0° altitude, $F_z = -206$ kips.	Z = −1.01 in.
4	Survival wind on dish at 30° azimuth and 0° altitude, $F_x = 628$ kips and $F_z = -362$ kips.	X = 3.19 in. Z = −1.77 in.

Summary and Conclusions

In general, the design of the 300-ft-diameter antenna was found to be satisfactory. The structural analysis showed that the deflections would not be excessive. Results indicated that no further stiffening of the superstructure would be needed but that cover plates should be added to the girders that connect the cone structure to the trunnion bearings for extra stiffness.

Numerous checks made throughout the analysis indicated that no serious errors were induced. With large structures, no matter which approach is used, precise data are not practically obtainable. We believe that the compromises made with this method are no more serious than those inherent in other and more costly methods.

Acknowledgments

The authors wish to acknowledge the contributions to this work of Austin Youmans and Clingman Emerson of the Naval Research Laboratory and David Eck, James Sorenson, and Miss Dolores Landreman of Battelle Memorial Institute.

This work was performed under Office of Naval Research Contract No. N00014-6-C0027.

H. C. Minnett, D. E. Yabsley, and M. J. Puttock
Commonwealth Scientific and Industrial Research Organization
Australia

STRUCTURAL PERFORMANCE OF THE PARKES 210-FT PARABOLOID

1. Introduction

The Parkes telescope has been described elsewhere[1,2] and only features relevant to the present paper need be recalled.

The 210-ft paraboloid (Figure 1) has an f/D ratio of 0.41 and is supported on a basic structure of 30 tubular ribs radiating from a central 23-ft-diameter hub and connected by four ring girders and two ring beams. The top chords of the ribs are interconnected by a system of left- and right-hand spirals to provide torsional and bending stiffness. The loads at the inner ends of the spirals are transferred to an assembly of welded steel panels which also form the central 54-ft-diameter section of the reflector surface.

The outer region of the surface consists of 1380 panels of tensioned high-tensile galvanized wire reflecting 98 per cent of the energy at $\lambda = 10$ cm. The panels are fastened to radial purlins supported by a series of screw adjusters mounted on the underlying spiral intersections. The 30 main purlins lie immediately above the radial ribs and extend from the plating edge to the rim. An equal number of shorter purlins, midway between the ribs, extend from 52-ft radius to the rim. Figure 2 shows the arrangement of panels on the purlins in a typical 12° sector of the surface (between ribs 25 and 26). There are 8 rings of 30 panels each out to radius 52 ft and 19 rings of 60 panels each beyond this. Also shown are the locations of the three connections to main ribs of the tripod structure supporting the equipment cabin above the focus.

Figure 1. Parkes 210-ft paraboloid with survey instrument at vertex.

2. Method of Surveying Surface

Since the telescope is in almost constant use, an instrument has been developed for the rapid measurement of surface deformations.[2] This instrument is essentially a 35-mm telescopic camera mounted at the vertex and capable of scanning a ring of up to 60 targets in 4 min in any attitude of the paraboloid. The targets are photographed relative to a fixed graticule in the telescope. Thus the angular displacement of a target when the dish is tilted or of the angular position of one target relative to another in the ring can be recorded and measured at leisure from the film.

The primary network of targets on the dish surface (Fig. 2) consists of one target on each mesh surface adjuster (678 total visible from vertex) together with 5 rings of 30 targets each on the solid plating. The distance of each of these targets from the instrument has been precisely taped.

During each dish survey the angular position of one target in each ring relative to a nominal paraboloid is accurately established by a Hilger-Watt ST-48 antenna theodolite, referenced by autocollimation onto a mirror fixed normal to the survey camera axis. These calibrations are conveniently made at

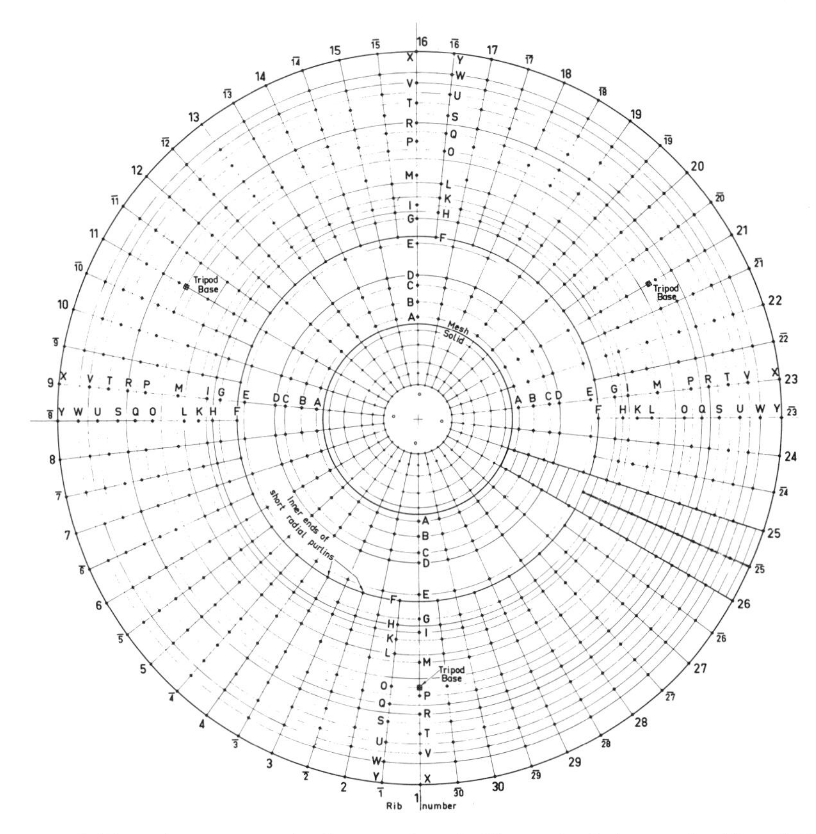

Figure 2. The primary network of survey targets. The sector between ribs 25 and 26 shows the layout of mesh panels.

zero zenith angle. The whole dish can then be scanned rapidly, ring by ring, at any zenith angle and the basic shape at every attitude computed from the records.

Interpretation of the angular displacement of a target caused by tilting the dish involves a knowledge of any changes in its radial distance with zenith angle. However, since the reflector is shallow the angular displacement measured by the instrument is relatively insensitive to radial target movement tangential to the surface. Furthermore, structural calculations confirmed by sample measurements show that target displacements are almost entirely normal to the surface. Consequently, although corrections based on computed or measured components of radial tangential displacements can be applied, these at most amount only to a few tenths of a millimeter. For most purposes it is therefore sufficient to assume that the angular changes represent target displacements entirely normal to the surface.

The scale of the pattern of gravitational distortions is large compared with the target spacing and the primary network therefore provides sufficiently detailed information. However, the network is not well related to the layout of the mesh panels (see Figure 2) and gives no information on the fabrication errors and gravitational distortions within individual panels. Such errors are substantially uncorrelated with those of adjacent panels and the panel system therefore provides a more natural way of subdividing the surface for analytical purposes.

For these reasons a secondary survey network consisting of a point in the center of each panel has been developed. The height of each point relative to reference points on the purlins at the intersections with the panel center line has been measured, normal to the surface, using a special stretched-wire instrument. This is fitted with telescopic sighting so that the presence of the observer does not disturb the reading obtained (Figure 3). The heights of the

Figure 3. Panel measuring instrument.

purlin reference points have been related to the nearest purlin targets of the primary network by linear interpolation. Gravitational displacements of targets mounted at the secondary network points can be measured in the usual way with the survey camera.

3. Analytical Procedure

In Ruze's statistical theory of the effect of surface errors on performance, a uniform distribution of errors is assumed and the result is then independent of illumination function. Since the survey technique at Parkes provides detailed information on the surface errors, their effect on performance can be computed directly from the radiation integral.* This avoids the need for assumptions about the error distribution and allows the effect of varying the illumination function to be analyzed.[5]

In this procedure the dish surface is subdivided into elementary areas each of which is replaced by a point source located at the phase center of the element. Each source is assigned an amplitude proportional to the product of the element area and the value of the feed illumination function at its center. For reasons already given, the natural element of the mesh portion of the Parkes dish is the individual panel. This portion of the surface is therefore represented by 1380 sources.

On the central solid-plated region the fabrication errors and the gravitational distortions of the surface both show correlation over distances greater than the target spacings. It is sufficient therefore to divide the region into 150 elements, each centered on a primary survey point, and the geometrical center of each element can also be taken as its phase center. The region is then represented by 150 sources at positions coinciding with the primary survey points.

The aperture efficiency and radiation pattern of the dish are degraded if the equivalent sources do not lie on a paraboloidal surface (not necessarily the nominal one). This degradation is found by a least-squares determination of a best-fit paraboloid to the array of sources. In this process the vertex coordinates, the direction of the axis and the focal length of the paraboloid (six parameters in all) are adjusted to minimize the loss of efficiency. The computations have been programed for a CDC 3600 computer. A key feature of the program is the use of a direct minimization technique,[6] providing for the systematic and simultaneous variation of all six parameters and for variation of search step size to achieve convergence to the proper minimum. The use of this search procedure and the setting of stringent convergence criteria combine to reduce the chances of convergence on a local minimum of higher function value (e.g., a saddle point).

In the fitting process, the surface efficiency is computed at a convenient wavelength (6 cm in the present case). The surface efficiency may be defined as the ratio of the actual aperture efficiency to the value that applies for a perfectly shaped paraboloid with perfect reflectivity, assuming the same

*A similar approach has been used by Dion (Ref. 4). However, in the method described here the homologous components of distortion are removed by best-fit analysis before the integral is evaluated.

illumination function in each case. The equivalent rms (half-path) error of a uniform random distribution having the same efficiency may also be calculated.

The computed radiation pattern is valid only out to the first few sidelobes because the relatively broad patterns of the finite-sized panels have, for simplicity, been disregarded. However, the restriction is not serious because this is the region of the radiation pattern of greatest interest. Previous determination of the paraboloid of best fit ensures that the computed pattern represents optimum axial and lateral focusing of the distorted paraboloid.

4. Gravitational Distortions

It is appropriate to consider first the performance of the dish supposing the mesh panels to be perfectly shaped at all zenith angles and adjusted so that the entire mesh region conforms exactly to the nominal paraboloid at zero zenith angle. When the dish is tilted the surface displacements will then be due solely to gravitational distortions of the main structure, and can be determined by camera survey of the primary target network alone. From this information the deviations of the equivalent sources from the nominal paraboloid can be obtained.

The effects of tilting the dish to zenith angles of 10°, 20°, 40°, and 60° have been studied. Figure 4 shows the surface displacements caused by a 60° tilt. The contour patterns have good symmetry about the vertical plane, and tripod leg loading shows clearly. Scalloping in the region of rib 16 is due to the lower stiffness of the backup structure below the intermediate purlins. The repeatability of the observations is good.

Least-squares analysis over the zenith angle range 10° to 60° shows that the best fit is obtained by progressively shifting the vertex, by rotating the axis in the rib 1 rib 16 plane of symmetry and by shortening the focal length. The combined effect of the first two changes is approximated by rigid-body rotation about the nominal vertex by the amounts shown in Figure 5, which also shows the change of focal length with zenith angle. The values plotted are for a 10-dB Gaussian illumination taper, but there is relatively little variation in the curves for illumination tapers ranging from zero to 20 dB.

The equivalent rms residual deviation from the best-fit paraboloid is a measure of the nonhomologous behavior of the structure, as defined by van Hoerner.[7] Although the maximum surface displacement due to tilting to 60° is more than 10 mm, the rms residual deviation is only 1.7 mm for uniform illumination and 1.4 mm for a 20-dB illumination taper (Figure 6). For smaller angles of tilt the rms error is lower but does not fall to zero at the zenith because the fabrication errors of the solid central region have been included in the analysis. These errors are significant only at small zenith angles. For angles exceeding 40° the rms residual error would be improved by less than 0.1 mm if the solid region were perfectly shaped at 0°.

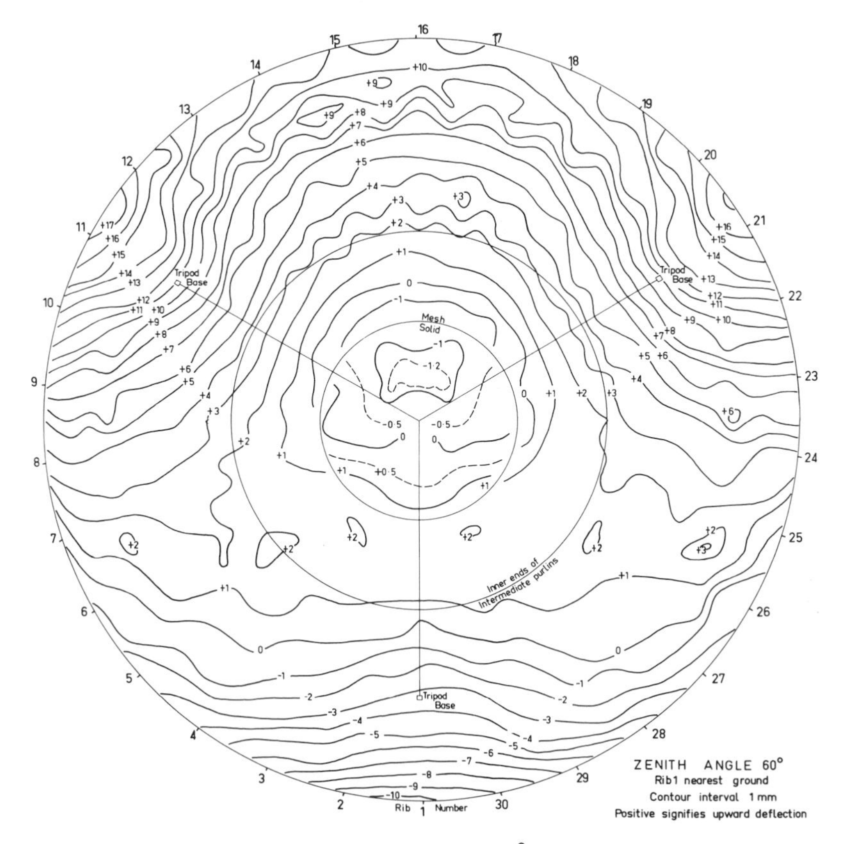

Figure 4. Displacement contours due to tilting dish 60° from zenith. Rib 1 is nearest to ground; contour interval 1 mm; positive signifies upward deflection.

The relative losses of efficiency caused by different parts of the dish surface are shown in Figure 7 for zenith angles of 20°, 40°, and 60°. The losses shown are for the solid region in the center and three annular zones making up the mesh region. The contribution from the solid region is almost constant with zenith angle, while those from the mesh zones increase with zenith angle and with radial distance from the center of the dish. Thus at both 40° and 60° zenith angle, the predominant contribution is from the mesh zone beyond the feet of the tripod (rings 18–27).

Further analysis shows that the losses at the higher zenith angles are dominated by distortions caused by tripod leg loading. From the shape of the dish deflection patterns an approximate estimate can be made of the effect of removing this loading. Least-squares analysis of the residual deflections then shows that the losses attributed to rings 9 to 27 are substantially decreased. At 60° zenith angle the equivalent rms error of the whole dish for a 10-dB taper is reduced to 0.8 mm.

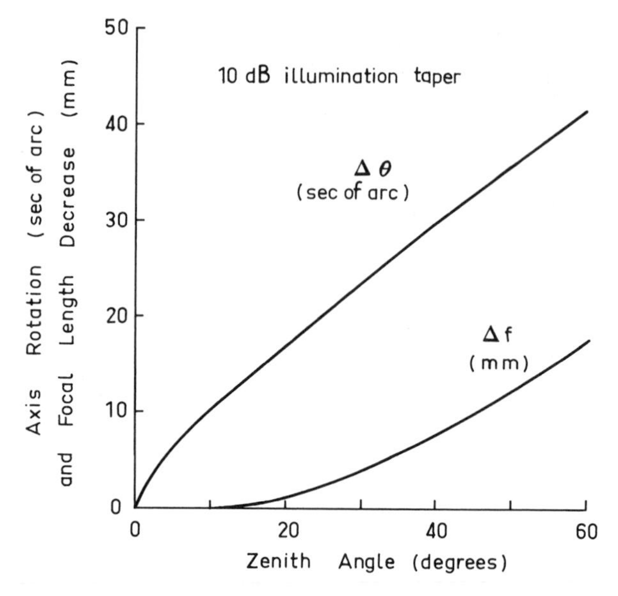

Figure 5. Rigid-body rotation $\Delta\theta$ towards rib 1 and shortening of focal length Δf versus zenith angle.

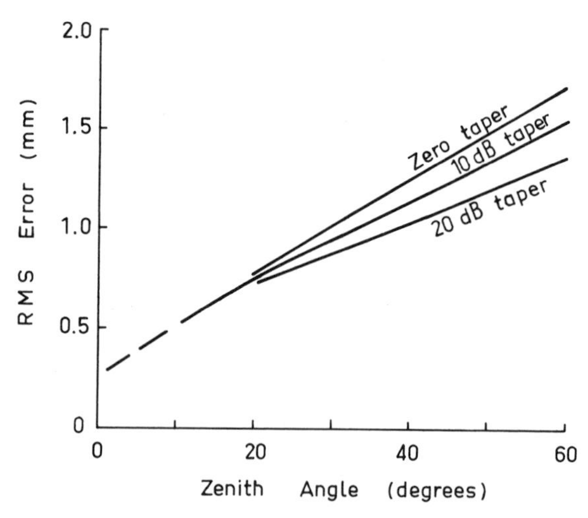

Figure 6. Root-mean-square deviation from best-fit paraboloid with mesh surface adjusted to perfect nominal shape at 0° zenith angle.

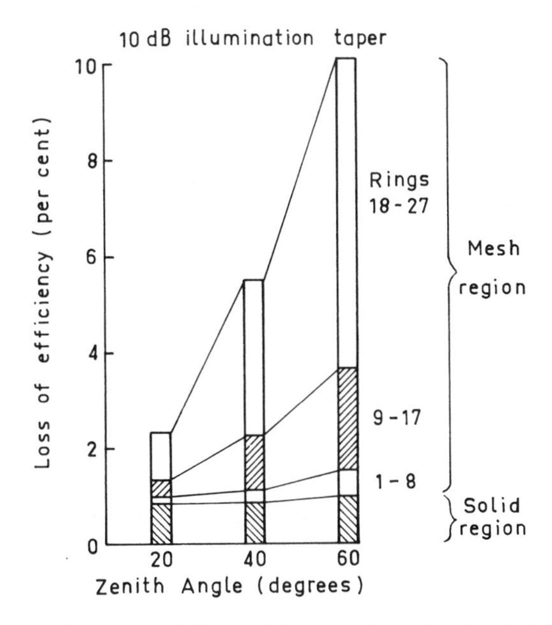

Figure 7. Contributions of different dish regions to surface loss before biasing.

5. Optimum Adjustment of the Surface

Adjustment of the shape of the mesh-panel surface must be carried out at zero zenith angle; initially the shape there was set as closely as possible to the nominal design paraboloid. With the theodolite-controlled adjustment procedures adopted, the resulting rms deviation from nominal shape was about 1.4 mm.

However, since an altazimuth mounting is used, the dish structure is always symmetrically oriented about a vertical plane. It is therefore possible to reduce gravitational distortions by biasing the surface adjustments. Having measured the dish shape at zero zenith angle and the deflections due to tilting, each adjuster can be biased so that the dish approaches the perfect nominal shape in the middle of the zenith angle range (excluding the solid region, which is not adjustable).

A slightly different approach was adopted when the surface was readjusted at the end of 1965. The measured deflections caused by tilting are not in general linear functions of zenith angle and the shape of the curve varies considerably over the surface. The bias for each adjuster was therefore chosen to minimize the maximum deviation at that point from the nominal paraboloid over the full zenith angle range. As the dish tilts, different points return to the nominal position at different zenith angles and the shape is not precisely nominal at any zenith angle. However, the distortion at the ends of the range are somewhat less than for the first method of biasing.

The predicted rms departure from a best-fit paraboloid over the zenith-angle range is plotted in Figure 8. Curves (a) and (b) are for uniform illumination

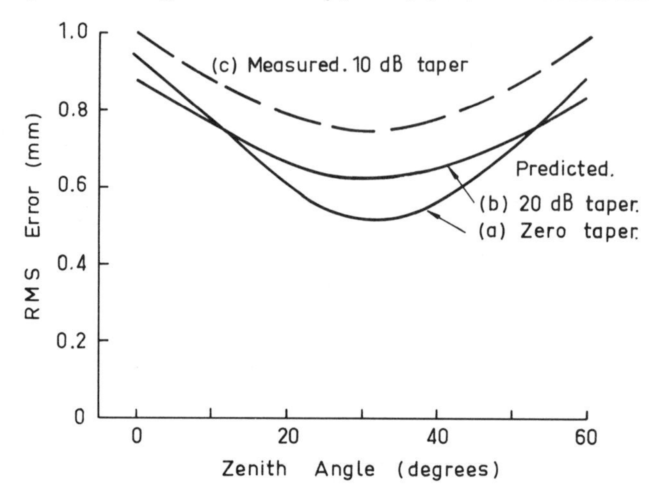

Figure 8. Root-mean-square deviation from best-fit paraboloid with mesh surface biased.

(zero taper) and 20-dB taper, respectively. The rms error remains below 1 mm in both cases and for uniform illumination falls to 0.55 mm in the middle of the zenith angle range.

After the biasing operation, the measured rms residual errors were about 0.15 mm higher than the predicted values. Curve (c) in Figure 8 shows these results for an intermediate value of illumination taper, 10 dB. Part of the error increase is probably due to unfavorable weather conditions during the period when the adjustments were made. Nevertheless, comparison with Figure 6 shows that a substantial improvement in the performance of the dish was achieved.

Confirmation of this improvement has been obtained from radio observations at both 11 and 6 cm wavelengths. Whereas the losses of surface efficiency at 60° zenith angle were previously about 7 and 20 per cent, respectively, the new values are reduced by a factor of about 3. For more exact comparison with the survey data, better receiver stability and facilities for lateral focusing of the feed (in the plane of tilting) are needed.

A feature of the predicted performance is that uniform illumination, although giving higher surface losses than tapered illumination at zenith angles of 0° and 60°, should give less loss at intermediate angles, as shown in Figure 8, curves (a) and (b). (This improvement is additional to the increase in feed illumination efficiency that would result from uniform illumination.) Examination of the radial distribution of the loss contributions shows the reasons for these results (Figure 9). The losses due to the distortions in the outer zones of

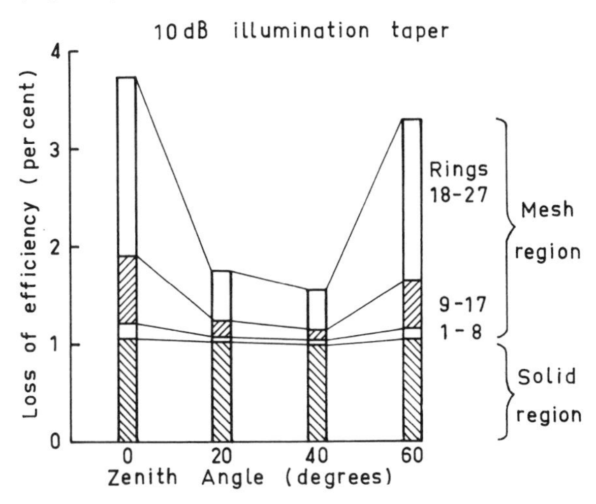

Figure 9. Contributions of different dish regions to surface loss after biasing.

the mesh region are high at 0° and 60° zenith angle, but lower at 20° and 40°. However, the losses attributed to the solid region (which cannot be biased) are almost constant, and provide the predominant component of the total at 20° and 40°. Thus uniform illumination, which gives greater weight to the outer regions of the surface, increases the surface losses only at the ends of the zenith angle range.

The important effect of tripod-leg loading on the gravitational distortions of the unbiased dish was mentioned in Section 4. With this loading removed in the case of the biased dish the rms error is reduced from 0.9 mm to an estimated 0.6 mm (for 10-dB taper) at 0° and 60° zenith angle. However, little improvement can be expected in the middle range of zenith angles since the losses in this case are mainly due to errors of the central region of the reflector.

6. Mesh-panel Characteristics

Because of their dimensions and method of fabrication and fixing, each of the present mesh panels on the Parkes dish tends to be constrained to a parabolic shape radially, but departs systematically from the correct circumferential curve. The circumferential error profile can be represented by a simple function that has been adjusted to fit the central error of each panel as determined by secondary target survey. The fractional reduction of the radiated field relative to a perfect panel and the position of the phase center relative to the nominal paraboloid can be determined simply from this model for every panel. This information, when combined with the primary survey data, modifies the equivalent source array and resulting best-fit analysis.

In addition to the systematic panel error, there is a small-scale random roughness which appears to be fairly uniform within a panel and from panel to panel, with an rms value of about 1.5 mm. Because of this uniformity, its effect is simply to reduce the over-all gain by an additional factor calculable from Ruze's statistical theory.

Further small changes of panel shape occur when the dish is tilted. The deflection at the center of a panel may be determined with the survey camera and to date such measurements have been made on 210 panels. The rms values of the deflections around the panel rings that have been sampled are 1–2 mm when the dish is tilted through 60°. Additional targets are currently being installed for the investigation of these errors in more detail.

7. Over-all Performance

The surface efficiency of the Parkes paraboloid after biasing is shown as a function of wavelength in Figure 10. A 10-dB illumination taper is assumed and appropriate mesh transmission losses are included. The upper curve in each case indicates the performance in the middle of the zenith angle range (30°) while the lower dashed curve applies at the extremes of the range.

The curves labeled (a) show the calculated efficiency with the present mesh panels. The isolated points show the decrease of performance with wavelength deduced from radiometry measurements. At $\lambda = 6$ cm the observed performance is about 0.4 dB below the value calculated from the survey result which is within the experimental uncertainty of the radio observations.

When surface distortions due only to deflections of the main structure are taken into account (i.e., assuming perfect mesh panels with the same rigidity), the performance is shown by curves (b). This indicates that the losses due to the errors of the present surface panels greatly exceed those due to gravitational bending of the main structure and is a measure of the scope for future improvements. A hypothetical surface assuming mesh panels with a combined rms deviation due to manufacturing and setting errors of 1.5 mm is shown by curves (c).

It is of interest to scale the performance to larger diameters assuming gravitational deflections increase as the square of the diameter. If the minimum operating wavelength λ_{min} is taken to correspond to a decrease in surface efficiency of 3 dB, the rms surface error is $\lambda_{min}/15$. The values of λ_{min} plotted as a function of diameter in Figure 11 are based on an estimate of the

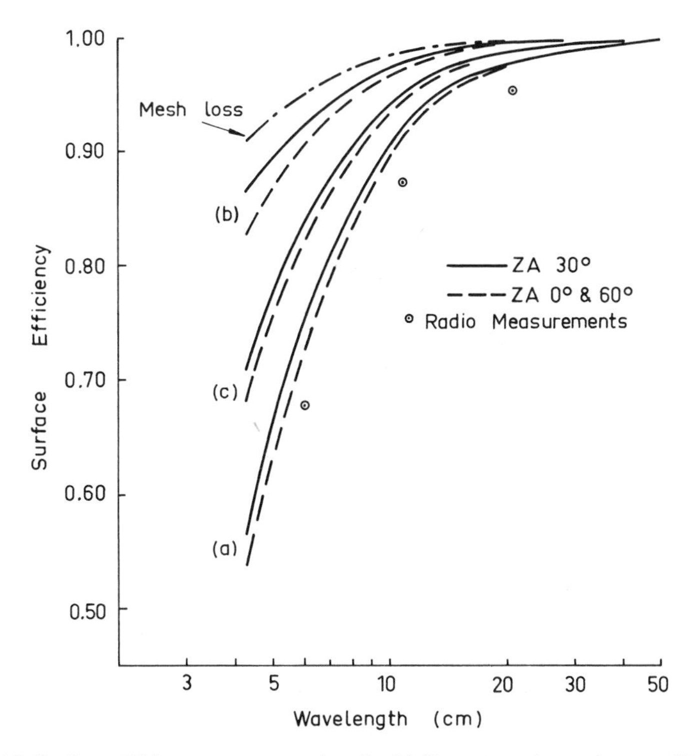

Figure 10. Surface efficiency versus wavelength. (a) Present mesh panel error. (b) Zero mesh panel error. (c) 1.5-mm mesh panel error.

performance of the Parkes dish at the zenith after biasing, with the zenith angle range extrapolated from 60° to 90° to allow comparison with the "gravitational deflection limit" estimated by van Hoerner.[7]

In curve (a) the total mesh-panel error is kept constant at the value that applies now at Parkes and the Parkes mesh dimensions are scaled with diameter to keep the leakage and porosity of the mesh constant. Curve (c) is also for a similarly scaled mesh but with the rms mesh-panel error reduced to 1.5 mm. In (b) it is assumed that mesh-surface errors are due only to gravitational distortion of the backup structure and leakage is zero.

The latter curve falls very close to curve (d) showing van Hoerner's gravitational deflection limit adjusted slightly to correspond with the 3 dB definition of λ_{min}. Thus the results predict that operation of a 400-ft-diameter dish down to a wavelength of 10 cm with a surface loss of less than 3 dB should be feasible. Use of the estimate given in Section 5 for the effect of removing tripod loading allows curve (b) to be replaced by curve (e). The predicted limit for a 400-ft-diameter dish is then reduced to 5 cm.

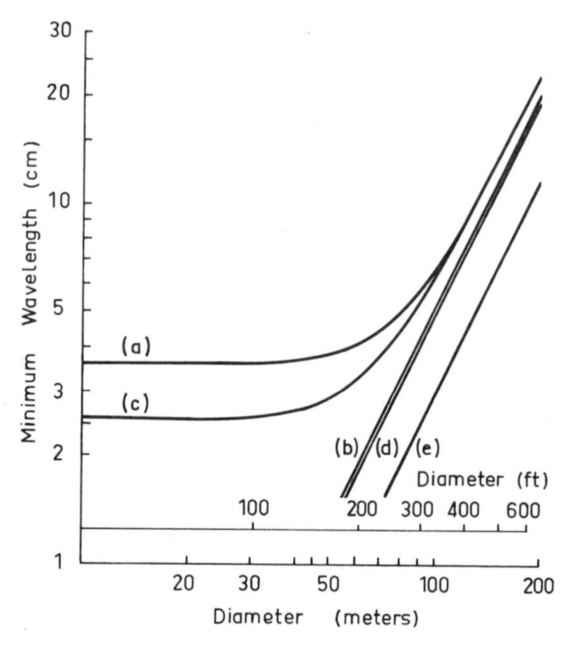

Figure 11. Minimum operating wavelength for 3-dB loss at various diameters. Panel errors as in Figure 10.

8. Conclusions

Extensive measurements with a special rapid-survey instrument have been made of the surface of the Parkes 210-ft paraboloid. These data have been used to study sources of on-axis gain loss by direct computation from the radiation integral using a least-squares best-fit technique to determine optimum focusing of the distorted dish. This method of analysis allows the effects of radial illumination functions and nonuniform error distributions to be evaluated.

The measured gravitational deflections of the main structure have been used as a basis for biasing the dish to achieve best performance in the middle of the zenith angle range. Subsequent observations have confirmed that a substantial improvement in performance has resulted.

Although tripod-leg loading contributes significantly to the errors of the outer portions of the dish surface, the rms residual error of the basic dish structure is less than 1 mm over the zenith angle range. More accurate panels would result in greatly improved performance at the shorter wavelengths. However, when the dish is scaled to a diameter of 400 ft, basic structural deflections dominate the panel losses. Nevertheless, the analysis suggests that operation down to a wavelength of 10 cm should be feasible for a total surface-error loss of less than 3 dB.

Acknowledgments

Part of this work was carried out under Research Grant NsG-240-62 from the U. S. National Aeronautics and Space Administration. We wish to acknowledge the encouragement given to the project by Dr. E. G. Bowen, Chief of the Division of Radiophysics and F. J. Lehany, Chief of the Division of Applied Physics.

It is also a pleasure to acknowledge the work of Dr. R. T. Leslie and D. E. Shaw of the Division of Mathematical Statistics on the development of least-squares best-fit programs for the analysis.

The authors are also grateful for the enthusiastic assistance of K. J. Loughry, of the Division of Applied Physics, and of P. W. Butler and Miss L. A. Martin of the Division of Radiophysics, in the collection and analysis of the survey data.

References

1. E. G. Bowen and H. C. Minnett, "The Australian 210-ft Radio Telescope," *Proc. IRE (Aust.)* **24**, 98 (1963).
2. M. H. Jeffery, "Construction and Operation of the 210-ft Radio Telescope at Parkes, Australia," Large Steerable Radio Antennas – Climatological and Aerodynamic Considerations, *Ann. N. Y. Acad. Sci.* **116**, 62 (1964).
3. M. J. Puttock and H. C. Minnett, "Instrument for the Rapid Measurement of Surface Deformations of a 210-ft Radio Telescope," *Proc. IEE* **113**, 1723 (1966).
4. A. R. Dion, "Investigation of Effects of Surface Deviations on Haystack Antenna Radiation Patterns," Lincoln Laboratory Technical Report 324, MIT Lincoln Laboratory (1963).
5. R. T. Leslie, D. E. Shaw, and D. E. Yabsley (to be published).
6. J. A. Nelder and R. Mead, "A Simplex Method for Function Minimization," *Computer J.* **7**, 308 (1965).
7. S. von Hoerner, "Design of Large Steerable Antennas," *Astron, J.* **72**, 35 (1967).

H. G. Weiss, W. R. Fanning,
F. A. Folino, and R. A. Muldoon
*Lincoln Laboratory,**
Massachusetts Institute of Technology
Lexington, Massachusetts

DESIGN OF THE HAYSTACK ANTENNA AND RADOME

Introduction

The Haystack Microwave Research Facility was designed for studies of long-range communications techniques and for radio and radar astronomy. The design objective was to obtain a versatile system with high sensitivity at frequencies up to 10 GHz. This was accomplished by the use of a very precise 120-ft steerable antenna, a transmitter system capable of radiating 500 kW of average power, and an efficient microwave configuration compatible with the use of low-noise receivers.

The design of the antenna evolved after extensive computer analyses and model testing. In the controlled environment within the radome, the surface and pointing precision could be greater than that of any existing large exposed antenna. Due to conservative engineering design and to the protection afforded by the radome, the system has provided essentially trouble-free operation during the past three years. The initial design objectives have all been realized, and the precision of the structure has permitted useful operation at frequencies as high as 36 GHz.

The Haystack facility employs a fully steerable, radome-enclosed paraboloidal antenna 120 ft in diameter. The radome is of the metal space-frame type with 1/32-in.-thick fiberglass panels. The surface accuracy (3σ) and pointing precision are ± 0.060 in. and ± 0.005°, respectively. This antenna system is used for many different experiments at numerous frequencies in both the radio and radar modes of operation. This usage is facilitated by the use of

* Lincoln Laboratory is operated with support from the U.S.Air Force

complete electronics systems within removable, interchangeable equipment rooms. The antenna is pointed, and collected data are reduced by means of two digital computers located at the site.

The radome is a 150-ft-diameter, seven-eighths sphere, aluminum space-frame surface structure with 1/32-in.-thick fiberglass panels. The radome is designed to withstand hurricane winds of 130 mph without damage.

The antenna employs an aluminum space-frame backup structure interacting with an integral thin-shell surface. This backup structure is composed of five concentric ring trusses interconnected by pretensioned diagonal and radial rods. The surface is composed of 96 aluminum honeycomb sandwich panels preloaded to act as a thin shell and to add considerable stiffness to the structure. In addition, the structural concept employs simple, gravity-operated, open-loop deflection control devices to reduce distortion.

The antenna is supported by a Y-shaped yoke assembly that houses all the mechanical-hydraulic-electrical and electronic components required to rotate and position the antenna to any point in the hemispheric sky. The yoke assembly floats on a 14-ft-diameter azimuth, friction-free, oil-film bearing located on top of a 35-ft-high concrete tower.

To direct the very narrow beam (0.05° at 10 GHz) of the Haystack antenna to the required positional accuracy, a digital control and data system is employed in a closed-loop positioning servo system. The specified maximum position error of 0.005° has been met.

An optical measurement system was integrated into the over-all design of the antenna system for the purpose of erecting, measuring, checking and monitoring the accuracy and precision of the reflector surface and the axes of rotation.

Electromagnetic and Electronic Characteristics

A Cassegrain configuration was selected to provide efficient electromagnetic performance and permit the placement of electronic components near the center of gravity of the reflector structure. In this configuration the energy spillover from the primary feed horn is largely in the direction of the "cold" sky, a desirable attribute for a low-noise receiving system. The hyperbolic subreflector size was chosen to be 9 ft, 4 in. in diameter to permit operation at wavelengths as long as 21 cm.

To permit the convenient use of the antenna in a variety of experiments, the electronic equipment is installed in a room that is "plugged in" directly behind the reflector. This design approach eliminates the need for rotary microwave joints and long runs of waveguide. A number of different plugin rooms containing low-noise receivers, high-power radar transmitter components, and antenna primary feed systems have been instrumented. The rooms, which are 8 x 8 x 12 ft in size and weigh 7000 lb, are raised from the ground and positioned by an integral hoist system.

To achieve an antenna with a high-aperture efficiency at 10 GHz, the overall reflector rms must be maintained to within 0.030 in. of a paraboloid contour and the reflector surface cannot be perforated. Since this precision with a solid-surface reflector could not be maintained out of doors, a radome has been used to provide a controlled environment. The particular radome employed at Haystack is not an optimum design but was utilized because it became available without cost from another program. It was designed to survive a 200-mph wind in the Arctic, and aluminum was used for the space frame to minimize the brittleness that occurs in steel at very low temperatures. The 0.032-in. fiberglass panels are also substantially thicker than would be necessary for a contemporary design. As a result of these factors, the radome is relatively inefficient and the space frame introduces a loss of about 1.1 dB, while the membrane contributes an additional 0.2 dB loss (at 10 GHz). Even this comparatively opaque radome introduces only relatively minor perturbations on system performance except for the loss in effective aperture and noise-temperature contribution of approximately 5° K. The effect upon pointing angle and polarization appears to be negligible.

As a result of the absence of variable environmental loads and the low "stiction" hydrostatic azimuth bearing, the antenna system performs with a pointing accuracy of ± 0.005°. The performance of the system at 10 GHz, including the radome, is equivalent to that of an "ideal" reflector having a diameter of approximately 105 ft but with an angular beamwidth corresponding to that of a 120-ft antenna.

The electronic instrumentation, control, and data processing system,[1] have been in active use at wavelengths of 21, 16, 3.8, and 2 cm, and some preliminary testing has been accomplished at 8 mm. Measurements at 8 mm have shown that the antenna beamwidth (half-power) is less than 0.02°, and the sidelobe levels are acceptably low for many experiments (Figure 1). The high sensitivity has permitted many exciting scientific experiments[2-5] to be conducted, and many others are currently in progress.

Structural-Mechanical Concept

The complex of antennas at Lincoln Laboratory's Millstone Field Site can be seen in Figure 2. This is an aerial photograph showing the 60-ft X-band Westford antenna within its 90-ft air-supported radome with the 84-ft L-band Millstone tracking antenna somewhat above it. To the left of the Millstone antenna is the 220-ft fixed ionospheric research antenna, and, finally, at the upper left of the figure is the Haystack facility.

A closeup photograph of the Haystack site is shown in Figure 3, where a

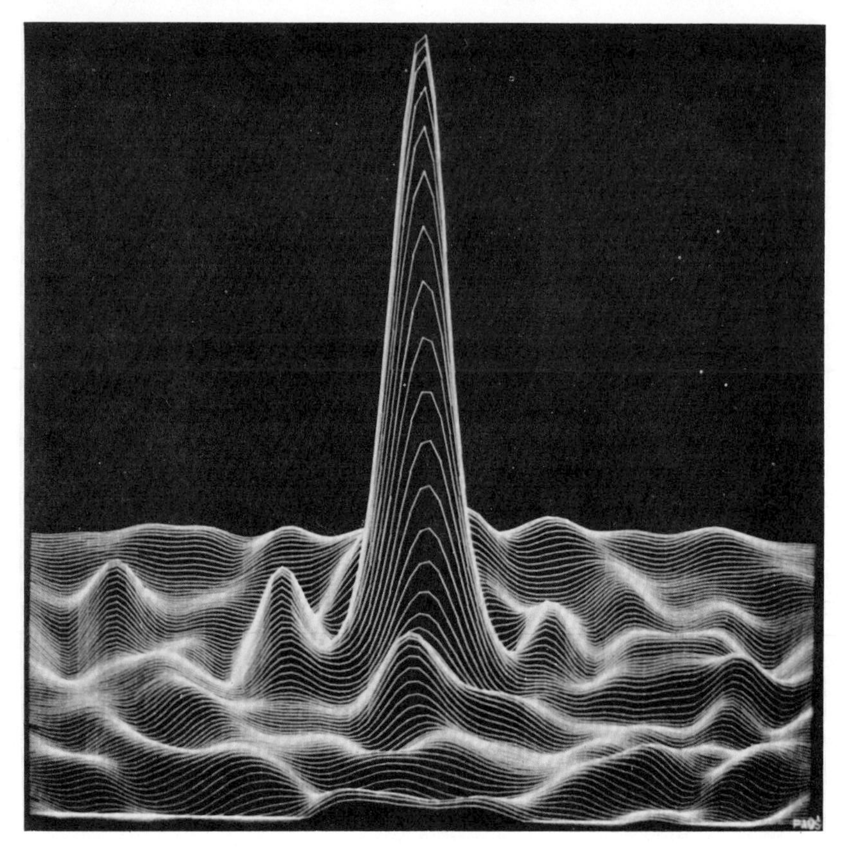

Figure 1. Antenna pattern at 8-mm wavelength (photographed from computer display). Contour lines at 0.5 dB intervals. Half-power beamwidth ≈ 0.17 ≈ 1 min of arc.

rendition of the antenna structure has been superimposed onto an actual photograph of the site. The pedestal for the antenna is a tapered octagonal concrete tower with 18-in.-thick walls and is anchored into bedrock and heavily reinforced with steel. On top of the concrete tower is a 6-ft-high steel weldment that serves as a load distribution ring for the hydrostatic pressure pads of the azimuth bearing. Inside the tower is a "Maypole" cable wrap that permits ± 300° of antenna motion for the 3000 pairs of wires and two 4-in.-diameter cooling water lines. Above the bearing is a 3-ft-high load distribution ring that contains the machined runner for the hydrostatic azimuth bearing and also serves as a transition piece between the bearing and the yoke structure (Figure 20). This yoke is the steel plate structure between the load distribution ring and the elevation bearings. The two sets of elevation bearings are standard, self-aligning roller bearings that incorporate a sliding Teflon pad on one shaft to allow for thermal growth between the aluminum antenna and the steel

Figure 2. Millstone field site.

supporting structure.

The elevation bearings support two parallel planer trusses ("trunnion beams") that in turn support the primary reflector from the middle (No. 3) ring truss (63 ft in diameter) at four points 90° apart. The trunnion beams also directly support the secondary reflector, the electronics equipment room (rf box) structure, and the main counterweight. In this way, these critical weight components are isolated from each other, and by means of this concept, the weights of any of these components can be varied with little effect upon the others. For example, added weight in the rf box has a negligible effect on the antenna's surface accuracy or the position of the secondary reflector; although it would require retrimming the main counterwegiht.

The secondary reflector with its quadripod support can be seen in a cutaway in the antenna. The quadripod legs are planar trusses 5 in. wide by 20 in. deep, which are braced at the third points with 1/2-in.-diameter rods. Figure 3 shows the removable (plugin) rf box in the process of being hoisted to the level of the rf axis. It will then be moved forward on a self-contained dolly to its operating position just behind the vertex of the paraboloid. At this point it will be attached to the antenna structure and electronic wiring will be hooked

Figure 3. Photographic rendition of the Haystack facility.

up by means of plugin connectors. The two cantilever beams protruding from the top of the center (main) counterweight are the pendulum ballast, surface control devices. These beams apply load to three cables which are attached to the outer ring trusses. The load in these cables, resulting from the pendulum or pivoting action of the beams is such that the load is zero when zenith pointing and then varies sinusoidally in moving toward the horizon. These three cables plus the two outboard counterweights are the gravity-operated surface control devices used in the Haystack concept. The effect of these devices was determined by computer analysis, and the compensation loads can be varied simply by changing weights if one wishes to trim the surface differently. These compensation devices considerably reduce surface distortion of the antenna.

The basic design philosophy of the Haystack concept is further illustrated in Figure 4. This figure shows the numbering system for the ring trusses and, in a pictorial sense, the stiffness of the ring trusses normal to their plane. The ring trusses are basically compression members which are supported by the pretensioned radial and diagonal rods. These rods are not shown in the figure for clarity and because of their small size (1/2–1 in. in diameter). In most movable structures, such as antennas, all members are designed for both a

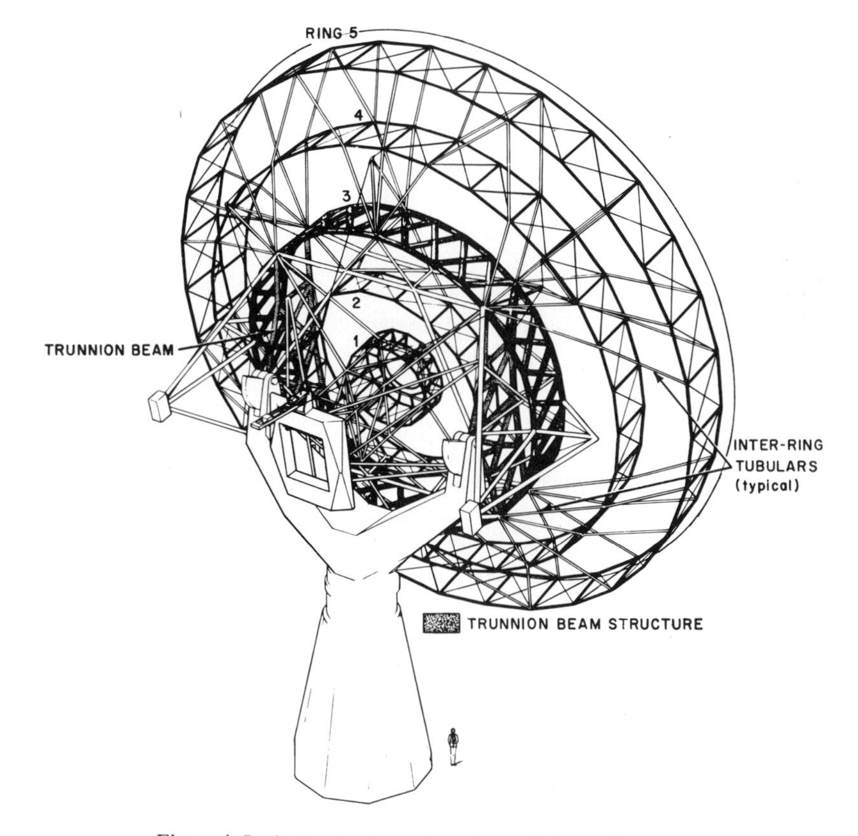

Figure 4. Basic structural configuration of Haystack antenna.

maximum compression and tension load; and since compression buckling is more severe than tension yield, most members must be sized for the compression load. This is not true in Haystack. Here, the prestressing force has been chosen such that most of the tension members (radials and diagonals) always remain in tension and compression members (circumferentials) always remain in compression.

The rather high circumferential stiffness of the ring trusses is an important and unique feature of this antenna. Many existing antennas use a radial system of heavy rib trusses with a fairly light system of circumferential members. The rib-truss-type antennas do seem to have smaller absolute deflections, but higher surface distortions; the difference here being accounted for in the large quasi-rigid body motion typical of ring-truss-type antennas. This is the contribution of the relatively stiff circumferential trusses since they are quite effective in resisting out-of-plane surface distortions. Further description of the antenna may be found in References 6 through 9.

Radome Considerations

The inclusion of a radome confers many major benefits on the antenna structure. The most important is a reduction in the required stiffness and rigidity and thus the total weight and cost of the antenna structure itself. When an antenna is exposed to the environment, increased stiffness and rigidity are essential in order for the structure to survive climatic conditions that may occur but once in many years. This survival criteria dominates and controls the complete design of many antennas; it demands larger and more costly structural components which are rarely utilized at their design capacity; higher power requirements are necessary to drive a heavier antenna against the prevailing low-velocity winds; moderate wind and snow storms compel a shutdown of the whole system; exposure to the elements increases the frequency of maintenance and repair time with a corresponding increase in shutdown time. All of these factors degrade the usefulness and efficiency of the installation; with the growing tendency toward larger and more accurate and precise reflectors, where surface deflections must be kept within a few hundredths of an inch, a controlled environment of the type that can easily be provided by a radome is essential.

Analytical Methods

Wind loads constitute the single most severe loading imposed on the radome structure by the environment. A loading of this nature can cause the radome to fail in any of three different modes. The maximum stress of a beam member may be exceeded, an individual beam may buckle, or a substantial section of the radome frame might suffer a general buckling collapse. The first two of these are treated as beam-column failure while generel instability criteria govern the last situation.

Beam-column Failure

The pressure distribution to be expected from the wind on the Haystack radome was determined from wind tunnel measurements using 27.5-in. scale models of the Haystack structure. A Fourier series was fitted to the experimental pressure distribution and, using this result, a membrane stress analysis on an equivalent thin shell was performed. The membrane stresses were than transformed into axial loads in the beam members. The transverse loads acting on individual beams were taken to be one-third of the load imposed on each of the adjacent triangular panels by the local overpressure. It was further assumed that this load varies linearly from zero at the end points of the beam to a maximum at the midpoint. With the axial and transverse loads for the beams established in this fashion, a conventional beam-column analysis was performed. The dimensions of the beam cross section were then selected to prevent failure due to beam column action.

General Instability

The buckling capacity of the frame is determined by transforming the space-frame radome into an equivalent, homogeneous, isotropic shell and then calculating the buckling resistance of this shell. The transformation is accomplished by equating the extensional stiffness and bending rigidity of the frame to the equivalent shell. This method indicated that for the Haystack radome the beam dimensions required by the beam-column analysis would yield a large margin of safety in resisting general instability.

Radome Construction

The design effort described above resulted in 3 x 5 in. aluminum (6061-T 6) beams with a 1-in. wall thickness near the base ring and ½ in. above this. The beams are connected to each other by means of hubs which consist of 6061-T 6 aluminum plates and 4340 steel bars (fingers) inserted into the core of the beam and bolted through the web. The dielectric panels were fabricated from three plies of 0.010-in. fiberglass cloth and an epoxy resin. These panels are made thicker at the edge to allow mechancial attachment to the outside of the beam by means of a cap strip bolted to the beam.

The erection started with the pouring of a duo-decagonal concrete base ring supported about 20 ft above grade by two concentric rows of columns and footings anchored to rock. Anchor-bolted to this were a series of special base plates with the beam connecting fingers projecting upward. A special central mast and elevating arm platforms were then erected on the center line of the dome. This consisted of an old oil derrick modified to support a rotating platform at the equator of the radome. From this rotating platform, two elevating arms extend to follow the curvature of the radome structure. These arms could be raised or lowered from the base ring almost to the top of the radome (except for a small cone of silence) just as a crane boom would be. On the end of those arms, an upside-down U-shaped bracket is connected with a work platform on each leg. This structure can be seen in Figure 5 at which time the erection was almost to the equator. Iron workers on these platforms would erect the beamhub structure in an area and then attach the panels. In the uncompleted form, particularly below the equator, a radome structure is very flexible at its free edge and must be restrained by an extensive network of temporary guy lines.

Erection as shown in Figure 5 proceeded until the arms and platforms approached instability near the zenith, at which point the erection was com-

Figure 5. Radome erection with central mast.

pleted (Figure 6) by means of a 100-ton truck crane with a 190-ft boom and a 40-ft jib.

Antenna Analysis Effort

The single most critical and analytically demanding dimensional specification for this antenna was the requirement that the peak surface error be less than ± 0.075 in. under all operating conditions. This goal was to be achieved under the worst environmental conditions and pointing position, and not just for a certain percentage of the time or under some average conditions. This error must include all contributors to surface distortion which include the changing direction of gravity, differential temperatures, noncorrectable surface panel manufacturing tolerances, and measurement errors during erection. The error budget which acocmpanied this specified peak error is shown in Figure 7. This is a substantially higher degree of precision than has been obtained in previous antennas.

Very early in the design program it was shown that hand calculations would be inadequate and that newly developed computer programs would have to be used. This inadequacy was caused by the complexity of the structure with

Figure 6. Radome closure with crane.

±0.040 in.	0° – 90° GRAVITY
±0.017 in.	THERMAL
±0.009 in.	MEASUREMENT (0.020-in. peak)
±0.009 in.	MANUFACTURING (0.015-in. peak)
±0.075 in.	PEAK ERROR
0.030 in.	RMS (0.4 × peak)

Figure 7. Surface accuracy budget (original objections).

its resulting simplifying assumptions which were necessary to perform a hand analysis. In addition, it could have provided only the maximum deflection for the over-all surface and not the distortion of many individual points. The STAIR[10] program, with its capability to analyze structures with up to 4000 joints, was able to calculate accurately these local deflections. North American

Aviation (NAA), the prime contractor for the Haystack antenna, developed a similar large-capacity computer program for use in their design effort.

A very precise detailed analysis was required to permit the effective use of open-loop compensation techniques consisting of control cables, distributed counterweights, and the introduction of preplanned distortions "bias rigging" in the reflector when the surface is set in the "face up" position. It requires the calculation of a target offset value for each of the surface measurement points, such that the surface will be distorted an appropriate opposite amount when it is in the "face-side" position. The calculated values must be accurate to within ± 0.01 in. to achieve significant over-all improvements in the already high-precision design.

One of the more significant structural aspects of the Haystack effort was the development and use of improved techniques for the analysis of complex three-dimensional frame-shell structures. Successful structural analysis of the Haystack antenna required the development of computer programs and analytical techniques capable of handling with unusual accuracy a large number of deflection calculations for an integral space-frame and shell structure. The computer programs used by Lincoln throughout this design effort were STAIR and FRAN.[11] STAIR and FRAN are both large-capacity structural analysis programs. STAIR is restricted to the analysis of three-dimensional trusses (pinned joints) with up to 4000 joints. FRAN is an almost completely general structural analysis program for framed structures (rigid joints), as well as trusses, with up to 2000 joints. STAIR was used during the early design phases, the one-fifteenth structural scale model tests and the prototype in-plant and on-site tests. FRAN was used during the prototype in-plant and on-site tests and for the final analyses of the operational conditions. This usage resulted from the availability and efficiency of the individual programs and the demands of the particular analysis effort. The STAIR analyses were for the backup structure only because no reasonable shell analogy was found for use with STAIR. The FRAN program was used later in the analytical effort for the combined frame-shell analyses. The generality of the FRAN program allowed the development of a good shell analogy using a combination of beams and bars. Subsequent to this time a shell analogy for STAIR was developed (Reference 12) and although it does require numerous joints, it might prove useful in other analytical problems.

Test Program

Due to the complexity of the analysis, the need to employ previously unevaluated digital computer programs, and the necessity for very accurate deflection calculations, three series of structural load tests were incorporated into the Haystack program. The first of these load test efforts was carried out on a structural model, built to a one-fifteenth scale. Considerable care was taken in building the model and numerous loading conditions (with and without the shell) were applied with the deflections measured at numerous points. The accuracy of the STAIR calculations was initially verified by comparisons with these test results.

The reference coordinate system for this series of test programs is depicted in Figure 8. The X axis is the elevation axis about which the antenna rotates;

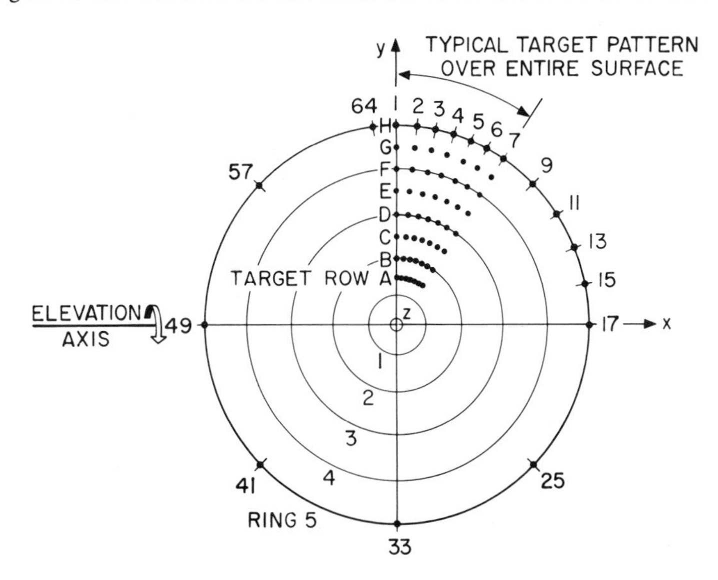

Figure 8. Reference coordinate system.

the Y axis is the vertical centerline of the figure and the Z axis is the axis of revolution for the paraboloid. The pattern of optical targets is shown by the dots with the radial rows lettered from A to H and numbered circumferentially from 1 to 64. Alternate target rows have the same radius as the ring trusses beneath the surface panels. These targets are located about 3 in. from a radial edge of each panel. There is also a series of secondary targets located between these targets in both the radian and circumferential directions. The rows corresponding to the ring trusses were used in the model and in-plant tests while the complete set of targets was used during erection and on-site tests.

The Y axis of this coordinate system is in all respects an axis of both structure and load symmetry. The left half is in all respects a mirror image of the right half. This allows one to calculate deflections for only one-half the structure provided the appropriate boundary condition is maintained along the axis of symmetry. Test results, which will be presented later, will show only the right half with the measurement results from both halves superimposed.

Figure 9. One-fifteenth scale model.

A photograph of the Haystack one-fifteenth structural scale model is shown in Figure 9. All member sizes, properties, and length are scaled down by a factor of 15. The model represented a fairly early design configuration since it was built at the start of the over-all design effort. Any discrepancies between this model and the actual antenna are immaterial since it was not used to predict deflections of the prototype but only to make analytical comparisons with the computer results. The prototype computer analyses were not used for this comparison; instead separate runs were made for the model with appropriate changes in input data.

Some of the actual comparisons of analytical versus test data for loads in the Z direction are shown in Figure 10. The original test data have been corrected for deflections of the trunnion beams by using as a reference plane their four

LOAD AT	RESULTS FROM	MEASURED POINTS								
		H-1	H-5	H-9	H-13	H-17	F-5	F-13	D-1	D-9/D-25
H-1	EXPERIMENTAL	4786	1905	−124	−1160	−1646	1073	−669	426	0
	COMP. NAA	5625	2421	−18	−1555	−2241	1415	−913	712	0
	COMP. STAIR	5656	2436	−42	−1604	−2299	1432	−900	733	0
H-5	EXPERIMENTAL	1924	2753	344	−706	−1258	1065	−413	340	0
	COMP. NAA	2426	2812	351	−1123	−1821	1084	−609	490	0
	COMP. STAIR	2442	2858	336	−1150	−1851	1092	−623	499	0
H-17	EXPERIMENTAL	−1440	−1144	−184	1626	7280	−614	912	−300	0
	COMP. NAA	−2098	−1685	−258	2258	8976	−965	1467	−552	0
	COMP. STAIR	−2165	−1722	−224	2340	9074	−999	1518	−589	0
F-5	EXPERIMENTAL	1438	1120	203	−450	−746	1314	−286	253	0
	COMP. NAA	1431	1094	156	−682	−1091	1370	−437	371	0
	COMP. STAIR	1448	1101	144	−705	−1117	1377	−450	380	0
F-13	EXPERIMENTAL	−544	−372	144	733	1154	−263	1075	−146	0
	COMP. NAA	−775	−520	183	937	1422	−357	1129	−242	0
	COMP. STAIR	−807	−537	203	983	1476	−373	1158	−260	0
D-1	EXPERIMENTAL	664	457	−12	−332	−514	343	−244	362	0
	COMP. NAA	734	504	−25	−456	−673	376	−317	399	0
	COMP. STAIR	753	514	−38	−481	−702	385	−331	409	0

COMPARISON BETWEEN CALCULATED AND MEASURED DEFLECTIONS

1. FACE-UP TYPE LOADS
2. DEFLECTION VALUES ARE IN MICROINCHES AND ARE REFERENCED TO THE FOUR HARD POINTS
3. THE LOAD IS 1 POUND PER QUADRANT −4 POUNDS TOTAL LOAD

Figure 10. Computed versus model. Test comparison.

end points. Points D-9 and D-25, because of this correction, show all results to be zero. The load applied for each of the tests was 1 lb per quadrant for a total of 4 lb. The tabulated values are in microinches.

The first point of comparison that can be made from this figure is the consistently good agreement between the two computer programs. Although two different methods of analysis were used with independently generated input data, the agreement at points with measurable deflections is generally within 2 per cent.

The other comparison that can be made from Figure 10 is between the test results and these calculated deflections. The model results are generally approximately 15 per cent less than the calculated values, which was considered good agreement since the model construction involved many compromises. Recently, NAA has reviewed this analytical work and has found much better ($\simeq$ 5 per cent) correlation.

The next series of tests was conducted after the entire antenna space-frame structure (no panels) was trial erected in NAA's Columbus, Ohio, plant. The results of these load tests and the corresponding computer calculations are shown in Figure 11 for loads in the Z direction and Figure 12 for loads in the Y direction. As with the model tests, the data has been referenced to the four hard points (the ends of the trunnion beams) but the load in these cases was 1000 lb per quadrant for a total of 4000 lb. These measurements were made with a theodolite and optical scales and they were normally repeatable within ± 0.01 in.

In both types of tests, measurements were taken at two load levels so as to show linearity and point out possibly bad data. The load levels in each test were 50 and 100 per cent, but the zero reference for the face-up tests was no load whereas it was 10 per cent of full load for the face-side tests. This 10-per cent load was used to take the slack out of the modified support structure used in these tests.

The differences shown between STAIR and FRAN are the effect of assuming pinned joints in STAIR versus rigid joints in FRAN, and they are not simply the results of different analytical techniques. A separate FRAN analysis, not shown here, which used exactly the same data as STAIR with pinned joints, gave results within 0.001 in. of the STAIR results shown here.

A review of these results indicates that there is very good correlation between measured and calculated deflections since, with only few exceptions, all of the STAIR deflections agree with the test results within the measurement accuracy. Over all, 94 per cent of the differences are less than 0.010 in. and 69 per cent are less than 0.005 in. In addition to this, FRAN is even closer to the test results than STAIR for more than 70 per cent of the points. STAIR gives very good results for this structure, but FRAN is even better and by tests such as this, it was decided to use FRAN for all of the final analyses and the preparation of bias rigging tables.

The only places that show significant discrepancies are at the point of load application in cases 1 and 2 on Figure 11. In case 1, the difference (at 100 per cent) is 0.042 in. and in case 2, it is 0.018 in. The 50-per cent values would indicate that better correlation existed at lower loads and it is felt that, in

TEST	LOAD	RESULTS FROM \ NODE POINT (Target Row and Number)		H-1	H-7	H-9	H-17 H-49 (avg)	H-33	F-1	F-17
1	FULL LOAD 2000 lb H-1 AND H-33	TEST	50 %	−0.105	−0.028	0.015	0.038	−0.110	−0.047	0.017
			100 %	−0.265	−0.052	0.042	0.081	−0.234	−0.102	0.050
		COMP.	STAIR	−0.223	−0.054	0.040	0.088	−0.241	−0.101	0.053
			FRAN	−0.223	−0.051	0.037	0.082	−0.232	−0.095	0.049
2	FULL LOAD 2000 lb H-17 AND H-49	TEST	50 %	0.029	0.011	−0.020	−0.164	0.027	0.019	−0.049
			100 %	0.066	0.039	−0.025	−0.363	0.075	0.049	−0.099
		COMP.	STAIR	0.076	0.046	−0.033	−0.345	0.078	0.053	−0.104
			FRAN	0.070	0.043	−0.029	−0.333	0.071	0.047	−0.097
3	FULL LOAD 1000 lb H-9, H-25, H-41 AND H-57	TEST	50 %	−0.006	−0.023	−0.011	0.002	−0.005	0.002	0.002
			100 %	−0.012	−0.033	−0.032	0.009	−0.006	−0.003	0.007
		COMP.	STAIR	−0.005	−0.037	−0.036	0.010	−0.004	−0.002	0.003
			FRAN	−0.006	−0.037	−0.035	0.011	−0.005	−0.003	0.003

COMPARISON BETWEEN CALCULATED AND MEASURED DEFLECTIONS

Figure 11. Face-up backup structure. Test results.

TEST	LOAD	RESULTS FROM \ NODE POINT (Target Row and Number)		H-1	H-7	H-9	H-17 H-49 (avg)	H-33	F-1	F-17
1	FULL LOAD 2000 lb H-1 AND H-33	TEST	40 %	−0.016	−0.005	0.003	−0.003	0.022	−0.004	−0.002
			90 %	−0.040	−0.009	0.000	0.003	0.043	−0.017	0.000
		COMP.	STAIR	−0.041	−0.009	0.003	0.000	0.051	−0.017	0.000
			FRAN	−0.040	−0.009	0.002	0.001	0.049	−0.016	0.001
2	FULL LOAD 2000 lb H-17 AND H-49	TEST	40 %	−0.004	−0.006	0.000	−0.002	0.001	−0.001	0.000
			90 %	−0.015	−0.004	0.002	−0.003	0.012	−0.008	0.000
		COMP.	STAIR	−0.015	−0.010	0.001	0.001	0.016	−0.009	0.000
			FRAN	−0.014	−0.010	0.001	0.001	0.015	−0.009	0.001
3	FULL LOAD 1000 lb H-9, H-25, H-41 AND H-57	TEST	40 %	0.004	−0.002	−0.001	0.002	0.008	0.000	0.001
			90 %	−0.001	−0.004	−0.017	−0.001	0.008	0.001	−0.003
		COMP.	STAIR	−0.007	−0.010	−0.018	0.000	0.008	−0.004	0.000
			FRAN	−0.007	−0.010	−0.018	0.001	0.007	−0.005	−0.001

COMPARISON BETWEEN CALCULATED AND MEASURED DEFLECTIONS

Figure 12. Face-side backup structure. Test result.

these two cases, one or more of the pretensioned rods in the vicinity of the load point went slack during the second half of the load tests. Since the typical operational joint loads are less than 10 per cent of these test loads, there should be no degradation due to this cause during actual use.

The next series of load tests followed the erection of the entire antenna (with surface panels) within the radome. These took place just prior to removal of the work platforms, scaffolding, and initial operation of the antenna. The analytical versus test comparisons for row H (ring 5) for one of these cases are shown in Figure 13. The analysis, in these cases, included a representation

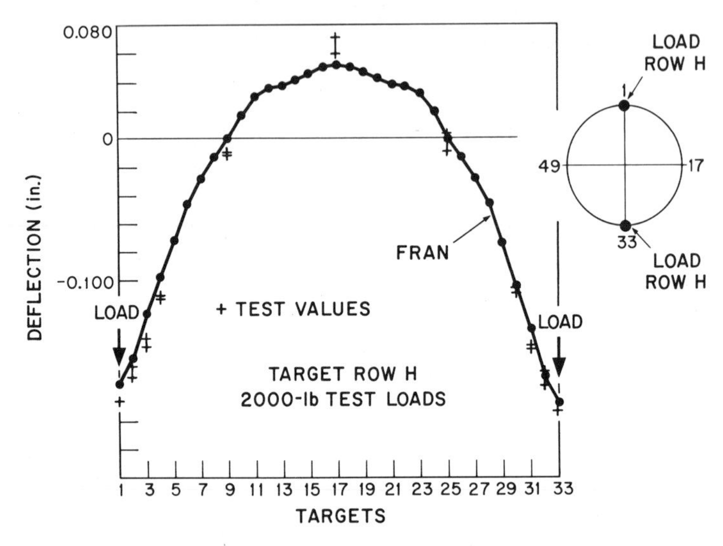

Figure 13. On-site tests--two Z loads on Row H (ring 5).

of the shell, which will be discussed in the next section. The type of load applied is depicted in the plan view in the upper right and consisted of two 2000-lb loads applied to ring 5 at points 180° apart (H-1 and H-33). Vertical deflections have been plotted versus target number and thus in plan view this is the deflection curve for a semicircular ring.

The extremely good agreement that has been achieved in this plot is readily apparent. The analytical value is generally closer to the test results than the maximum variation of test measurements at symmetrical points. This variation and over-all repeatability of measurements indicates that these values have an accuracy of approximately ± 0.010 in. Additional plots of a similar nature for different points of load and direction are presented in Figure 14 and 15. These support the previous agreement obtained in Figure 11. These plots are three of approximately 200 that were made from these on-site tests. The over-all conclusion that was reached from this series of tests as well as the two previous ones is that the analytical work can be considered extremely accurate and generally as precise as can be measured in most cases.

Structural Idealization and Shell Analogy

The FRAN program was used to analyze the behavior of the complete integral shell and the space frame as a composite structure. Although the shell

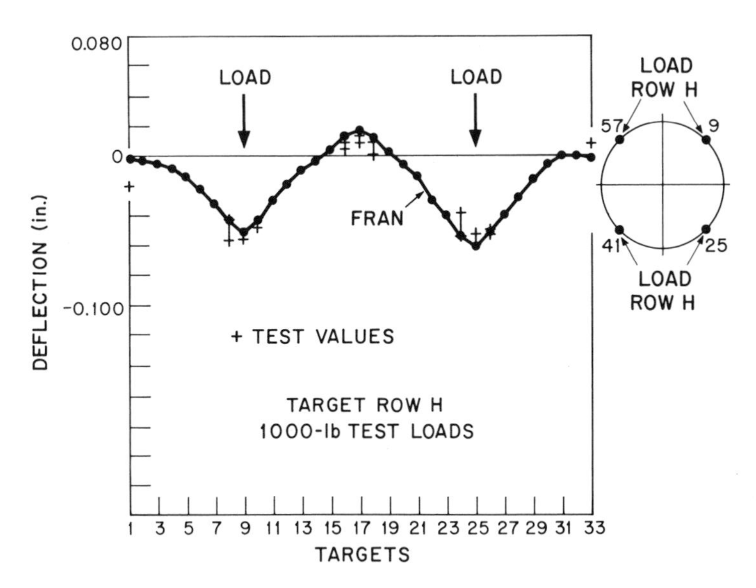

Figure 14. On-site tests--four Z loads on Row H (ring 5).

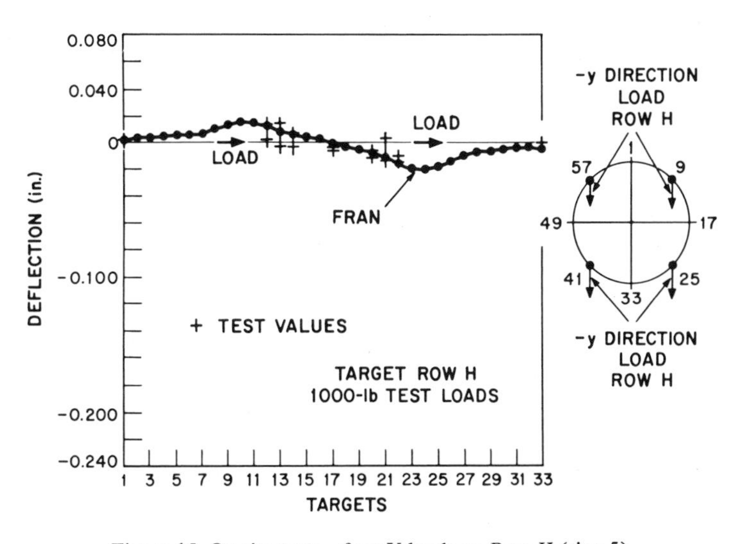

Figure 15. On-site tests--four Y loads on Row H (ring 5).

represents only 10 per cent of the reflector weight, it adds approximately 20 per cent to the over-all structural stiffness. In most other reflectors, the surface is a load on the reflector, which contributes no structural stiffness.

A lattice analogy of finite elements was developed which satisfactorily represented the stiffness of the shell and this was used in conjunction with the data that identified the space-frame backup structure and the FRAN program. This reflector grid analogy was derived as a network of beams and bars with appropriate axial, bending and torsional stiffness (area and inertias) from the usual thin-shell equations. The reflector shell representation used in the FRAN analyses divided the surface into trapezoidal grid elements as shown schematically in Figure 16. Each of these elements is composed of six members which

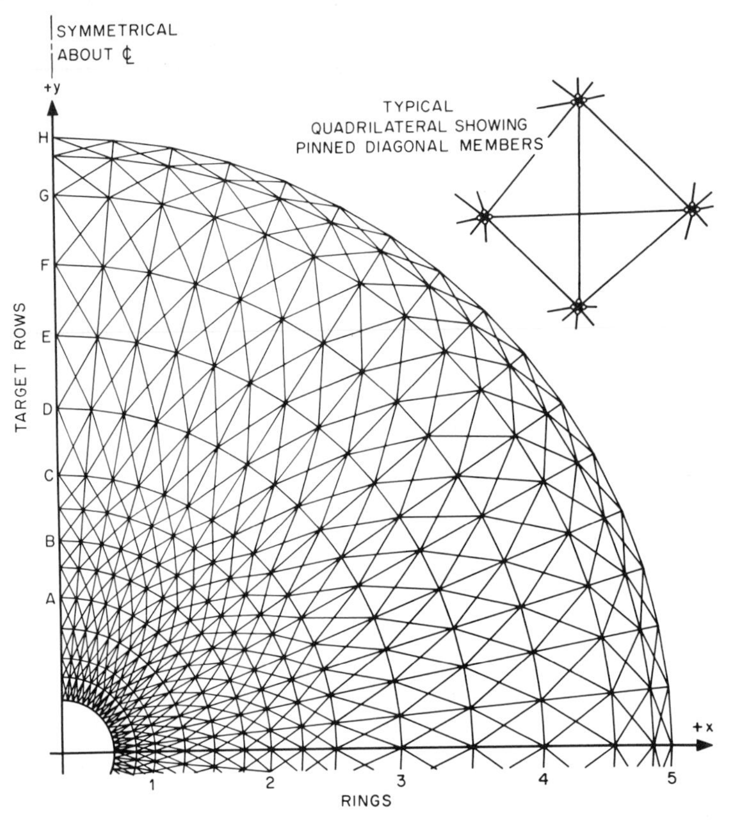

Figure 16. Grid analogy for reflector surface.

interconnect the four node points. The radial and tangential members of each element are rigidly connected to the nodes while the diagonals are pinned. There were 495 joints and 1911 members in the shell grid analogy which connected to 80 joints in the backup structure by members that are normal to the shell surface. These members have ball and socket joints at each end as shown physically in Figure 17. This is a photograph of the top chord of ring

Figure 17. Shell to backup structure connection.

3 with a section of the surface attached to it by the members normal to the shell. The chord members are 10-in.-diameter tubes with fully welded butt joints. These can be compared visually with the clevis pin connections on the diagonal rod members in the lower right.

The complete antenna structure, including the shell, backup structure, counterweight supports, rf box supports, and the secondary reflector support made up a total analytical model of 840 joints and 2700 members. These FRAN analyses for the final round of analysis involved approximately 4000 simultaneous equations per loading condition and a total of 18 loading conditions.

Operational Design Conditions

The ultimate use of the programs developed and then validated by the load tests was, of course, to predict the operational behavior of the reflector under gravity and thermal loads. The destortion pattern of the reflector, the point-by-point deflection plot from an ideal parabolic contour was determined for all positions of the antenna (face-up to face-side). The minimum surface distortion envelope would be then achieved by adjusting the shell surface (bias rigging), control devices, and counterweight system such that all points have two equal and opposite maximum surface distortions (generally at 0° or 90°). This results in an antenna with its best surface near 45° even though it was erected and adjusted face up. Figure 18 shows a view of the present

Figure 18. Present antenna surface.

Haystack surface. Visual reference and size can be gained from the man in the picture, who is standing over the main ring at a radius of 32 ft. The picture was taken from the edge of the reflector. The black radial sections are simply walkways that are supported over the rings. The surface can be walked on but this is allowed only in emergencies. The black ring is the 1-in.-thick aluminum splice plate which connects to the inner and outer rows of panels. It has been painted black to lower its thermal time constant. The accuracy of the surface can be judged visually by noting the clarity and straightness of reflections in the picture particularly the radome beams and secondary reflector supports.

The measured distortions of this surface is shown as a contour plot in Figure 19. The surface errors indicated on this plot are the normal deviations, in thousandths of an inch from a best-fit paraboloid. This data was obtained from a set of 240 optical target readings taken at noon on a thermally unstable day and thus represents a worst-case contour plot. The peak surface error in this plot is 0.135 in. and over-all it yields an rms error of 0.048 in. A more

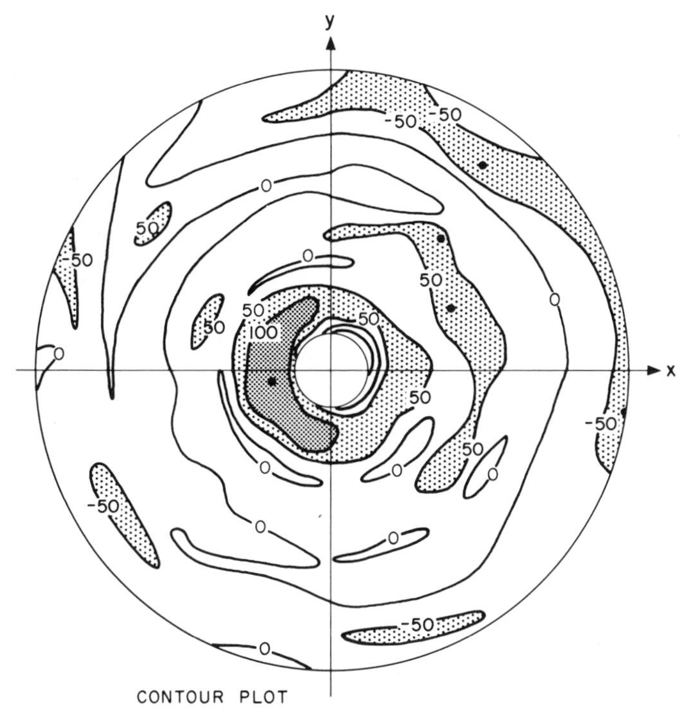

CONTOUR PLOT
HAYSTACK REFLECTOR-ZENITH POSITION
1. Normal deviations from best fit paraboloid in thousanths of an inch
2. Obtained from the measured position of 240 targets
3. RMS surface error 0.048 in
4. Peak surface error +0.135 in

Figure 19. Antenna contour plot.

typical value for day and night operation would be 0.040 in. rms. Measurements on thermally quiet nights indicate that the antenna has a basic surface rms error of 0.036 in. face-up and 0.039 in. face-side with a minimum of 0.034 in. at an elevation angle of approximately 30°. These surface distortions in the present instance are primarily caused by temperature distortions and an erroneous adjustment of the surface during construction. There were several reasons for this erroneous adjustment. The most significant reason was a serious degradation of the optical instrumentation as a result of corrosion in the mounting of the pentamirrors. The basic mirrors and equipment had a 1 sec of arc capability; however, as a result of this corrosion, one of the mirrors developed a peak error of 42 sec of arc.

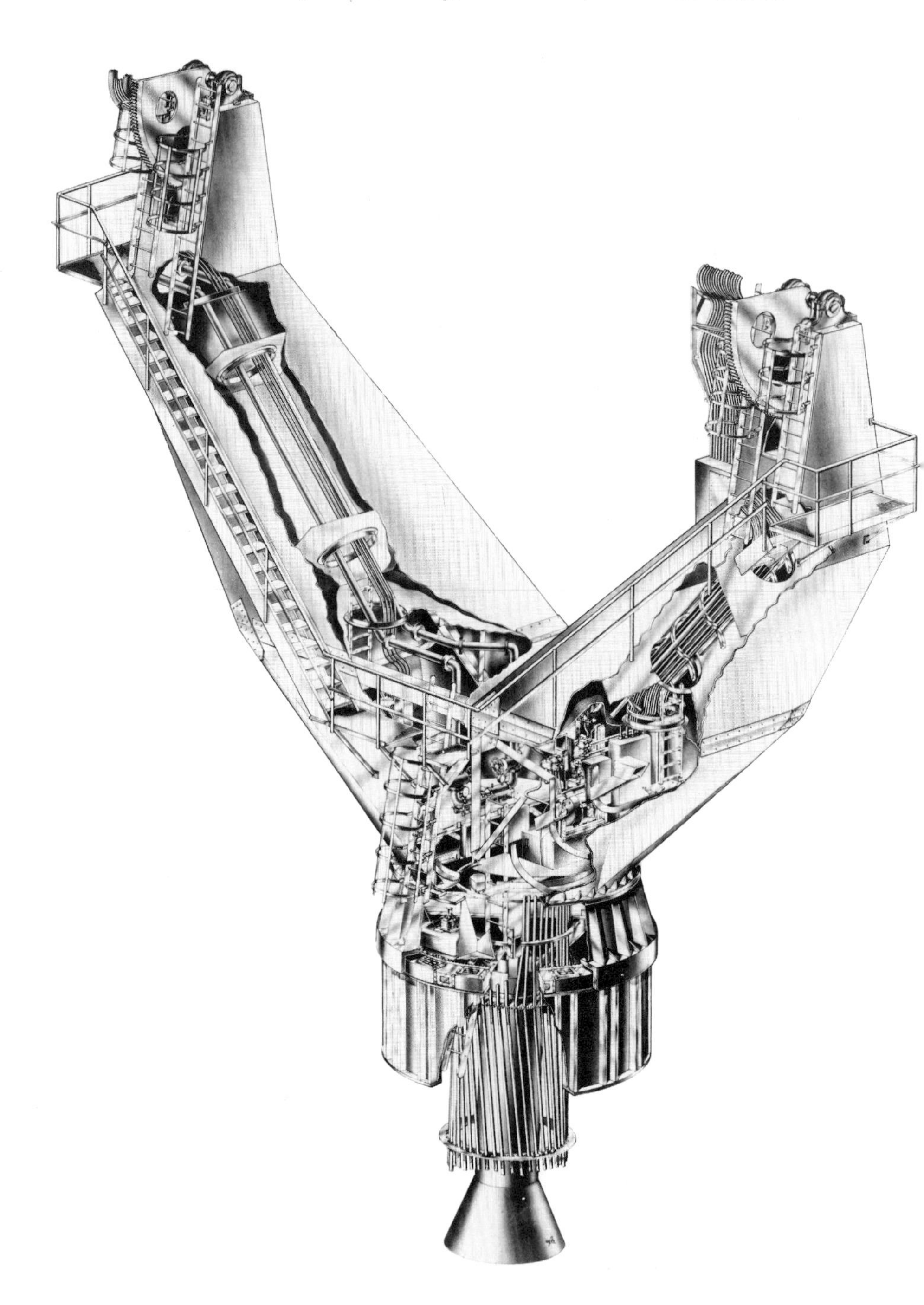

Figure 20. Yoke assembly.

Mechanical Description

The yoke assembly illustrated in Figure 20 provides an elevation over azimuth rotating support structure for the antenna.

The azimuth hydrostatic bearing, bottom of photograph, is a combined thrust and radial fluid film bearing which, when fully loaded with 190 tons of antenna and yoke, operates at an average radial pad-to-runner clearance of 0.006 in. and at an average thrust pad-to-runner clearance of 0.005 in. Nominal operating pool pressures of 250 psi, with a total flow of 26.5 gpm, are supplied by a 50-hp hydraulic pump. Temperature-compensated oil flow control valves are inserted between the pump and each bearing pad to insure constant flow control and eliminate any "downstream" effects.

Housed within the tower and yoke and passing through the azimuth hydrostatic bearing is the azimuth cable wrap. Included in the wrap are over 3000 pairs of electrical conductors, eight microwave cables, and two 4-in.-diameter water lines delivering about 250 gpm at 200 psi, a high-voltage line, and all utilities. These cables run up both sides of the yoke arm to two elevation cable wraps from which they proceed to the rf equipment "plugin" boxes located directly behind the reflector vertex.

The elevation bearings located atop the two yoke arms consist of two pairs of conventional, self-aligning, spherical roller bearing assemblies. The 10.5-in.-diameter elevation shafts in these bearings support the trunnion beams of the reflector assembly. Each pair of pillow blocks is designed to have one bearing fixed and the other float. The pairs of bearing are 47 ft apart across the arms of the yoke.

Figure 21 illustrates the principal components of the azimuth drive assembly. Two gear reducers, each driven by a 10-hp hydraulic motor, mesh with an internal bull gear that provides the torque required to rotate the antenna. The performance characteristics of the antenna will be proportional to the load inertia since the radome eliminates wind effects and the hydrostatic bearing, for all practical purposes, eliminates "stiction" and friction torques. Backlash is eliminated in the dual pinion parallel gear train by a device that preloads the parallel gear trains and removes backlash in every mesh. At the center of the bearing, rotating at the same speed of the antenna, is a phasolver transducer. This analog device, with the proper conversion circuitry in the digital control system, will produce azimuth angle data with a resolution of 2.47 sec of arc (19 bits).

The elevation drive system, Figure 22, illustrates one of the two elevation drives located on each of the yoke arms. The azimuth and elevation drives, except for actual gear reduction, are very similar. The same motors and phasolver transducer are used. The spherical roller bearings are small enough in diameter so that their "stiction" and friction torques can be neglected.

Hydraulic buffer stops will mechanically stop the antenna in elevation and azimuth at the limits of travel. The stops were designed and successfully tested

HYDRAULIC DRIVE

POWER TRAIN			TACH DRIVE		
GEAR	NO TEETH	G.R.	GEAR	NO TEETH	G.R.
1	21	.266	7	96	4.57
2	79		8	21	
3	21	.250	9	105	4.38
4	84		10	24	
5	26	.046			
BULL	564				

DATA BOX DRIVE

CAM DRIVE			SYNCHRO DRIVE		
GEAR	NO TEETH	G.R.	GEAR	NO TEETH	G.R.
BULL	564	21.7	BULL	564	21.7
1	26		1	26	
2	30	.250	2	30	.250
3	120		3	120	
4	28	.199	4	28	.199
6	141		5	141	
7	64	.500	4	28	.199
8	128		6	141	

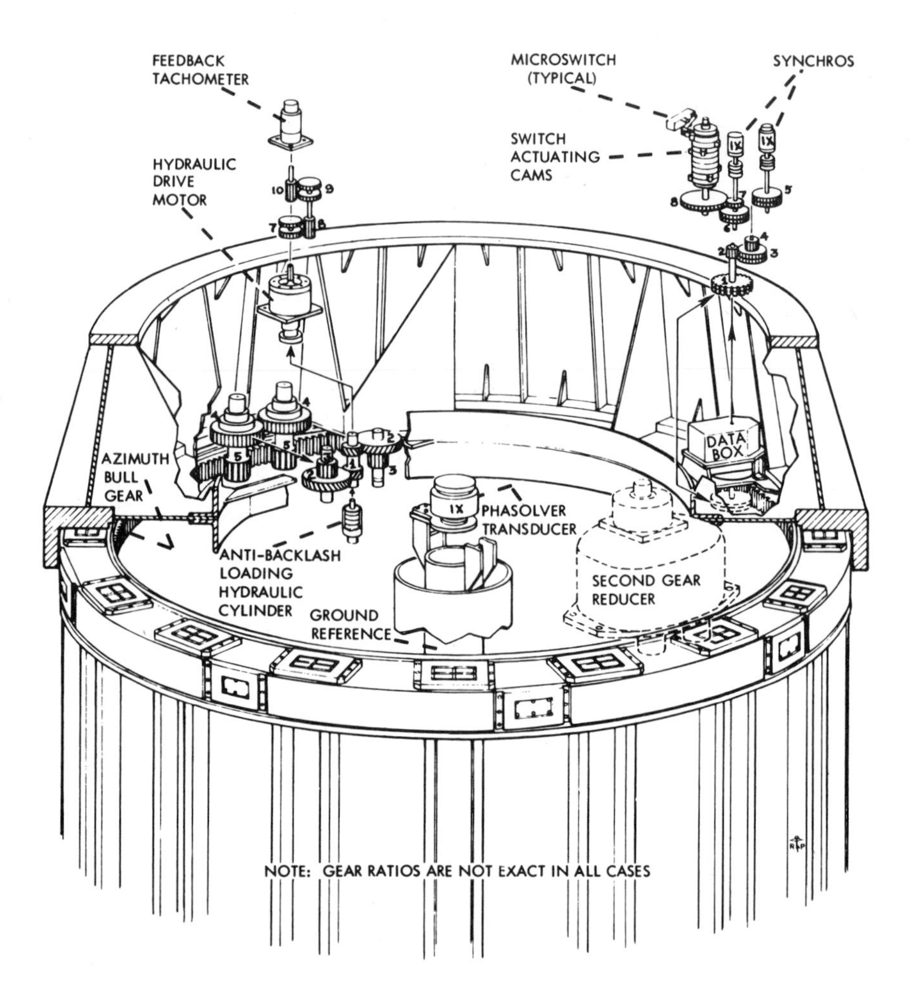

Figure 21. Azimuth drive assembly.

to stop the antenna at maximum velocity with motors delivering maximum torque. Stow locks, remotely controlled, can be used to lock the antenna at 0°, 45°, and 90°.

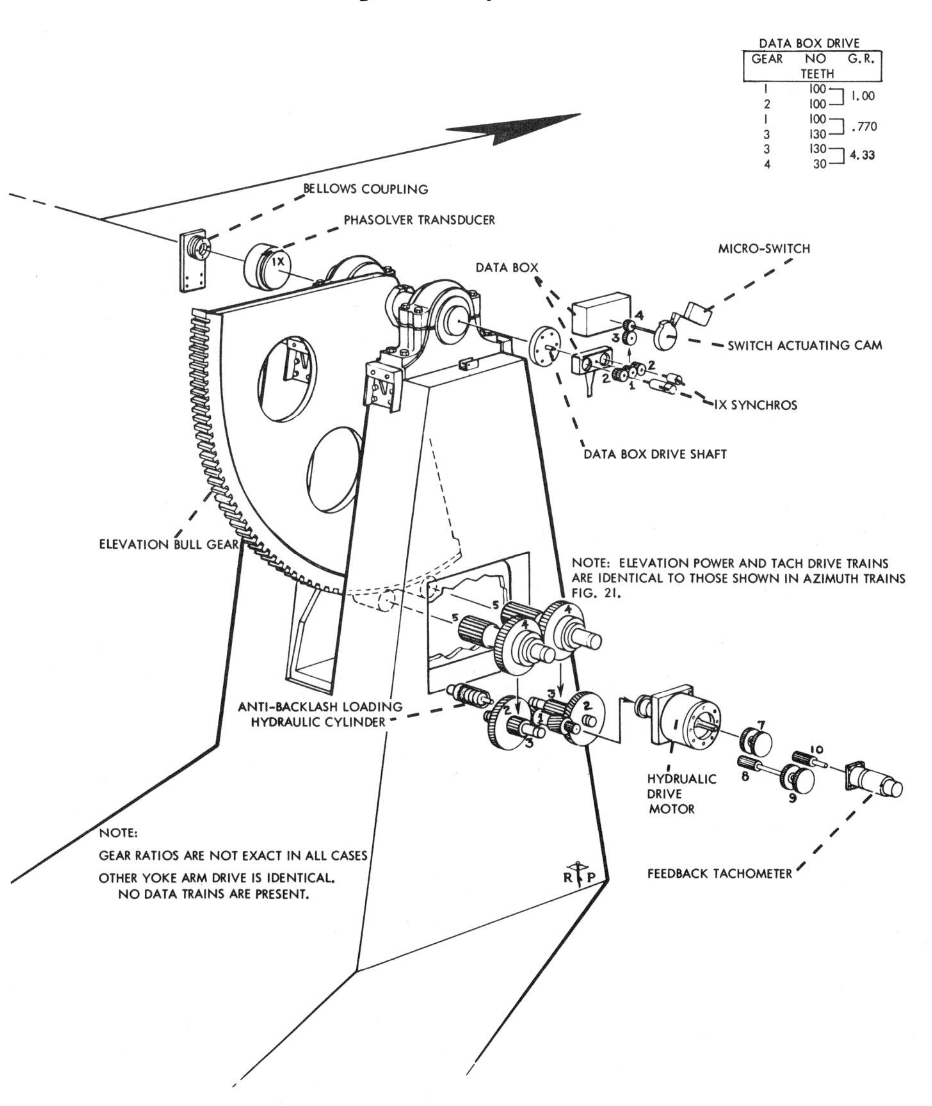

Figure 22. Elevation drive assembly.

One unique feature of Haystack is its ability to interchange rf electronic "plugin" boxes. Figure 23 illustrates the radiometer box being hoisted into place. The turn-around time for removing one box and replacing it with another can be as little as 1 hr.

The servo control system is a hybrid, closed-loop, positioning system utilizing a digital computer for pointing instructions and an analog angle data system for measuring antenna position. Basically, a digital-to-analog conversion of the

Figure 23. Radio-frequency equipment hoist.

input command signal is compared with analog angle position data to provide analog error signal for controlling the drive systems of each axis. The least significant bit of the position control system, which is designed for 19-bit accuracy, is equal to 2.47 sec of arc. The total static error permitted in the angle data system is ± 3.6 sec of arc.

Optical Measurement and Alignment Considerations

To accommodate the many alignments, location and measurement requirements of the antenna system, an integrated optical system was designed by the Keuffel and Esser Company of New Jersey. Figure 24 is a schematic of the

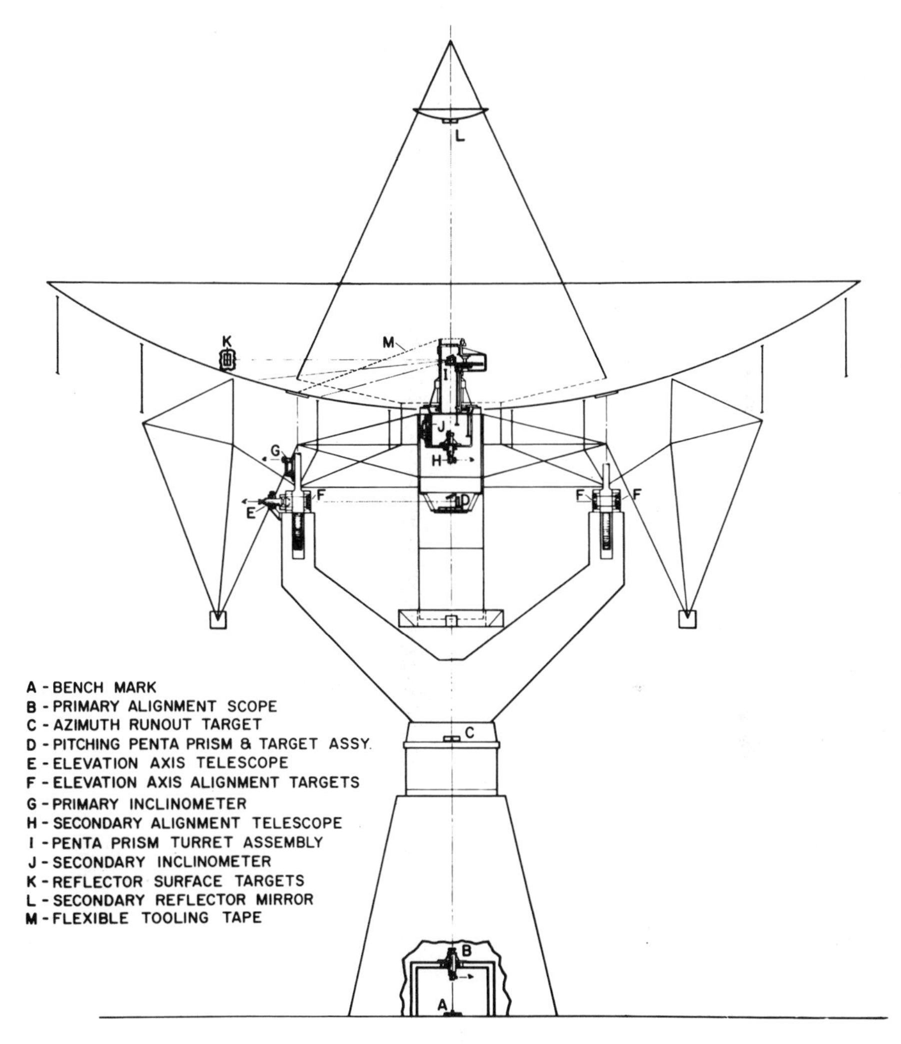

Figure 24. Optical measurement schematic.

various elements comprising the optical system.

Basically, a primary alignment scope B established coincidence of the azimuth axis with the focal axes of the primary and secondary reflectors. Orthogonality of the elevation axis is established by the 45° pitching target assembly D and the primary inclinometer G. Alignment of the four elevation bearings across the yoke structure and its squareness with the azimuth axis is accomplished by the alignment scope E, targets F, and the pitching pentaprism target D.

An optical probe I, consisting of eight pentamirrors in a turret assembly, is used to locate and measure 484 targets K located on the primary reflector panels. A calibrated tape (accuracy 0.003 in. in 32 ft) was used to locate the splice plate from which in turn the inner and outer rows of panels are located.

An autocollimating target mirror L, precisely positioned during the final machining operation of the hyperbolic surface, is used for alignment of the secondary reflector.

Rerigging Effort

The antenna was monitored throughout its early operational use to determine its actual time and position deformation history. During this period the above-mentioned optical errors were detected. In addition, although the radome with its heating and blower system does substantially mitigate against temperature changes and differentials, the antenna was found to be sensitive to diurnal temperature changes. After a considerable analytical effort this was traced to the thermal interaction of the 1-in.-thick splice plate and the remainder of the honeycomb panel shell. The honeycomb panels are very thin and respond almost instantly to temperature changes, whereas the splice plate lags several hours behind and thus introduces considerable distortion into the shell surface. This effect has been considerably reduced by using the heating system to damp out the diurnal temperature change and the splice plate has been painted black to reduce its thermal time constant.

Part of this investigation involved a brief series of gravity deformation tests using a stretched wire and a linear potentiometer. This device was attached to two points on the antenna and measured the relative displacement of these points as the antenna was rotated 90°. The test results along with the corresponding analytical deflections are shown in Figure 25. As shown in line 1 of this figure, the relative displacement of the secondary reflector and the rf box structure was measured to be 0.238 in. when the antenna was rotated 90°. The corresponding calculated value is 0.211 in. The other comparisons are similar and in view of the estimated accuracy of the test equipment, this series of tests as with the previous ones further verified the accuracy of the analysis techniques being used.

As a result of this study effort, the error budget shown in Figure 26 was established as the objective in a planned rerigging effort. These values are, in contrast to those in Figure 7, firm values resulting from analyses and investigations of the actual antenna over a period of five years. A limited, six-week

POINTS MEASURED	MEASURED VALUE	CALCULATED VALUE
(1) SECONDARY REFLECTOR TO RF BOX	+0.238	+0.211
(2) SECONDARY REFLECTOR TO INNER PANEL (top)	−0.032	−0.034
(3) SECONDARY REFLECTOR TO OUTER PANEL (bottom)	+0.201	+0.181
(4) SECONDARY REFLECTOR TO TARGET ROW H (top)	−0.188	−0.202
(5) DIAMETER CHANGE (top to bottom)	+0.030	−0.014

(1) ELEVATION MOTION FROM −0° TO 90°

(2) ESTIMATED REPEATABILITY OF MEASUREMENTS − ±0.02 in.

(3) VALUES ARE IN INCHES AND PLUS IS EXTENSION

Figure 25. Stretch wire measurements.

rerigging effort is running concurrently with this conference (October 1967) and present optical and radiometric measurements indicate that we have achieved the objective shown in Figure 26.

Conclusions

The Haystack antenna system has been in almost continuous operational use since its dedication in October 1964. After its initial survey and panel setting, which was carried out with an optical system having uncertain errors, and before the temperature in the radome was fully controlled, an antenna with an rms surface tolerance of 0.040 in. was obtained for use at its 8 GHz design frequency. The pointing system consistently operates with a peak pointing angle uncertainty of 0.005° and a pointing "jitter" of only 0.002°. This capability is consistent with the known rigging, optical, thermal, and gravity errors. Radiometric, analytic, and limited measurements indicated that this 0.040-in. rms error is relatively constant at all elevation angles and all times of the year. The residual uncompensated gravity errors contribute only

±0.034 in.	20° – 70° GRAVITY
±0.030 in.	THERMAL
±0.015 in.	MEASUREMENT
±0.015 in.	MANUFACTURING
±0.010 in.	ERECTION
0.051 in.	ROOT SUM SQUARE
0.020 in.	RMS (0.4 × RSS)

Figure 26. Rerigging error budget.

slightly to the initial surface tolerance.

The operation of the hydraulic drives and hydrostatic bearing has proved to be exceptionally trouble free. The present system utilizes a common source of hydraulic supply and current plans are to separate this sytem, thereby reducing "cross-talk" and in this way improve the present performance characteristics even further.

The three years of operation to date have shown the interconnect system (electrical cables and water lines) to be more than adequate and that the removable rf boxes provide extremely flexible operational usage approaching almost continuous utilization.

The radome has required little maintenance since 1960. The top half was painted once and the entire dome is now being repainted. No radome panels have ever been replaced, although a few have been repaired.

With this busy operational schedule, it has only recently been possible to plan and carry out a short rerigging effort to materially improve the surface tolerance of the reflector. Early indications (optical measurements) are that a tolerance of between 0.017 in. and 0.020 in. rms have been achieved, and this performance has been confirmed by radiometric measurements at wavelengths as short as 8 mm. In achieving this goal, Haystack has pioneered in several notable areas such as a complete deflection design implemented by digital computer, integration of a radome into a research antenna system, application of a hydrostatic bearing in azimuth to reduce "stiction" and to improve pointing accuracy, and the use of removable electronics modules resulting in a quick-change flexibility in over-all operations scheduling.

The future application and improvement of these techniques to the logical and economical design of much larger projects can be readily foreseen and, in fact, is taking place at the present time in the planning stages of the 440-ft antenna study effort being conducted by the Northeast Radio Observatory Corporation (NEROC).

Acknowledgments

The planning and implementation of this program is the result of the efforts of many different organizations and individuals. The authors wish to acknowledge the contributions of North American Aviation, Inc., the principal antenna contractor, and the funding agency, Electronic Systems Division of the United States Air Force. The radome was designed by Lincoln Laboratory and constructed by the H. I. Thompson Fiberglass Company of Gardena, California. Dr. Howard Simpson of the firm, Simpson, Gumpertz & Heger, of Cambridge, Massachusetts, has worked very closely with Lincoln Laboratory as a structural consultant throughout this effort. Professor Dudley Fuller or Columbia University served as design consultant on the hydrostatic azimuth bearing.

The success of this program is due directly to the diligence and competence of the many Lincoln Laboratory staff who have participated in various phases of the over-all program.

References

1. H. G. Weiss, "The Haystack Microwave Research Facility," Institute of Electrical and Electronic Engineers Spectrum, New York (February 1965).
2. J. V. Evans *et al.*, "Radar Observations of Venus at 3.8 cm Wavelength," *Astron. J.* **71**, 902 (1966).
3. T. Hagfors, "Some Recent Lunar Radar Observations and their Interpretation," NEREM, Boston, Massachusetts (November 1966).
4. G. H. Pettengill, "Recent Radar Results on the Rotation and Surface Features of the Inner Planets," Colloquium, Goddard Space Flight Center, Greenbelt, Maryland (May 1966).
5. G. H. Pettengill, "3.8-cm Radar Observations of the Surface and Atmosphere of Venus and Mars," International Astronomy Union, Prague, Czechoslovakia (August 1967).
6. W. R. Fanning, "The Haystack Antenna – Mechanical Concept, Design and Testing," American Society of Mechanical Engineers publication "Deep Space and Missile Tracking Antennas" (December 1966).
7. W. R. Fanning, "Mechanical Design of the Haystack Antenna," *Inst. Elec. Engrs. Proc. (London)* Publ. No. 21 (1966).
8. J. Banche and D. N. Ulry, "Axial Load Design for High Performance Radar Antenna Structure," American Society of Mechanical Engineers, New York (November 1962).
9. H. Simpson and J. Antebi, "Space Frame Analysis and Applications to

Other Types of Structures," Proceedings of the First Annual Workshop of the SHARE Design Automation Commitee (June 1964).

10. "STAIR (Structural Analysis Interpretive Routine) Instruction Manual," Lincoln Manual No. 48, MIT Lincoln Laboratory, Lexington, Massachusetts (March 1962).
11. "IBM 7090/7094 FRAN (Framed Structures Analysis Pragram) 7090-EC-01X," International Business Machines Corporation, White Plains, New York (September 1964).
12. K. S. Parikh, "Analysis of Shells Using Framework Analogy," thesis presented to the Massachusetts Institute of Technology, Cambridge, Massachusetts (June 1962).

Supplementary Readings

A. H. G. Weiss, "The Haystack Experimental Facility," Technical Report No. 365, Massachusetts Institute of Technology, Lincoln Laboratory, Lexington, Massachusetts, (September 1964).

B. G. R. Carroll, "A Hydrostatic Bearing for Haystack," American Society of Mechanical Engineers, New York (November 1962).

C. A. O. Kuhnel, "Haystack Antenna System Status Report," Technical Report 324, MIT Lincoln Laboratory (September 1963).

D. H. G. Weiss, "Performance Measurements of the Haystack Antenna," *Inst. Elec. Engrs. Proc. (London)* Publ. No. 21 (June 1966).

M. Smoot Katow
Jet Propulsion Laboratory
California Institute of Technology
Pasadena, California

TECHNIQUES USED TO EVALUATE THE PERFORMANCE OF THE NASA/JPL 210-FT-REFLECTOR STRUCTURE UNDER ENVIRONMENTAL LOADS

1. Introduction

The Jet Propulsion Laboratory (JPL) is the cognizant agency of the Deep Space Network for the National Aeronautics and Space Administration (NASA). Based on planned increases in the capability of the Deep Space Network, JPL determined that the increased performance levels of 210-ft-diameter antennas operable at S band or 2300 MHz frequency would be the next upgrading step over the present 85-ft antennas.

In June 1963, NASA awarded a contract to the Rohr Corporation for the final design, fabrication, and erection of the prototype 210-ft antenna to be located at Goldstone, California. The 210-ft has been in use since April 1966.

The completed 210-ft antenna, photographed from a helicopter, is shown on Figure 1. The main features are the concrete pedestal, an alidade rotatable around an azimuth axis, and supporting elevation bearings that carry the tilting reflector assembly. Also shown are the solid and 50 per cent porous surface panels and the truss-type quadripod supporting the subreflector. The rf feed horn protrudes from the 40-ft-high Cassegrain cone assembly mounted on the vertex of the reflector. The wind shield that houses the independent "optical tower" supporting the master controlling unit is visible between the pairs of the elevation gears.

This paper describes the techniques employed by JPL to evaluate the performance of the 210-ft-reflector structure under the environmental loads. The environmental loads considered were (1) the thermal loads due to differential

Figure 1. The completed 210-ft antenna, photographed from a helicopter.

heating from exposure to sunlight, (2) the steady wind loads, and (3) the gravity loads due to changes in the direction of the gravity force with respect to the reflector structure. The influence on performance by rf feed deflections was also considered in the total performance picture.

The evaluation process required the calculation of the deflections due to the environmental loads, and the resolution of the resulting deflections to determine the performance of the reflector structure.

2. Calculation of Deflections

During the feasibility study period, it became obvious with the development of the computer as a structural analysis tool that only computer-organized analysis would produce structural deflections with acceptable accuracy. With the JPL Structural Program (STIF-EIG)[1] limited to an analysis of a maximum of 40 to 50 joint truss, the STAIR[2] computer program was transferred for use to JPL in 1962, with the cooperation of Lincoln Laboratory. STIF-EIG was returned in an exchange for its capability of solving dynamic problems.

Also in 1962, recognizing the lack of wind data on antenna structures for design purposes, JPL's Aerodynamics Section was commissioned to conduct a series of wind tunnel tests on antenna models.[3-7] To compute panel loads on reflector structures, thin paraboloidal shells fitted with static pressure tubes were tested in the wind tunnel. The test data were then presented as pressure difference coefficients over the face of the paraboloid for a particular direction of the wind. Figure 2 shows a photograph of the 25 per cent porous model mounted in the test section of a wind tunnel. Other tests produced force data on many configurations as well as on a complete model.

For computing the effect of thermal loading on the reflector structure, a temperature difference between the parts of the structure exposed to sunlight and in the shade was required. Tests made by the Rohr Corporation, JPL, and others resulted in 10°F as an acceptable temperature difference for a structure properly painted.

The problem of evaluating the deflection vectors computed by STAIR developed. To solve this, in the summer of 1962, a program[8] was coded at JPL to best fit a paraboloid to the deflection vectors computed by the structural program and output the rms of the deviations in ½ rf pathlength errors together with position information of the fitting paraboloid for boresight calculation purposes.

After use at JPL and by contractors, additional coding was added to the rms program[9-11] to control restraint conditions of the best-fitting paraboloid and to plot contour level maps to the ½ rf pathlength errors with respect to the fitting paraboloid.

When the size limitations of the STAIR analysis applied to the 210-ft-reflector structure were considered, it became apparent that only one quadrant of the structure could be readily analyzed. With data for one quadrant, only three load conditions could be computed per run. The data card count was large (about 6000), and each computer run required about 1½ hours. Also, at that time, the additional cost and difficulties of commencing an analysis of half of the reflector could not be justified because only limited experience had been gained with computer analyses of this magnitude.

The need to consider the half-reflector analysis arose because the reflector backup structure was not perfectly symmetrical about the XX plane. The lack of symmetry was caused by the open section required between the elevation gears for rotational clearance with the "optical tower." A simulation by a symmetric structure was considered to adequately represent the actual structure with the computed load conditions and on a sound structural basis.

Figure 3 shows the plan view of one quadrant of the 210-ft-reflector structure that is symmetric about the X and Y planes. The truss arrangement is of the radial rib type. It should be noted that this type of structure is

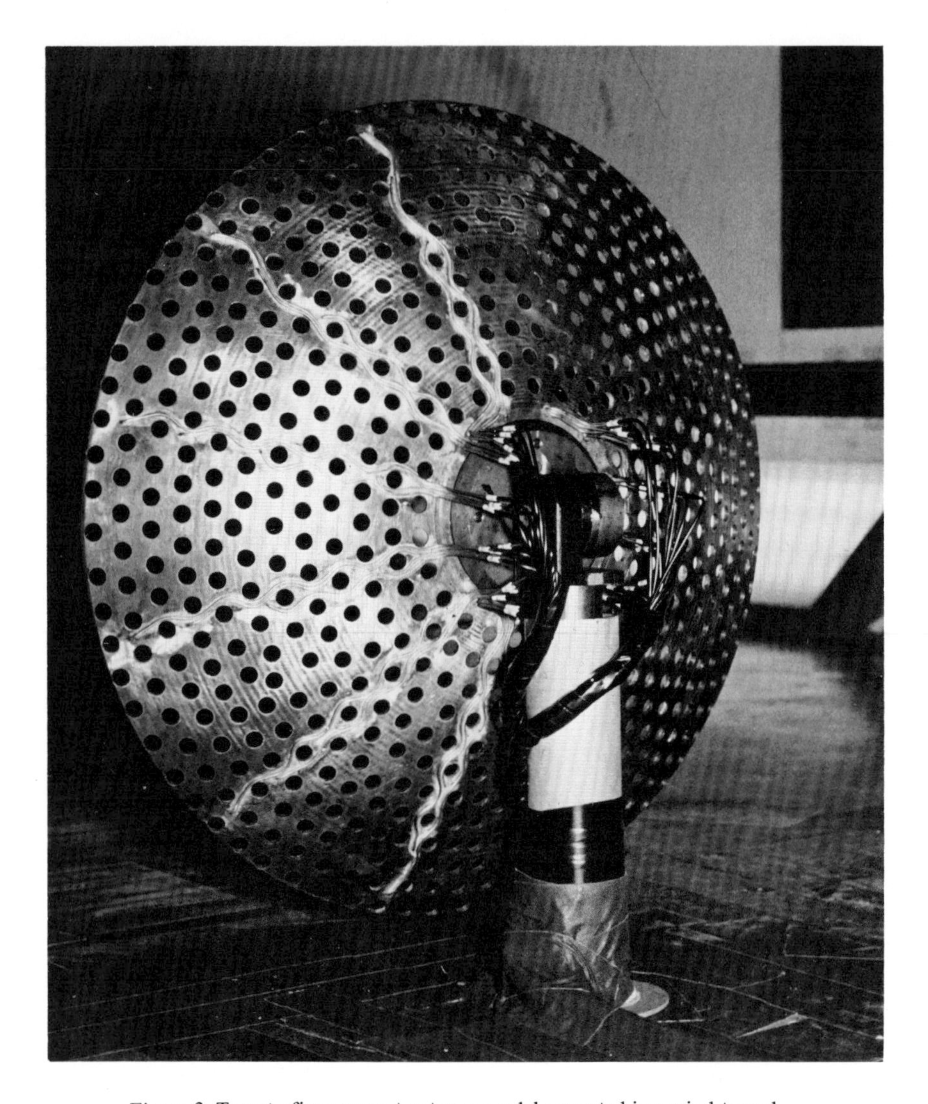

Figure 2. Twenty-five per cent antenna model mounted in a wind tunnel.

actually a box with triangular trusses in the top and bottom surfaces.

Figure 4 illustrates the method used to compute deflections of a reflector consisting of quadrants that are structurally symmetric to each other. Only data for one quadrant, as shown shaded, is required. Applying the rules for the arrangement of loads as outlined by Newell[12] in 1939 for symmetric structures, asymmetric wind and temperature loads about the YY plane

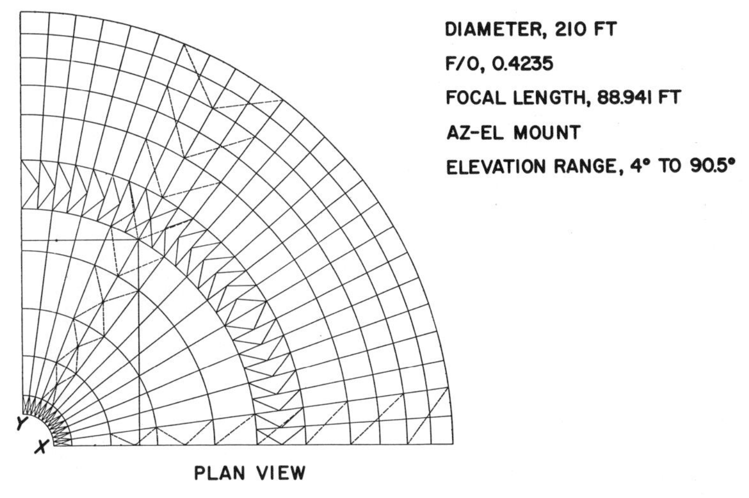

Figure 3. Quadrant symmetric 210-ft-reflector structure.

require antisymmetrical-type restraints (Y and Z direction restraints) in the YY plane. Symmetric loadings about the XX plane require symmetrical restraints (Y restraints) in the XX plane as shown.

If one air load L_s is applied to a joint in the first quadrant as shown in section XX with symmetrical-type restraints on both the XX and YY planes, the computed deflections for the data for one quadrant would simulate a complete reflector with four symmetrically located loads L_s. With one load L_a applied to analysis of one quadrant using restraints as pictured in Figure 4, the simulated complete reflector will be loaded with L_a in the positive direction in quadrants I and IV and L_a in the negative direction in quadrants II and III.

When the results of the above loads L_a and L_s are summed, the computed deflections will be for the asymmetric loading case of $+ L_s$ plus $+ L_a$ in the first and fourth quadrants. The $+ L_a$ and $- L_s$ in the second and third quadrants cancel each other.

The asymmetric temperature loads can be handled in the same fashion Also, one can proceed to the analysis of one load in any quadrant by adding the four answers from the four different restraint conditions for the structural data for one quadrant. Bar stresses may be computed by the same method.

3. The Contour Maps

The specifications for the antenna selected the orientation of the reflector structure for the worst case of the combined environmental loads. The selection

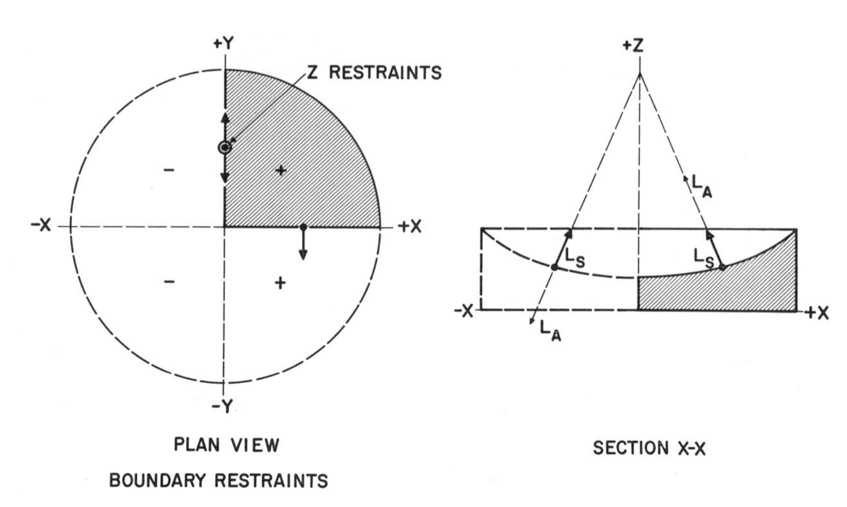

Figure 4. Temperature and wind loadings diagram. Quadrant symmetric structure.

resulted in the sum of the gravity deflection with the antenna pointing at the horizon added vectorially to the thermal load from the sun from one side, plus the 30 mph 120° yaw wind load from the same side. To illustrate the contribution to the total distortion by each load, the STAIR computed individual results; the summed results will be illustrated with contour maps.

All of the contour maps are duplicated exactly as produced by the SC-4020 plotter connected to the 7094, except for the addition of enlarged numbers showing the values of the contour levels, the enlarged zero (0.0) level lines, and the special titles for slide projection reasons. The ½ rf path-length error levels are organized by a plotting subroutine in the rms program[10] by totally restraining the fitting paraboloid. Complete circular data were generated to show the contours as clearly as possible.

Figure 5 shows the STAIR computed deflections in the ½ rf path-length errors to the original paraboloid. The captions note the temperature and wind conditions.

Figure 6 shows the vectorially added deflections summed for the three environmental loads. The left contour map shows the summed deflections and the right contour map shows the errors to the best-fit paraboloid, which has changed its focal length. It should be noted here that it is the writer's opinion that the wind loads used in the design were conservative because of the aerodynamist's decision to apply the pressure cofficients to the total area rather than to a reduced area which accounted for the porosity.

The deflections for gravity loads were analyzed by first dividing the gravity force into two components. One component was designated parallel to the

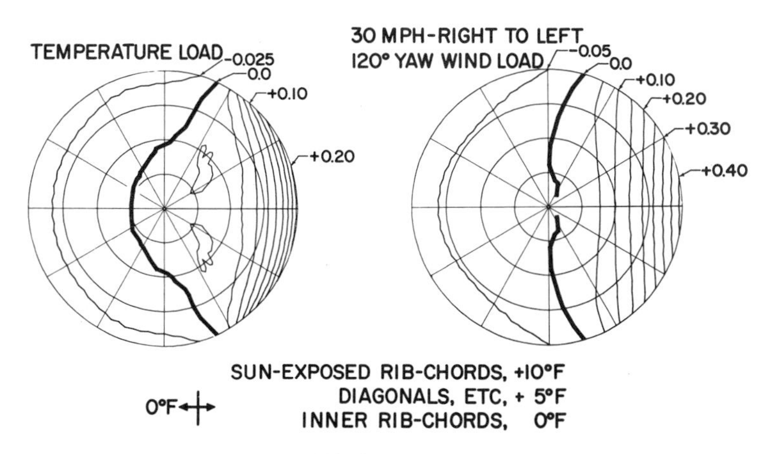

Figure 5. Contour maps of ½ rf path-length errors (in.). STAIR outputs.

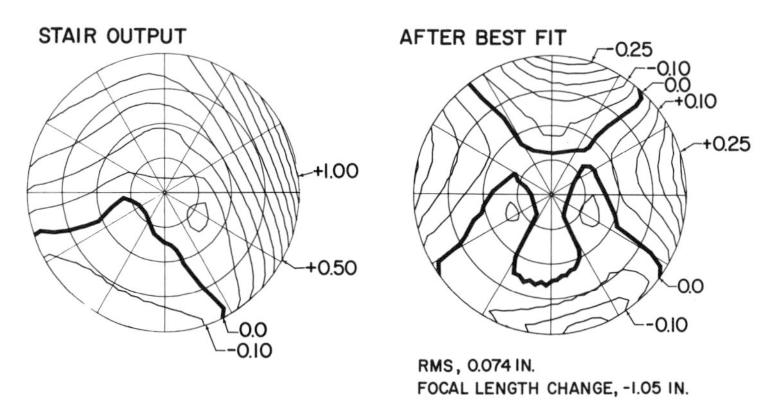

Figure 6. Contour maps of ½ rf path-length errors (in.). Temperature, 30 mph wind, and horizon-look gravity loads vectorially combined.

symmetric axis of the paraboloid and the other at right angles to it. As shown in Figure 7, the maximum value of the first component, designated as "ZEN," is 1.0 g when the antenna is pointed at the zenith. The other component value is zero when the first is a maximum and increases to 1.0 g when the antenna is pointed at the horizon. The maximum value of this component was named "HOR." At 45° elevation, both components have a value of 0.7 g.

It follows that, if the surface panels are set to a perfect paraboloid shape at 45°, the final deflection vectors will be the summation of the incremental deflection vectors of the two gravity load components. Thus, with the panels set at 45°, the final deflection vectors with the antenna pointed at the horizon is equal to the sum of −0.7 ZEN plus 0.3 HOR.

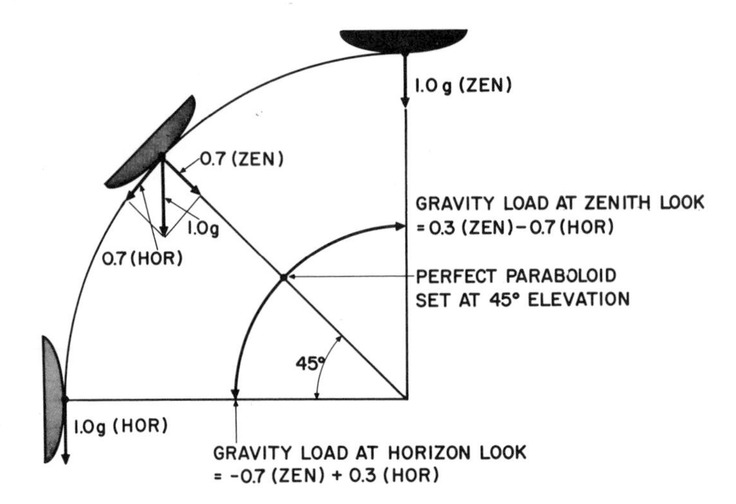

Figure 7. Gravity loading diagram for deflection computation.

The previous designations were made so that normal computer analysis will compute the deflections in trusses for the ZEN and HOR cases. Thus, for the ZEN case and data for one quadrant, the boundary restraints must be symmetric as required by rules for symmetric loads on symmetric structures.[12] The common description for this load case is "zenith look–gravity off-on." For the "horizon look–gravity off-on" or the HOR case, the restraints arrangement is shown on Figure 8 with the antisymmetric restraints in the XX plane as required by the above noted rules.

Shown also are the directions of the deflection vectors on each side of the antisymmetric restrained plane with the curvature change in the deflections above and below the same antisymmetric restrained plane.

4. The Prediction of the Deflections

Based on the data presented and developed to this point, definite predictions on the shape of the deflections for the individual HOR and the ZEN cases can be made. These predictions should then lead to a true picture of the actual deflections since ZEN and HOR are additive, as shown previously.

The predictions are as follows: (1) In the ZEN case, one can easily picture a deflection shape to be another paraboloid of longer focal length in an axisymmetric reflector structure. Neglecting the other requirements for antenna use, the distortion errors for the ZEN load can be reduced to a negligible value by a change in focal length of the paraboloid. (2) For the HOR case, the restraints arrangement prevents a change in the horizontal diameter. In fact, the parabolic shape across the section XX remains stable with this load case.

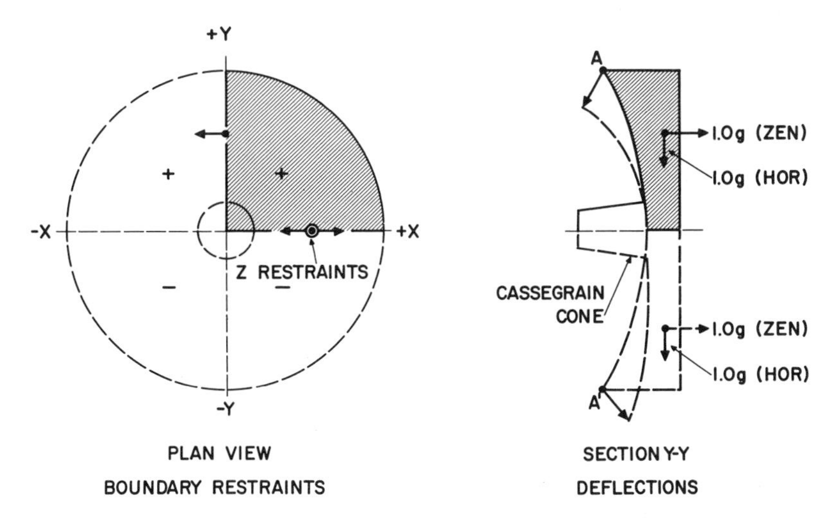

Figure 8. Antisymmetric gravity loading and deflection diagram. Quadrant symmetric structure.

In the vertical YY plane, the diameter also has not measurably changed because the deflection vector at A is equaled at A′ in the counter direction. Up to this point we have considered only the problem of compensating for changes in focal length in the ZEN case. However, it will be difficult to solve the problem of asymmetric deflections in the reflector surface. (3) Thus for the HOR case, the top half of the reflector surface deflects to a deeper concave shape while the lower half becomes less concave; this situation is shown by the dotted lines in Figure 8.

Prediction (3) plus the concentration of steel near the center hub of the radial rib type of reflector structure made it possible to mount a heavy Cassegrain cone on the vertex of the 210-ft. Radio-frequency considerations such as the mounting position of the feed horn about 42 ft from the vertex and short waveguide runs to the maser and transmitters are thus made possible.

Although it was obvious for some time that the bending moment reaction in the reflector holding the Cassegrain cone would tend to compensate for the deflection shape of the normal HOR load, actual computations to isolate this effect on the deflections were made only recently. STAIR analysis showed that although compensation did not occur the Cassegrain cone contributed only 0.006 in. rms for the HOR case and 0.001 in. rms in the ZEN case.

To provide a clearer deflection picture of the ZEN and HOR, let us examine the deflection pattern for the 210-ft ZEN case in Figure 9. On the left map, the STAIR deflections show that a near-ideal symmetric deflection pattern was obtained for a large part of the total area of the reflector. The low deflection areas that always occur around the elevation bearings are minimized.

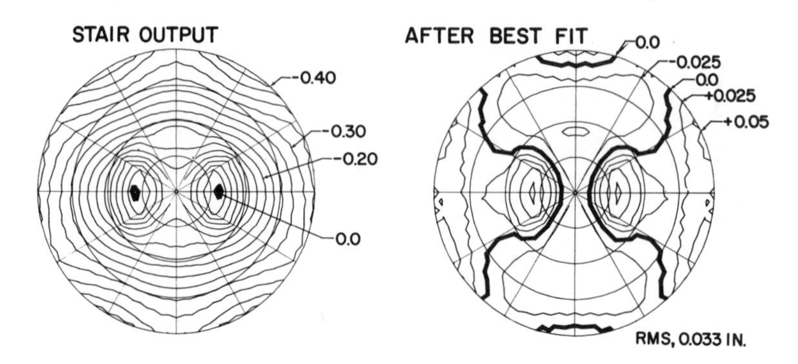

Figure 9. Contour maps of ½ rf path-length errors (in.). Zenith look--gravity "off-on."

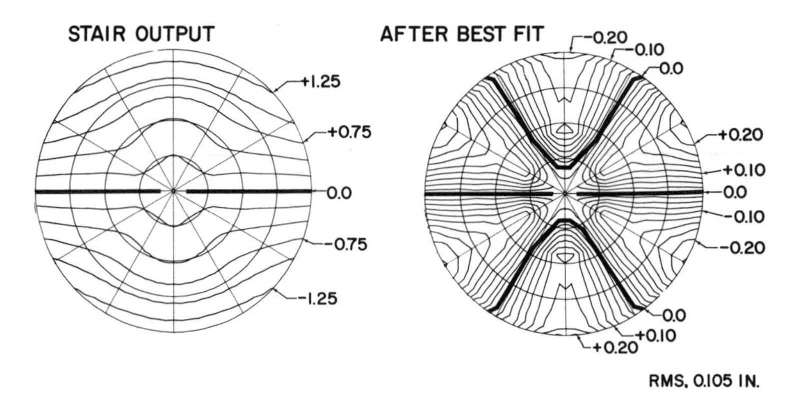

Figure 10. Contour maps of ½ path-length errors (in.). Horizon look--gravity "off-on."

The contour map "after best fit" shows small deviations or errors of 0.033 in. rms with a focal length change of + 0.98 in. The quadripod load (45,000 lb) is apparently absorbed into the heavy main girder structure with little effect on the surface errors.

The HOR case is shown on Figure 10. In the STAIR computed map on the left, the rotation component masks the deformation pattern. But, after best fit, the residual produces contour levels that clearly define the deflection pattern. On the vertical centerline, the greater concavity above the horizontal centerline and less concavity below is clearly pictured.

Peculiar to the 210-ft structure, there are deflections along the 45° lines caused by the moment reactions of the extended arms of the reflector structure. These extended arms were necessary as a result of clearances required for the master controlling unit on the elevation bearing centerline and for

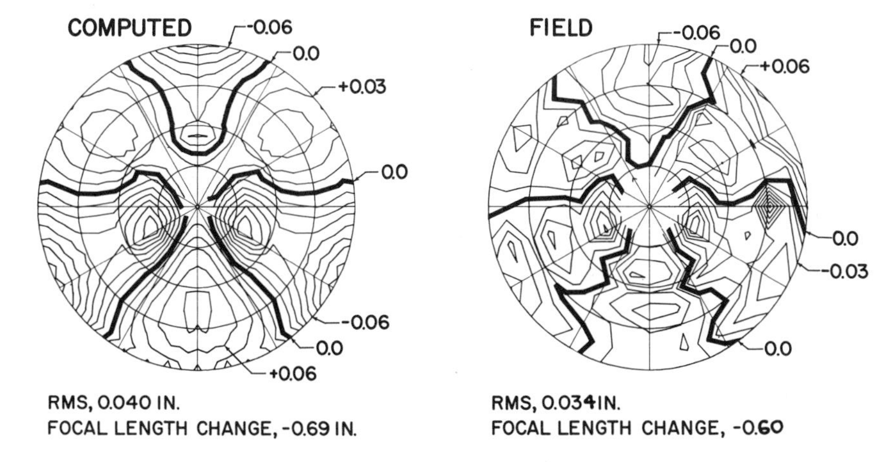

Figure 11. Contour maps of ½ path-length errors (in.), after best fit. Horizon look. Surface set at 45° elevation.

full elevation angle travel down to 4° of the horizon. Again the quadripod's concentrated loads do not cause local errors.

Figure 11 on the left shows the predicted computed deflections in terms of errors "after best fit," when the two cases of ZEN and HOR were added in the manner previously worked out, namely, the horizon look deflection with the panels set at 45° elevation = -0.7 ZEN + 0.3 HOR.

5. Correlation with Field Data

On the right of Figure 11 is the contuor map after best fitting the field data using only the differentials in the deflection readings resulting from the movement of the reflector structure from 45° to 6° elevation. The similarity of the deflection patterns is apparent. The field data were obtained by using a theodolite, 15 in. above the vertex, reading targets on the surface placed accurately in terms of arc lengths from the center. The targets were in close proximity to the surface panel mounting points and to the work points in the reflector truss structure.

The measured errors are presented in Tables 1 and 2 with estimated total surface errors for the 45°, zenith, and horizon looks. From the individual field readings, the conclusion might be reached that a higher elevation angle than 45° might have been selected for the setting of the panels. However, the distortions caused by the displacement of the rf feed due to the change in focal length of the reflector reverses the conclusion.

It should be noted that large equivalent distortion is caused by the rf feed movement or shift in the axial direction.

Figure 12 shows the equivalent distortion values for the two directions of

Table 1.

At 45° Elevation–Gravity Only		
Panels support position (R/J)[a]		0.019 best fit, rms
Panels manufacturing error (R/J)		0.035 std dev, rms
Subreflector mfg error (R/J)		0.027 rms
	Total	0.048 rms
At Horizon Look (6°)–Gravity Only		
Structure gravity deflection (R/J)		0.034 best fit, rms
Surface at 45°		0.048 rms
Effective focal length change (0.77 in.)[b](J)		0.058 rms
rf feed lateral shift (1.1 in.) (J)		0.020 rms
	Total	0.085 rms

[a] R/J, Rohr measured/JPL monitored;
J, JPL computed or estimated.
[b] Not refocused.

Table 2.

At Horizon Look		
With 10°F Thermal + 30 mph Wind + Gravity		
10°F + 30 mph Wind + Gravity (J)		0.074 best fit, rms
Surface error at 45°		0.048 rms
Effective focal length change[a] (J)		0.080 rms
rf feed lateral shift (J)		0.040 rms
	Total	0.126 rms
At Zenith Look–Gravity Only		
Structure gravity deflection (R/J)		0.063 best fit, rms
Surface error at 45°		0.048 rms
Effective focal length change (0.31 in.)[a] (J)		0.023 rms
rf feed lateral shift (J) (2.8 in.)		0.050 rms
	Total	0.097 rms

[a] Not refocused.

shift as calculated by the JPL/Radiation Program[13] and the application of the Ruze equation.[14] A cam-driven servo drive could compensate for the gravity load case by driving the subreflector to refocus. However, a more exotic device would be needed to sense changes in cuvature of the reflector due to the wind or temperature loads and to compensate for the focal length. A computer working with wind speed, direction, and the sun position could also supply the correction data.

With the availability of the displaced position of the best-fit paraboloid, it followed that rf boresight direction could be computed by ray tracing as shown in Figure 13. It was only necessary to bounce the ray from the deflected position of the rf feed horn off the vertex of the best-fit paraboloid.

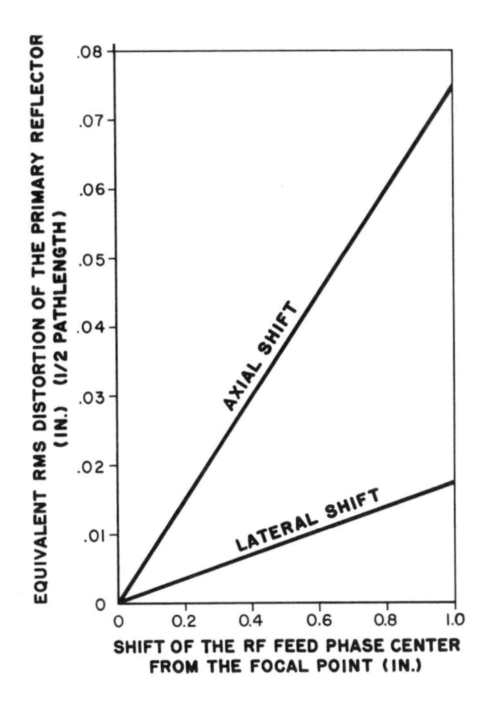

Figure 12. Two hundred ten-ft antenna, 2300 MHz frequency. Root mean square (equiv) versus rf feed point shift.

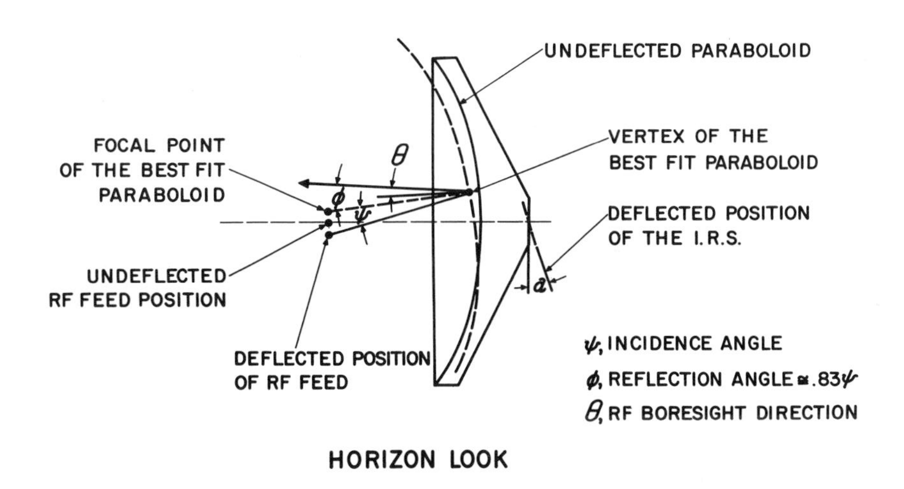

Figure 13. Ray tracing for determining rf boresight direction.

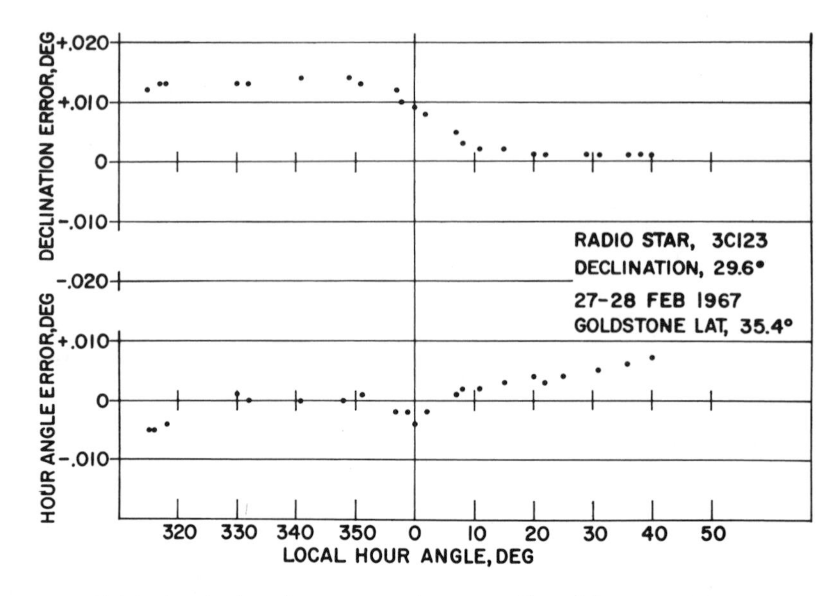

Figure 14. Two hundred ten-ft-antenna total system--rf boresight error and field measurement.

The computations made during the design period by Rohr engineers resulted in low rf boresight errors due to gravity loads and steady wind loads. This was confirmed by field tests as shown by a star track graph in Figure 14. The rf boresight direction was determined by a bracketing method using the sides of the main listening lobe. As the radio star was continuously tracked by the master controlling unit (master equatorial mount), the deviations were read from the encoders of the master equatorial mount. Its pointing error, in its present stage of adjustment, is less than 24 sec of arc (peak).

The above field data include any boresight shifts due to normal winds from the sides if it was windy on the particular day. However, the Servo Group, with limited test periods assigned to them, has not yet taken wind into account. In fact, they ignore the common everyday winds. Compensation in rf boresight direction seems to occur for wind loads similar to the gravity loads. Limitedfield data seem to show little wind effect on the rf boresight directions.

6. Summary

In summary, the field rf performance of the reflector can be gauged by everyday mission operational data listening to and commanding Mariner IV, Pioneer VI and VII. The AGC voltage of the receiver varies a maximum of 0.3 dB throughout a pass from horizon to horizon for the ordinary operation day. This decibel difference includes the change in maser calibration.

Acknowledgments

The development of the techniques was made possible only by full support and encouragements from R. Stevens, W. D. Merrick, and Phillip Potter. Appreciation is expressed to R. Bamford and B. Wada for structural analysis methods, to K. Bartos for the STAIR computations, to R. Jirka and L. Schmele for the computer program developments, to W. Peterschmidt for the Star Track Data, and to C. Valencia and the Rohr engineers for the field reflector measurements.

This paper represents one phase of research carried out or the Jet Propulsion Laboratory, California Institute of Technology, under Contract NAS7-100, sponsered by the National Aeronautics and Space Administration.

References

1. B. Wada, "Stiffness Matrix Structural Analysis," Technical Report No. 32-774, Jet Propulsion Laboratory, Pasadena, California (31 October 1965).
2. "STAIR (Structural Analysis Interpretive Routine)", Lincoln Manual No. 48, MIT Lincoln Laboratory, Lexington, Massachusetts (March 1962).
3. N. L. Fox and B. Dayman, Jr., "Preliminary Report on Paraboloidal Reflector Antenna Wind Tunnel Tests", Internal Memorandum, JPL CP-3, Jet Propulsion Laboratory, Pasadena, California (28 February 1962).
4. N. L. Fox, "Load Distributions on the Surface of Paraboloidal Reflector Antennas," Internal Memorandum CP-4, Jet Propulsion Laboratory, Pasadena, California (July 1962).
5. N. L. Fox, "Experimental Data on Wind-Induced Vibrations of a Paraboloidal Reflector Antenna Model," Internal Memorandum CP-5, Jet Propulsion Laboratory, Pasadena, California (January 1963).
6. R. B. Balylock, "Aerodynamic Coefficients for a Model of a Paraboloidal-Reflector Directional Antenna Proposed for a JPL Advanced Antenna System," Internal Memorandum CP-6, Jet Propulsion Laboratory, Pasadena, California (1 May 1964).
7. R. B. Balylock, B. Dayman, Jr. and L. N. Fox, "Wind Tunnel Testing of Antenna Models," from conference on "Large Steerable Radio Antennas—Climatological and Aerodynamic Considerations," *Ann. N. Y. Acad. Sci.* **116**, 1 (26 June 1964).
8. S. Utku and S. M. Barondess, "Computation of Weighted Root-Mean-Square of Pathlength Changes Caused by the Deformations and Imperfections of Rotational Paraboloidal Antennas," Technical Memorandum 33-118, Jet Propulsion Laboratory, Pasadena, California (March 1963)

9. H. Christiansen, "RMS-Paraboloid Fitting Program," Western Development Laboratories, Philco Corporation, Palo Alto, California (September 1964).
10. M. S. Katow and L. W. Schmele, "Antenna Structures: Evaluation Techniques of Reflector Distortions,"JPL Space Programs Summary No. 37-40, Jet Propulsion Laboratory, Pasadena, California (31 August 1966) Vol. IV, pp. 176-184.
11. L. Schmele and M. S. Katow, "UTKU/Schmele Paraboloid RMS Best-Fit Program," A. Ludwig, Ed., part of Technical Report No. 32-979, Jet Propulsion Laboratory, Pasadena, California (15 April 1967).
12. J. S. Newell, "The Use of Symmetric and Anti-Symmetric Loadings" *J. Aeronaut. Sci.* (1939).
13. D. Bathker, "Radiation Pattern Programs," A. Ludwig, Ed., part of Technical Report No. 32-979, Jet Propulsion Laboratory, Pasadena, California, Ed. (15 April 1967).
14. J. Ruze, "Physical Limitations on Antenna," Technical Report 248, Research Laboratory of Electronics, Massachusetts Institute of Technology, Cambridge, Massachusetts (30 October 1952) (ASTIA/AD-62351).

T. G. Butler
NASA, Goddard Space Flight Center
Maryland

ROSMAN I DYNAMIC ANALYSIS

This paper is devoted to dynamic analyses of the Rosman I antenna. The Rosman antenna has an 85-ft-diameter reflector and is used in collecting data from earth-orbiting satellites. A satellite will be in view anywhere from 5 min to 30 min only. Relatively fast changes of antenna position are needed to track the satellite. Consequently dynamics can be an important factor in the antenna design. The Rosman antenna was designed principally on the basis of static considerations. In operating the antenna, an error characteristic was found in signals from the control circuits which indicated the need for structural vibration information. It was suspected that the structure was rich in natural frequencies over the control system band of 0–30 Hz, but to find out how rich and how responsive, seemed to the control engineers at the time to be an impossible analytical task.

Fortunately a computer program SB-038 was at hand, incorporating finite element statics and Givens–Householder dynamics. Our experience in solving large dynamics problems, at the time that we were approached to do the Rosman analysis, had only been in the 300 to 400 degree-of-freedom range. Our confidence was growing sufficiently in the Givens–Householder eigenvalue routine that we bargained to do the Rosman analysis as a 1000 degree-of-freedom problem. However, we tailored the problem down to 645 dynamic degrees-of-freedom. It was not just the Givens–Householder theory that made it possible for us to attack this problem but also the skillful management of this theory on the computer by the Martin Company programmer, Thomas L. Clark.

The major part of the analysis, of course, was the eigenvalue problem from which we could volunteer to tackle three related dynamic problems: (1) transfer function spectrums between drive gears and feedbox, (2) transfer function spectrum between drive gears and bore sight, and (3) dynamic history at any point of the structure, due to the drive system inputs from satellite-tracking commands.

Figure 1. Rosman I tracking antenna.

This report will treat item (1) only, but anticipation of items (2) and (3) influenced our structural modeling.

The arrows in Figure 1 identify the focal points of interest pertaining to the transfer function problem. The lowest arrow indicates the drive system for the X wheel; next above, the arrow indicates the drive system for the Y wheel and the highest arrow indicates the feedbox. The antenna was stroked sinusoidally at the X wheel bull gear or the Y wheel bull gear with a steady state amplitude of 1-lb force and the steady state response was measured in inches at the feedbox.

The Martin Company SB-038 program solves the finite element problem by the force method and has these limitations: (1) a maximum of 1225 primary elements, (2) a maximum of 1225 redundant elements, (3) a maximum of 1225 dynamic degrees of freedom, and (4) all relations must be linear.

With this extensive structure the program's limitation could rapidly be exceeded. Fortunately SB-038 is provided with the feature of structural partitioning. In effect, this allows the large structure to be solved in segments confined to the static limitations of the program. These segments can be combined after the influence coefficients are obtained for all interior points in terms of all interface and constraint points.

The anatomy of the analysis is a nine-part affair: (1) six individual static solutions of the segments, (7) combining the six static segments into a single large influence coefficient matrix for the entire structure, (8) eigenvalue solution of the reduced model and (9) transfer function solution of the reduced model.

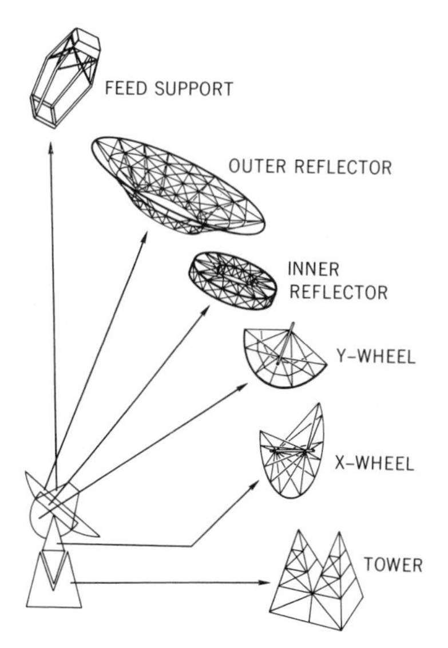

Figure 2. Rosman I static analysis segments.

Figure 2 displays the static analysis segments into which the antenna was divided: earth, pilings, concrete and tower first, X wheel next, followed by the Y wheel, then the parabolic reflector divided into an inner and outer portions, and finally the feedbox.

A catalogue of the analysis is given in Table 1. The headings are: segment

Table 1. Rosman I Analytical Model.

Segment	No. grid points	Degrees of freedom	Elastic elements			Solution time (min)
			Total	Redundant	Primary bending	
Tower	92	267	317	50	20	15
X wheel	103	290	358	68	56	30
Y wheel	52	141	165	24	15	10
I. reflector	324	700	722	22	0	30
O. reflector	168	632	800	168	96	90
Feedbox	130	395	429	34	20	20
Total static	869	2425	2791	366		195
Dynamic	215	645				900

identification, number of grid points, degrees of freedom, elastic elements and solution time.

The segments vary in complexity from the Y wheel with 52 grid points, 141 degrees-of-freedom, and 165 elastic elements to the outer portion of the reflector with 218 grid points, 632 degrees of freedom, and 800 elastic elements. The sixth column is added to give a kind of measure of how markedly the segment departs from a true truss. It gives the number of bending elements that are necessary to include in the model to obtain a well-conditioned primary structure. Proportionately the X wheel is the poorest truss with 56 of its 290 primary elements required to be bending. The effect of both the amount of bending and number of redundants can be seen by comparing the same size. The X wheel takes twice as long to solve as the tower. The real impact of redundancies is seen in the outer reflector segment. To solve structures with a large number of redundants is a costly operation by means of the force method. The inner reflector of about the same size, but a perfect truss and practically no redundants takes 1/3 as long as the outer reflector for solution.

When the total structure was put together it embraced 869 grid points, 2425 degrees of freedom, and 2791 elements. The assembly of the 2791 x 2791 influence coefficient matrix from the six segments requires an additional 4 hr. The most costly time was the eigenvalue solution of the coarser-meshed model of 645 dynamic degrees of freedom. More than 14 hr were consumed in, first, transforming the matrix to tridiagonal form; second, solving the eigenvalues; third, determining the eigenvectors, and fourth, making orthogonality checks. The transfer function program and the structure cutter took an additional 2 hr which totals $24^1/_3$ hr for a complete solution without mistakes from statics through the transfer function.

To provide an idea of the difficulty with the X wheel, a detailed view is given in Figure 3. Heavy black lines represent the members with large cross-

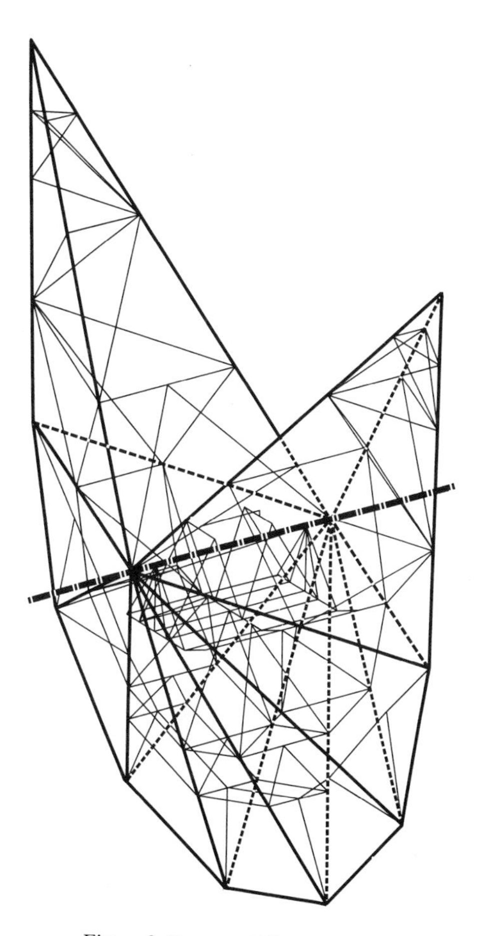

Figure 3. Rosman I X wheel.

sectional area. There is a pair of triangles in the horizontal plane sharing the axle for their common base with spokes radiating out to the vertex at the bull gear on the rim. This is the substructure, which must carry the lion's share of the load to resist bending acting on the principal plane of the X wheel perpendicular to the axle and passing through the bull gear and Y axle bearings. The importance of this moment-carrying ability will be evident when the elastic behavior of the mode shapes is examined. The X wheel was sufficiently different from a true truss that it was necessary to use an automatic redundant selector routine in order to establish a well-conditioned matrix of primary elements.

An added difficulty in making a dynamic analysis of this nonaxisymmetric structure is that for every different mating position between the tower and the X wheel and between the X wheel and the Y wheel there is in effect a new elastic path and thus a new potential energy function in the dynamic equations. This implies an infinity of sets of eigenvalues, each set of which contains an infinity of its own eigenvalues. To avoid this monstrous condition it was assumed that an eigenvalue determination was sufficiently representative of the structure over an arc of 30° such that no appreciable error would result. Implementing this assumption required the determination of eigenvalues for each of the three positions, or a total of nine. But for purposes of brevity this paper will be confined to the zenith position of the antenna only.

The suspicions of the controls people about the richness of structural resonances were amply sustained. One hundred twenty-five modes were found in the 0–30 Hz range. Six of the more important modes are shown in Figures 4 through 10. They are X-axis foundation mode at 1.21 Hz, automatic plotter, X wheel out of plane mode at 1.36 Hz, feedbox torsional mode at 1.42 Hz, Y-axis foundation mode at 1.51 Hz, Y-direction foundation mode at 2.84 Hz, longitudinal foundation mode at 3.12 Hz, and reflector ripple mode at 21.28 Hz.

Because of the nature of the control system we can ask four things about the structural behavior in the modes: (1) In how many of these modes is the participation of the feedbox important? (2) In how many of these modes is the participation of the reflector important? (3) In how many of these modes is the participation of the Y wheel important? (4) In how many of these modes is the participation of the X wheel important?

We will eliminate No. 2 immediately because we are in no position to assess the importance of reflector deformation until radiation analyses are able to predict the boresight pattern from nonclassical geometric shapes. This is a current research project at Goddard Space Flight Center. The structural aspects of this problem will be investigated in about a year and a half.

Arbitrarily this paper will ignore Nos. 3 and 4 in the interests of time. We confine our attention to the feedbox problem only as being representative of the detailed interpretations that are necessary to advise control systems analysis regarding dynamics of structures.

The amplitude at the feedbox due to gear drive forcings is a function of

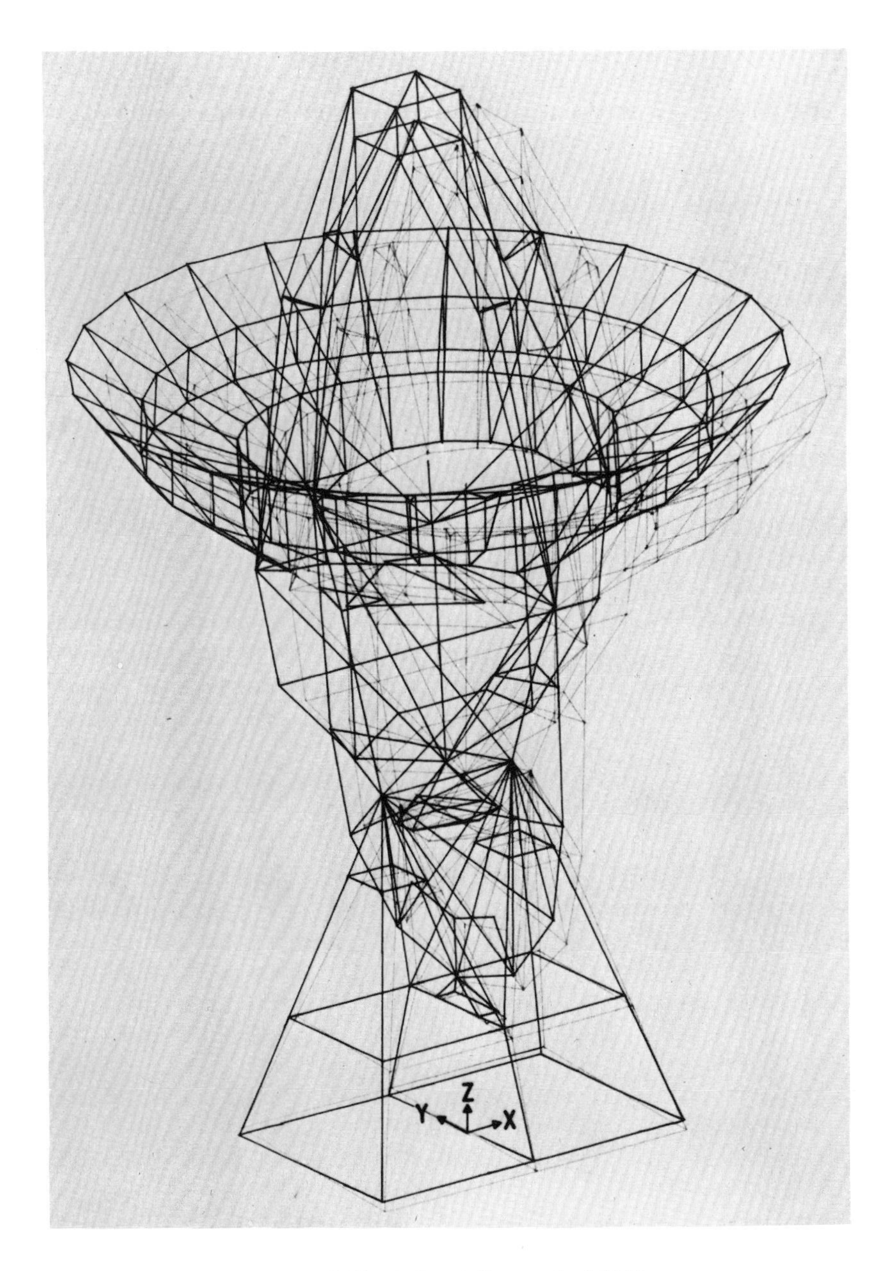

Figure 4. X axis foundation mode. 1.21 Hz.

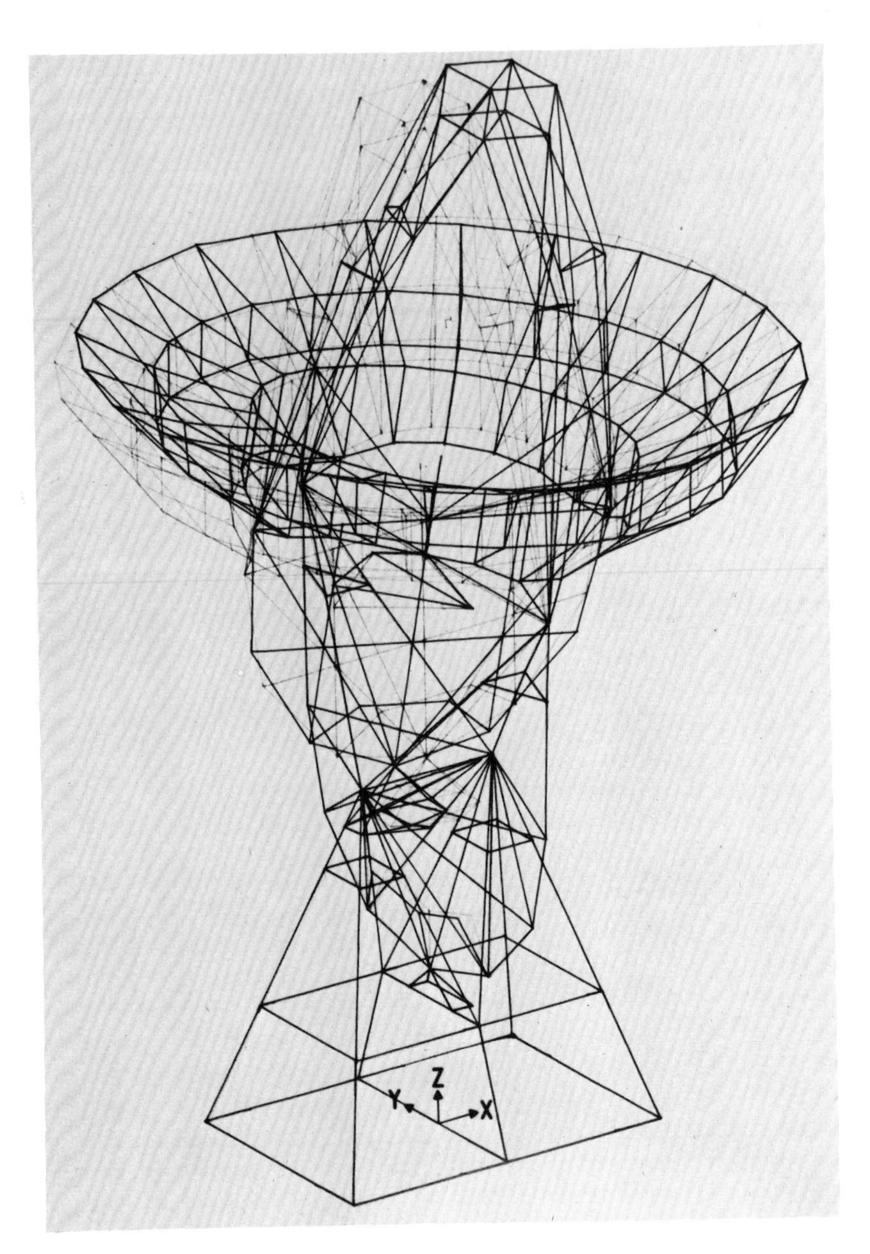

Figure 5. X wheel out-of-plane mode. 1.36 Hz.

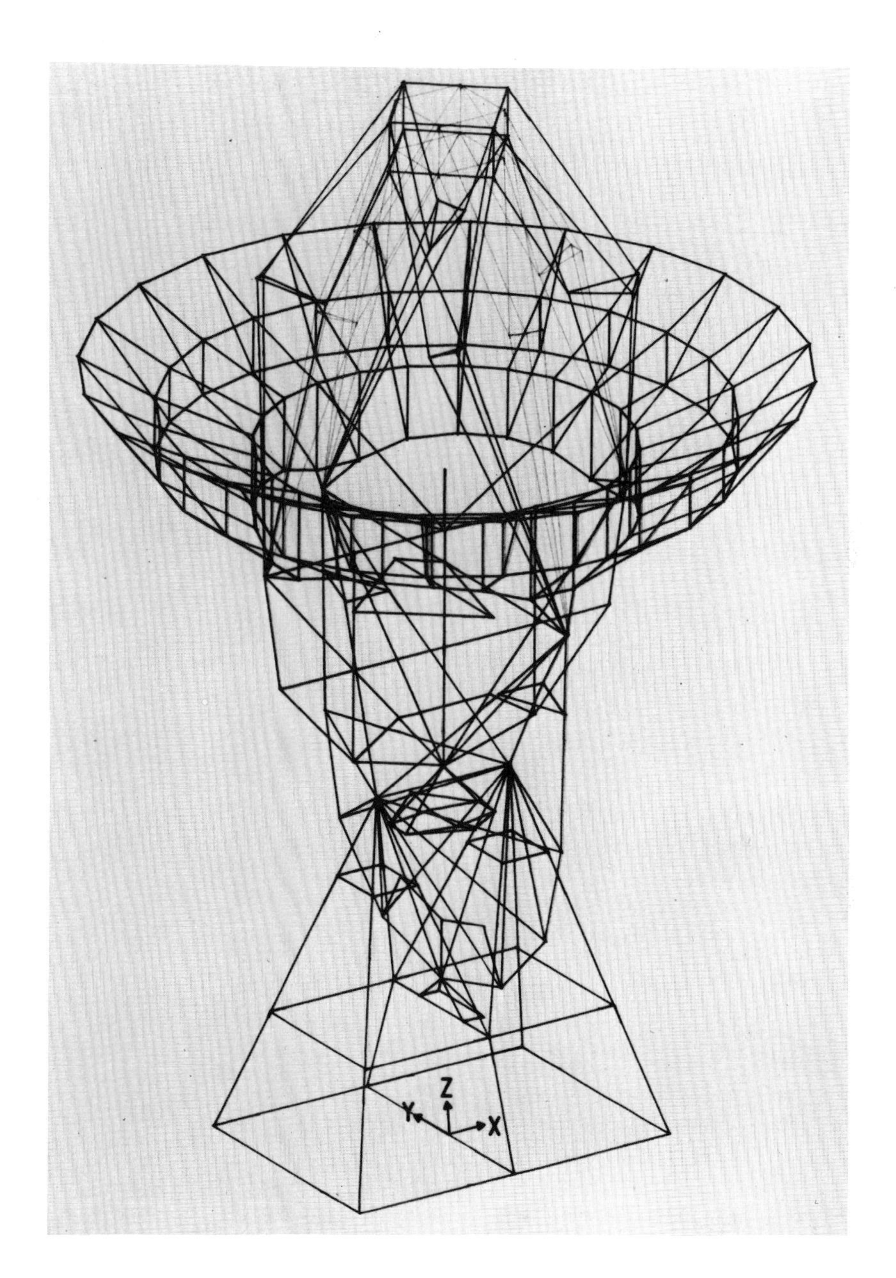

Figure 6. Feed box torsional mode. 1.42 Hz.

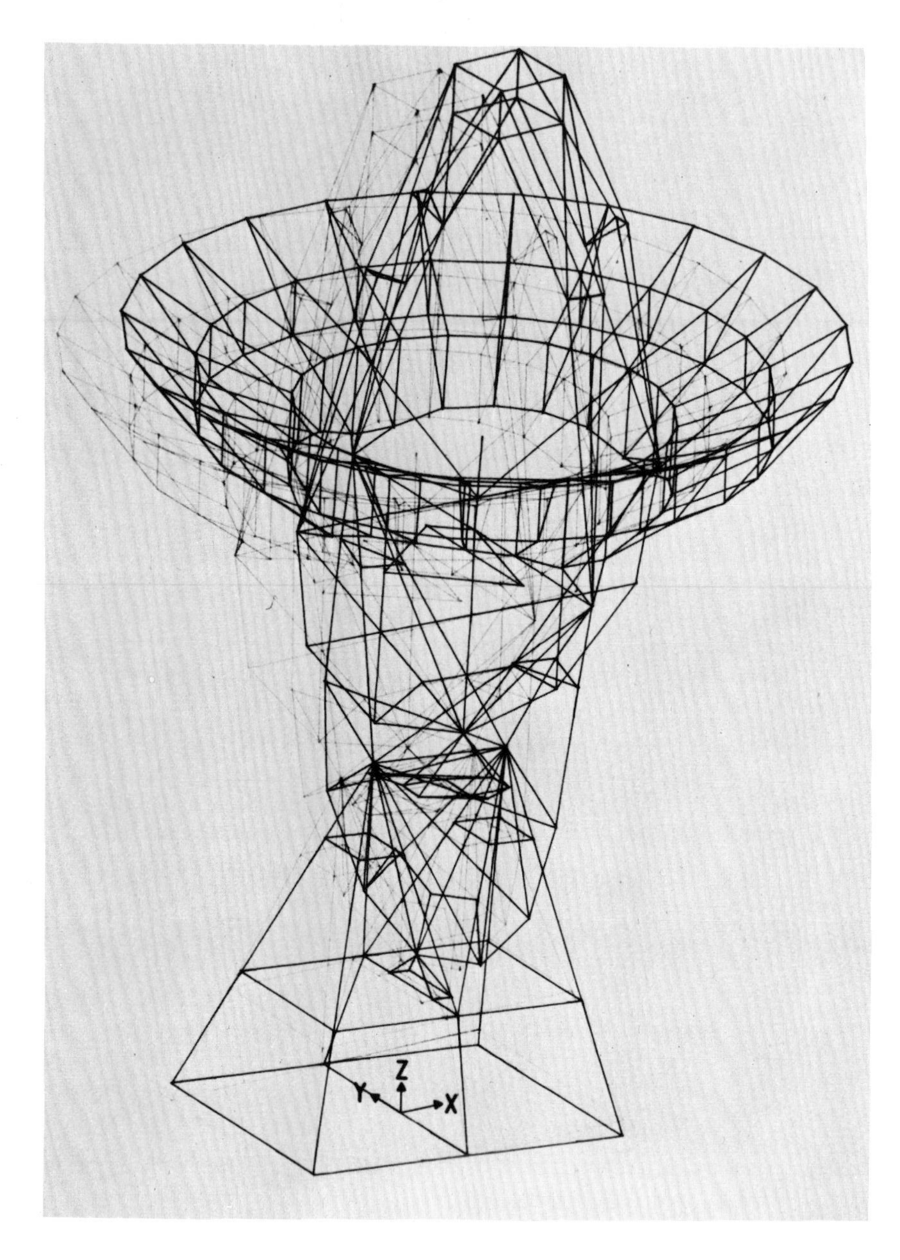

Figure 7. Y axis foundation mode. 1.51 Hz.

Figure 8. Y direction foundation mode. 2.84 Hz.

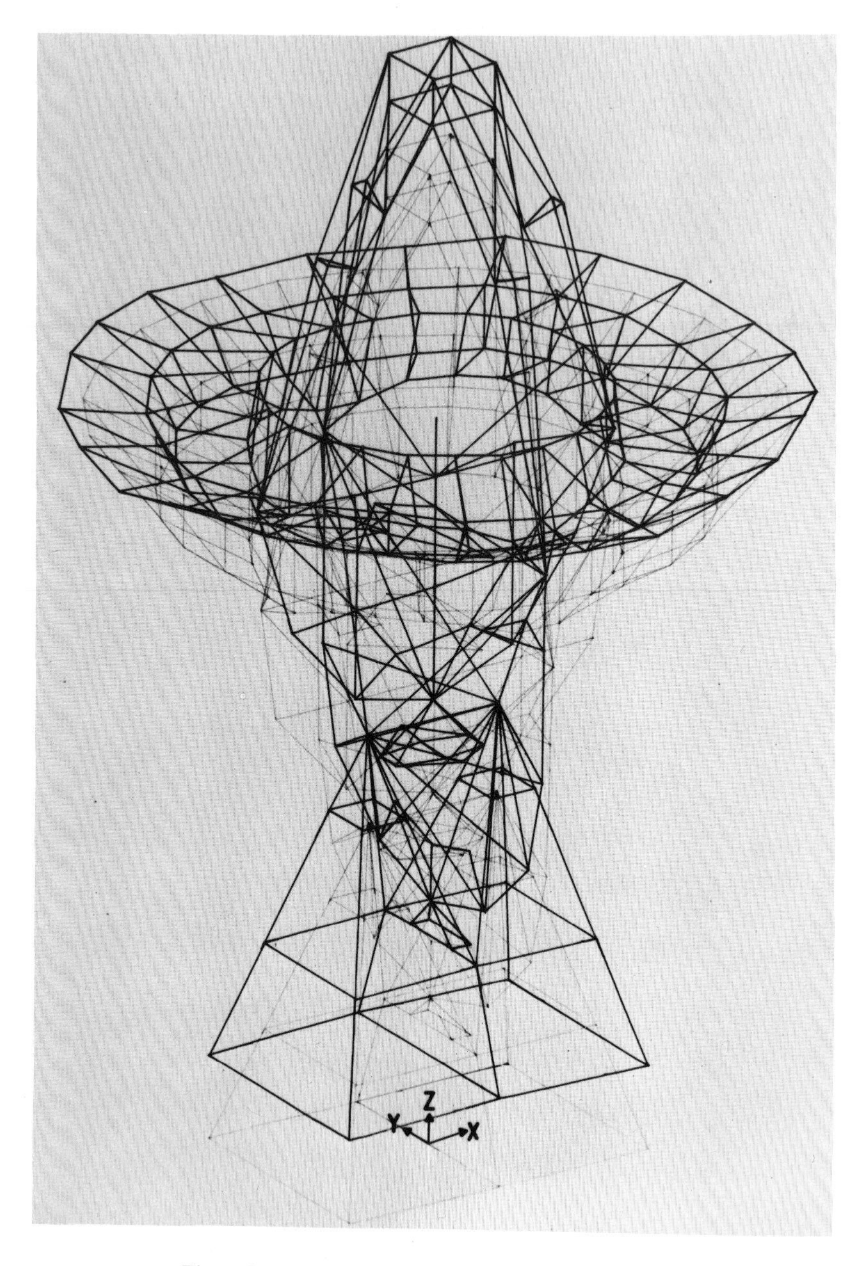

Figure 9. Longitudinal foundation mode. 3.12 Hz.

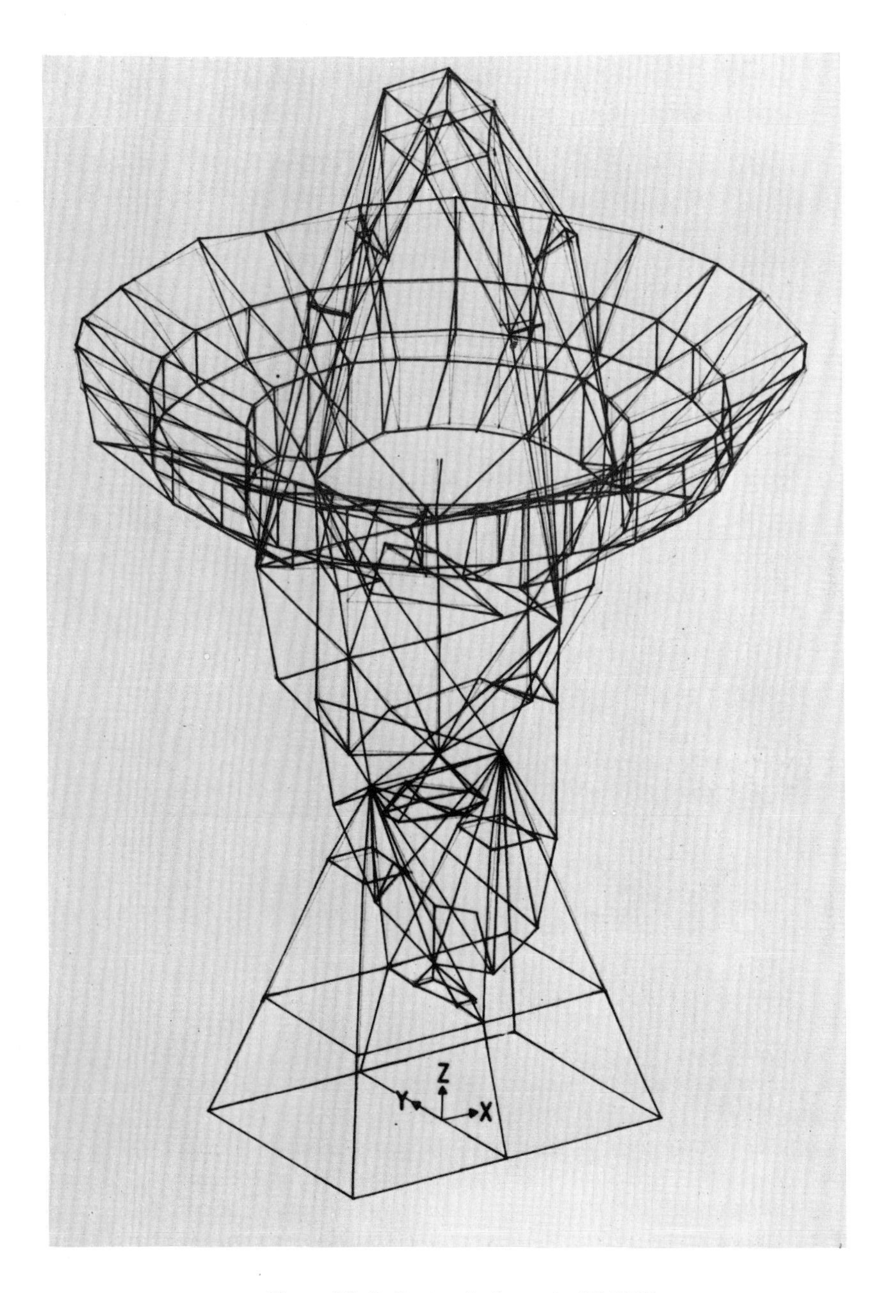

Figure 10. Reflector ripple mode. 20.28 Hz.

damping. Lacking any experience with bolted structures, we were fortunate to obtain experimental data on damping versus frequency. Data were reduced to critical damping and plotted as in Figure 11. Notice that damping is greatest

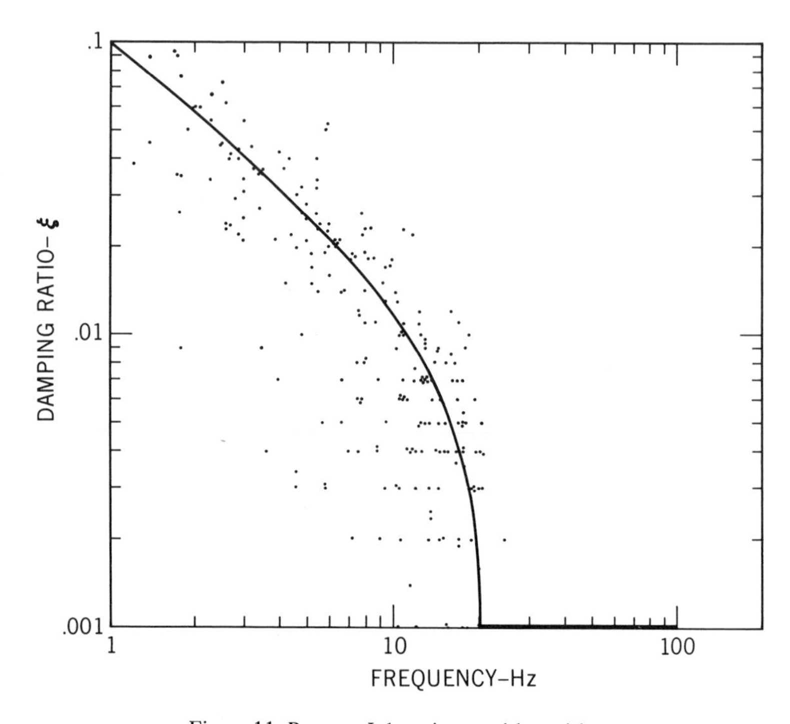

Figure 11. Rosman I damping, zenith position.

in the lower-frequency band and diminishes rapidly to a low level in the higher modes.

Transfer functions were obtained by a program distinct from SB-038. The output of SB-038 is the collection of modal data representing the dynamic characteristics of the structure. Transfer functions are obtained by taking the Laplace transform of the differential equation and solving the resulting set of simultaneous algebraic equations in matrix form.

Figure 12 and 13 are the spectra of the feedbox structural transfer functions

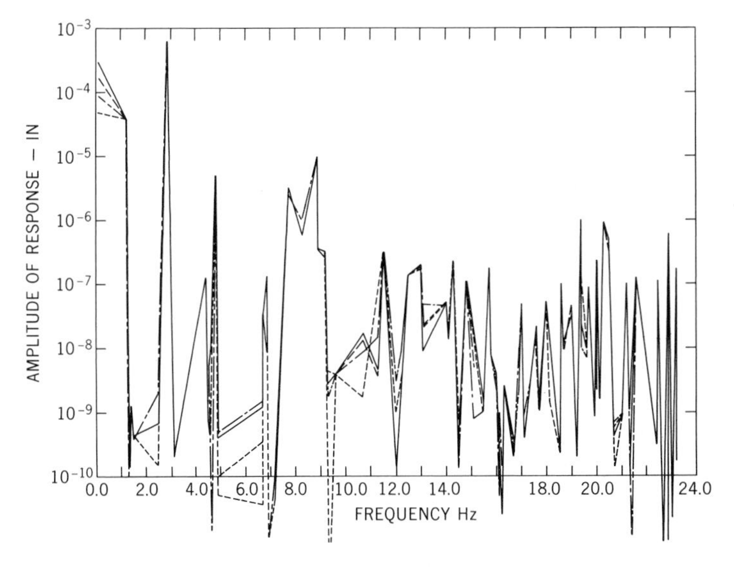

Figure 12. Rosman I transfer spectrum (X-wheel forcing).

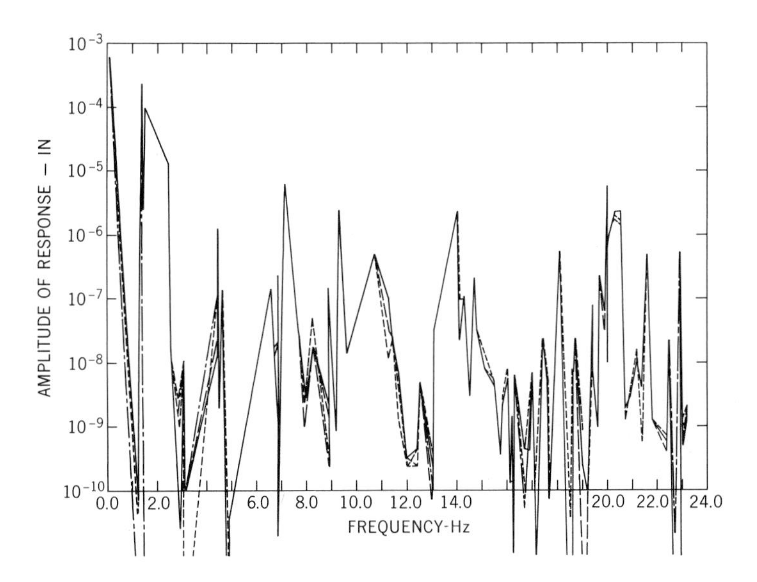

Figure 13. Rosman I function spectrum (Y-wheel forcing).

in inches due to 1-lb forcing amplitudes at the X gear and Y gear, respectively. Notice that except for a spike at 14 Hz and a cluster at 20 Hz the important responses take place in the 0–11 Hz range.

Table 2. Rosman I Summary of Mode Types.

Type of mode	Mode	Freq. (Hz)	Order of magnitude
Translation in X direction	3	1.35	10^{-4}
	5	1.51	10^{-4}
	6	2.45	10^{-5}
	12	4.44	10^{-6}
	14	4.65	10^{-7}
	17	6.70	10^{-6}
	23	7.16	10^{-5}
	30	9.30	10^{-6}
	32	10.7	10^{-5}
	41	14.0	10^{-6}
	77	20.0	10^{-6}
	82	20.5	10^{-6}
Translation in Y direction	2	1.21	10^{-4}
	8	2.84	10^{-4}
	15	4.82	10^{-5}
	25	7.77	10^{-6}
	27	8.89	10^{-5}
	38	12.5	10^{-7}
Rotation about Z axis	4	1.42	10^{-6}
	7	2.55	10^{-7}
	9	3.09	10^{-8}
	18	6.75	10^{-8}
Longitudinal in Z direction	10	3.12	10^{-8}
Skew translation (in X–Y plane)	11	4.43	10^{-7}
	20	6.87	10^{-7}
	28	8.90	10^{-7}
	81	20.3	10^{-6}

Table 2 classifies these responses in terms of only how the feedbox quadripod behaves locally. At first one might think that torsional modes would not concern the control system, but because incoming signal sums and differences are taken, the torsional response consequently becomes important.

Pictorially this data is displayed in Figure 14 which is a bar graph of important responses with line links to those of the same type. In summary, if responses less than 10^{-5} in. are ignored the frequency range can be paired down from 30 to 20 Hz and the number of modes can be truncated from 125 to 15.

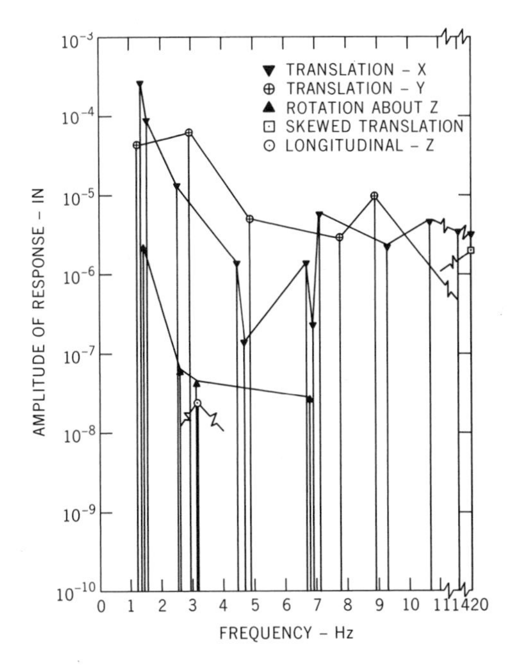

Figure 14. Rosman I transfer function spectrum summary.

It is apparent that finite element analysis by the force method is a useful tool in predicting the dynamic behavior of large complex structures. It is urged that, in the future, designers take dynamic factors into account earlier in the design of complex structures to avoid having to make modifications to existing structures that are liable to dynamic malfunction.

The author wishes to acknowledge the assistance of B. Bata and T. L. Clark of Martin Company, Michael Oien of Cornell University, and W. L. Cook of NASA.

M. H. Jeffery
Freeman Fox & Partners,
London, England

CONSTRUCTION AND PERFORMANCE OF THE 150-ft NRC ANTENNA AT ALGONQUIN RADIO OBSERVATORY, ONTARIO, CANADA

Introduction

The National Research Council of Canada has been operating a 150-ft radio telescope at the Algonquin Radio Observatory since May 1966. The telescope is generally very similar to the 210-ft telescope operated by CSIRO at Parkes, Australia, which was also designed by Freeman Fox & Partners. This paper is concerned only with the fabrication, erection, adjustment and survey of the telescope dish, and its actual structural performance compared with its computed behavior. The principal defference between the two telescopes, and the only one relevant to this paper, is that the 210-ft CSIRO dish is mainly mesh covered, only the central 54-ft diameter being plated, where the 150-ft NRC dish is plated out to 120-ft diameter, only the outer 15-ft radius being mesh.[1] A basic feature of the NRC dish construction, which we believe to be unique at this order of diameter, is that the plated reflector surface panels are connected together to form a single membrane which is then rigidly connected to the backup structure at zero zenith angle, so that it participates fully in resisting gravity stresses as the dish is tilted off zenith, together with wind and thermal stresses at any attitude. A sketch of the general arrangement is shown in Figure 1 and a photograph of the completed antenna is shown in Figure 2.

The solid reflector surface consists of ¼-in.-thick mild steel plate, stiffened at 1-ft–9-in. spacing with 6 x 2½ x $^5/_{16}$-in. angles, running circumferentially. The plates are bolted at 4-in. pitch to each other and to 36 radial purlins of

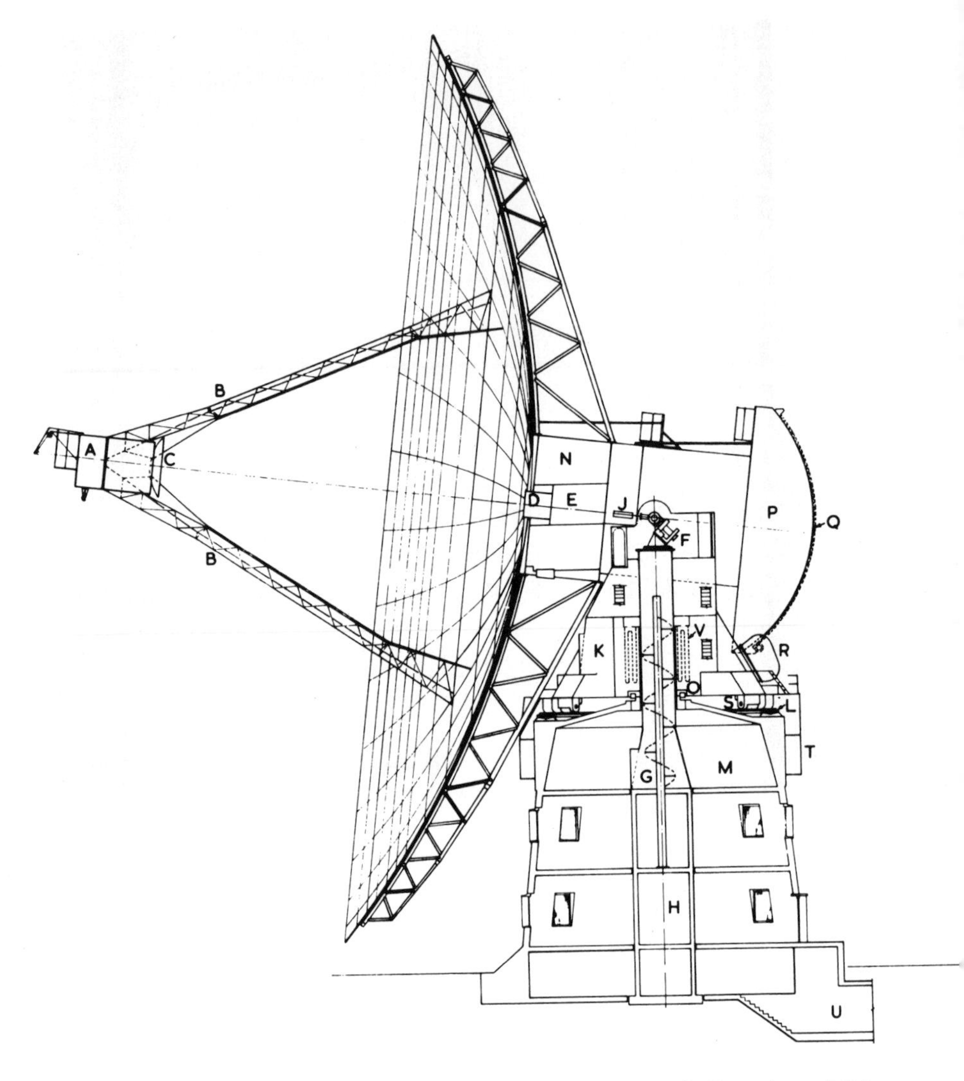

Figure 1. General arrangement. A–Aerial cabin. B–Tetrapod. C–Gregorian subdish. D–Vertex tube for Gregorian feeds. E–Receiver room. F–Master equatorial control unit. G.H–M.E. unit pillar. J–Error detector. K–Air-conditioning package space. L–Azimuth track. M–Concrete base tower. N–Cylindrical hub. P–Counterweight. Q–Altitude drive rack. R–Altitude gearbox. S–Azimuth roller and gearbox. T–Canopy. U–Tunnel to control room. V–Cable twister.

4 x 4-in. I section, formed to the required parabolic profile. Each purlin is then attached to a rib at its eight truss node points by means of a device combining the function of shear connector and height adjuster, accessible from the dish surface. The shear connection is supplied by two steel blocks

Figure 2. The completed instrument. June 1966.

that are clamped against the web of the purlin by two high-strength grip bolts.

The 36 ribs cantilever radially from a cylindrical steel hub, 23 ft in diameter. The ribs are interconnected with two ring girders at 66- and 48-ft radius (where the tetrapod feed support legs are attached) and with two bays of bracing around the rim of the dish, in the plane of the top chords of the ribs.

Structural material is mild steel throughout. Ribs, ring girders, and bracing are all of tubular construction, round or rectangular, and every connection is made by direct welding, no gusset plates being used.

Shop Fabrication

The hub (Figure 3) was fabricated for shipment in three sections only; at the points of attachment of the ribs, two strips were machined truly circular within 0.020 in., to provide an accurate starting point for the rib erection.

The rib trusses (Figure 4) were all fabricated on one shop jig. Top chord sections of 4½-in.-o.d. tube were first hot-formed to the parabolic profile,

Figure 3. Hub base section. Machining for rib root connection.

then butt-welded together in the jig. Bottom chords, made up from straight sections, $8\frac{5}{8}$, $6\frac{5}{8}$, and $4\frac{1}{2}$-in.-o.d. tubular with tapered reducers, were then jigged in place and butt-welded. The diagonal web members, their ends having first been shaped in a Steffan automatic tube-profiling machine, were then inserted and all joints between them and the top chord welded up. Since all these welds were to the underside of the top chord, it was initially set in the jig with an extra upward curvature of about 1/8 in. at each end, distorting to the correct profile after welding. The bottom welds on each diagonal were then made, jamming the diagonals into tight contact with the chord to minimize weld size and hence weld distortion. The result of this carefully devised welding sequence was that, after the ribs were released from the jig, peak error was 0.105 in. at any node point, 90 per cent of all node points were within 0.060 in., while half the ribs were within 0.025 in. at all nodes.

The I section purlins were first machined on one flange to a plane perpendicular to the web, to remove the effect of any twist in the section, then hot-formed to within 0.040 in. of the desired parabolic profile. Each purlin was then attached to a rib, with the latter still in the jig, the shear connector devices being then welded in place.

Figure 4. Rib fabrication in shop jig.

The ¼-in. reflector plates were rolled in both directions until, when laid on a template, they were within 1/8 in. of it at all points, under their own weight alone. The angle stiffeners for these plates were first rolled to approximate profile, then planed to within 0.005 in. of the desired shape. It was found that this could not generally be achieved with one cut, a second skim being needed, on most of the 1080 pieces. The stiffeners were not welded to the reflector plates but attached with special bolts having 1/8-in.-thick cheese heads, fitting into 1/8-in. counterbored holes, so as to leave a flush finish on the reflector surface. Each stiffened reflector plate was checked on another jig (see Figure 5), a point being measured on every square foot, all such points to be within 0.025 in. of theoretical profile. A few plates falling outside this tolerance were tweaked into shape by judicious application of small hydraulic jacks.

The mesh panels forming the outer portion of the reflector surface are very similar to those used on the 210-ft CSIRO dish. They were made up of sheets of hot-dip galvanized woven H. T. 19-gauge steel wire mesh, the galvanizing being carried out after weaving to ensure good electrical conductivity at all intersections. The mesh edges were then sheared to size, and selvedged with

Figure 5. Checking reflector panel.

galvanized steel strip. The panel frames, from 10-ft – 6-in. long to 13-ft long x 2-ft – 6-in. wide in the radial direction, were made up out of cold-rolled 12-gauge steel sections, the long sides being curved to the correct tangential profile. The sheets of selvedged mesh were then tensioned over the frames, and attached by pop rivets driven through the selvedge at 2-in. intervals. The heavy tension served to pull out any wrinkles formed in the galvanizing process, and to prevent permanent deformation of the mesh under subsequent foot traffic. Peak fabrication errors of the mesh panels were about ± 0.06-in. at 120-ft radius going up to ± 0.10-in. at the rim.

Site Fabrication and Erection

The ribs were just small enough to permit transport to site in one piece, by road, stacked flat on a low loader. To provide structural stability in erection, however, they had to be hoisted and attached to the hub in pairs. A jig (see Figure 6) was therefore made to hold two ribs in correct alignment while the connecting ring girder and bracing material was welded on (see Figure 7). The jig was erected on trial in the shops, then re-erected on site. Some of the main

Figure 6. Site jig for assembling rib pairs.

features of the erection process are shown in Figures 8 – 10.

Before erection, the two innermost reflector plates were bolted onto each pair of ribs, to give some support to the top chords laterally. Since no lateral bracing is provided between ribs, except near the rim, a guyed boom was set up, to which the first pair was attached, to provide temporary lateral stability; even with all ring girders in place, the torsional strength of the whole assembly was very low until most of the reflector plating had been installed. When all 18 pairs of ribs had been erected, each was surveyed individually for level, which was adjusted by inserting shims between the hub and the bottom chord root. On completion of this operation, all adjusting points were within ± 0.2 in. of desired level.

Half the ring girders and lateral bracing material had, of course, been welded in place in the jig. The other half then had to be welded up *in situ,* 120 ft or so above the ground. Furthermore, the necessity of keeping to program meant that almost the whole of this phase took place in the Canadian winter, welding being successfully carried out in ambient temperatures down to 10°F.

The horizontal distance between each pair of ribs had to be kept to the nominal amount ± 0.06 in. to ensure good fit of the reflector plates on the

Figure 7. Typical rib to ring-girder connection.

purlins. One-third of the reflector plates were therefore installed, to provide as much uniform dead loading as possible, then in each successive unwelded bay, the bracing material was fitted and ground to exact length before welding, checking that the horizontal angle from each successive rib, relative to a datum, was within ± 15 sec of arc of the nominal value. We had calculated that during the welding of the bracing, the weight of scaffolding material was reasonably equivalent to the weight of the reflector plating omitted. Thus any cumulative buildup of error was avoided in the final closing bay.

With ring girders and bracing completed, the remaining reflector plates were installed; the dish was then tipped for the first time to permit erection of the tetrapod feed support structure and apex cabin. This operation completed, the dish was tipped back to zero zenith angle, and the feed cabin then aligned on axis by means of jacking points provided, while the mesh panels were installed.

Reflector Survey

The principle of the survey method is by now well established. Placing a

Figure 8. Beginning dish erection, with temporary props between pairs of ribs. Note boom providing lateral stability of unfinished dish.

suitable target on each adjuster, the distance of each target from the vertex is first measured. Any variation of this distance from the nominal is then converted to a correction to the nominal elevation angle of the target. The angle is then measured by theodolite, and any error found is converted back to a linear correction to be applied at that point, normal to the dish surface. Knowing the pitch of the adjuster thread, any desired correction is applied by rotating the adjuster sleeve appropriately.

Figure 9. Welding in ring girder and bracing material between erected pairs of ribs.

At the vertex of the dish, a heavy steel cone was installed, bolted to a circular flange projecting from the hub, already provided for subsequent attachment of Gregorian feed horns. On the cone was mounted a Hilger and Watts ST .561 spherical theodolite base, which, once aligned, was left unmoved throughout the entire dish erection and adjustment. The base carried either a Hilger and Watts ST .48 antenna theodolite (which is a modified version of a Microptic No. 2) or a device to hold the end of a steel tape.

Figure 10. Installing reflector plating. Note the radial purlins with shear connection to rib top chords.

The targets installed on each adjuster were of bronze, painted matt black, having a conical top to which azimuth angles and vertex distances are measured, and a V-shaped groove below, to which elevation angles are measured. The groove provides an ideal target for a horizontal hairline, and defines a constant height above the adjuster, whatever the angle of inclination of the axis of the target to the theodolite line of sight.

Target distances were measured using an ordinary steel tape, but to avoid setting up and tensioning the tape 288 times, and making sag corrections for the more distant targets, a device was employed to enable all eight targets per rib to be taped at once. A small nylon roller was mounted over each target, the tape passed under these rollers, and then out over another larger roller on the rim of the dish to a 10-lb tensioning weight hanging down below the rim. By this means, no sag corrections were necessary and the taping could be carried out very quickly, only about 2 hr being required for the whole set of 288 targets. The peak error in taping target distances was estimated at about ± 0.03 in. corresponding to a peak error in the displacement of any target normal to the dish surface of about ± 0.009 in. Furthermore, computations showed that changes in radial distances of targets when the dish was tilted over were small enough to be negligible. Thus no taping was necessary with the dish tilted away from zero zenith angle.

When using the theodolite, a check on its alignment and level was available by sighting down the dish axis, permanently defined by two illuminated glass disk targets mounted rigidly to the hub structure about 5 and 14 ft below the theodolite.

All taping and theodolite surveys were, of course, carried out at night, to obtain as near uniform temperature conditions as possible. For the theodolite measurements, a spotlight mounted at the focus was moved around to illuminate each target in turn from above. In order to complete the surveys as quickly as possible, only face-right readings were made, a series of corrections being already prepared for face right to left error at each target distance. To reduce personal error, two observers made each survey, one person always setting the theodolite, the other always reading the scale. To save more time, the observers wore headsets plugged into the telescope intercom system and gave their readings verbally to a third person who carried out all the paperwork indoors in comfort. Since the azimuth settings of the hairline are much less critical than the elevation settings, it was quicker to read the targets a ring at a time rather than a rib at a time. Using all these aids to speedy observation, it was possible to survey all targets in about 3 to 4 hr without undue fatigue, even with the dish tilted at 70° zenith angle. The peak error in reading any target was about 3 sec of arc, which corresponds to a normal displacement error of 0.013 in. at the rim, reducing to 0.004 in. on the innermost ring of targets.

Adjustment of the Reflector Surface

With the dish at zero zenith angle, after all reflector plating and mesh panels were in place, together with the tetrapod feed support structure and apex cabin, all of the grip bolts (see Figure 11) forming the shear connections at each adjuster point were loosened. On each reflector plate panel, all except eight of the bolts connecting the panel to the purlins, and to adjacent panels, were also slackened off. Thus the total weight of reflector surfacing material, plus purlins, was temporarily being carried on the ribs alone. Having thus as far as possible removed stresses locked up in the plating or purlins during erection, the first survey of the dish was made. Adjusting points were then raised or lowered as required the next day in normal working hours, no attempt being made to survey and adjust at the same time. After this primary adjustment, in which peak errors up to about ± 0.35 in. maximum had been found, another survey and adjustment was carried out, after which all bolts were tightened up at night, under uniform temperature conditions.

The procedure during subsequent adjustments was to slacken off again all grip bolts, move the adjusters during the day, and retighten the grip bolts

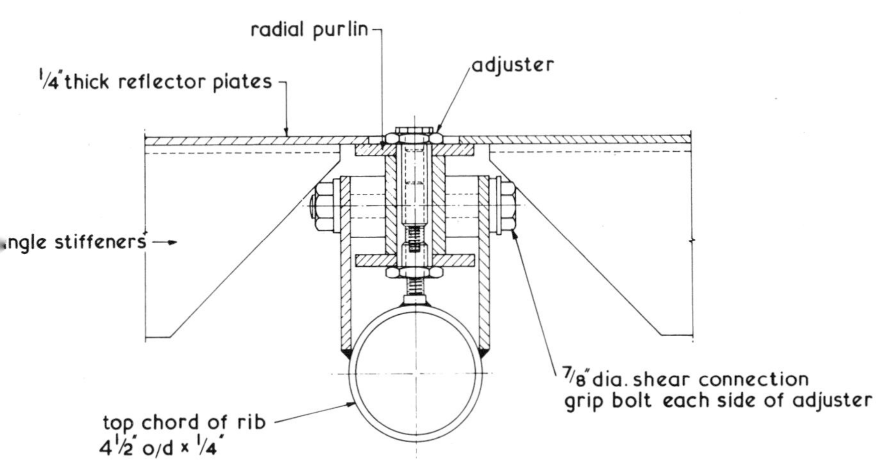

Figure 11. Section through adjuster and shear connector.

during the next night. It had been thought that the reflector plating bolts would also need to be slackened off and retightened, but this proved, most fortunately, to be unnecessary. We found we could release 288 shear connectors in about 3 hr, reset the adjusters in 2hr, and retighten all the shear connectors in another 3 hr, with a squad of eight people. Since the slackening-off can commence before sunset, working from the rim inward, it is quite feasible to make a complete dish readjustment in one short summer night. However, if loosening and retightening about 15,000 plating bolts were also required, an extra 200 or so man hours would be needed, and the whole operation could not possibly be completed overnight.

The dish was adjusted four times in all. The rms error of the target points, weighted appropriately for illumination tapering off at the rim, was found to be about 0.030 in. after the third survey, and only a little less, about 0.028 in., after the fourth. This is considerably greater than would be expected from the theodolite and taping errors alone. Some error is unavoidable in setting the adjusters themselves. These have 18 threads per inch, and are difficult to set to better than, say, one-tenth of a turn, which corresponds to 0.006 in. Variations in torque applied to the adjuster may add an error of comparable magnitude. But the largest single component of error in setting up the dish seems undoubtedly to be due to ambient temperature changes during survey and adjustment.

Structural Performance of the Reflector

In the previous section, we considered the errors in adjustment of the dish when held static at zero zenith angle. The structural performance of the design, with the reflector surface participating integrally with the ribs and ring girders, was irrelevant. This performance can be examined in isolation by studying the differences between the dish contours at 0°, 40°, and 70° zenith angle, surveyed without intervening adjustment, so that errors in setting the dish are canceled out.

Figures 12 and 13 thus compare computed with measured deflections in the two ranges of 0° to 40°, and 40° to 70° zenith angle, respectively. (N.B. All contours in these and subsequent figures are in thousandths of an inch, positive upward.) Note the apparent rotation of the plane of bending away from thevertical, seen in the measured contours. No convincing explanation of this phenomenon is available.

A computer program to determine the best-fit paraboloid (BFP) has been applied to these deflected shapes, with the results shown in Figures 14 and 15. Deflections have been weighted to correspond with 10-dB taper of the illumination at the rim. (It may be noted that the effect of increasing this taper to 15 dB was examined, and found to be negligible; to the nearest thousandth of an inch, the rms deviations from the BFP were unchanged.) Note that inFigures 8 and 10, the apparent rotation of the plane of bending has disappeared simply because the computer program inserts not only a vertical, but also a horizontal, rotation of the BFP axis.

The last two contour diagrams, Figures 16(a) and (b), show differences between two pairs of surveys, all carried out at zero zenith angle. Figure 16(a) shows the difference between two surveys made only three days apart, but in one case, the temperature varied from 26° to 23°F, and in the other from 63° to 60°F. Figure 16(b) shows the difference between the survey made in an ambient of 26° to 23°F and a third survey, carried out about nine months previously, in an ambient of 34° to 30°F. These and other surveys, not recorded here, show no discernible change of dish shape with time, but changes in shape seemingly more or less proportional to changes in ambient temperature are very clearly present.

One specific feature of this telescope design may account, partly at least, for this phenomenon. The cylindrical hub of the dish is designed to contain a great deal of equipment, especially the receivers used in Gregorian operation. To operate and maintain this equipment in the Canadian climate, it had been deemed necessary to air-condition this hub space. Thus, while the outside of the hub remains at ambient temperature, being insulated on its inner surface with 2½ in. of sprayed polyurethane foam, interior diaphragms of the hub may be maintained up to 70°F. It is not yet possible to tell how much this may be contributing to the observed changes in the shape of the dish in changing ambient temperatures.

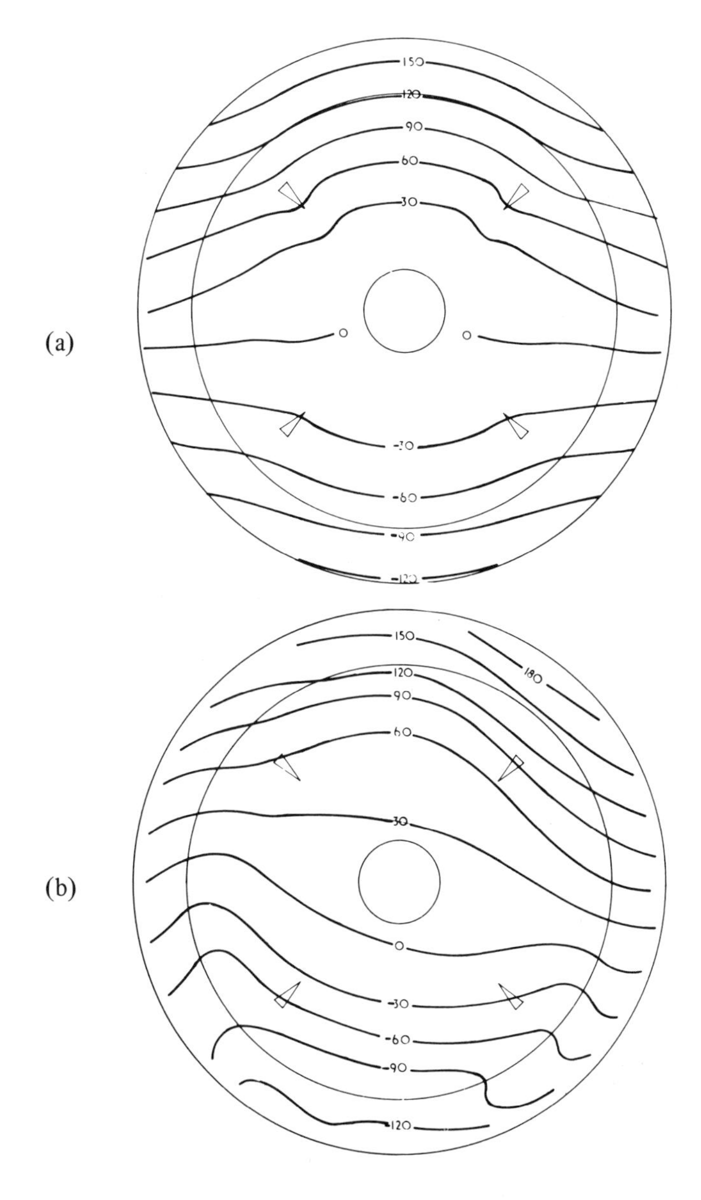

DEFLECTIONS POSITIVE UPWARDS

INCHES × ·001

Figure 12. Contours of deflection of target points due to tilting from 0° to 40° zenith angle. (a) Computed. (b) Measured.

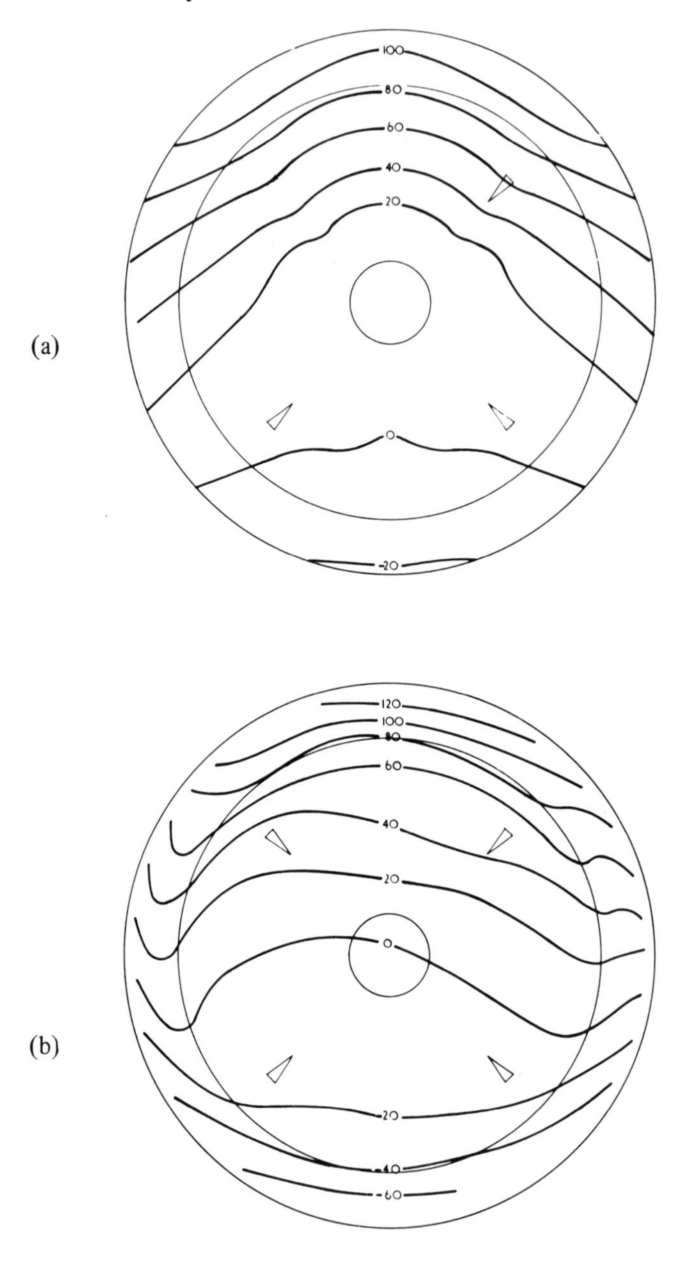

Figure 13. Contours of deflection of target points due to tilting from 40° to 70° zenith angle. (a) Computed. (b) Measured.

(a)

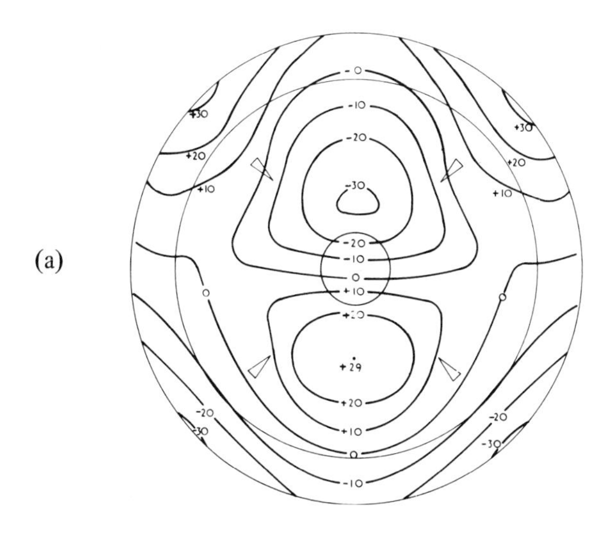

(b)

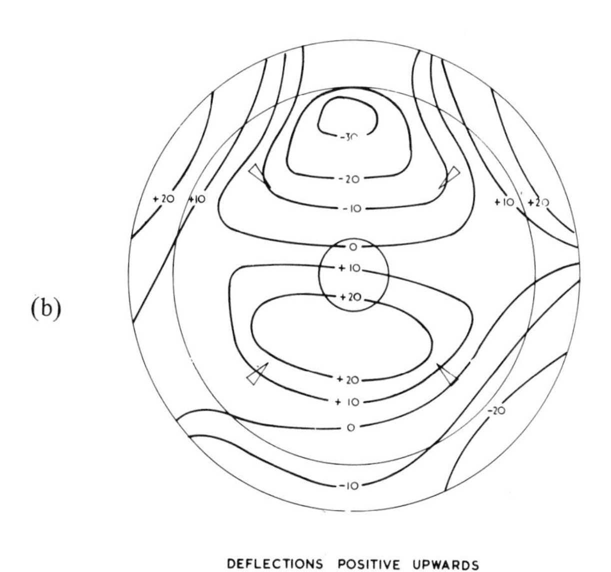

DEFLECTIONS POSITIVE UPWARDS
INCHES x ·001

Figure 14. Contours of residual deflection of target points due to tilting from 0° to 40° zenith angle, relative to BFP at 40°. (a) Computed. (b) Measured.

(a)

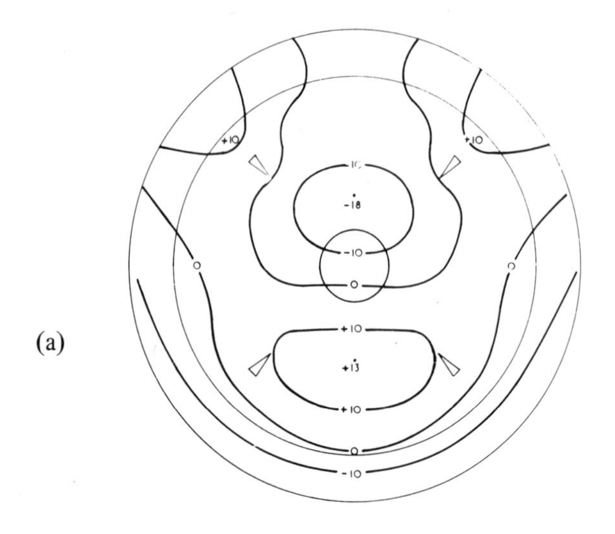

(b)

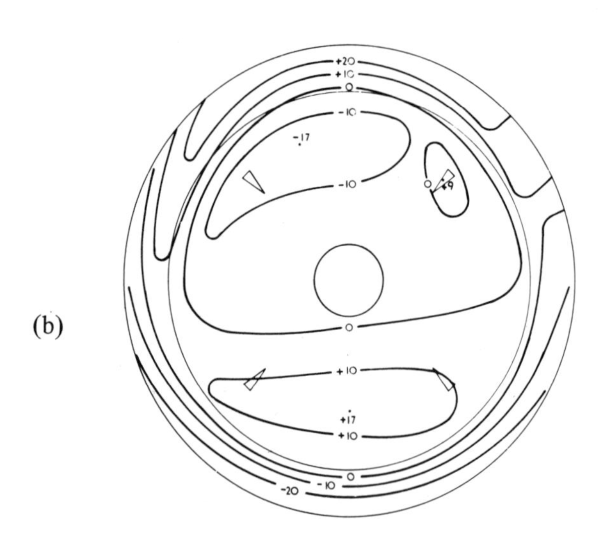

DEFLECTIONS POSITIVE UPWARDS
INCHES x ·001

Figure 15. Contours of residual deflection of target points due to tilting from 40° to 70° zenith angle, relative to BFP at 70°. (a) Computed. (b) Measured.

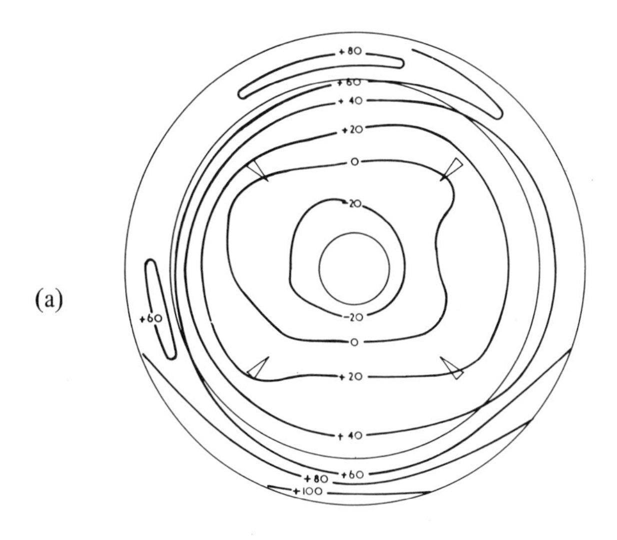

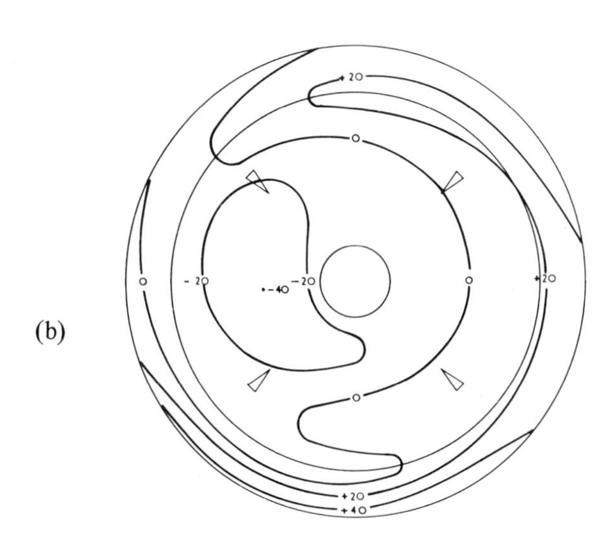

DEFLECTIONS IN INCHES × ·001

Figure 16. Contours of difference between pairs of surveys at 0° zenith angle. (a) 36°F ambient change. (b) 8°F ambient change.

Table 1. Properties of Dish Deflection Contours.

θ_x lateral rotation of BFP axis. θ_y vertical rotation of BFP axis, positive downward. f change in focal length, positive increasing. z change in vertex, positive towards focus. rms (weighted for 10-dB illumination taper to rim).

		θ_x (sec of arc)	θ_y (sec of arc)	f	z (in.)	rms
150-ft-diameter dish						
0 to 40°	computed	0	30	−0.087	0.002	0.015
	measured	9	35	−0.110	−0.006	0.015
40° to 70°	computed	0	14	−0.158	−0.004	0.007
	measured	1	19	−0.062	0.005	0.013
120-ft-diameter dish						
0 to 40°	computed	0	25	−0.105	0	0.012
	measured	9	32	−0.046	0.004	0.013
40° to 70°	computed	0	12	−0.190	−0.001	0.006
	measured	2	15	−0.094	0	0.009
Comparison between pairs of 0° zenith angle surveys						
36°F ambient temperature change		2	6	−0.312	−0.026	0.013
8°F ambient temperature change		4	7	−0.130	−0.024	0.012

Fortunately, the increased curvature of the dish, due to a lowering of the ambient, is reasonably uniform over the dish, so that any increase in rms error from a BFP is small.

Table 1 shows the results of computations carried out on all these cases. Comparing computed and measured behavior during change of zenith angle, note particularly the identical rms errors of 0.015 in. from 0° to 40°, and the small increase of measured over computed error from 40° to 70°, 0.013 in. compared with 0.007 in. It is to be noted that the unexplained lateral rotation of the BFP axis, 9 sec of arc from 0° to 40°, has not been confirmed by any radio measurement to date.

Comparative figures are given for the inner 120-ft diameter of the dish, as well as for the whole dish, because the design was made initially on the assumption that below 10-cm operation only the 120-ft plated portion would be illuminated. In fact, the accuracy of the outer 15-ft-radius mesh-covered portion has bettered the specification such that it is advantageous to use the whole dish even at 3 cm.[2]

The computation on the temperature deformation shown in Figure 16(a) reveals, as one might expect, a relatively large change in focal length, 0.31 in., but the weighted rms error of the deformation is only 0.013 in., about the same as Figure 16(b).

In summary, rms surface fabrication errors are, for radial purlin profile between adjuster points 0.004 in., for the solid panels 0.013 in. off the templates, plus variations in plate thickness 0.008 in., and for the mesh panels 0.035 in. Weighting these figures for area, 64 per cent solid, 36 per cent mesh, gives a total surface fabrication error of 0.023 in. rms.

The rms components of error in setting up the adjuster points are target distance measurement 0.004 in., theodolite instrumental plus reading error 0.003 in., and adjuster setting error say 0.003 in., but these are all small compared with errors due to different ambient temperatures during survey, slackening, adjustment, retightening, and resurvey of the adjusters. The total rms error of adjusters as now set up is 0.028 in. We consider this might possibly be reduced to 0.020 in. or even 0.015 in., given very good ambient conditions, and faster survey techniques to take advantage of them.

Finally, rms gravity deflection error about the mean 40° zenith angle is 0.015 in. at the adjuster points, plus gravity deflection of the surface between adjusters, which is negligible on the solid panels, and about 0.007 in., area weighted, on the mesh panels. This yields a total equivalent rms error of not less than 0.017 in., say, about 0.020 in. because these deflections are not all randomly distributed.

Structural and Radio Performance Compared

The rms reflector errors may thus be summarized, for fabrication 0.023 in., for setting up 0.028 in., possibly in future reducible to 0.015 in., and for gravity deflection 0.020 in. The sum of these, the total rms error, is 0.041 in., possibly in future reducible to 0.034 in. This compares very favorably with the NRC specification, which was for 0.040 in. over the central 120-ft-diameter plated surface only.

Latest radio measurements have indicated at 4.6-cm wavelength an aperture efficiency of 50 ± 3 per cent, from zero to about 40° zenith angle. The theoretical efficiency, assuming 0.040 in. rms surface error, is 52 per cent.

At 2.8 cm, aperture efficiency is 42 ± 2 per cent from zero to 25° zenith angle, compared with theoretical efficiency of 45 per cent. At larger zenith angles, the measured efficiency steadily falls to 46 ± 2 per cent at 70°. The beamwidth throughout the range is about 2.4 min of arc.

Finally at 0.8 cm, the aperture efficiency is 6 ± 1 per cent at small zenith angles, which is still commensurate with 0.040 in. rms surface error. The beamwidth is 1.0 min of arc from zero to 50° zenith angle.

In conclusion, it is to be noted that the National Research Council are now in process of shop testing an automatic survey instrument, which they have devised and manufactured themselves. Its working principle follows closely that of the instrument devised by CSIRO for survey of the 210-ft Parkes telescope, which is described elsewhere.[3] When this instrument is installed, it will be capable of making a complete survey of all 288 target points, at any zenith angle, in 30 min or less. By its ability to take repeated surveys, several in one night if necessary, it should eventually be possible to elucidate the nature of differential temperature effects on the dish with some precision. With this information, there is little doubt that the setting accuracy of the adjusters can be improved, as suggested earlier in this paper.

Acknowledgments

The author is grateful to Sir Gilbert Roberts, FRS, the partner in charge of the project in Freeman Fox & Partners, for permission to publish this paper.

The main contractors, who carried out all the structural fabrication and erection described in this paper, were Dominion Bridge Co., Ltd. of Montreal, Canada.

References

1. C. R. Blackwell, "The Reflector Dishes of the 210′ Diameter Radio Telescope at Parkes, Australia, and the 150′ Diameter Radio Telescope at Lake Traverse, Ontario," *IEE Conference,* Publication No. 21, London (1965), pp. 335–343.
2. W. A. Cumming, "The Design of a Multi-Band Feed System for a 150′ Diameter Radio Telescope," *IEE Conference,* Publication No. 21, London (1965), pp. 130–134
3. M. J. Puttock and H. C. Minnett, "Instrument for the Rapid Measurement of Surface Deformations of a 210′ Radio Telescope," *Proc. IEE* **113,** 1723 (1966).

William Weaver, Jr.
Stanford University
Stanford California

COMPUTER-AIDED DESIGN OF A LARGE STEERABLE ANTENNA STRUCTURE

Introduction

The preliminary structural design of large, fully steerable azimuth-elevation antenna has been completed for the Stanford Research Institute. The antenna consists of a 400-ft-diameter paraboloidal reflector structure to be fabricated of aluminum and steel and mounted upon a steel support structure. The total configuration contains approximately 525 tons of 6061-T6 structural aluminum and 675 tons of A36 structural steel. Structural properties and performance characteristics under a variety of service loading conditions have been ascertained. Background studies, framing details, computer techniques, and deflection control are described in the paper.

The reflector structure was designed and analyzed as a space truss using a computer program devised especially for steerable antennas. In addition to the usual features of such a program, it is capable of performing automatically a dead- or live-load analysis of the reflector in any position. As an added feature, a perspective line diagram of any geometric layout of the structure may be drawn by a plotter linked to the digital computer. The antenna was designed in cyclic fashion with the capability of varying the geometry, member properties, support conditions, and load in each cycle.

The design of the reflector is governed primarily by deflections occurring at various elevation angles due to its own dead weight. These deflections are minimized by specifying a favorable geometric layout, a quadripod framed integrally with the reflector, appropriate initial rigging, steering adjustments, deflection control devices, and a movable feed mount. With these provisions the maximum deflection of the reflecting surface can be restricted within an envelope of ± 1/2 in. (approximately 1/6 in. rms) in the axial direction. Because the reflector and its backup framing are balanced with respect to the elevation axis, no counterweights are required.

Strength and geometric considerations govern the design of the support structure. Member sizes in this assemblage are based upon a series of dead-load and survival wind-loading conditions. The total configuration is balanced with respect to a central azimuth bearing. An approximate frequency analysis of the combined structure was conducted to serve as a guide for designing the drive systems.

Reflector Structure

Geometry of Reflector

Delineation of the geometry of the reflector must begin with the properties of the paraboloidal surface. This surface of revolution is generated with a parabola having a focal ratio specified to be 0.42. That is, the focal distance f for the parabola shown in Figure 1 is taken to be 42 per cent of the aperture distance d.

$$f = 0.42d. \tag{1}$$

This relationship serves to define the equation of shape for the parabola as

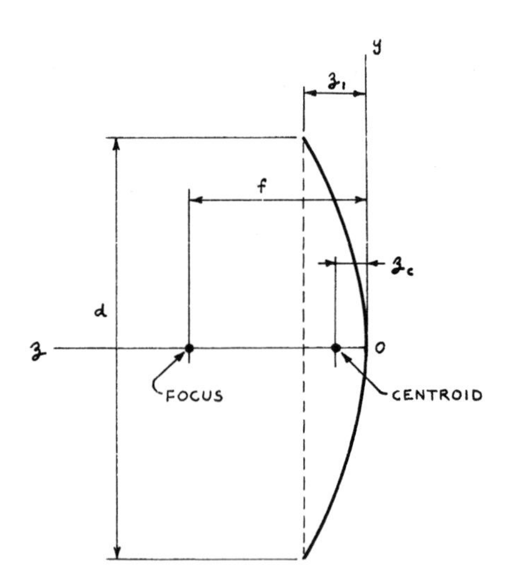

Figure 1. Parabola.

follows:

$$y^2 = 4fz = 1.68dz. \tag{2}$$

If d is 400 ft, then f is 168 ft, and the coordinates of any point on the surface may be computed using Equation 2. A computer program was used to compute other properties of the paraboloidal surface, and some of these are as follows:

$$z_1 = 59.524 \text{ ft}, \quad z_C = 30.510 \text{ ft},$$

$$\text{Surface area} = 136{,}212 \text{ ft}^2.$$

Figures 2 through 8 consist of line drawings of the reflector structure, which

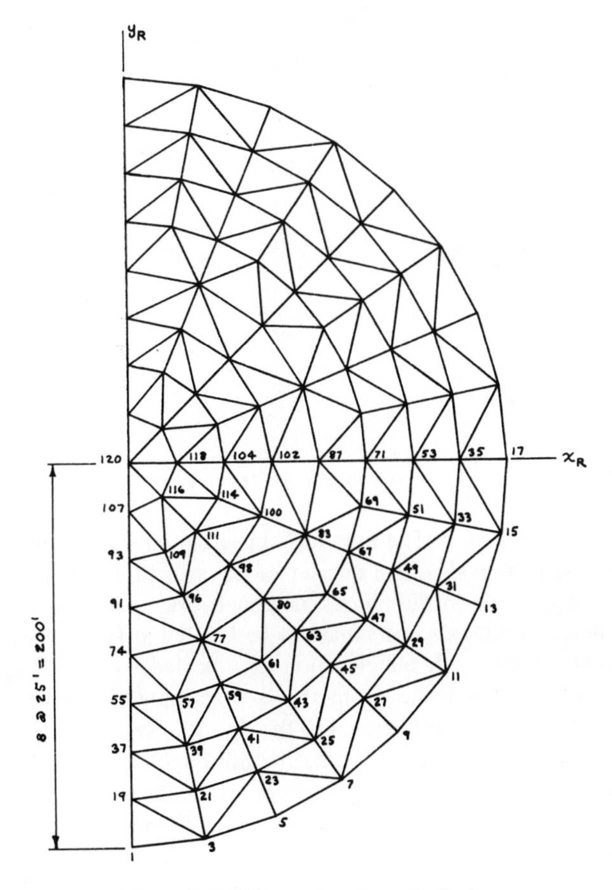

Figure 2. Half front elevation of reflector.

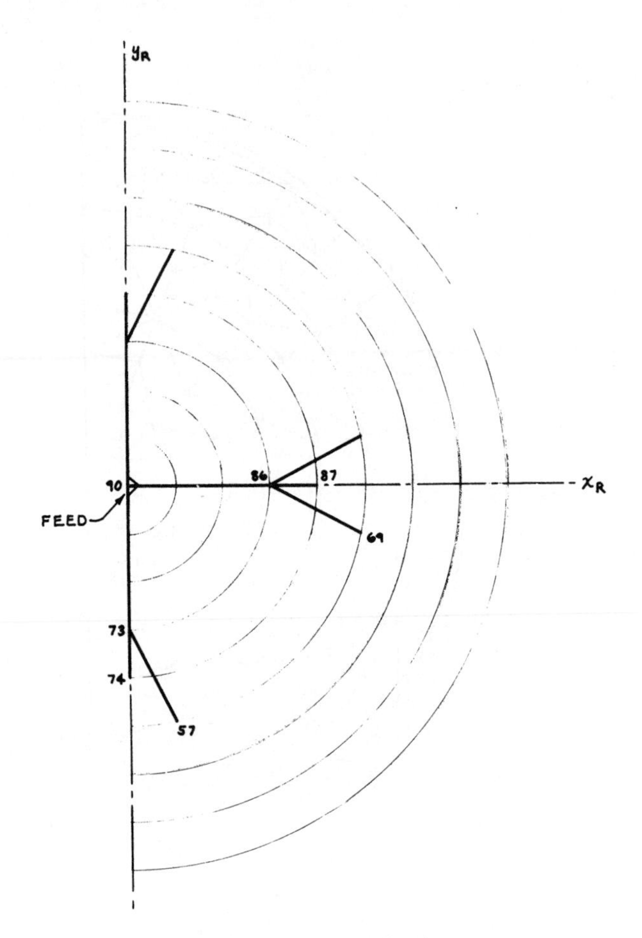

Figure 3. Quadripod framing.

is framed as a space truss. The numbers appearing on these figures are joint designations used in analyzing a quarter of the structure. The reflector axes x_R, y_R, and z_R shown in the figures have their origin at the joint numbered 121, and the elevation axis is assumed to be located at ($y_R = 0$, $z_R = 20$ ft). An alternative scheme with the elevation axis located at ($y_R = 0$, $z_R = 30$ ft) was also considered, but the geometry is practically the same. Since the reflector is to be balanced with respect to the elevation axis, the actual location of this axis cannot be determined precisely until the final phase of the design.

A scale model of the reflector and its support was constructed, and two views of the model are shown in Figures 9 and 10. In addition, perspective drawings of the total congifuration were prepared on a plotter linked to a digital computer, and two drawings of this type are included as Figures 11 and 12.

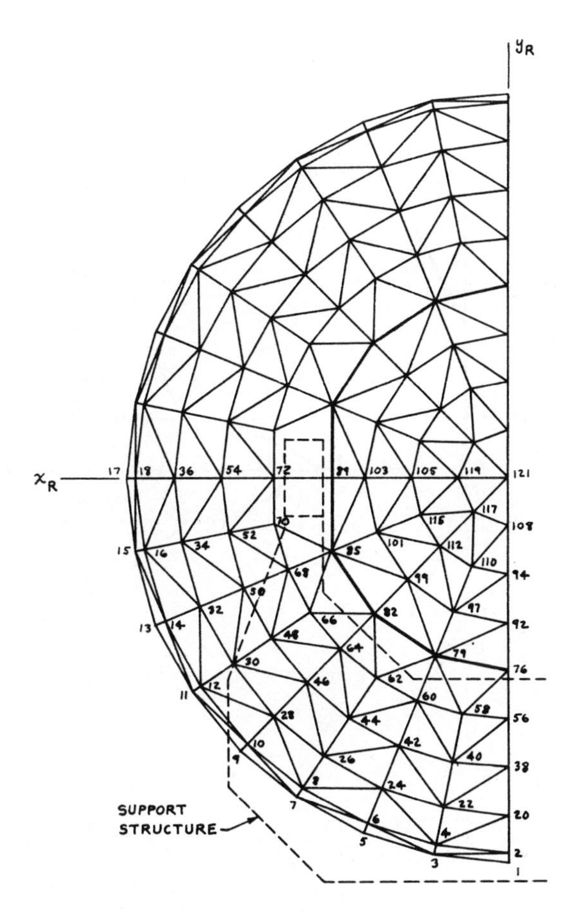

Figure 4. Half rear elevation of reflector.

The reflecting surface is to be expanded aluminum mesh, estimated to have a unit weight of approximately 0.33 psf of surface area. For the purpose of supporting the mesh, radial stringers, annular stringers, and truss grids were considered and compared; but radial stringers proved to be the most desirable choice with respect to deflection and dead weight. Figure 13 shows the general layout of stringers for one-eighth of the reflector. Each stringer is to be continuous over at least four spans, and the spacing between stringers is 2.5 ft minimum, 5 ft maximum, and approximately 3.75 ft average. Two supporting rings (at radii 2.5 ft and 12.5 ft) are required near the center of the reflector, and supplementary bracing struts are also needed to bolster the annular members in the front face of the reflector structure. These struts are indicated by dashed lines in Figure 8.

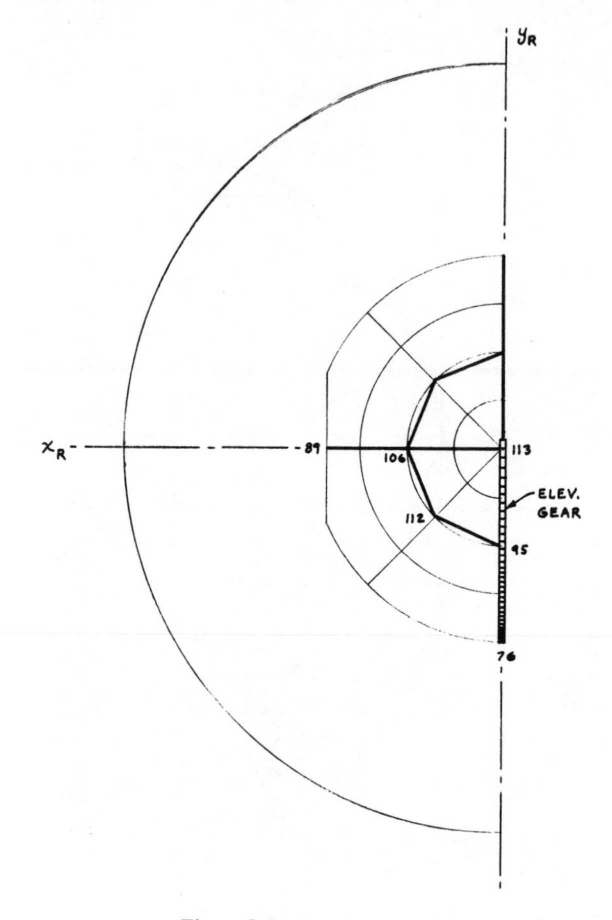

Figure 5. Backup framing.

Details of a typical radial stringer are given in Figure 14. The bottom chord may be straight, but the top chord must be curved in the shape of a parabola. The stringers are to be fabricated of structural aluminum 6061-T6 and welded either as shown or in an equivalent fashion. The dead weight of the radial stringers is approximately 0.38 psf of reflector surface area.

Design of Reflector Framing

Structural aluminum 6061-T6 was chosen for most of the reflector members and the quadripod. This material has the desirable features of low unit weight, fairly high strength, and good weldability. Furthermore, this type of aluminum is readily available and is practically maintenance free.

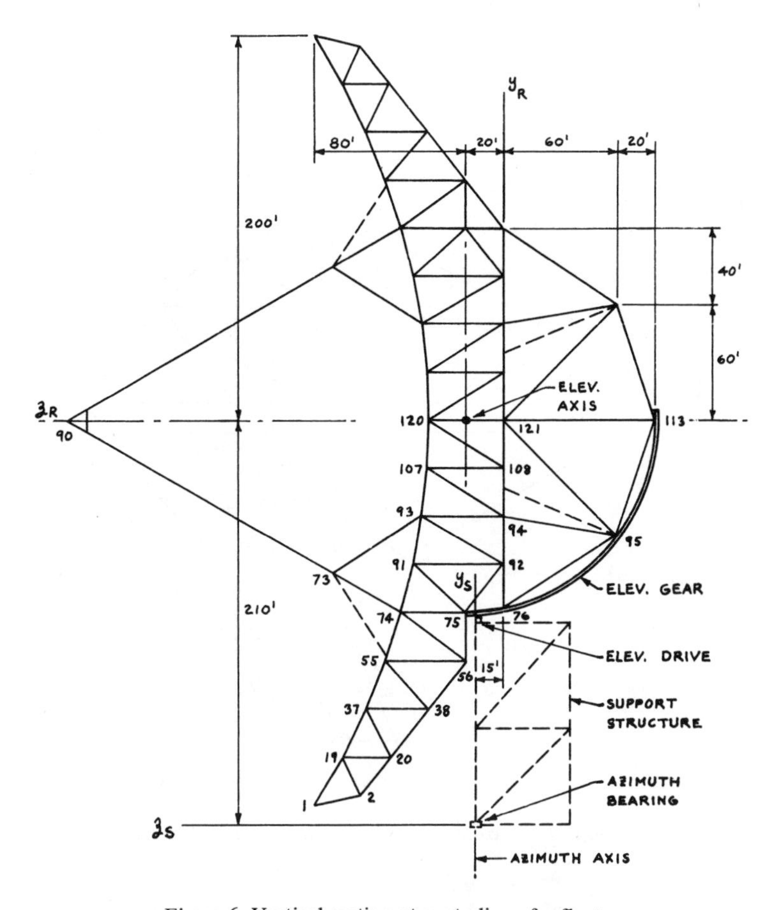

Figure 6. Vertical section at centerline of reflector.

Members of an antenna structure inherently operate at a low level of stress because deflection is the primary design criterion. Therefore, initial member sizes may be selected on a rather arbitrary basis. Since the members of a movable space truss may be subjected to either tension or compression, a limiting value of L/r (ratio of length to radius of gyration) was selected as the most suitable criterion. The ASCE Specifications for structural aluminum[1] allow a working stress of 8 to 10 ksi for welded members, and a stress of 8 ksi corresponds to a value of L/r equal to approximately 95. This relationship is indicated by the dashed lines in Figure 15(a), which contains a graph of allowable compressive stress for axially loaded compression members versus L/r. This particular plot applies to an assumed condition of partial restraint at the ends of the member which produces an effective length of 0.75L. The selections of 8 ksi as the limiting stress also determines the maximum

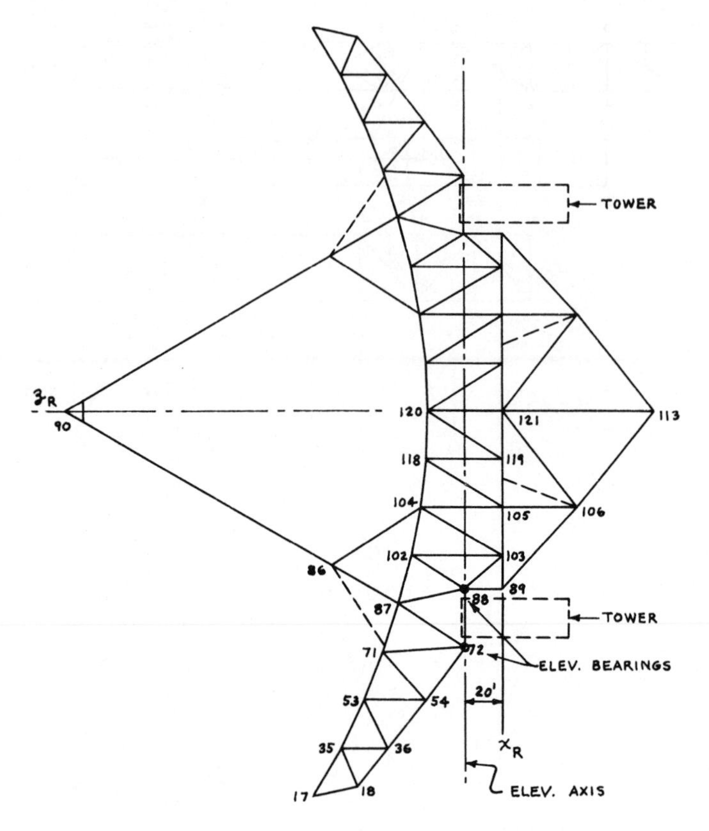

Figure 7. Horizontal section at centerline of reflector.

b/t ratios for angles (b/t = 14) and square tubes (b/t = 60), as shown in Figures 15 (b) and 15 (c).

Square tubes were chosen for most of the members of the reflector, and their minimum sizes were based upon satisfying the conditions $L/r \leqslant 95$ and $b/t \leqslant 60$. Next, many of the minimum member sizes were supplanted by nominal sizes in an effort to produce greater orderliness and repetition in the over-all layout. These nominal sizes of square aluminum tubes ranged from 10 x 10 x 1/4 in. to 16 x 16 x 1/4 in.

The quadripod acts as an integral part of the reflector structure, and its design is based upon satisfying $L/r \leqslant 95$ in each leg. A quadripod leg consists of four aluminum chord angles laced with light aluminum tubing as shown in Figure 16. Each individual piece satisfies $L/r \leqslant 95$, and the chord angles satisfy the condition $b/t \leqslant 14$. Lacing members are to be slit and welded to the chords either as shown or in an equivalent manner.

Members of the reflector that are located to the rear of the plane $z_R = 0$

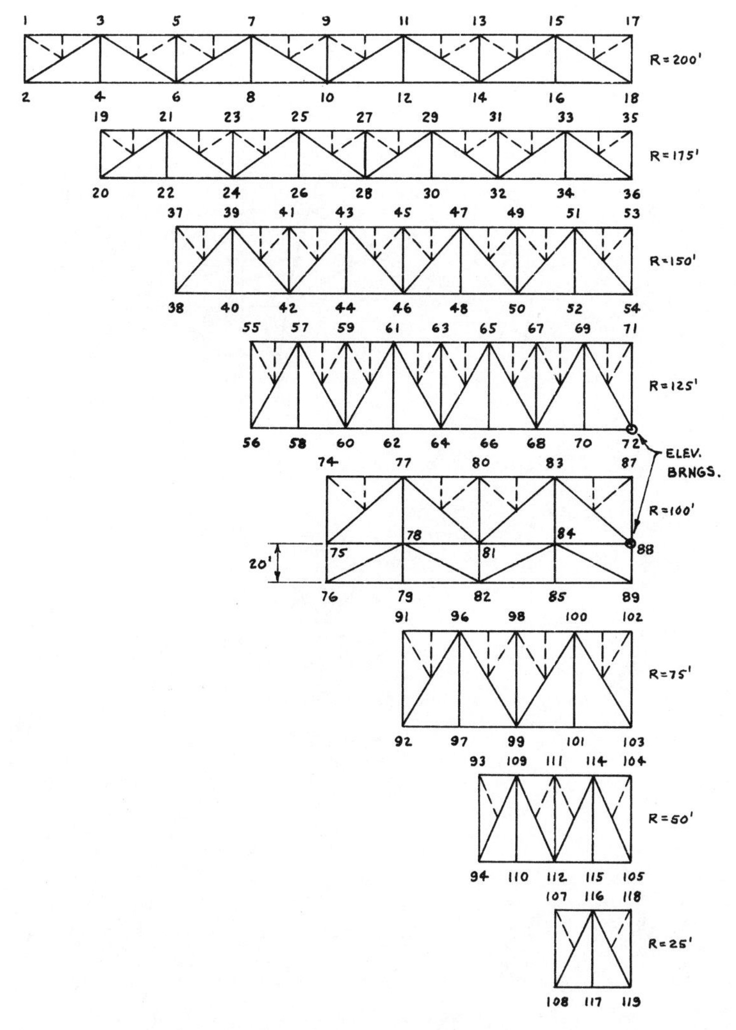

Figure 8. Developed views of annular bracing.

are designated as backup framing. These backup members appear in Figures 5, 6, and 7, and their lengths range from 60 to 80 ft. In order to provide both stiffness and weight to strengthen and balance the reflector, A36 steel was chosen as the material for these members. In this instance an $L/r \leqslant 100$, a K value of 0.75, and $KL/r \leqslant 75$ were chosen as the basis for design. The 1963 AISC Specifications[2] allow a compressive stress of 16 ksi under these conditions, provided that $b/t \leqslant 42$ for square tubes. A nominal tube of dimensions 24 x 24 x 1/2 in. approximately meets these standards, and

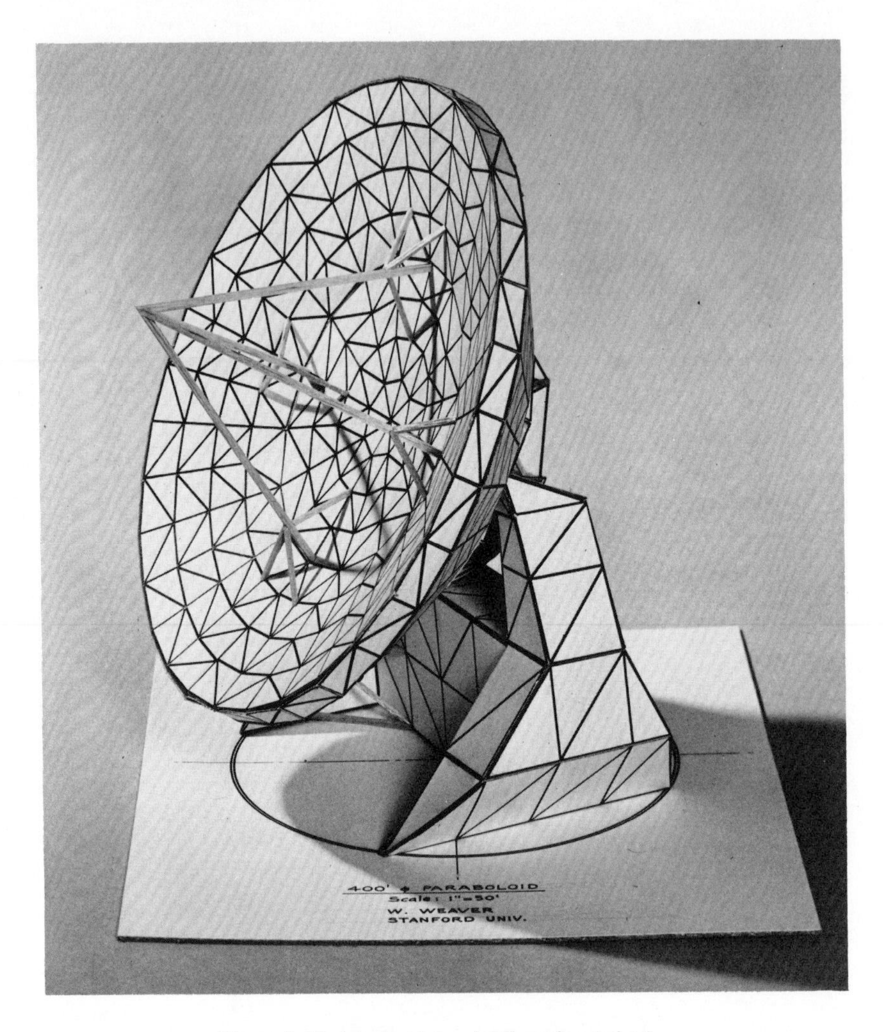

Figure 9. Model of antenna (oblique front view).

this size was used for many of the backup members. However, some of the members were specified to be 24 x 24 x 1 in. to give greater stiffness and weight in favorable locations. The cross-sectional areas of backup members were transformed to equivalent aluminum for purposes of analyzing the reflector structure as a space truss of one material.

After initial member sizes were selected as explained above, the reflector structure was analyzed cyclically for dead loading, wind loading, and several cases of deflection control. The computer program used for these analyses is a space truss analyzer devised especially for steerable antennas. It takes

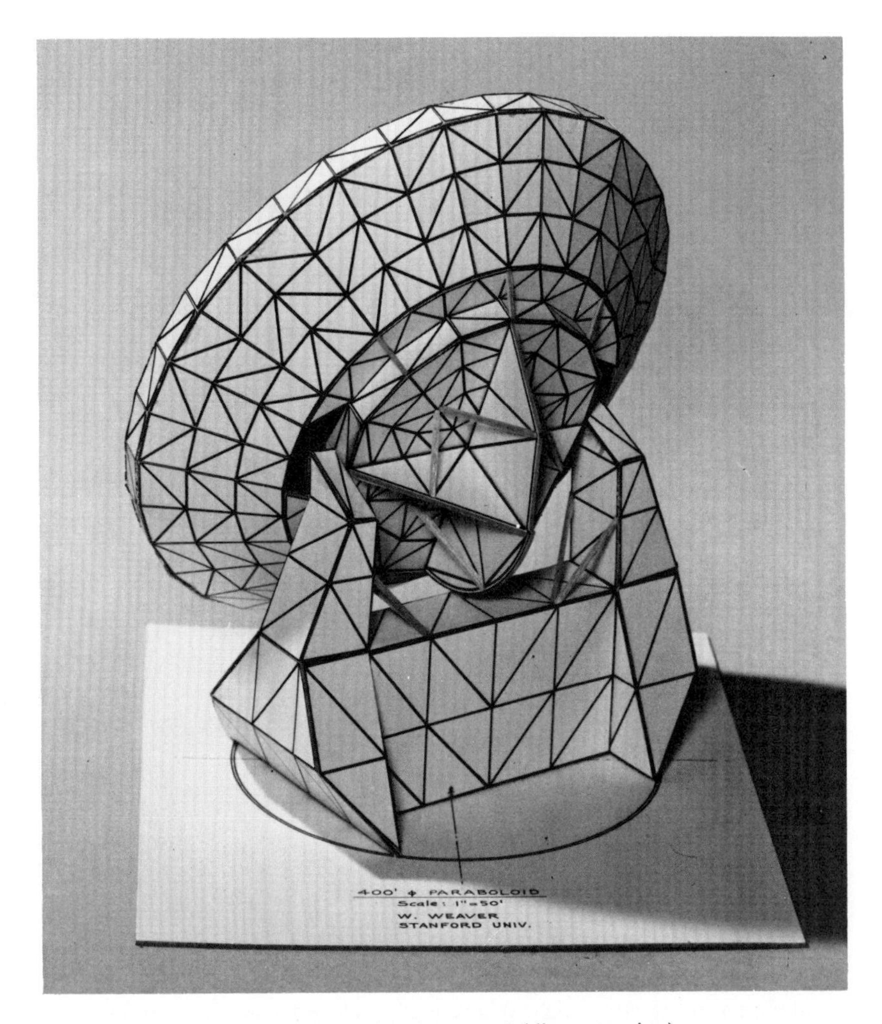

Figure 10. Model of antenna (oblique rear view).

advantage of the symmetry and bandwidth of the joint stiffness matrix and performs calculations using only the core storage of a large digital computer.[3] In addition to the usual features of such a program, it is capable of performing automatically a dead- or live-load analysis of the reflector in any position.

After each cycle of analysis, new member sizes were selected on the basis of stresses due to the dead weight of the reflector in the zenith position. Since deflections in this position are of paramount importance (as demonstrated later), member sizes were proportioned to reduce dead-load stresses to

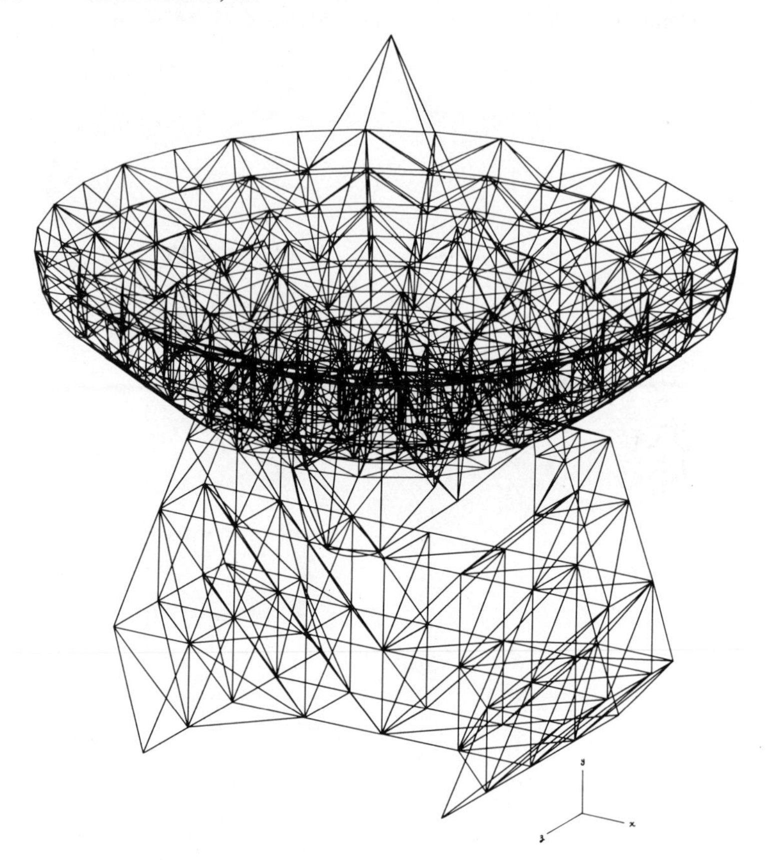

Figure 11. Perspective drawn by plotter.

approximately 2 ksi maximum throughout the structure. This technique may be characterized as a nominally stressed design with certain side constraints (minimum member sizes, lightweight quadripod, and heavy backup framing). Although this procedure will not result in a structure of minimum weight, it will produce a high stiffness-weight ratio.[4] Final sizes of square aluminum tubes selected on this basis range from 10 x 10 x 1/4 in. to 16 x 16 x 13/4 in.

Table 1 shows sample results from certain analyses of the reflector structure with final member sizes. Translations (in the z_R direction) of selected joints on the front face of the reflector and the quadripod appear in columns (1) through (5) of the table. Column (1) contains deflections due to the dead weight of the reflector in the zenith position. Those in column (2) are caused by an operational wind velocity of 30 mph on the front face of the reflector in the horizon position. The latter deflections are only about 9 per cent of the former because the wind loading is small compared to the dead weight of the structure.

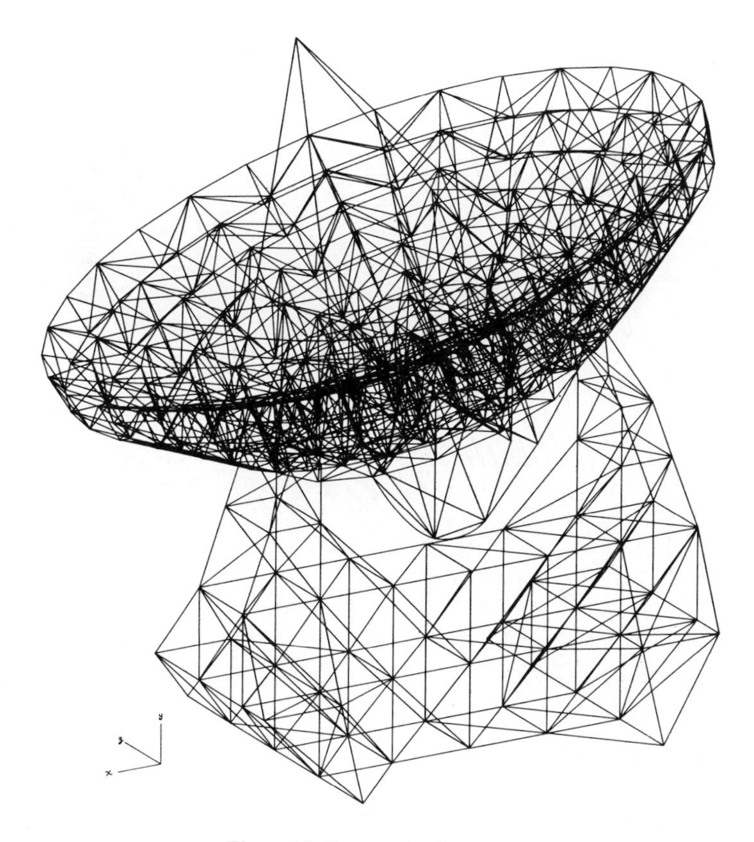

Figure 12. Perspective drawn by plotter.

Table 1. Translations (in inches) of Selected Joints in z_R Direction.

Joint	Loading pattern (1)	(2)	(3)	(4)	(5)
1	−1.699	−0.147	−0.551	0.510	−0.145
9	−1.215	−0.111	−0.190	0.166	−0.043
17	−0.699	−0.075	0.114	−0.236	0.111
74	−0.823	−0.056	−0.401	0.375	−0.108
80	−0.721	−0.055	−0.121	0.210	−0.093
87	−0.111	−0.009	−0.006	−0.006	0.006
120	−0.572	−0.028	0.146	0.543	−0.385
90(Feed)	−0.549	−0.037	−0.176	0.161	−0.045

An extensive study of feasible methods for controlling deflections revealed that a substantial correction could be achieved with a modest number of devices built into members of the backup framing. For this reason, three representative cases of deflection control are listed in columns (3), (4),

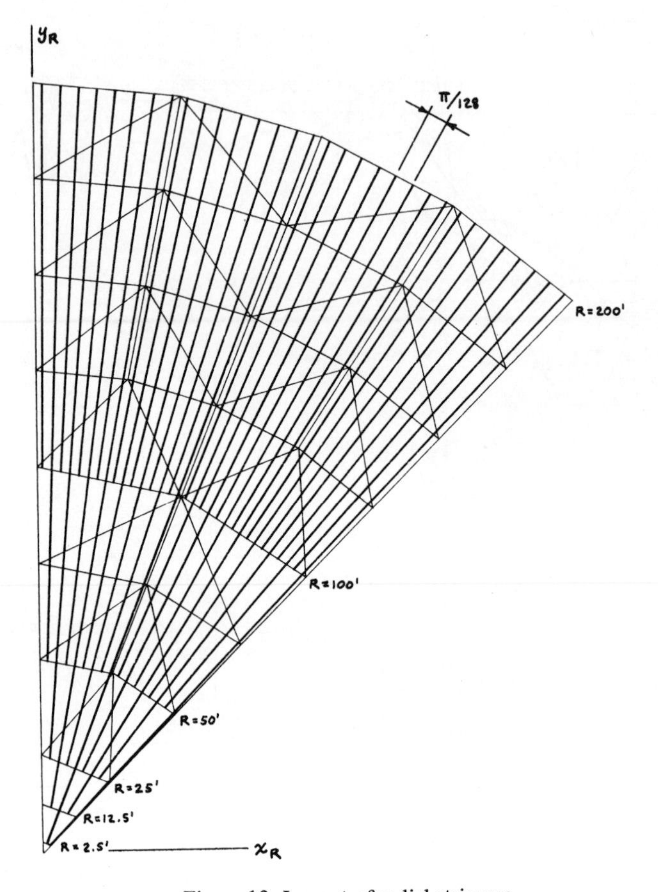

Figure 13. Layout of radial stringers.

and (5) of Table 1. These cases involve shortening back up members 95-113, 106-113, and 113-121 one inch, respectively, by means of a screw jack built into each member. The combination of loading patterns (1) and (3) in Table 1 is discussed at the end of the paper.

Properties of Reflector

Over-all physical properties of the reflector configuration (including mesh, stringers, quadripod, etc.) were obtained by a computer program written for this purpose, and the results are summarized in Table 2. The weight of steel in the backup framing constitutes approximately 25 per cent of the total weight, which is about 1400 kips. The location of the center of gravity of the reflector indicates that the elevation axis should be moved forward about 10 ft in

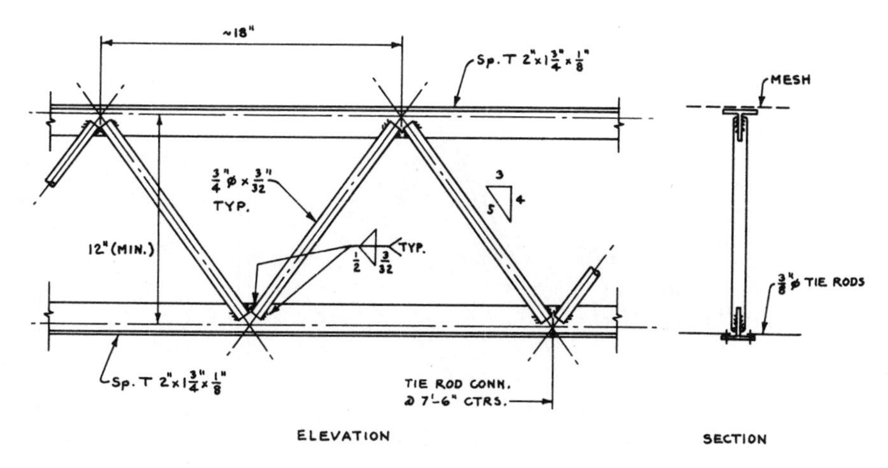

Figure 14. Details of radial stringers.

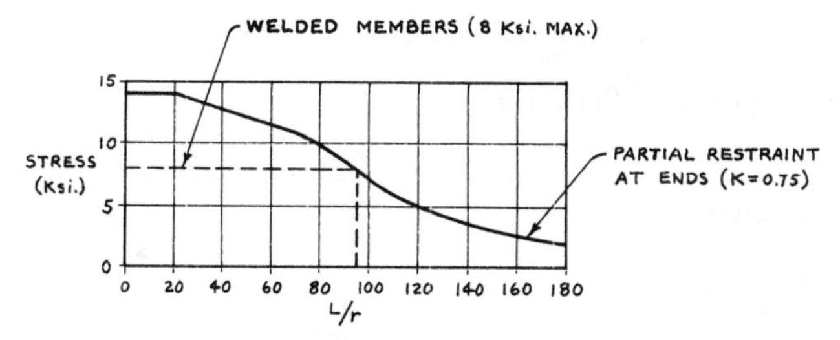

(a) Axially loaded compression members.

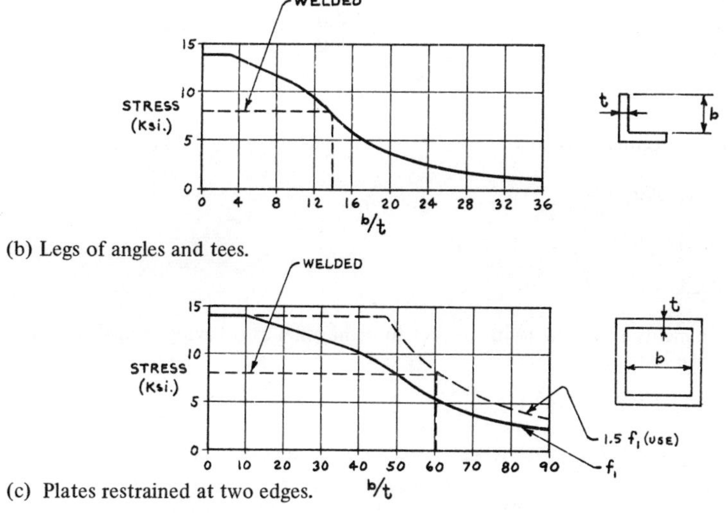

(b) Legs of angles and tees.

(c) Plates restrained at two edges.

Figure 15. Allowable compressive stresses for 6061-T6

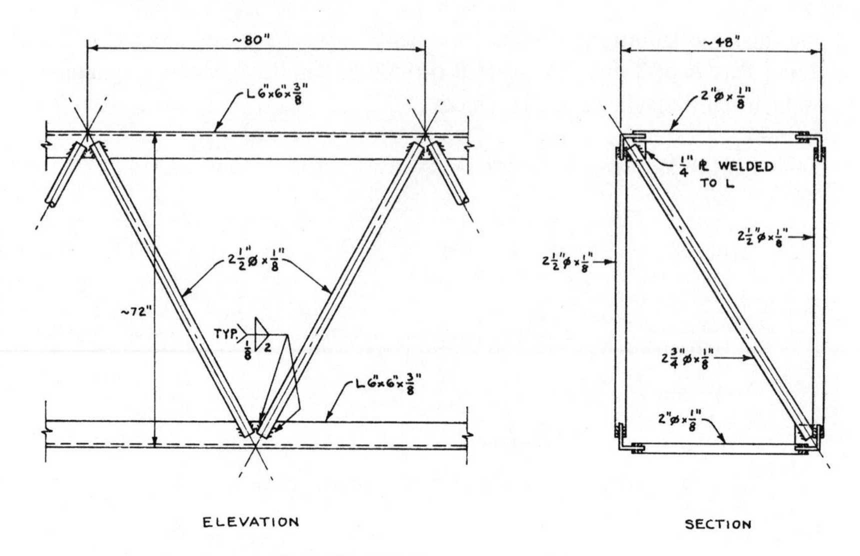

Figure 16. Details of quadripod legs.

Table 2. Properties of Reflector.

A. Dead Weights (kips):

Reflector members		
Aluminum		955.2
Steel		348.0
Stringers		51.8
Mesh		45.0
Feed		4.0
Quadripod lacing		8.0
	Total	1412.0

B. Location of center of gravity:

$(x_R)_{c.g.} = 0$

$(y_R)_{c.g.} = 0$

$(z_R)_{c.g.} = 31.1$ ft

c. Mass moments of inertia (with respect to orthogonal axes having their origin at the c.g. of the reflector):

$(I_x)_{c.g.} = 0.452 \times 10^6$ k-ft-sec^2

$(I_y)_{c.g.} = 0.460 \times 10^6$ k-ft-sec^2

$(I_z)_{c.g.} = 0.580 \times 10^6$ k-ft-sec^2

the final configuration. The mass moments of inertia given in Part C of Table 2 and Part A of Table 3 are useful for computing drive system requirements and natural frequencies of the system.

Table 3. Mass Moments of Intertia[a] with Respect to Azimuth Axis.

A. Reflector	
1. Horizon position:	0.472×10^6
2. Zenith position:	0.591×10^6
B. Support structure:	0.346×10^6
C. Combined system	
1. Horizon position:	0.818×10^6
2. Zenith position:	0.937×10^6

[a] Units are k-ft-sec^2.

Support Structure

Geometry of Structure

The primary functions of the support structure are to provide full steerability for the reflector, to support the weight of the reflector as well as its own weight, and to resist overturning wind and earthquake forces. The geometry of the support structure is determind to a great extent by the desired locations of bearings and drives. In addition, the reflector must be supported clear of the ground and clear of the support structure itself.

Figures 17, 18, and 19 contain line drawings of the support structure, which has also appeared in previous figures. The structure is framed as a space truss and is symmetric with respect to a central vertical plane. The joint numbers appearing in the figures were assigned for the purpose of analyzing half the structure with a space truss computer program. Reference axes x_S, y_S, and z_S for the support structure have their origin at the joint numbered 31, which is the location of the azimuth bearing. Azimuth drives are located at other points of support, such as joints 35 and 39. Elevation bearings are indicated at joints 1 and 2, and a single elevation drive is mounted at or near joint 11.

Design of Support Framing

Structural steel A36 was selected as the most economical material for all members in the support structure. A standard member size of 24 x 24 in. outside cross-sectional dimensions was chosen on the basis was chosen on the basis of $L/r \leqslant 100$, $K = 0.75$, and $KL/r \leqslant 75$ for the longest member. Under these conditions, the 1963 AISC Specifications allow a compressive stress of

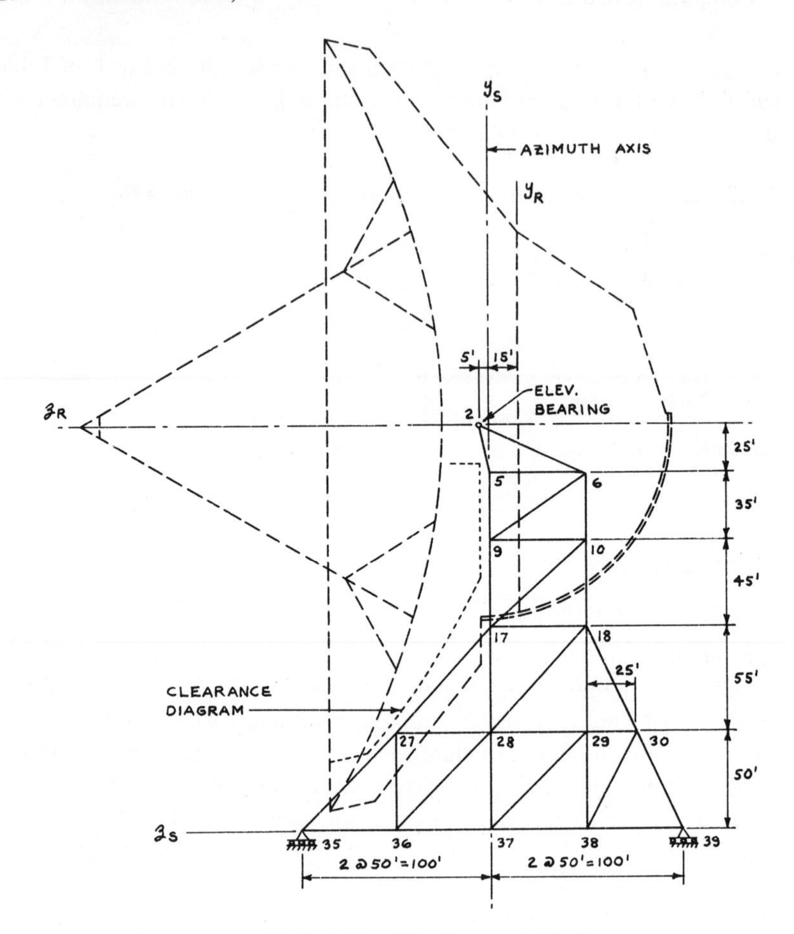

Figure 17. Side elevation of support structure.

16 ksi, provided that b/t ⩽ 13 for legs of angles and b/t ⩽ 42 for plates forming a closed section. Most of the members in the support structure are proportioned to satisfy these standards. Figure 20 shows the minimum feasible member size ascertained on this basis. It consists of four chord angles laced with tubing in a manner similar to that for the quadripod legs described earlier. Stronger members were provided by changing the characteristics of the cross section, resulting in a series of laced, semilaced, and tubular cross sections varying in area from 6 to 60 sq in. Nominal sizes for individual members of the support structure were chosen from this series on the basis of the following considerations: (1) results of a dead-load analysis with all members having equal cross-sectional areas, (2) attainment of approximately 12 ksi maximum stress in members due to dead load only, and (3) an orderly

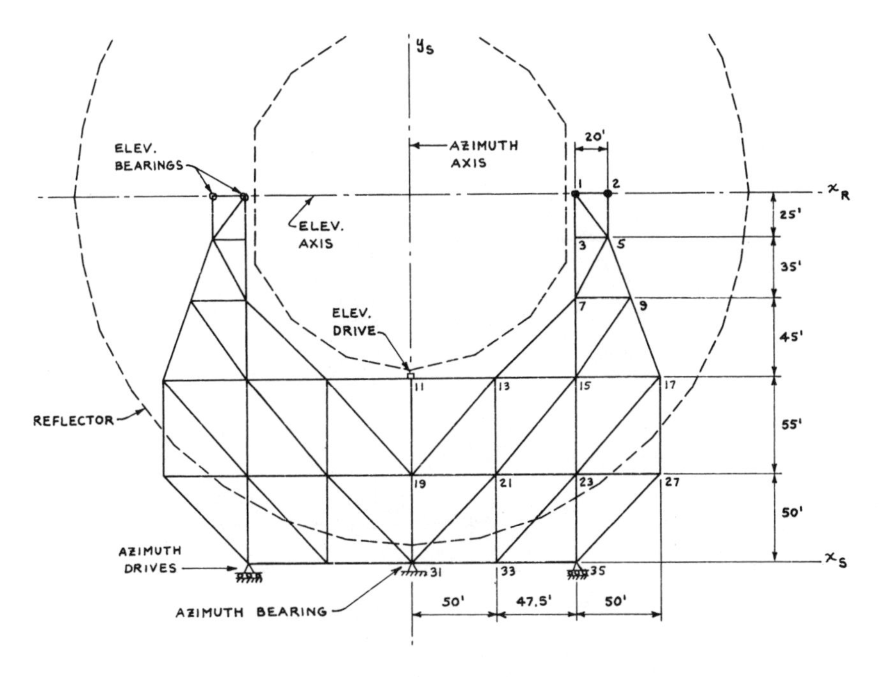

Figure 18. Front elevation of support structure.

arrangement of member sizes. The structure was then analyzed cyclically under the following conditions of dead load plus survival wind forces in various directions, modifying member sizes in each cycle:

(1) Dead load of support structure only.
(2) Dead load of support structure plus reflector.
(3) Dead load of case (2) plus survival wind force in +x direction.
(4) Dead load of case (2) plus survival wind force in −x direction.
(5) Dead load of case (2) plus survival wind force in +z direction.
(6) Dead load of case (2) plus survival wind force in −z direction.

A survival wind velocity of 90 mph was assumed for cases (3) − (6) with the reflector in the zenith position. The strongest members (square tubes 24 x 24 x 5/8 in.) were required in the vicinity of bearings and points of support, where large dead-load and wind forces must be transmitted.The final member sizes are subject to revision at a later date because of the fact that response to earthquake ground motion was not investigated.

Properties of Support Structure

Physical properties of the support structure were determined by computer. The total weight of steel framing is approximately 1000 kips, and the center

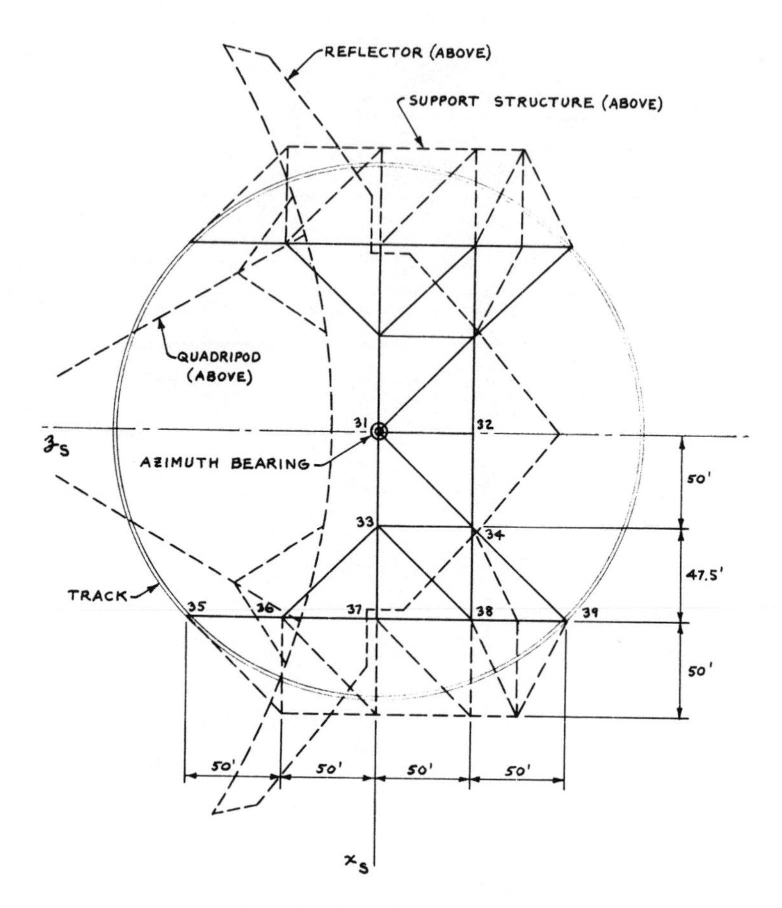

Figure 19. Plan of support structure.

of gravity of the structure lies at the coordinates ($x_S = 0$, $y_S = 85.8$ ft, $z_S = -16.3$ ft), as indicated in Figure 21. The mass moment of inertia of the framing was computed with respect to the azimuth axis and is given in Part B of Table 3.

Combined Structure

Properties of Combined Structure

The combined configuration has a total weight of approximately 2400 kips, consisting of about 1050 kips (525 tons) of 6061-T6 aluminum and about 1350 kips (675 tons) of A36 steel. These figures must be augmented, of

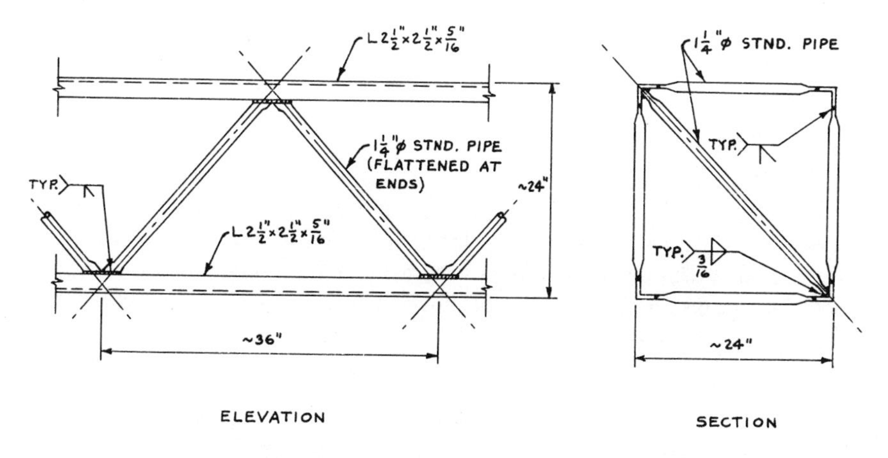

Figure 20. Minimum member size for support structure.

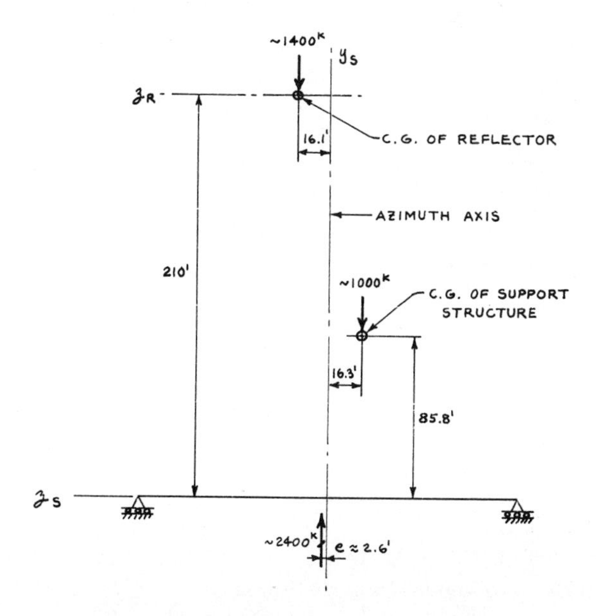

Figure 21. Balancing of combined structure.

course, by the weights of bearings, drives, supplementary framing, and other appurtenances added to the structure.

The center of gravity of the combined structure lies approximately 2.6 ft forward of the x_S - y_S plane, as indicated in Figure 21. This small amount of eccentricity may be easily eliminated in the final design to produce a completely balanced system.

Mass moments of inertia of the combined structure with respect to the azimuth axis are given in Part C of Table 3. The maximum value of this property is obtained with the reflector in the zenith position.

Frequency Analysis

An approximate frequency analysis of the combined configuration was conducted to serve as a guide in designing the drive systems. For this purpose a crude analytical model was devised by assuming the reflector to be a rigid body on elastic supports. Figure 22 shows this analytical model, in which the

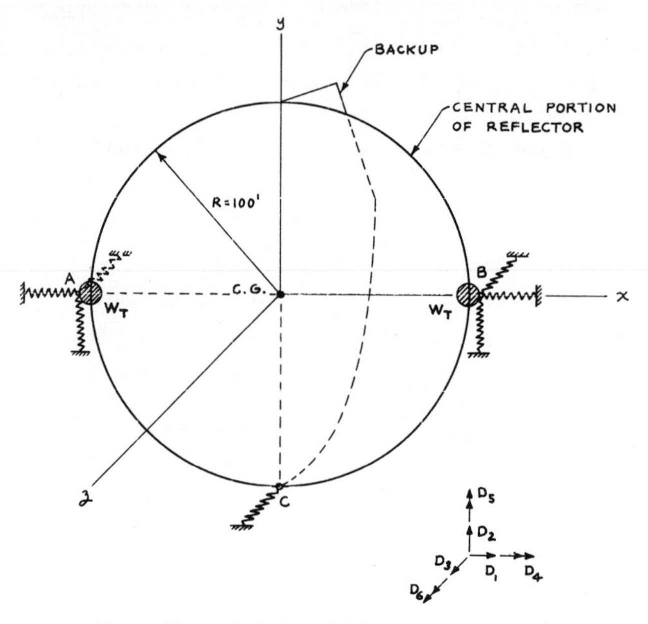

Figure 22. Analytical model for frequency analysis.

elastic restraints at points A, B, and C represent stiffnesses of the support structure at the elevation bearings and drive. The rigid body has six degrees of freedom, as indicated by the displacement vectors D_1 through D_6 in Figure 22.

Mass, inertia, and stiffness properties of the analytical model were obtained from previous analyses of the structure. The weight of the reflector (1400 kips) is assumed to be augmented by contributions from the support structure, consisting of two weights (W_T = 100 kips each) lumped at points A and B. Frequencies and mode shapes for this analytical model were obtained by computer[5] with the reflector in both the horizon and zenith positions. The lowest symmetric and antisymmetric mode frequencies were calculated to be approximately 0.46 and 0.60 cps, respectively, with the reflector in the zenith position.

The analytical model used for this analysis is admittedly a poor representation of the actual structure. However, it is only intended to provide a rough estimate of natural frequencies for the preliminary design of the drive systems.

Discussion of Deflections

The primary criterion for satisfactory performance of a steerable antenna is the deflection of the reflector under its own dead weight for elevation angles from 0° to 90°. It is convenient to resolve this deflection problem into two constituent parts. If the reflector is doubly symmetric, its dead-load deflections in the zenith position are also doubly symmetric. On the other hand, deflections in the horizon position are antisymmetric. Moreover, the deflections at any elevation angle α may be expressed as

$$\mathbf{D}_\alpha = \mathbf{D}_0 \cos\alpha + \mathbf{D}_{90} \sin\alpha \tag{3}$$

The symbols $\mathbf{D}_\alpha$, $\mathbf{D}_0$, and $\mathbf{D}_{90}$ in Equation 3 represent displacement vectors for the joints of the reflector for elevation angles of α, 0°, and 90°, respectively. Equation 3 shows that if dead-load deflections are determined for the horizon and zenith positions, deflections in any other position can also be obtained. The maximum displacement at a given joint can occur at any elevation angle from 0° to 90°, depending upon the relative values of elements in the vectors $\mathbf{D}_0$ and $\mathbf{D}_{90}$.

However, if the reflector is constructed and precisely aligned in the zenith position, the following equation applies:

$$\mathbf{D}_\alpha = (\mathbf{D}_0 - \mathbf{D}_{90}) \cos\alpha. \tag{4}$$

In this case the maximum joint displacements will occur when the reflector is rotated to the horizon position.

$$\mathbf{D}_{max} = \mathbf{D}_0 - \mathbf{D}_{90}. \tag{5}$$

The Appendix contains the results of a study that demonstrates the fact that the antisymmetric displacements (in the z_R direction) of joints on the front face of the reflector may be compensated almost entirely by a small rigid-body rotation about the elevation axis. This small compensating rotation constitutes a steering adjustment that varies as the cosine of the elevation angle.

$$\theta_\alpha = \theta_0 \cos\alpha, \tag{6}$$

in which θ_α is the correction for elevation angle α and θ_0 is the correction for case of zero elevation angle. If this type of correction is utilized, Equation 4 simplifies to

$$\mathbf{D}_\alpha = -\mathbf{D}_{90} \cos \alpha. \tag{7}$$

Thus, the symmetric deflections $\mathbf{D}_{90}$ under dead load in the zenith position remain as the displacements of primary concern. Displacements of this type (in the z_R direction) may be controlled to a certain extent by tightening screw jacks built into the backup members of the reflector. If such controlled deflections are denoted by $\mathbf{D}_C$, Equation 7 becomes

$$\mathbf{D}_\alpha = (\mathbf{D}_C - \mathbf{D}_{90}) \cos \alpha. \tag{8}$$

Note that the magnitudes of these controlled deflections also vary as the cosine of the elevation angle and that the maximum displacements still occur when the reflector is rotated from zenith to horizon.

$$\mathbf{D}_{max} = \mathbf{D}_C - \mathbf{D}_{90}. \tag{9}$$

Therefore, the reflector surface should be initially rigged in the zenith position to have initial displacements $\mathbf{D}_I$ according to

$$\mathbf{D}_I = - C(\mathbf{D}_C - \mathbf{D}_{90}) = C(\mathbf{D}_{90} - \mathbf{D}_C). \tag{10}$$

The constant C in Equation 10 may be specified arbitrarily to be any number between 0 and 1.0. If an elevation angle α_n of normal usage could be chose, then C would be selected as the cosine of that angle.

$$C = \cos \alpha_n. \tag{11}$$

With the reflector in this position the displacements $\mathbf{D}_{\alpha_n}$ are theoretically equal to zero. In lieu of this criterion, the constant C may be set to the average value of 0.5, and Equation 10 becomes

$$\mathbf{D}_I = 0.5\,(\mathbf{D}_{90} - \mathbf{D}_C). \tag{12}$$

In this case the initial displacements for the reflector in the zenith position are as shown in Equation 12, and those for the horizon position are equal and opposite. The position for zero displacements corresponds to an elevation angle of 60°. If it were practically feasible, the reflector could be rigged in the 60° position with zero initial displacements.

A measure of the deviation of the reflector from a true paraboloid may now be estimated on the basis of Equation 12. For this purpose the displacements in the z_R direction of several key points on the reflector will be considered. A deflection control study divulged the fact that a significant correction could be realized by shortening only member 95-113 (and its counterpart above the elevation axis) by the amount 1.5 in. Using this correction, $\mathbf{D}_C$ in Equation 12 is computed as 1.5 times column (3) of Table 1. The calculation of key displacements according to Equation 12 is illustrated in Table 4. Neither

Table 4. Initial Translations (in inches) of Selected Joints in z_R Direction.

Joint	D_{90} (1)	D_C 1.5x(3)	D_I 0.5(D_{90}-D_C)
1	−1.699	−0.826	−0.436
9	−1.215	−0.285	−0.465
17	−0.699	0.171	−0.435
74	−0.823	−0.601	−0.111
80	−0.721	−0.181	−0.270
87	−0.111	−0.009	−0.051
120	−0.572	−0.219	−0.396
90 (Feed)	−0.549	−0.264	−0.142

the absolute displacements $\mathbf{D}_I$ nor the relative displacements in the table exceed 1/2 in. The position of the feed apparatus can be further improved if desired by utilizing a movable mount attached to the quadripod. This type of device would allow the antenna operator to shift the feed apparatus to the focal point of the paraboloid of "best fit."

In summary, the deflections (in the z_R direction) of the joints on the front face of the reflector can be restricted to an envelope of ± 1/2 in. (approximately 1/6 in. rms) by means of the following features:

1. Steering correction as given by Equation 6.
2. Initial rigging at zenith according to Equation 12.
3. Deflection control devices for shortening member 95-113 and its counterpart approximately 1.5 in.
4. Movable feed mount.

Under these conditions the maximum dead-load displacements will occur at the zenith and horizon positions, and the reflector will theoretically take the form of a true paraboloid at an elevation angle of 60°. Alternatively, the reflector surface could be rigged in the 60° position, in which case length changes of approximately ± 0.75 in. would be required in the control devices.

Acknowledgment

The preliminary design described in this paper was developed by the author while acting as a consultant to the Radio Physics Laboratory, Stanford Research Institute, Menlo Park, California.

References

1. *Aluminum Construction Manual,* The Aluminum Association (1959), Part 4 revised 1963.
2. *Manual of Steel Construction,* American Institute of Steel Construction (1963), 6th ed.
3. W. Weaver, Jr., *Computer Programs for Structural Analysis* (D. Van Nostrand Company, Inc., Princeton, N. J., 1967).
4. J. W. Young, and H. N. Christiansen, "Synthesis of a Space Truss Based on Dynamic Criteria," *J. Structural Div. ASCE* **92** (No. ST6), Proc. Paper 5021 pp. 425-442 (December 1966).
5. W. Weaver, Jr., "Dynamics of Elastically-Connected Rigid Bodies," Third Southeastern Conference on Theoretical and Applied Mechanics, University of South Carolina, Columbia, South Carolina (31 March – 1 April 1966).

Appendix: Background Study of Unsubdivided Paraboloidal Reflector

Geometry

The 400-ft-diameter paraboloidal reflector studied in this phase of the project is shown in Figures 23 and 24. It was intended that this configuration would subsequently be subdivided to the extent necessary to provide reasonable support for the mesh and stringers on the front face of the reflector.

An elevation bearing is assumed to be located behind joint 45 at a distance to be determined later when balancing the reflector. As a result of a previous study, quadripod legs framing into the structure at joints 37, 45, and 53 were considered to be an integral part of the structure.

Dead-load Analyses

Half of the reflector structure was analyzed under its own dead weight in the horizon and zenith positions. For purposes of these analyses, unit cross-sectional areas were assigned to all members in the structure, and the material was assumed to be structural aluminum 6061-T6. In addition, the elevation bearing was assumed (in this instance) to be located at joint 46, which was taken to be 30 ft forward of the rear face of the reflector structure. Figure 25 summarizes the significant results of these analyses. Displacements (in the z_R direction) of joints on the front face of the reflector are plotted

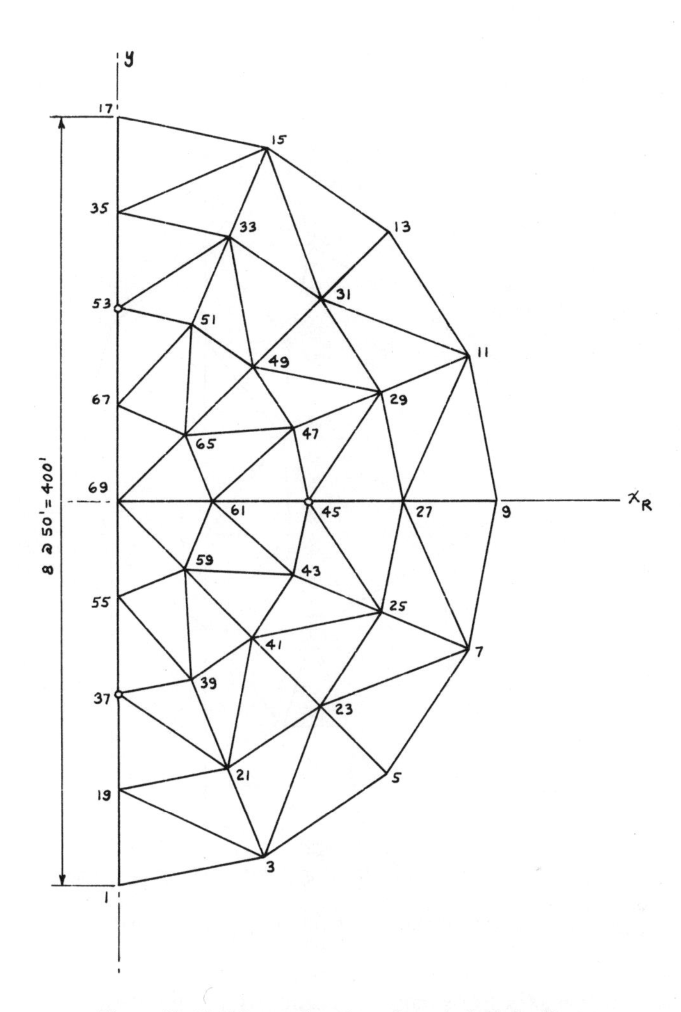

Figure 23. Half front elevation of paraboloidal reflector.

for both the horizon and zenith positions. The joints represented in Figure 25 are located at the centerline and at the edge of the reflector. Displacements in the horizon position are antisymmetric, and the maximum values of ±4.37 in. occur at joints 1 and 17. Symmetric displacements in the zenith position reach a maximum value of − 2.28 in. at joints 1 and 17.

Deflection Control

Inspection of Figure 25 reveals the fact that antisymmetric displacements in the z_R direction can be corrected by a rigid-body rotation of the reflector

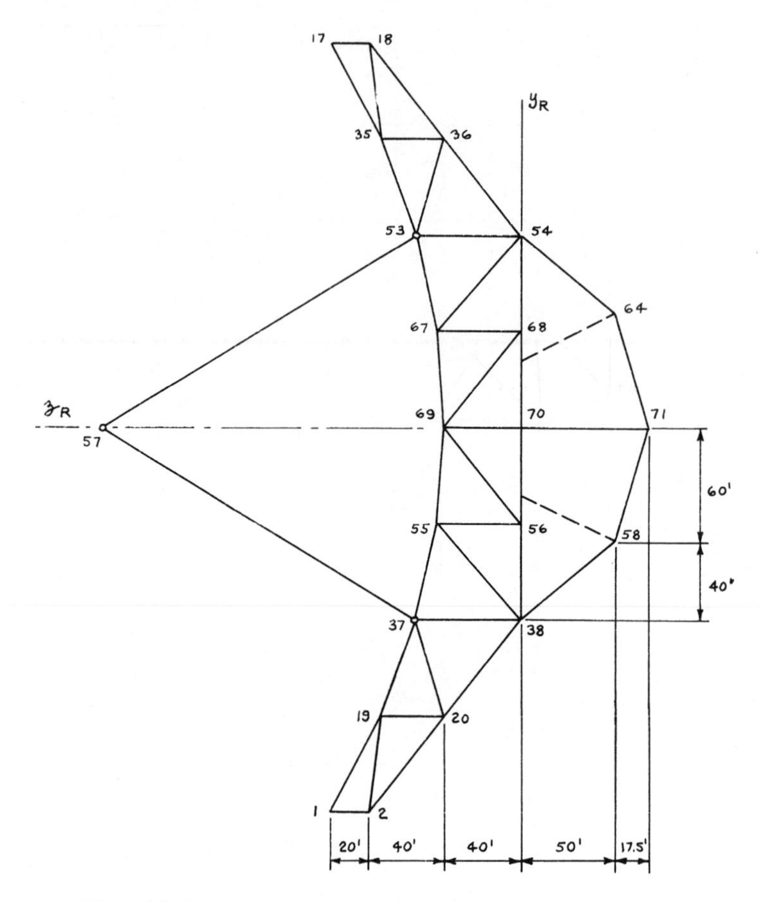

Figure 24. Vertical section at centerline of paraboloidal reflector.

about the elevation axis. For example, in order to return the z_R displacements at joints 1 and 17 to zero, the required correction angle θ is calculated as follows:

$$\theta = \frac{4.37}{200 \times 12} = 0.00182 \text{ rad} = 0.104 \text{ deg} = 6.25 \text{ min.}$$

The effect of a rigid-body rotation of this magnitude upon the y and z displacements of selected joints is shown in Table 5. The resulting net values of z displacements are considered to be negligible in comparison with the original values.

On the other hand, symmetric displacements in Fig. 25 must be corrected by some deflection control system built into the structure. For this purpose, the

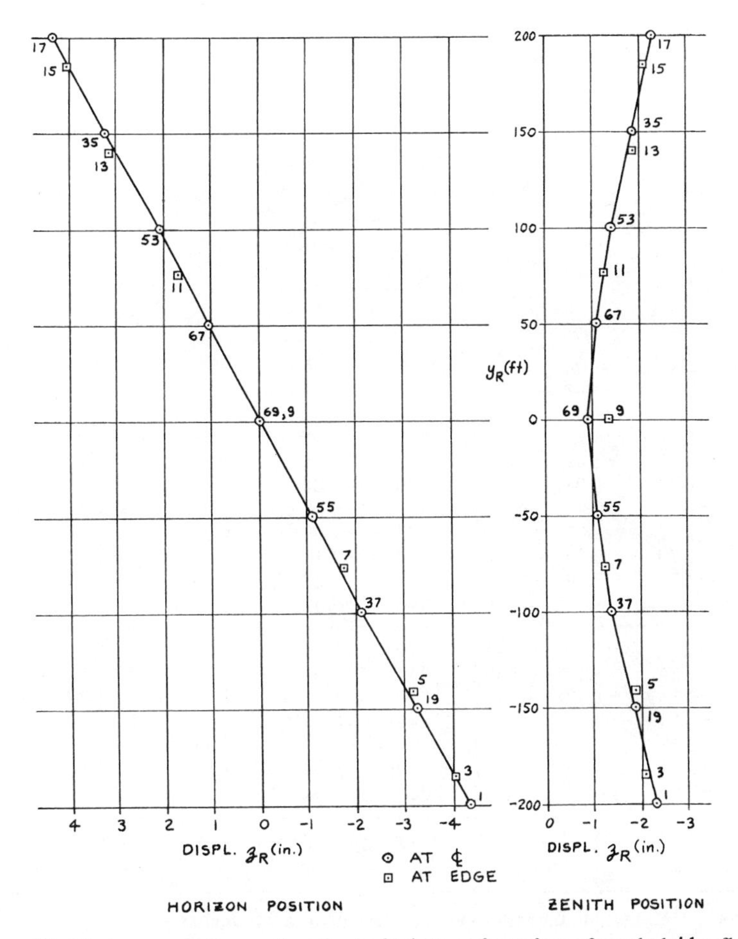

Figure 25. Displacements (in z_R direction) of joints on front face of paraboloid reflector due to dead load.

effects of contraction jacks in individual members of the backup framing were assessed for seven different members. The symmetric deflections in a paraboloid are difficult to control, but the study demonstrated that their magnitudes could be reduced to fall within an envelope of approximately ± 0.5 in. at all joints on the front face of the reflector.

Variation of Parameters

Several factors were varied in the paraboloidal reflector in order to test their effects upon the dead-load deflections. For example, when the quadripod legs were eliminated from the analysis, the displacements for the zenith

Table 5. Effect of Rigid-Body Rotation.[a]

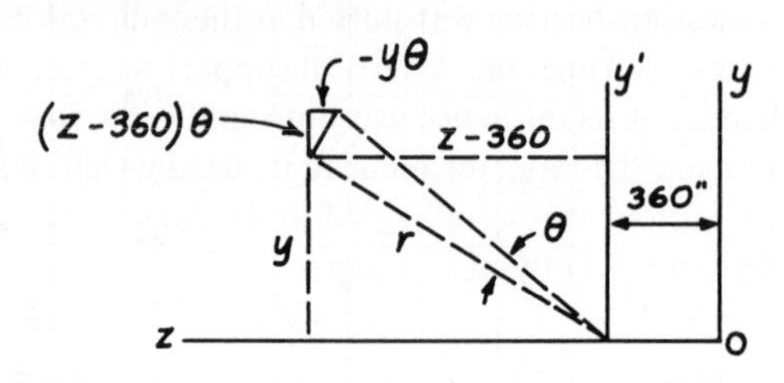

Joint	y displ.	z displ.	y	z-360	−yθ	(z−360)θ	Net y displ.	Net z displ.
1	−2.72	−4.37	−2400	840	4.37	1.53	−1.19	0
19	−2.14	−3.24	−1800	527	3.28	0.96	−1.18	0.04
37	−1.70	−2.09	−1200	304	2.18	0.55	−1.15	0. 09
55	−1.42	−1.08	−600	170	1.09	0.31	−1.11	0.01
69	−1.30	0	0	126	0	0.23	−1.07	0
67	−1.42	1.08	600	170	−1.09	0.31	−1.11	−0.01
53	−1.70	2.09	1200	204	−2.18	0.55	−1.15	−0.09
35	−2.14	3.24	1800	527	−3.28	0.96	−1.18	−0.04
17	−2.72	4.37	2400	840	−4.37	1.53	−1.19	0
3	−2.64	−4.04	−2217	840	4.04	1.53	−1.11	0
5	−2.64	−3.19	−1697	840	3.09	1.53	−1.11	−0.10
7	−2.53	−1.72	−918	840	1.67	1.53	−1.00	−0.05
9	−2.54	0	0	840	0	1.53	−1.01	0
11	−2.53	1.72	918	840	−1.67	1.53	−1.00	0.05
13	−2.64	3.19	1697	840	−3.09	1.53	−1.11	0.10
15	−2.64	4.04	2217	840	−4.04	1.53	−1.11	0
57 (Feed)	−5.07	0	0	2142	0	3.90	−1.17	0

[a] $\theta = 0.00182$ rad (all dimensions and displacements are in inches.)

position increased by approximately 30 per cent.

In an effort to reduce the number of framing members in the reflector, cross-sectional areas of certain braces were set to zero in a systematic manner. By this approach it was determined that certain interior braces could be eliminated without increasing deflections significantly or creating an instability. However, the removal of other members, such as the front face diagonals, caused a geometric instability in the structure which was detected by the advent of a singular stiffness matrix.

The location of the elevation bearing was varied in the z_R direction, leading to the conclusion that antisymmetric deflections can be corrected by a rigid-body rotation for any reasonable position of the elevation axis. The location of the elevation bearing was shifted in the x direction, but the effect upon deflections was detrimental. A vertical support reaction at the elevation drive was considered, but its influence was not significant. Decresing the depth of the main framing in the reflector reduced its dead weight but increased the deflections. These various studies all aided in the final choice of the reflector geometry described in the body of this paper.

H. Rothman and F. K. Chang
Ammann & Whitney, Inc.,
New York

MAINTAINING SURFACE ACCURACY OF LARGE RADIO TELESCOPES BY ACTIVE COMPENSATION

Introduction

In the design of large-diameter radio telescopes in the range of 400 ft or larger with small rms surface tolerance (less than a fraction of an inch) and high pointing accuracy (pointing error less than a fraction of a minute), the use of the brute force method or homology for deflection control may become uneconomical if not impossible. The use of partial compensation, either active or passive, may not solve the problem. In such a case, one has no choice but to rely on fully active compensation for a solution.

The Compensation Method

The most efficient method for active compensation is the separation of the reflection surface from the supporting structure. The uncoupled reflector surface is divided into individual triangular segments. The corner of each segment is kept in its correct position by reversible screw jacks interposed between the segment and the supporting structure.

This method of compensation is especially suitable for a nonenvironmental antenna. The reflector surface will be surveyed in a reference position, say face up, while the screw jacks are all in their home position. If computed or measured deflections at various attitudes are then superimposed on the surveyed shape, the uncorrected shape of the reflector is known for these attitudes. This is all that is necessary for the compensation of the jack extension needed to produce a perfect paraboloid. Since the best-fit paraboloids can be found for various attitudes, the movement of the jack screws will not be too excessive. These curves of jack extension (actually rotation) versus attitude, stored in a computer, will be used to monitor and adjust the jack positions.

For environmental antennas, the compensation method will become much more involved. The theoretical surface of the reflector should be established at all times by some means. One method proposed for a previous design is the use of reference tower made of Invar bars with negligible deformation due to its own weight as well as the optical equipment and temperature changes. A pair of projectors, one on the top of the tower and one on the bottom of the tower, are adjusted so that their light beams intersect at a point on the theoretical paraboloid over a jack. If 144 jacks are used, then 144 pairs of projectors are required. Once the reference paraboloid is defined, the correction of the actual structure can be made.

Photosensitive cells, mounted on the reflector panels, send signals to logic circuits that actuate the screw jacks.

The active compensation method also offers many other advantages such as: (1) the supporting structure can be optimized on cost rather than on deflection, and normal and therefore inexpensive techniques can be applied to the backup construction, (2) distortions due to temperature, added loads, and loads other than gravity can be compensated, provided they are predictable and vary slowly with time, and (3) the antenna can be returned by the use of the computer with no rigging.

The Hardware

The method of dividing the uncoupled reflector surface into triangular segments in shown in Figures 1 to 3. There are many advantages associated with this type of construction.

1. The system is statically determinate as shown in Figure 1. The B triangle is supported on the A triangle at two hinged upper corners, 1 and 2. There are six reaction components on this triangle and therefore it is statically determinate as far as external reactions are concerned. It can be seen that triangle A is also statically determinate. Since the panel system is statically determinate, if a jack freezes or fails at midstroke, there will be no damage to the panels.

2. The arrangement will permit the free expansion or contraction of the panels due to temperature difference between aluminum panels and steel backup structure.

3. The support pattern shown in Figure 2 will also minimize the panel clearance gaps. These gaps should be adequate to allow for jack motion, failure of adjacent jacks, structural deflections of the framework parallel to the surface under wind and dead load as well as the temperature difference

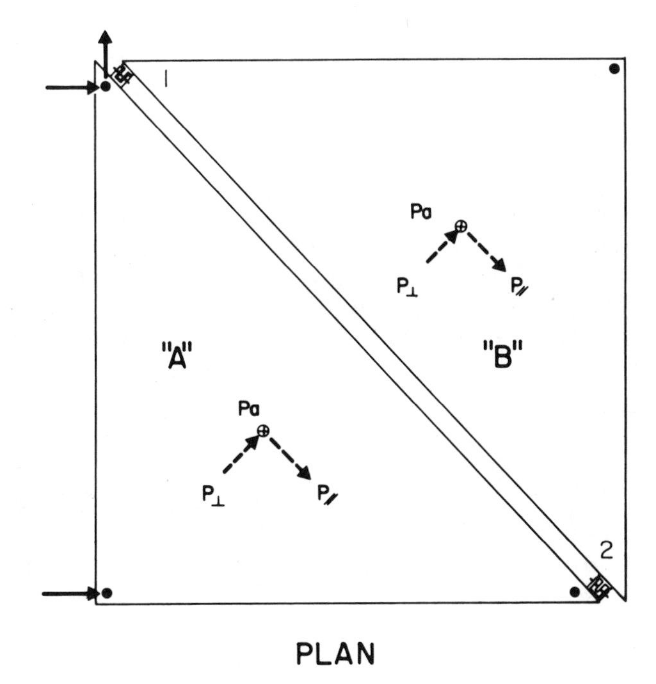

Figure 1. Force components.

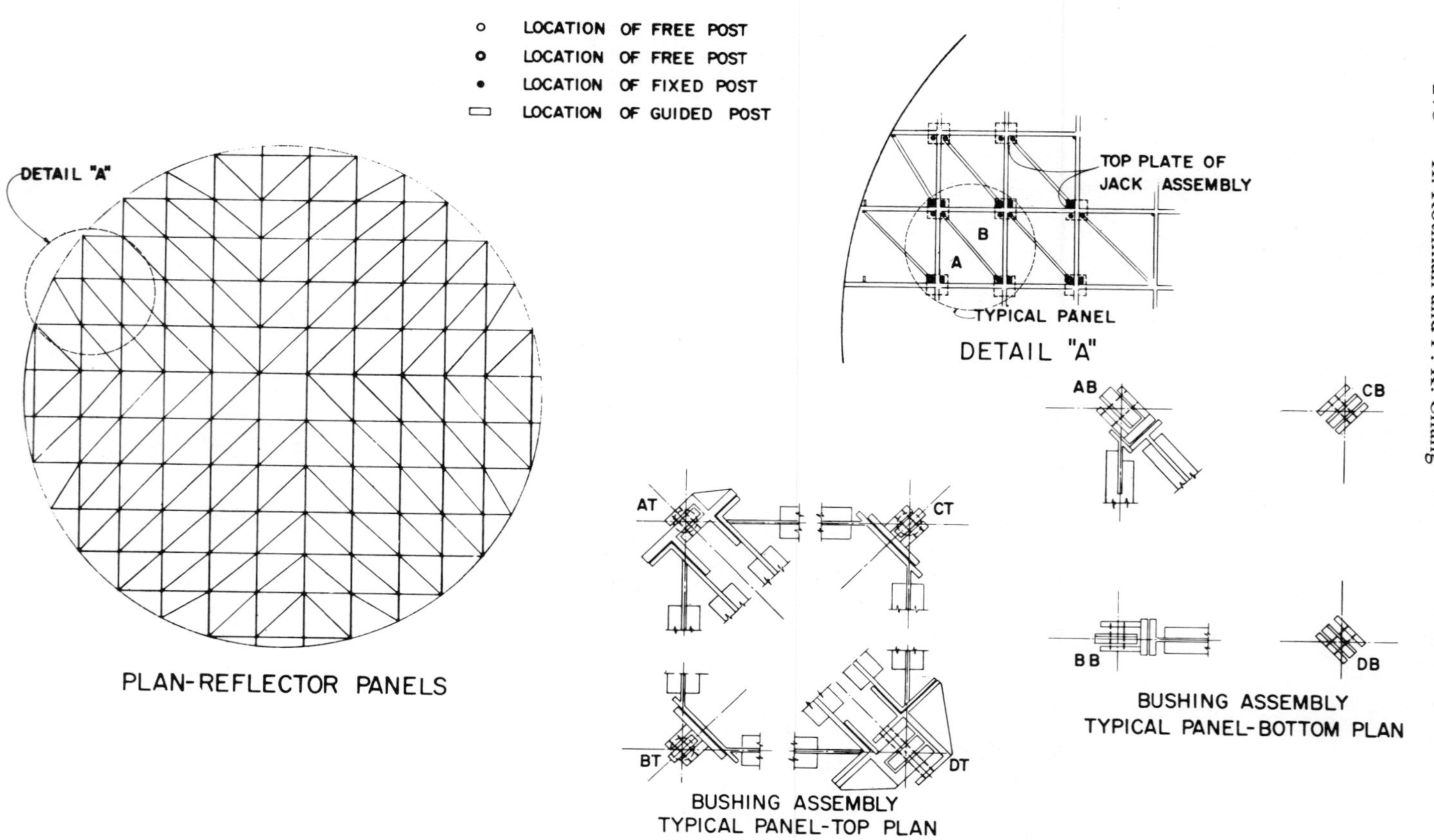

Figure 2. Reflector panels.

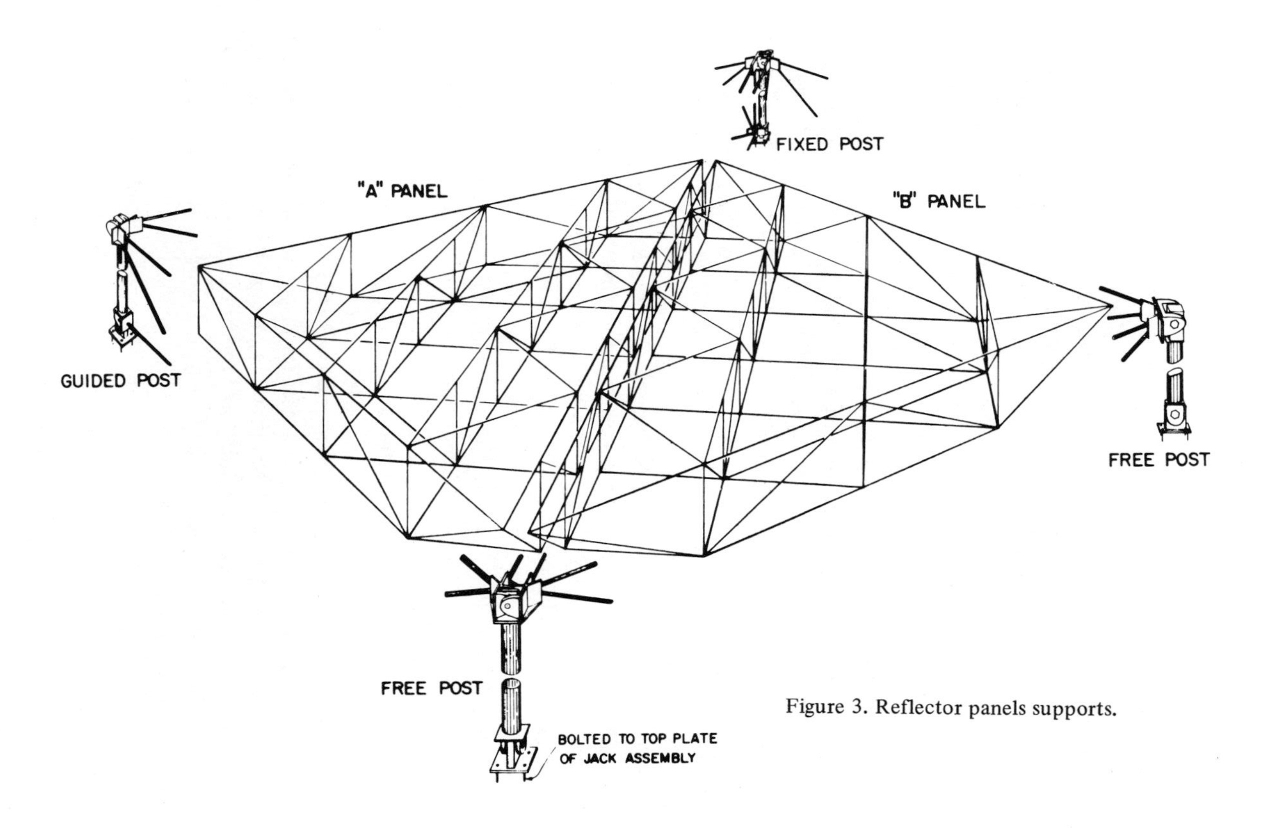

Figure 3. Reflector panels supports.

mentioned in 2. Since gap areas are nonreflecting shadow areas, it is important to minimize the gap dimensions.

4. The arrangement will permit all the jacks to be located parallel to the main axis of the paraboloid. The motions required for the jacks are along their axis only.

The corner fittings, bearings and post details are given in Figures 2 and 3. Each of the four steel corner posts has an upper and a lower spherical bearing; reactions to force components that are parallel to the posts are therefore easily solved. It is seen that no surface-plane reactions are possible at the C corner. A sliding pin along the diagonal line of the DT corner provides that force components on B panel in the direction of the diagonal are supported by the AT fitting and ultimately by the A panel. Force components perpendicular to the diagonal on the B panel are supported by both AT and DT fittings on A panel. Thus the B panel can be considered to be suspended on the A panel. Only one lower fitting can react forces in all directions: the AB corner.

One alternative scheme for the triangle panels is the use of pyramids as shown in Figure 4. It has been found that this type of construction is very efficient. This configuration also places the structural weight as far as practicable toward the trunnion axis. This results in a lighter counterweight and a smaller moment of inertia for the structure.

Figure 5 shows one type of construction for the jack assembly. Each of the jacks that support the reflector panels is a moving tube that slides within a backup structure fixed tube. The moving tube is activated by a drive screw revolving in a thrust bearing mounted on the fixed tube.

Each drive screw is coupled through a speed reducer to a two-speed fractional horsepower motor with magnetic brake. The combination may be made of modified stock items. The modifications include a special housing and oil seals for multiposition use.

Both the thrust bearing and the threaded drive screw have takeup details to eliminate backlash and both are pressure-lubricated.

The fixed tube in the backup connection is bored for a pressure-lubricated bronze bushing. Hand holes permit access for inspection of the details within the tube. An accordion-shaped Neoprene boot together with gasketed hand hole covers protect the screw assembly and prevent escape of excess lubricant.

Another type of jack assembly is shown in Figure 6. In this scheme, the jacks are placed inside the vertical members of the backup structure which are parallel to the axis of the paraboloid. The circular plunger is guided through upper and lower circular bearings of cast steel with spherical bronze bushings,split and bolted together. Torsion loads about the axis of the plunger were resisted by a shear plate key mounted through the plunger and extending into slots in the bearing. The upper bearing was supported on four or more structural legs which transmitted the loads to the upper plate of the weldment

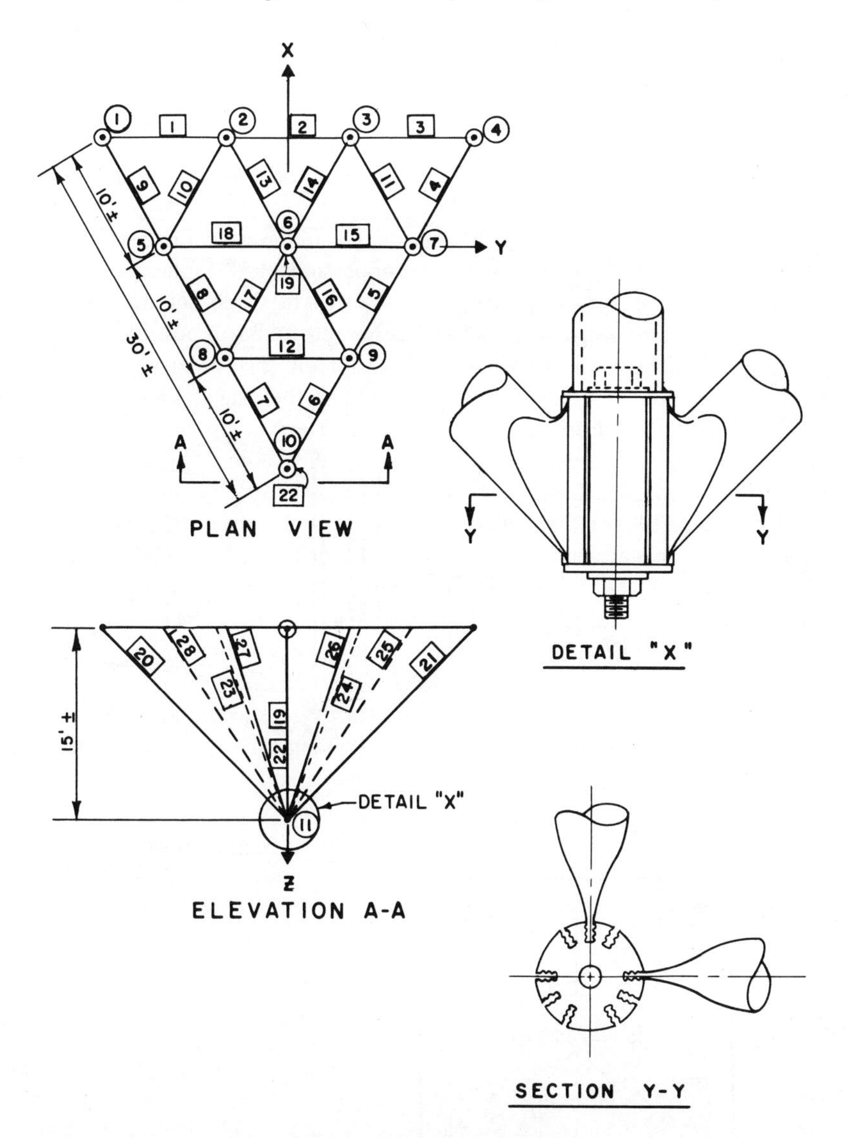

Figure 4. Pyramid assembly.

joining the truss members at the top joint. The use of an exterior fixed pipe for bearing support is shown in Figure 6.

Although the jack assemblies are rugged, they should be inspected periodically and provision must be made for repair and replacement. This can best be done when the radio telescope is in the face-up position. Access to all jack assemblies should also be provided.

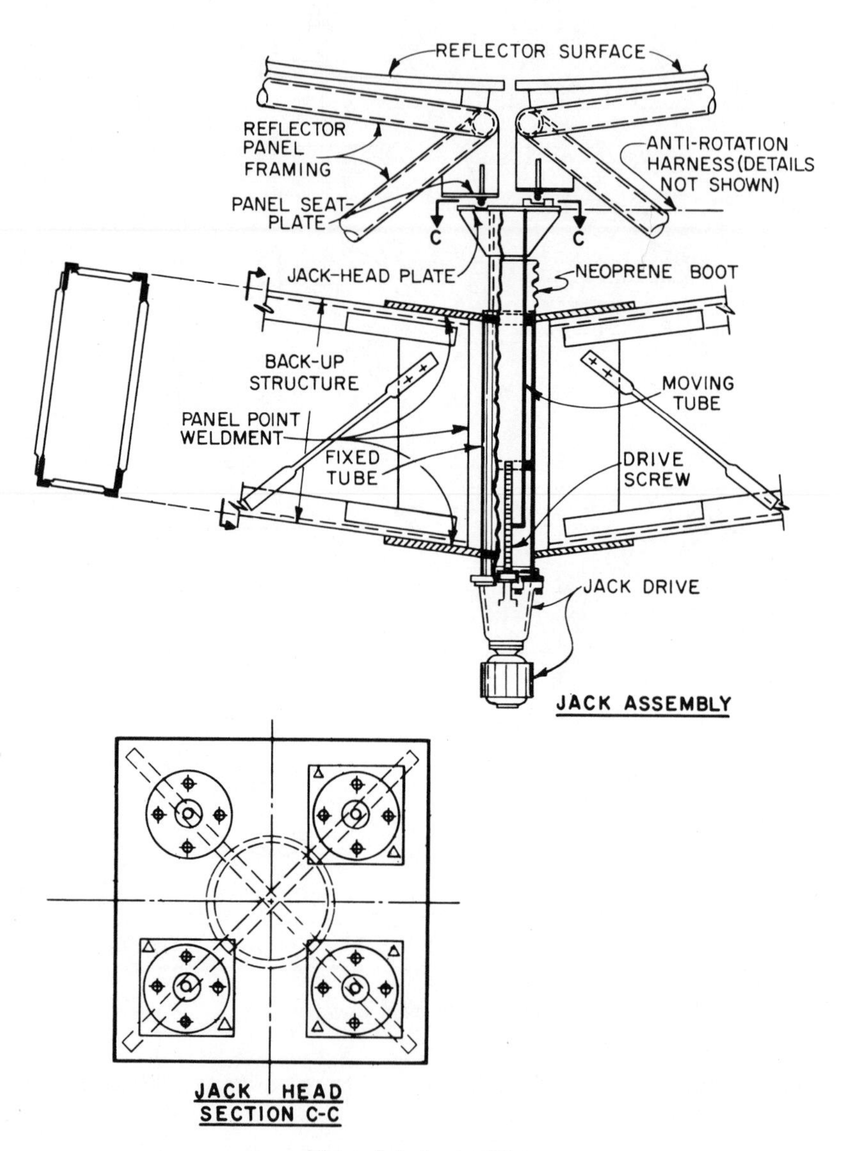

Figure 5. Jack assembly.

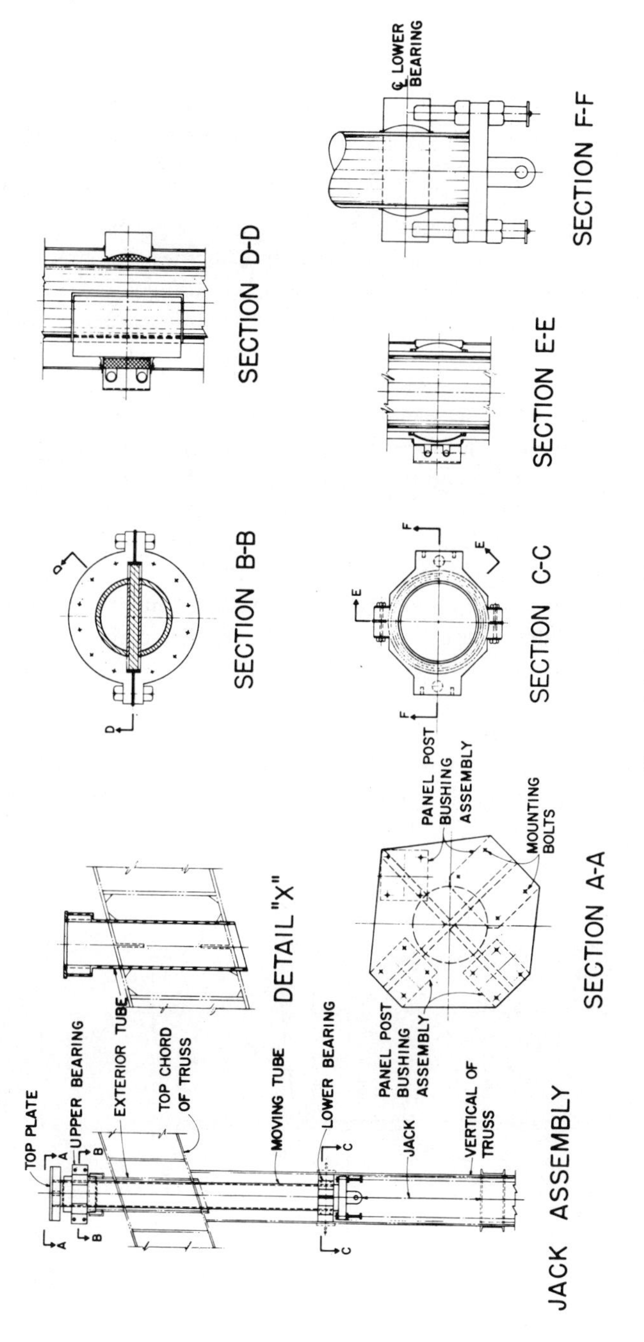

Figure 6. Jack assembly.

Compensation Control

For a nonenvironmental telescope, a semiclosed loop control can be used. One approach to the problem is outlined as follows:

A process control computer receives altitude, azimuth, and time information from the main computer, and program modification instructions from a console or remote typewriter. It also receives signals from a revolution counter within the speed reducer assembly. The difference between the measured and theoretical revolutions is computed, and one of the three possible signals depending on the difference is sent out. The three possibilities are: jack-forward, jack-backward, or do-not-jack. Because the signal opens or closes relays, a signal to an advancing jack to advance will permit the motion to continue. A jack set in motion will stop only when a subsequent stop signal is received.

The operating cycle of the computer-jack system is as follows: As shown in Figure 7, on discovering a panel point deflected more than 0.028 in. off its theoretical position, the computer signals the jack controlling the point to move toward the correct position. It will re-examine the jack position every 10 sec and will turn the jack motor off when the jack has overcorrected by more than 0.025 in. It continues examining this (and all other points) with a 10-sec period, responding to errors over 0.028 in. with an overcorrection of 0.025 in.

These intervals plus motions beyond these limits due to delayed computer discovery produce a dead band 0.030 in. each side of the mean. A random position within the band is ensured by the varying rate at which the points deflect, by a random monitoring sequence, and by a special starting sequence that positions the jacks randomly in the band.

The jacks operate in either direction, at a constant speed that is slightly

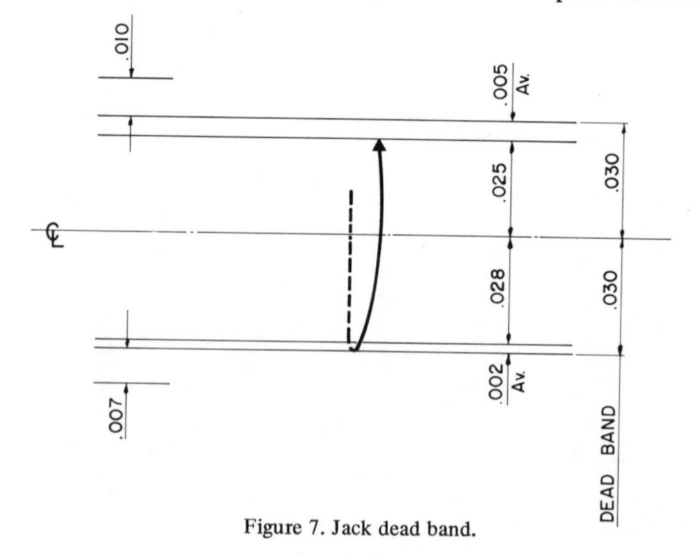

Figure 7. Jack dead band.

greater than the speed of the fastest deflecting panel point. This speed establishes the maximum computer return period per jack. Too long a period will permit the jacks to exceed the limits of the ± 0.030 dead band.

The dead-band width is a first cycle of optimizing the requirements of small deflections servo stability and random deflections. It produces an rms surface error of 0.017 in.

Assume that the tracking speed is 0.005°/sec and the maximum deflection of the backup structure over a 75° altitude motion is 10 in., the structure deflection speed is 0.005/75 x 10 = 0.0007 in./sec. A 1/2 rpm jack screw (8 threads per inch) rotation will be satisfactory since the jack speed will be 0.5/8 x 60 = 0.001 in./sec, which is faster than the deflection speed. For the jack to travel 0.055 in. as indicated in Figure 7, 55 sec minimum and 180 sec maximum are required. The jack need never restart in less than 80 sec. The computer must examine 144 jack points in 10 sec, allowing 70 msec to compute the error and signal a relay. The maximum distance which a jack overcorrects is 0.001 x 10 = 0.010 in. and the maximum distance which a jack permits to slide out of band is only 0.0007 x 10 = 0.007 in.

In the slewing mode, the logic and bandwidth will remain the same, but the jack will be sped up by a factor of about 15 by the use of a two-speed motor as mentioned previously to ensure that the antenna is ready to track shortly after the slew stops. During the slew, the jacks may overshoot the dead band by about 1/8 in. requiring at most 2 min to return to a suitable surface. Most of this delay will take place while the antenna is slowing down from slew to track and in practice should add less than 1 min to the slewing time.

For an environmental telescope, a completely closed loop control should be used. This type of configuration control is very complex and will not be examined here.

Error Analysis

Special treatment is required in the error analysis of an actively compensated reflector surface. In order to demonstrate the general approach to the problem, the following example is outlined: Assume the diameter of a nonenvironmental Cassegrain antenna is 400 ft, the secondary reflector, 40 ft. A reticulated shell-type backup structure is used with the maximum deflection in the order of 10 in. Some of the errors given in the example are actually computed values while some are assumed figures.

Let us examine first the various rms errors affecting the surface accuracy of the primary reflector:

1. Jack dead band—as mentioned previously, the jack dead band is ± 0.030 in. In other words, the errors are randomly between these limits. The rms error will be in the order of *0.017 in.*

2. Readout error of jack rotation–allowing a 10° error in the shaft rotation for an 8 threads to the inch screw, the peak error will be 1/8 x 10/360 = 0.0034 in. and randomly distributed. The rms error is about *0.002 in.*

3. Deflection calibration curve error–one major source of compensation error is the inaccuracy of the deflection calculations. Considering that the calculations can be augmented by direct measurement, the error may be conservatively taken as 1 per cent of the total deflection range about a mean position. The deflection range is far greater than the screw stroke since it includes the rigid body displacement of the telescope. Since the range of the worst panel deflection is 10 in., the peak calibration error will then be of the order of 0.10 in., and the random error will be conservatively taken as *0.050 in.* rms.

4. Panel deflection–although nonrandom, the jacks will operate to make the mean of the total panel deflection zero, rather than to make the panel corner deflections zero. This removes the bulk of the bias effect, permitting panel deflection to be treated as random.

The rms deflection of a bias-rigged panel on the center of the dish is 0.010 in. All other panels have smaller deflections. The rms deflection for the whole reflector surface is estimated to be 0.006 in. before the jack reduction of the mean taken place. This could reduce the deflection to a negligible value. However, in order to be conservative. and to allow for some bias, an rms value of *0.010 in.* will be used.

5. Initial survey–for the initial survey the error at jack points only will be assumed to be *0.010 in.* rms.

6. The panel manufacturing error–this error is assumed to be *0.010 in.* rms.

7. Miscellaneous errors–errors or at least part of the errors due to hysteresis, minor damages, bearing and roller imperfections can be correlated with altitude and azimuth position. If they can be measured, they can be considered a systematic error, and compensated. The rms error is assumed to be *0.005 in.*

For the secondary reflector, since it is a single structure with a diameter smaller than that of the primary reflector, rms errors due to various sources except that due to jack dead band and readout errors of jack rotation are substantially smaller than that for the primary reflector. These errors can be listed as follows:

8. Jack dead band–*0.017 in.* rms.
9. Readout error of jack rotation–*0.002 in.* rms.
10. Deflection calibration curve error–*0.025 in.* rms.
11. Deflection due to gravity–*0.008 in.* rms.
12. Initial survey–*0.005 in.* rms.
13. Manufacturing error–*0.005 in.* rms.

All the errors (1) through (13) can be taken as random, The total rms error of the system is $(0.017^2 + 0.002^2 + 0.050^2 + 0.010^2 + 0.010^2 + 0.010^2 + 0.005^2 + 0.017^2 + 0.002^2 + 0.025^2 + 0.008^2 + 0.005^2 + 0.005^2)^{1/2} = 0.064$ in. rms.

Deflections due to temperatures and acceleration have not been covered in the above analysis. Temperature deflections are caused by a nonuniform temperature state in the structure. This can be due to a nonuniform time response to a temperature change, to a stratified temperature environment, or to the absorption of solar radiation. These effects can be partially controlled by a radome heating-cooling system. Observations may show that they are partly predictable, and may possibly be a function of telescope position, time of day, date, and weather. If so, it can be compensated; if not, corrections based upon thermometer readings may be introduced into the computer a few times a day. Centrifugal forces due to both altitude and azimuth rotation and tangential forces due to acceleration cause small deflections. Compensation of these deflections can be performed by adding terms that are a function of the tracking parameters, and therefore, a function of time.

For telescopes with reflector diameters other than 400 ft, its total rms can be estimated by using the following plausible assumptions:

1. The survey errors are proportional to the antenna diameter.
2. Deflections and therefore, errors in the deflection prediction are proportional to the square of the diameter.

The rms surface error versus telescope size is plotted in Figure 8. It can be seen that for a 500-ft-diameter nonenvironmental telescope, an rms surface error in the order of 0.093 in. rms can be achieved; for a 300-ft-diameter, 0.043 in. rms. If the jack dead band can be reduced, a substantially lower surface error can be achieved for large-diameter telescopes.

Reliability

No estimate of mean time between failures of the jack system has been made. However, the over-all reliability appears to be very high. The loads, accelerations, and velocities are all so small that each element of the assembly can be oversized at a small cost making failure due to wear a minor factor. The assembly carries loads smaller than an automobile jack and is not much more complicated.

If the process control computer is merely a part of the main computer, it will make practically no contribution to a reduction in reliability. A computer failure will already have been included in the aiming system reliability. If it is a separate computer, its reliability is a factor that is manageable with redundancy.

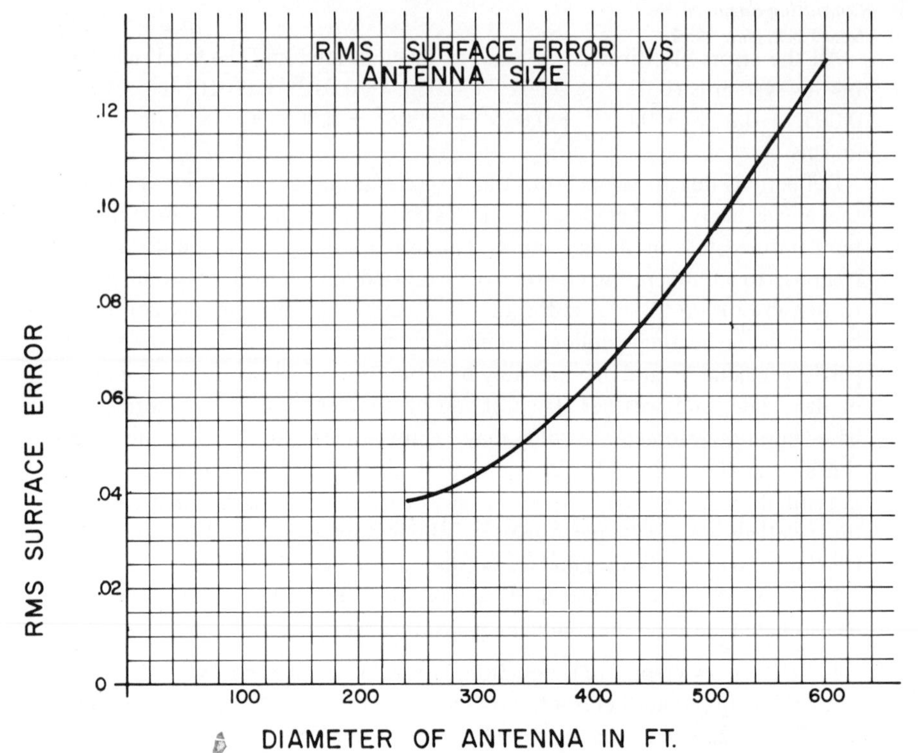

Figure 8. Root mean square error versus antenna size.

In no event can a failure of a jack assembly or electronic component cause structural damage to the reflector or backup. This is due to the statically determinate support system of the reflector panels. A low-probability motor or screw failure means at worst, a short delay for repairs, and at best, continued operation of the telescope with a 3 per cent loss of gain.

Conclusion

The uncoupling of the reflector surface from the backup structure and the use of active compensation is an effective method for maintaining the surface accuracy of large-diameter radio telescopes. The hardware required is inexpensive and readily available. The compensation control is not too difficult for nonenvironmental surfaces and complex but feasible for environmental conditions. A very favorable rms surface error can be achieved by compensating out the predictable errors and all the remaining errors can all be taken as random.

There is no doubt that active compensation will have its place in the future development of large-diameter radio telescopes. For a very large diameter surface with a very small rms tolerance and permissible pointing error, this may prove to be the only feasible method.

Paul Weidlinger
Consulting Engineer
New York, New York

CONTROL OF RMS SURFACE ERROR IN LARGE ANTENNA STRUCTURES

1. Introduction

This report contains a summary of the results obtained from a preliminary engineering feasibility study for a 400-ft-diameter radome-housed Cassegrain antenna. The study was executed under a contract (Purchase Order SR 10388) with Massachusetts Institute of Technology, Lincoln Laboratory, in behalf of the Cambridge Radio Observatory Committee.

The report presents a concept of an antenna structure satisfying the following requirements:

Sky coverage:	Elevation +15° to 90°
	Azimuth ±300°
Angular velocity:	Tracking 0.25°/sec
	Slewing 1.00°/sec
Angular acceleration:	0.04°/sec^2 max (at 87°)
	0.01°/sec^2 (typical)
Operational pointing accuracy	15 sec
Operational surface accuracy	rms = 0.10 in.
Mechanical resonant frequency	$<$ 0.05 cps
Ambient temperature:	-40°F to +140°F – survival
	+45°F to +85°F – operational
Gradient:	10°F top to bottom; 5°F side to side.

In the process of evolving the "force compensating" concept, numerous other approaches were investigated and reviewed. Some of the details and ideas found in the course of this review are incorporated in the design of the antenna structure and several of these are briefly described Section 2 of this report.

The antenna structure concept presented in this report consists of a parabolic dish supported on a series of parallel backup trusses. The gravity distortion of these trusses is actively compensated by externally applied forces during rotation around the elevation axis. The details of this force compensation concept are given in the body of the report together with drawings and description of the entire antenna structure and substructure. The report also contains detailed weight breakdown of the various components and an extrapolation of those weights in the 300- to 500-ft-diameter range. It is found that the weight is directly proportional to the square of the dish radius and significant increase in accuracy is achievable without comparable increase in cost.

2. Control of Distortions of Antenna Structure

The control of system tolerance due to all causes of a large-diameter fully steerable Cassegrain antenna structure is the principal development and engineering challenge, and it has a major influence on the cost and feasibility of these instruments. To achieve the design objectives, various techniques have been proposed, and many of these are considered and included into the design concept of all antenna structures. The most important of these follow.

1. *Elimination of source of distortions.* An important part of distortion can be eliminated by protecting the antenna structure from large ambient temperature variations and temperature gradients and by protecting it from the effects of wind and weather in general. This is accomplished by housing the antenna in a radome. The influence of the radome on the cost and efficiency of the antenna has been previously evaluated by CAMROC. It was found that because of noise and temperature effects, a radome-enclosed antenna will have to be 6 to 10 per cent larger in diameter to achieve an equivalent aperture efficiency. Since the cost of the radome appears to be approximately equal to the cost of the antenna, it was decided by CAMROC to restrict the present studies to radome-housed antenna structures. In view of the radome environment, wind and large temperature effects are, therefore, eliminated as a contributory cause of distortions.[1]

2. *Distortion control of the substructure.* Rotation around the azimuth axis does not produce static deformations and the substructure deformation can also be made largely independent of the elevation angle of the dish by bringing the elevation axis bearing to coincide with the center of gravity of the antenna dish. Further improvements can be made by appropriate details at these points to minimize torsional distortions caused by the torque of the elevation axis drive. This approach is used in the "hammerhead antenna" concept of Lincoln Laboratory.[2] A similar principle is used in the antenna concept presented

in this report.

3. *Control of panels and stringers.* These elements are usually of small dimensions as compared to the major antenna dish backup structure and consequently their contribution to the system error budget is minor. Nevertheless, because of the weight superimposed by these subassemblies to the backup structure, the weight stiffness of these elements merit serious studies. An excellent solution to this problem is proposed in Reference 2.

4. *Cambering and bias-rigging.* In controlling the surface tolerance of the antenna dish, cambering of the backup structure can bring about a significant reduction in the rms deformation of these major elements. Generally it could be anticipated that by optimum cambering the rms error can be reduced to approach the difference of the extremes of these values at the corresponding elevation angles ϕ_1 and ϕ_2.

$$\text{rms}_{\text{cambered}} = \text{rms}\,(\phi_1)_{\text{max}} - \text{rms}\,(\phi_2)_{\text{min}}\,. \tag{1}$$

The actual reduction obtained seems to approach a factor of 5 for large rms values in some cases that were investigated. It appears that reductions by a factor of 2 can be achieved in case of smaller rms quantities. On the other hand, cambering and bias-rigging, especially to optimal values can significantly increase both shop fabrication and erection costs and these must be compared to other means by which corresponding reductions of rms errors are obtainable.

5. *Best-fitting.* A further reduction of the rms error of the reflecting surface can be obtained by best-fitting the distorted surface at some or each elevation angle to a paraboloid of revolution of different focal length and by an appropriate rigid-body translation and rotation. The reduction of the rms error which may be achieved by best-fitting of a shell in the shape of a paraboloid of revolution is estimated at one order of magnitude[3] and for linear elements, i.e., plane trusses, it is found that a reduction by a factor of 5 may be attainable. (See Section 3 of this report.)

To achieve the required system tolerance it may be necessary to use all, or many, of the above-listed procedures. Since the weight contribution to the system error and, even more to the surface error, derives from the antenna dish backup structure, special effort, and techniques need to be applied to this component. At the present time the following approaches seem to be possible and promising:

6. *Maximum stiffness.* This approach is the conventional technique of design and it requires the achievement of maximum stiffness, consistent with minimum weight. This method requires stringent and, therefore, costly fabrication and erection tolerances. At present it appears that large high-precision antennas cannot be built without at least the partial application of some of the com-

pensating techniques described below.

7. *Control by compensation of displacements.* At the extreme application of this method, the antenna backup structure is designed without regard to distortion requirements and surface tolerance is achieved by controlled displacements of the panels of the surface itself. Each panel must be supported on jacks that are continuously adjusted to the required position during rotation around the elevation axis. This method requires elaborate servo systems and continuous monitoring of displacements.

8. *Control by homologous deformations.* This method, developed by S. von Hoerner, takes advantage of the fact that it is possible to define the geometry and the cross-sectional area of the members of a space truss in a manner that a paraboloidal surface defined by the nodes of the truss will be continuously deformed, unhindered, into another family of paraboloids, at all elevation angles. The method in some respects is a further exploration of the previously described best-fitting procedure. Its application to an actual antenna structure has, as yet, to be attempted.

9. *"Exotic" solutions.* There are a number of unusual approaches that have been proposed, such as the use of helium-filled structures, a spherical metal shell-radome combination floated on air, a reinforced concrete structure supported on torroidal shells floated on water. While the advantage gained from these solutions is evident, a number of technological problems remain to be solved before they can be evaluated.

10. *Force compensating concept.* Distortions of the antenna surface can also be controlled by applying one or more external forces that vary with the elevation angle. In the ideal situation, i.e., with sufficient number of compensating forces, the gravity deformations are independent of the elevation angle of the antenna and it approaches that of a structure operating in a gravity field rotating with the elevation axis. If the compensating forces are at an optimum level their effect approaches that of zero gravity. The most significant practical advantage lies in the relative simplicity of the control of the magnitude of compensating forccs that are independent of the magnitude of distortions; control, therefore, does not require continuous monitoring with respect to a fixed reference point. The preliminary design of the antenna structure presented in this report is based on the "force compensating" concept.

3. rms Tolerance Limit of Simple Structures

Since the major effort in the development of an economical antenna system is in the direction of obtaining an acceptable rms surface tolerance within the most economical solution, certain basic considerations leading to initial estimates of limiting quantities are desirable.

Apart from the more exotic solutions, it is safe to assume that the minimum weight (and approximately minimum cost) could be achieved if the antenna backup structure were designed by strength considerations only, i.e., without regard to stiffness. The attainable rms surface error of structures designed for strength only is therefore of fundamental importance in evaluating the efficiency of other designs. It turns out that the rms deformation of linear elements (such as plane trusses) of constant strengths can be readily calculated, without additional assumptions. Consider a plane cantilevered truss of span R, and depth h designed for constant strength in bending. This implies a constant extreme fiber stress and, therefore, constant maximum strain ϵ_m. If the depth h is also constant along the span R, the structure will deform with a constant radius curvature, which for small deflections is given by

$$\delta(x) = \epsilon_m(R/h)R(x/R)^2 \tag{2}$$

and the peak deflection δ_m at $x = R$ is

$$\delta_m = \epsilon_m(R/h)R, \tag{3}$$

where the span-depth ratio R/h is an essential geometric design parameter. The rms deflection can be calculated from Equation 2:

$$\text{rms} = (1/\sqrt{5})\epsilon_m(R/h)R\,. \tag{4}$$

The ratio of peak to rms deflection by Equations 3 and 4 is

$$\delta_m/\text{rms} = \sqrt{5} = 2.24\,. \tag{5}$$

In steel structures operating at 15 to 20 ksi, average peak stress levels, the strain is bracketed between

$$5 \times 10^{-4} \leqslant \epsilon_m \leqslant 6.7 \times 10^{-4}\,. \tag{6}$$

For the lower value of ϵ_m Equation 4 can be written as

$$\text{rms} = 2.24 \times 10^{-4}(R/h)R\,. \tag{7}$$

If the deflection curve of Equation 3 is best-fitted to a straight line, the best fit rms is given by

$$\text{rms}_{\text{best fit}} = (1/3\sqrt{2})\text{rms} = 0.23\ \text{rms}\,. \tag{8}$$

The customary range of values of the span depth ratio for backup trusses is

$$2 < R/h < 4\,, \tag{9}$$

and on Figure 1 approximate typical rms values attainable for least-weight designs are shown. The aim of further improvement is to produce lower rms errors with weights which are comparable to those obtained by strength design only.

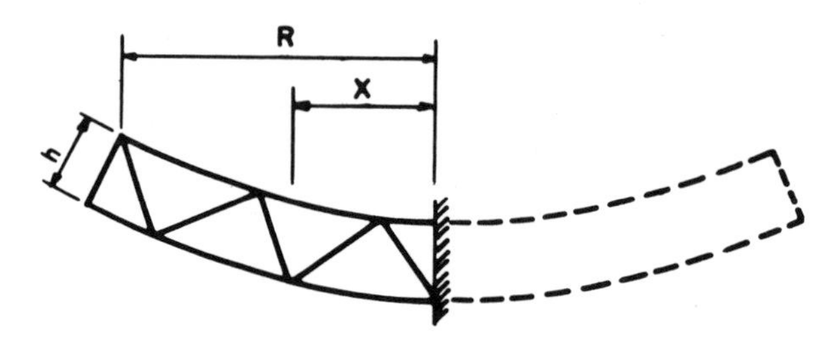

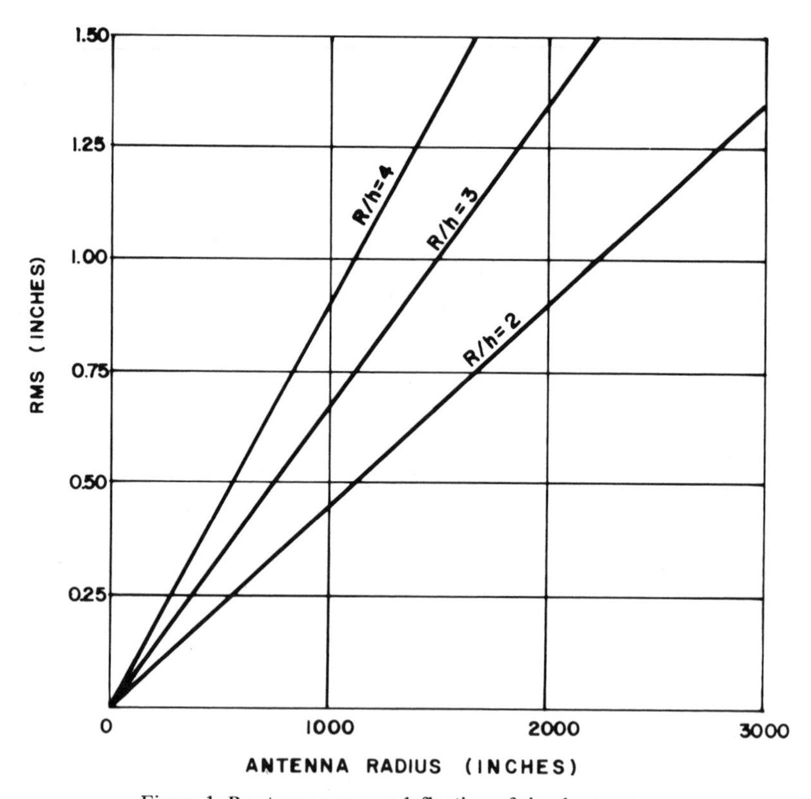

Figure 1. Root mean square deflection of simple structures.

4. Force Compensating Concept

Consider a large parabolic antenna surface supported on a backup structure, subject to gravity forces. We divide the backup structure into a number of segments with the weight **W** of each such subdivision assumed to be acting at the center of gravity. If the antenna is rotated by the angle ϕ from its original position, the gravity vector **W** will rotate by the same angle with respect to a coordinate system attached to the antenna. We now also introduce the external compensating force $F(\alpha,\phi)$ acting on the center of gravity at the angle α with respect to the coordinate system. The resultant of the two vectors **W** and **F** is

$$\mathbf{R} = \mathbf{W} + \mathbf{F} \ . \tag{10}$$

The value of **F** is continuously adjusted by control of the compensating system in such a manner that the resultant vector

$$\mathbf{R} = \text{const} \ . \tag{11}$$

The gravitational distortion of the structure is now independent of the elevation angle, i.e., the structure behaves during rotation as if it were in a field rotating with it. The constant distortion introduced by the force R can be eliminated by cambering. In practical application, with a finite number of compensated points, gravity distortion will exist in the areas between such points. Since our aim is to obtain a structure in which the radial component of rms deformations is to remain below a predetermined value, it is sufficient to control the radial component R_n of the resultant force only. For the condition

$$R_n = \text{const} \ , \tag{12}$$

it is sufficient to define for each elevation angle ϕ the value of $F(\phi)$ only, while its direction α can remain constant throughout the rotation.The compensating forces F acting on the backup structure can be reacted on some other part of the same structure and the entire system of truss and compensating forces constitute a prestressed structure in which the intensity of the prestress is varied as a function of the elevation angle ϕ only. Since this variable prestress is now an internal force with respect to the entire structure, the system remains in equilibrium. The variable prestress produces internal distortions, but as long as the variation of the force F at a constant angle α satisfies the condition

$$F(\phi) = [R_n - W \cos(\phi - \theta)] / \cos(\alpha + \theta) \ , \tag{13}$$

where R_n is a constant normal component at an angle θ with respect to the axis, the distortion of the structure remains controlled at selected points. (See Figure 2.) During rotation of the antenna the intensity of the force $F(\phi)$

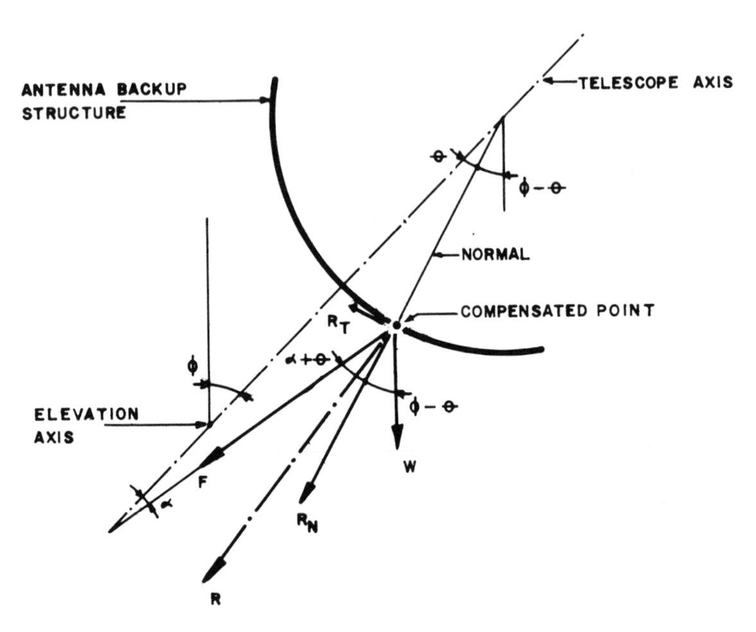

Figure 2. Compensating force geometry.

is continuously controlled at a predetermined value of each elevation angle and the force does not depend on displacements at the points of application or at the anchorage or displacements caused by the elongation of control rods through which the force is applied. This leads to a simple, open-loop control system, depending on the magnitude of the force only.

To apply the principles described, we proceed by designing an antenna backup structure, consisting of a series of parallel plane trusses, perpendicular to the elevation axis and symmetrical with respect to it. Each plane truss incorporates an anchorage point or pole (Figure 3), to which the compensating control rods are attached. The intensity of the force in each rod can be controlled either through a preprogrammed hydromechanical actuator or through gravity forces obtained through the truss counterweight. Each truss is designed for strength only, the approximate peak deflection can be obtained from Equation 3 and the rms deflection from Equation 7 or Figure 1. The actual calculated radial component of the deflection is shown in Figure 4. The rms radial deflection is 2.02 in.

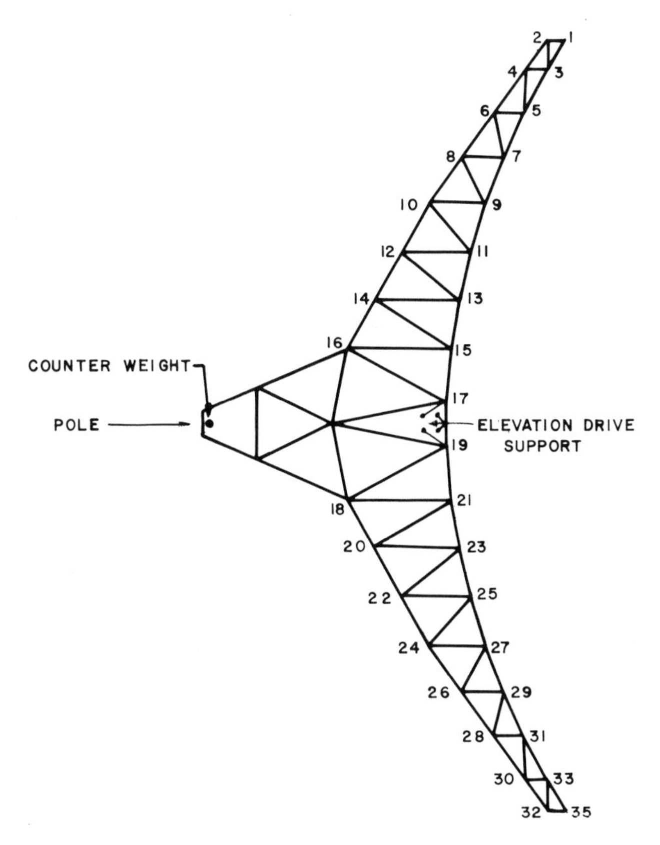

Figure 3. Typical truss.

In order to study the effect of the position and number of compensating forces influence coefficients I_{ik} were calculated denoting the radial deflection of the kth joint due to a pair of symmetrical compensating forces applied to the symmetrical points i and i'. The deflection of the truss at joint k due to a pair of forces F_i at joints i and i' is

$$\delta_k = F_i I_{ik} \tag{14}$$

and the compensated deflection δ_c of a joint at k, including the effect of gravity deformation δ_g and n pair of forces, is

$$\delta_c = \delta_g + \sum_{i=1}^{i=n} F_i I_{ik} . \tag{15}$$

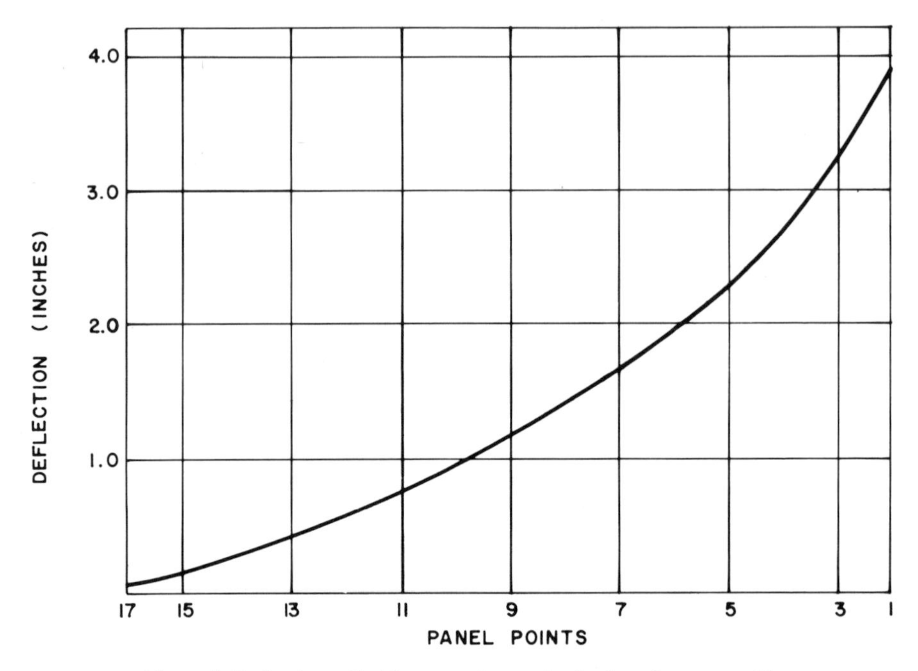

Figure 4. Deflection of half truss under gravity loads – face up position.

A deflection curve, for n = 3, applied to the main truss of the antenna structure is shown in Figure 9. In a symmetrical truss consisting of j pairs of joints the number of all possible points of compensation is

$$\sum_{n=1}^{n=j} \frac{j!}{n!(j-n)!} = 2^n - 1 \, .$$

In case of the truss shown in Figure 2, j = 8 and the number of combinations is 255. With the aid of the computer, these combinations can be scanned and the magnitude of the compensating forces and the rms deflections for elevation angles $0 \leqslant \phi \leqslant 90$ calculated. The result of such calculations is shown in Table 1 for a maximum of six pairs of forces showing best combination, i.e., the one that gives the least rms value together with the maximum value of any single compensating force and the sum of the absolute values of all compensating forces, measured in units of the total gravity load W on the truss. The sum of absolute values is given instead of the algebraic sum, because the compensating system as designed transmits tension only, and consequently the truss is bias-rigged to adjust to this requirement.

Table 1. Effect of Force Compensation on Typical Truss.

Number of pairs of forces	ϕ_m [a]	rms deflection (in.)	rms gain	Location of compensating forces at joint								F_{imax} [b,c] W	$\Sigma\|F\|$ [b] W
				2	4	6	8	10	12	14	16		
0	0	2.020	1									0	0
1	0	0.280	7		*							0.47	0.73
2	75	0.050	40		*				*			0.90	1.60
3	0	0.028	72	*		*				*		1.23	2.28
4	0	0.020	101	*	*		*			*		1.23	2.28
5	0	0.011	183	*	*		*		*		*	33.00	68.0
6	0	0.010	202	*	*	*	*		*		*	33.00	68.0
7	0	0.010	202	*	*	*	*		*	*	*	33.00	68.0
8	0	0.010	202	*	*	*	*	*	*	*	*	33.00	68.0

[b] $W = 1.5 \times 10^6$ lb.
[a] ϕ_m: Elevation angle at max rms.
[c] Maximum force occurs at $\phi \neq \phi_m$.

Table 1 also shows the rms gain, i.e., the ratio of the uncompensated rms to the compensated rms. As is to be noted in Table 1 and in its graphic presentation in Figure 5, the rms values approach a minimum value of 0.01

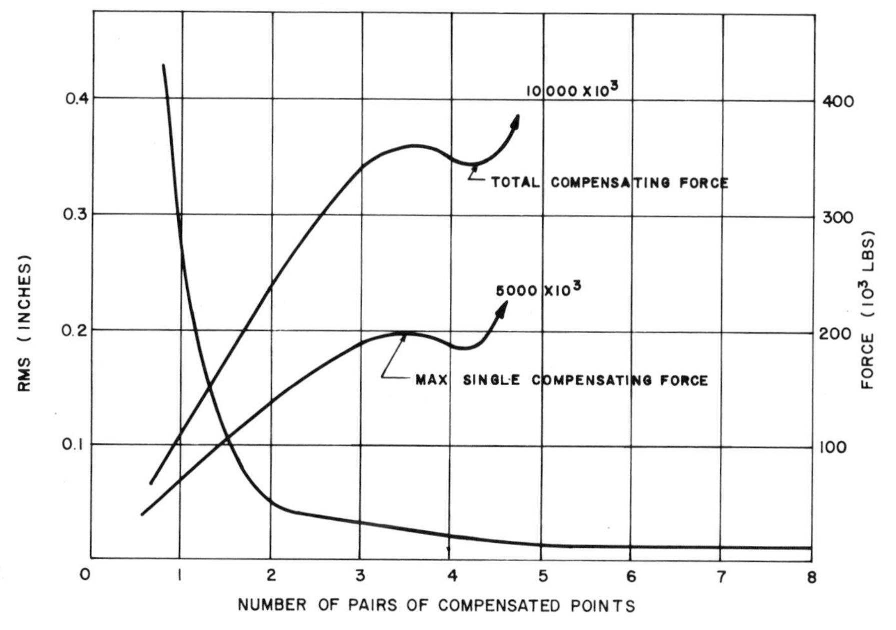

Figure 5. Root mean square deflection and force versus number of compensated points.

in. The magnitude of compensating forces increase nearly linearly up to three pairs, but increase very rapidly at five pairs. With these data it seems reasonable to select three pairs providing an rms = 0.028 in. although two pairs with an rms = 0.050 in. may also satisfy system tolerance requirements.

It is to be noted that while in Figure 5 the rms deflection approaches a nonzero limiting value with increasing number of pairs of compensating forces, the value rms = 0 is approached if nonsymmetrical combinations are also considered. The variation of the rms deflection for n = 3 for various elevation angles ϕ are shown in Figure 6, and the variation of the three pairs

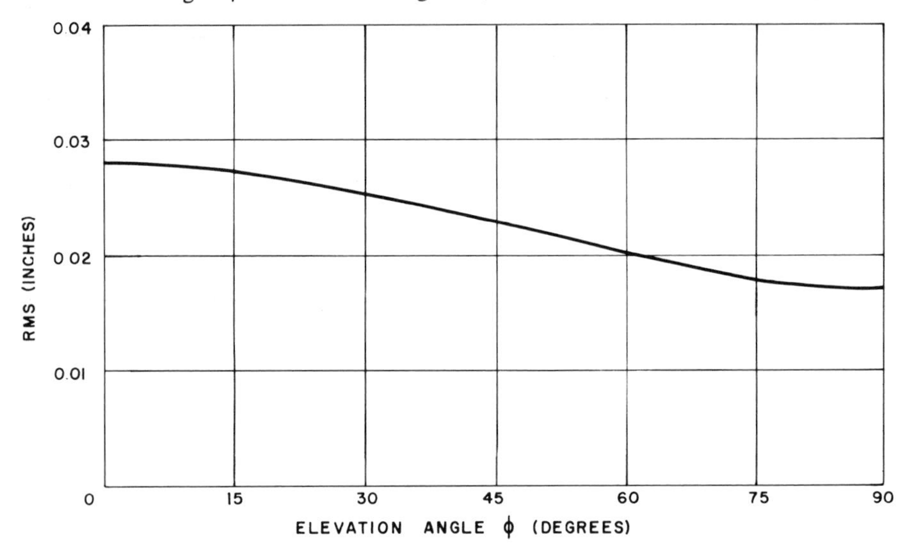

Figure 6. Variation of rms versus elevation angle. Compensating forces are applied at panel points 2, 6, 14, 20, 28, and 32.

of forces and the sum of their absolute values are given in Figure 7 and 8. The radial deflection of the half truss at $\phi = 0$ is shown in Figure 9.

From these studies it is concluded that the gravity rms deflection of the antenna backup structure can be controlled at any desired level and that very low surface distortion can be achieved with two or three pairs of compensating forces. It is also concluded that significant increase in efficiency through reduction of gravity distortions can be achieved without significant increase in cost.

5. System Description

Geometry

The method of controlling the rms deformation of an antenna dish is, in

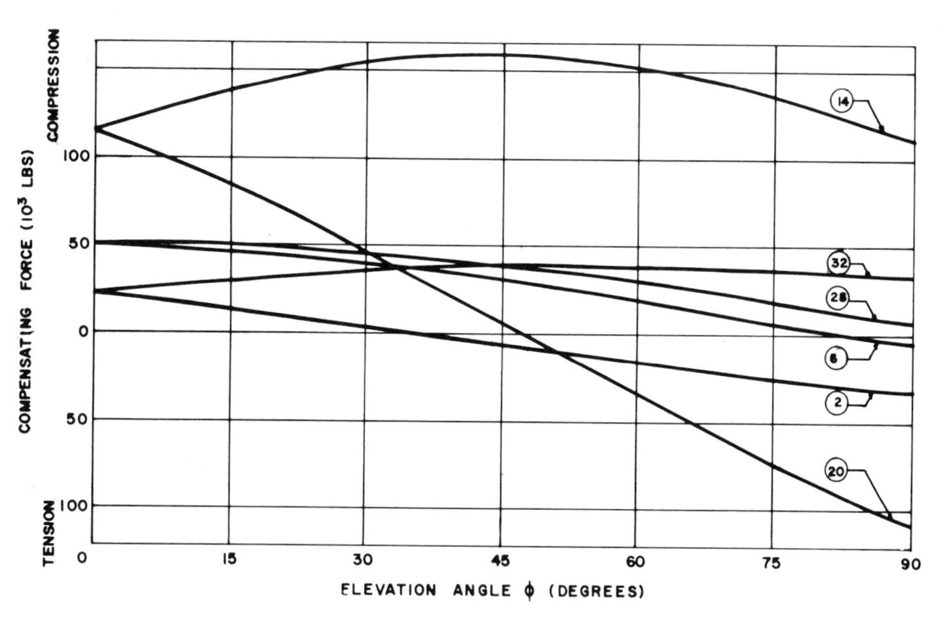

Figure 7. Compensating forces versus elevation angle.

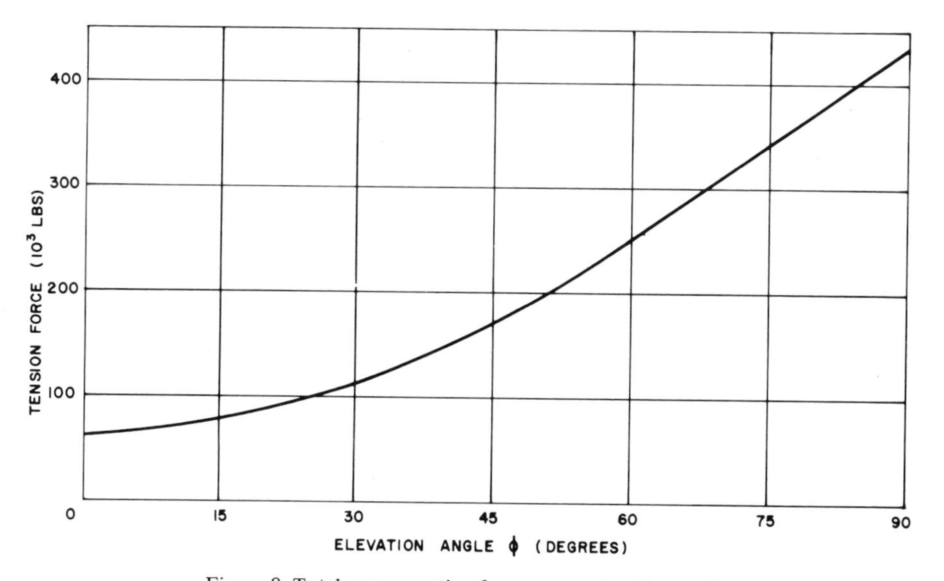

Figure 8. Total compensating force versus elevation angle.

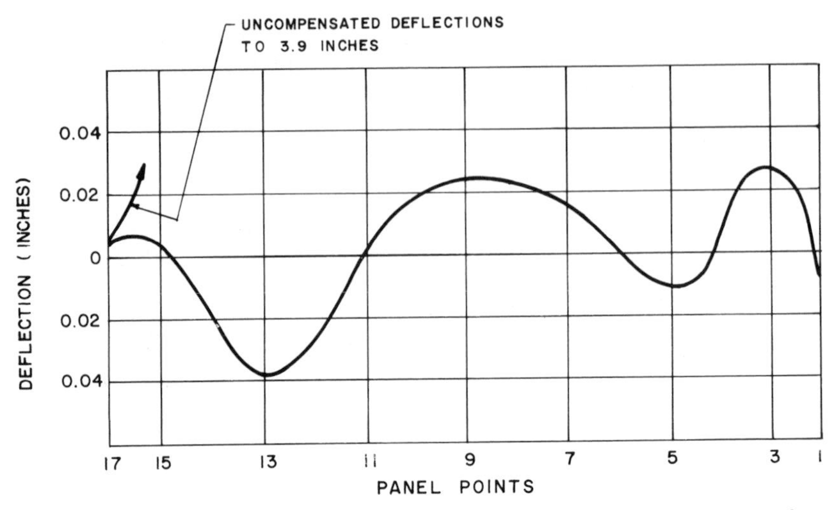

Figure 9. Deflection of half truss under gravity loads and six compensating forces—face up position.

principle, applicable to various types of structures. In its practical application to the CAMROC study, certain additional specific constraints are desirable. Since it is anticipated that a number of points on the antenna backup structure need to be compensated simultaneously, it is advisable from the point of view of stability of the control system to design the antenna structure consisting of essentially independent subassemblies which are uncoupled from each other, with respect to deformation. Since the distortion of the antenna structure is governed by the rotation around the elevation axis, a series of independent axisymmetric structures (instead of a centrally symmetric space structure) have been selected. In effect, the antenna dish backup structure consists of a series of parallel plane trusses supported on the elevation axis.

Compensating Force

There appear to be two practical methods by which the compensating force can be applied to the antenna backup structure. In both cases it is desirable that the compensating forces be applied in form of tension only; this is accomplished by appropriate bias-rigging and prestressing of the individual backup trusses.

In *one method,* the application of the compensating forces is accomplished by an external power source. This requires the use of a hydromechanical or hydraulic power device. Control will be accomplished by means of a nulling-type electrical system and appropriate amplifiers. The magnitude of the force is precalculated for each altitude angle. The shaft of the elevation axis drive will be used to make available noise-free voltage analog of the reflector altitude to control the compensating forces on the structure. It is estimated that the compensating forces will be obtainable within a required tolerance.

The *second method* does not require an external power source. In this approach the rolling weight on a fixed track is used to produce the required tension. In this system the counterweight required for the vertical truss itself will be used as the weight required to provide the compensating force through suitable mechanical amplifying devices. This passive method of compensation, (i.e., without external power source) can be used because the variation of the compensating force is a sine function of the elevation angle, as shown in Equation 13.

Structure

Specific application of the force compensated concept to a 400-ft-diameter parabolic Cassegrain antenna is shown in Figures 10–13. The structure is divided into two major components, namely that of the structure of the dish supported on an elevation axis and the substructure consisting of a yoke and supporting towers which, in turn, rest on azimuth bearing pads (Figure 10).

The antenna backup structure consists of purlins spaced at 13 ft on center and main vertical trusses spaced at 50 ft on center. The elevation axis on which the vertical trusses are supported is placed as closely as possible to the vertex of the paraboloid, thus reducing the eccentricity of the center of gravity with respect to this axis, consequently reducing the required counterweight and also producing a more compact design with favorable dynamic properties, and requiring minimum radome radius. Achievement of the above objective also suggests the use of a minimum shaft diameter and this, in turn, requires closely spaced support towers of the shaft. These conditions are met by the substructure consisting of the yoke and supporting towers (Figures 11 and 13).

The design of the vertical trusses was controlled by strength. They are balanced with respect to their support (elevation axis) by counterweights. The tension in the compensating wires is reacted at a central pole supported by a protruding part of the truss. The deflection of each truss in its own plane is independent of the deflection of an adjacent truss, and for this reason each truss has its own compensating system. The vertical trusses are supported on an elevation axis, which consists of a triangular space truss and hollow shaft. To minimize bending and torsion in the axis, bearings for the axis and support points for the trusses are close to each other. The entire axis rotates as one unit, thus simplifying the problems of bracing. Due to the truss action of the substructure supporting the dish, there is a horizontal thrust along the elevation axis, which is taken up by a single stationary rod, running between the end points of the yoke through the hollow shaft of the elevation drive axis (Figure 11).

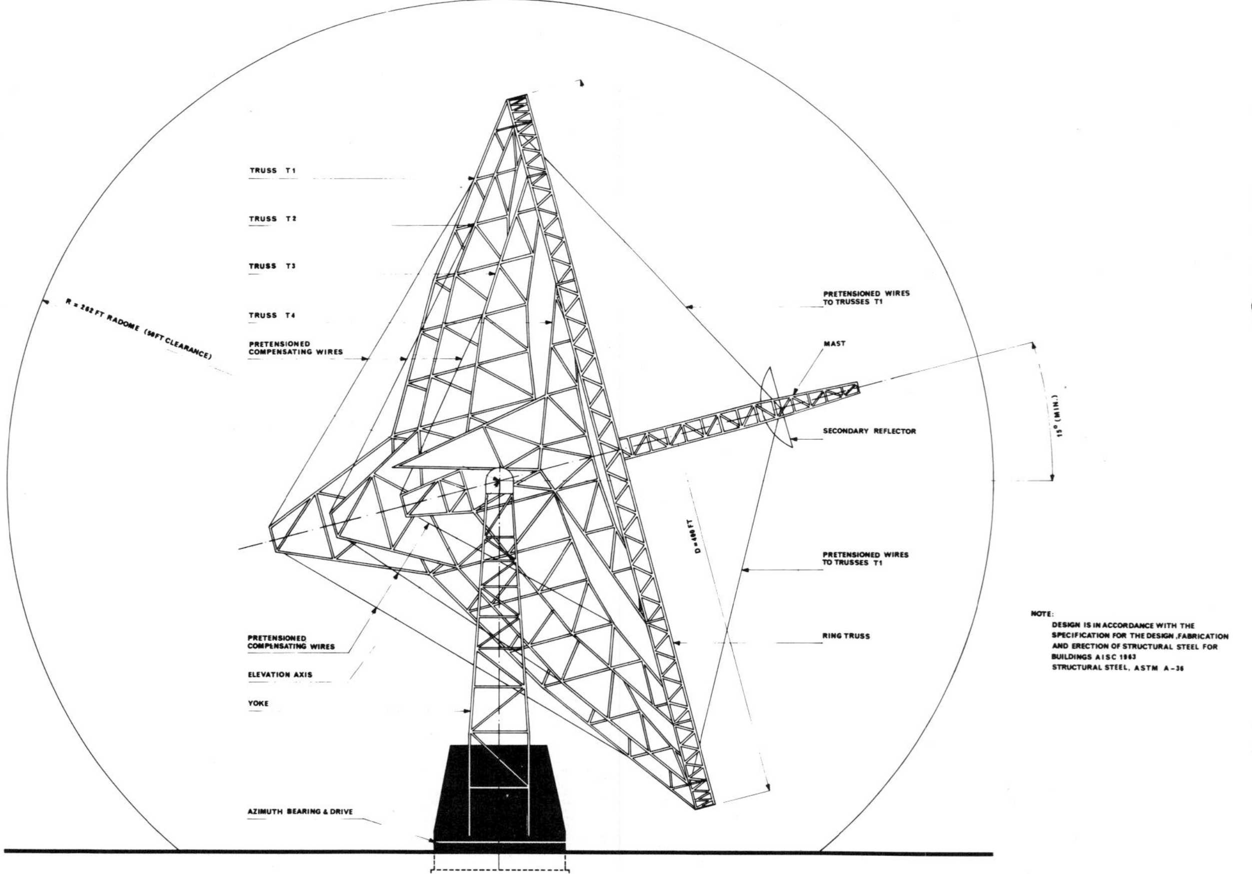

Figure 10. Force compensating concept, 400-ft-diameter antenna. Side elevation.

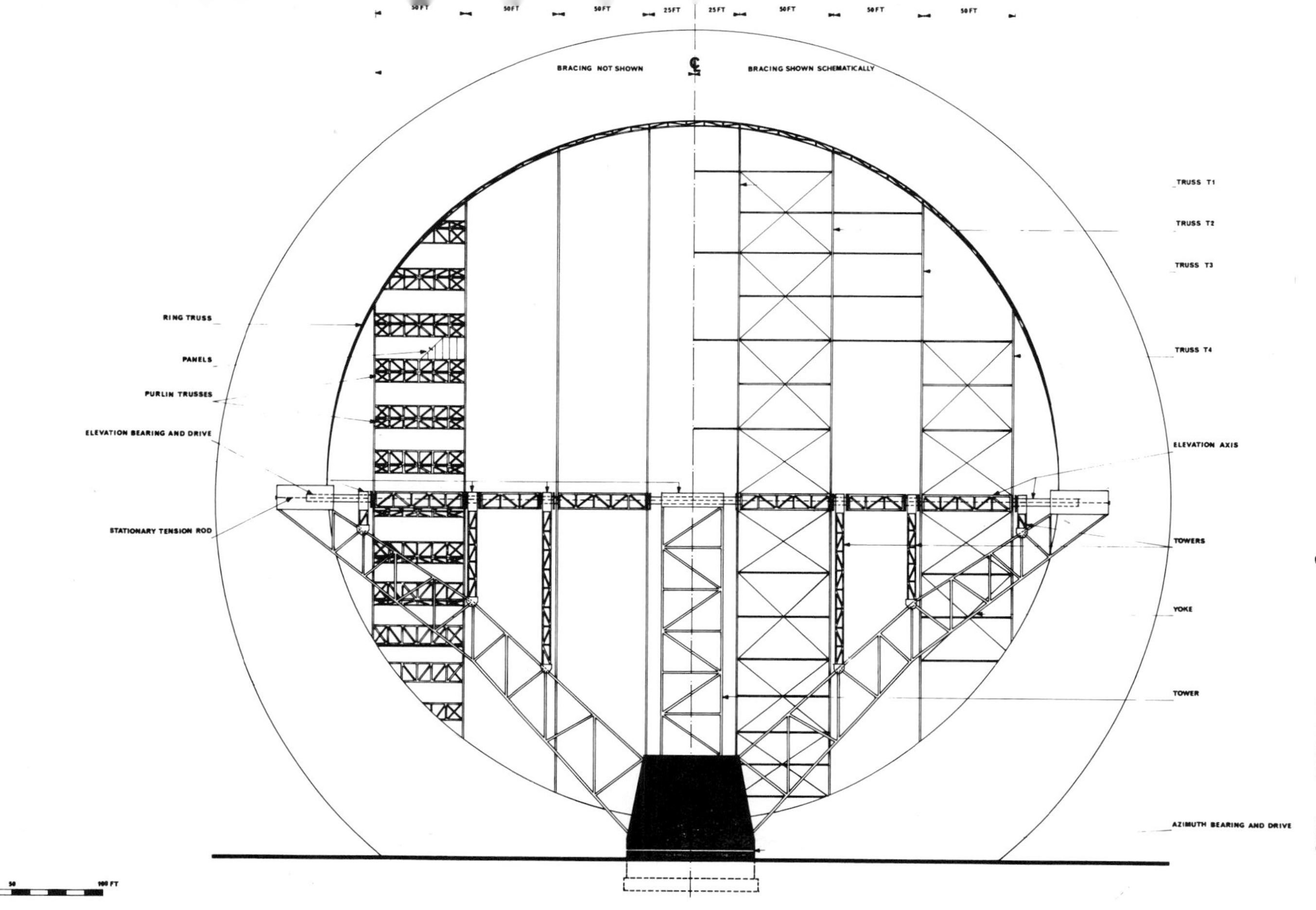

Figure 11. Force compensating concept, 400-ft-diameter antenna. Rear elevation.

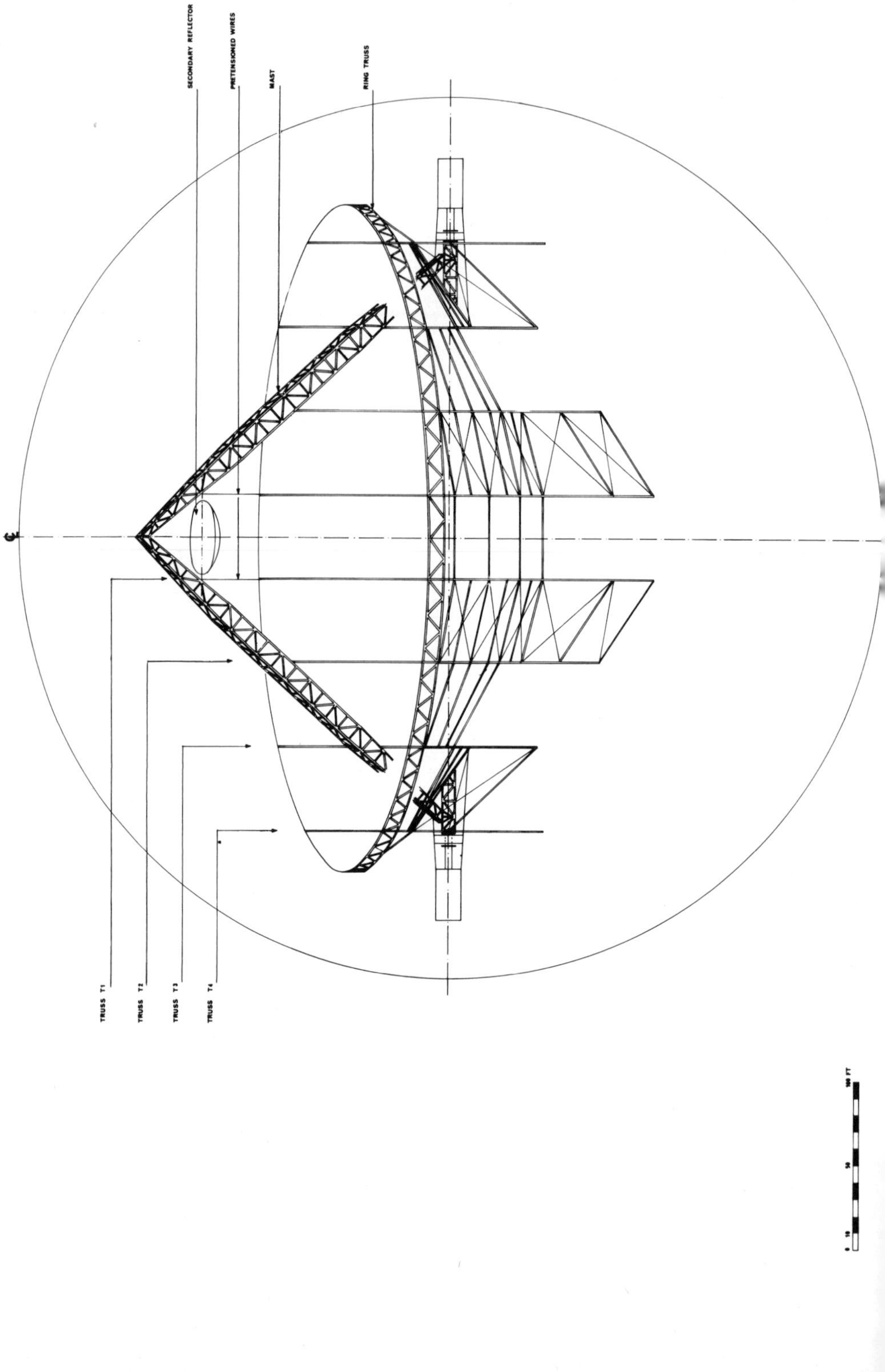

SECONDARY REFLECTOR
PRETENSIONED WIRES
MAST
RING TRUSS
TRUSS T1
TRUSS T2
TRUSS T3
TRUSS T4
0 10 50 100 FT

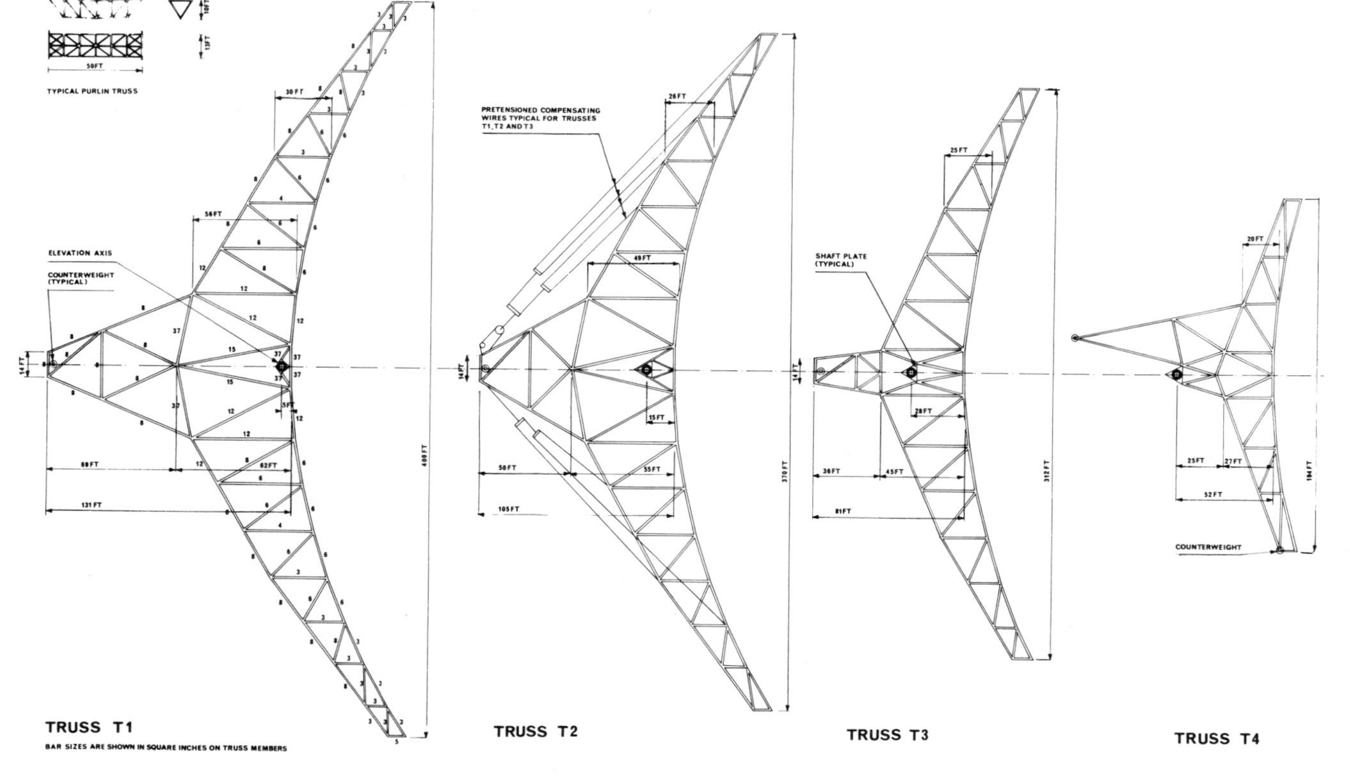

Figure 13. Force compensating concept, 400-ft-diameter antenna. Truss elevations.

The elevation axis is supported by towers. Each tower contains an elevation bearing and drive unit. Thus the axis is driven at a short distance from the support point, where the torque is transmitted to a vertical truss. Azimuth rotations cause a radial, outward-pointing, dynamic force (centrifugal) that causes tension in the elevation axis. This is a self-equilibrating force, and since the vertical trusses are rigidly connected to this axis, there is no need for thrust bearings.

The towers are designed to transmit vertical loads to the yoke. They also participate in resisting deformations due to tangential dynamic loads in case of azimuth accelerations. Some portion of these loads is reacted by the beam action of the elevation axis in a horizontal plane.

The towers are supported on the yoke. The shape of the yoke is chosen in a manner that moments due to eccentrically applied compression and vertical loads largely compensate. It transmits the torque for azimuth drive to the ends of the elevation axis.

The yoke is supported on a cellular reinforced concrete structure in the shape of a conoid with a square upper and a circular lower surface. It will house electronic equipment; the entire antenna is rotated on azimuth bearings at its lower circular surface.

Supports for the secondary reflector are provided in such a manner that a major portion of the load enters the elevation axis directly. Two compression masts and in a perpendicular plane two pairs of wires make up the supporting assembly. The variable force in the wires is compensated so that the resultant force acting on the truss is tangential rather than radial, resulting in negligible bending in the truss. Further advantage of this kind of support is that it allows the adjustment of the position of the secondary reflector on a vertical circular path. The vertical trusses are braced against the elevation axis, and each other. The topchords are held in position by the purlin trusses; adjacent bottom chords are cross braced (Figure 12).

The entire dish is "enclosed" by a ring truss, which also serves as a support for the purlin trusses in the peripheral areas of the paraboloid. In order to prevent interaction between adjacent vertical trusses, portions of the ring truss are connected by a combination of rigid and slip joints (Figure 10).

Preliminary calculations indicate that inertial forces have a negligible influence on deformation and that the frequency of lower modes is well below the resonant frequency specified.

6. Weight and Cost

The basic design of the CAMROC antenna is based on a 400-ft-diameter parabolic dish, and detailed calculations to establish the weight of the individual subassemblies were made. In order to assess the variation of the weight

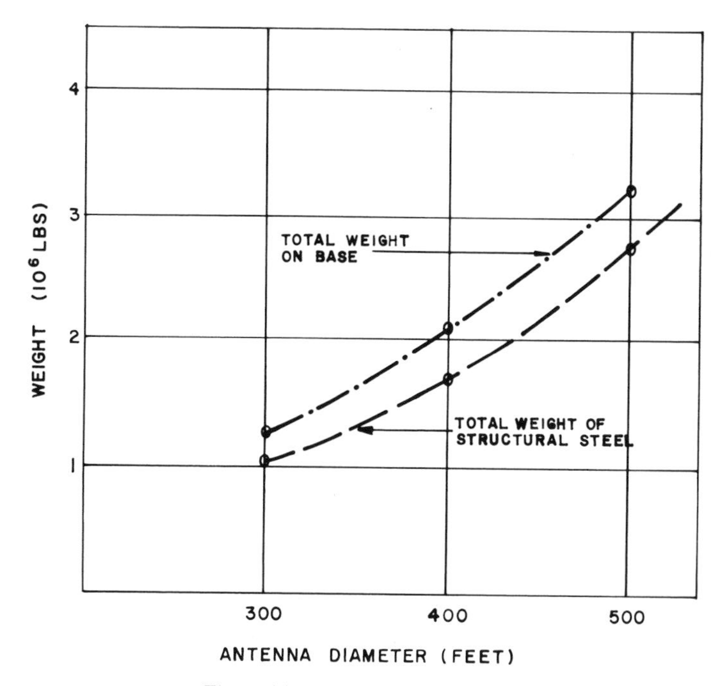

Figure 14. Weight versus diameter.

versus the diameter, additional calculations were obtained partially by extrapolating from the 400-ft-diameter basic design, and partially by redesigning certain major members for this purpose. In extrapolating the values obtained from the 400-ft-diameters dish, it was assumed that the weight per square foot of panels and purlins is constant. The weight of the subsequent subassemblies was calculated by superimposing on them the weight carried by the subassembly and calculating the required weight of the subassembly itself to carry the superimposed loads. In this manner the total weight of the antenna was obtained for various diameters.

The weights indicated do not include the electronic equipment and the conoidal reinforced concrete base on the azimuth bearing.

At this stage of the study it is not possible to obtain a reliable cost estimate; however, it is felt that the cost will probably be proportionate to the weight of the structural steel alone. This item is given separately on the weight breakdown in Table 2, and is also shown in Figure 14. The cost of the hydromechanical control system, assuming a maximum of 50 points controlled, has been estimated at $250,000. This cost will be substantially reduced if the passive counterweight control system is used.

Table 2. Weight Calculations.

Diameter	300 ft	Weight in 10^3 lbm 400 ft	500 ft
Counterweight (concrete)	190	260	380
Panels (aluminum)	40	75	120
Truss bracing	100	180	280
Purlins	100	175	270
Total Weight on Trusses	430	690	1050
Trusses	370	420	570
Secondary reflector	25	50	100
Weight on elevation axis	825	1160	1720
Elevation Axis	150	270	500
Towers	60	100	150
Yoke	230	500	900
Total Weight on azimuth bearing (excluding concrete base)	1265	2030	3270
Total weight of structural steel only	1035	1695	2770

7. Conclusions

As a result of the preliminary study the following conclusions have been reached.

1. A 400-ft-diameter radome-housed Cassegrain antenna can be designed and built to the CAMROC specifications.

2. Surface accuracy even beyond those specified is achievable without significant cost increase by the use of the "force compensation" system.

3. The entire weight of the 400-ft-diameter antenna, excluding reinforced concrete base, foundations and electronic equipment is 2×10^6 lb.

4. The weight, and probably cost, of similar antennas in the 300 to 500 ft range varies as the second power of the diameter.

5. The "force compensating" concept appears to offer an economic method to obtain extremely high surface accuracies. Further development of many crucial details of this method are required, however, to improve its practicability and explore its potential. In principle, the force compensating method could also be applicable to very large diameter antenna structures well beyond the range of the present study. It appears to be also well suited to large antenna structures built of materials of low modulus of elasticity, such as aluminum and, conceivably, of reinforced concrete.

6. The rms tolerance can be further decreased and/or the intensity of compensating forces decreased by best-fitting and bias-rigging.

7. Because of the compact design of the structure the radome size is reduced to a radius of 262 ft including a clearance of 50 ft.

8. Since sufficiently low values of rms deflections are achievable, the antenna structure can be fabricated within commercial structural steel tolerances, resulting in a significant cost saving.

9. The compensating system may be used, if this turns out to be desirable, to reduce the effects of thermal distortions. This can be accomplished by activating of the compensating control system, through temperature sensing devices placed on the backup structure.

Acknowledgments

This report was prepared in collaboration with the following members of the firm Paul Weidlinger, Consulting Engineer: Steven Varga, Ronald Check, Dr. Mario Salvadori, and Dr. John M. McCormick.

The author also gratefully acknowledges the helpful comments and suggestions of the CAMROC Committee, and the help of the following members of M.I.T. Lincoln Laboratory: Herbert G. Weiss, Paul Stetson, and F. George De Santis.

References

1. "Procurement Specification CAMROC-2," M.I.T. Lincoln Laboratory, Lexington, Massachusetts (8 March 1966).
2. P. Stetson, "CAMROC Hammerhead Antenna Concept," paper 5 in *Radomes and Large Steerable Antennas,* Conference Proceedings, the Cambridge Radio Observatory Committee (17–18 June 1966).
3. J. W. Mar and F. Y. M. Wan, "The Influence of Shell Behaviour on the Design of Large Antennas," Discussion.

Sebastian von Hoerner
National Radio Astronomy Observatory
Green Bank, West Virginia

HOMOLOGOUS DEFORMATIONS OF TILTABLE TELESCOPES

A telescope tilted in elevation angle must deform under its dead load, and this sets a lower limit to the shortest wavelength of observation once the diameter is chosen. A most natural way of passing this limit is by designing a structure which deforms unhindered, but which deforms one paraboloid of revolution into another, thus yielding a perfect mirror for any angle of tilt. Focal length and axial direction are permitted to change (to be servo-corrected by focal adjustments).

1. The Gravitational Limit

There are three natural limits (as opposed to financial ones) for diameter D and shortest wavelength λ, for tiltable conventional telescopes[1]:

Stress limit $D \leqslant D_{st} \approx 600$ m; (1)

Thermal limit $\lambda \geqslant \lambda_{th} \approx 2.4$ cm $(D/100 \text{ m})$; (2)

Gravitational limit $\lambda \geqslant \lambda_{gr} \approx 8.0$ cm $(D/100 \text{ m})^2$. (3)

The stress limit is reached when the dead load of the structure produces at the bearings the maximum allowed stress of the material; at present we are far below this limit. The thermal limit applies to a telescope with good protective paint in full sunshine ($\Delta T = 5°C$); this limit can be passed by a factor 2 – 5 in a radome, or an open dome (36-ft telescope at Kitt Peak), or at night. The gravitational limit arises from the deformations under dead load if the telescope is tilted from zenith to horizon; the value given, 8.0 cm, applies to an

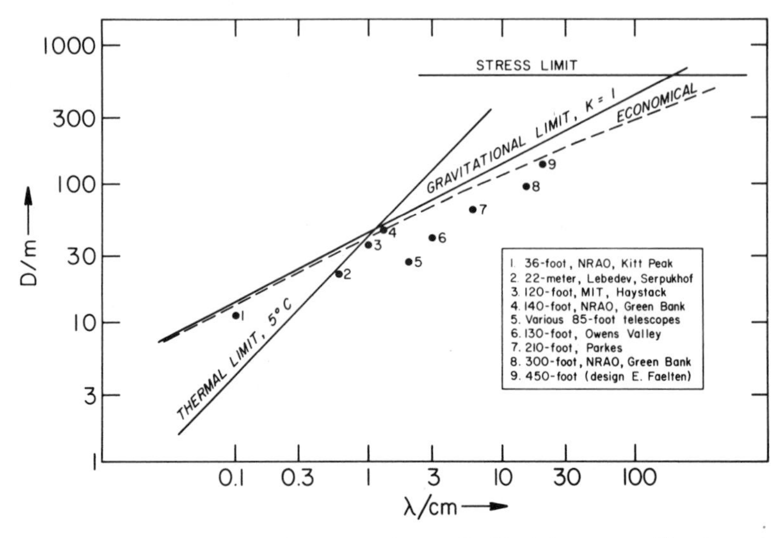

Figure 1. Three natural limits for conventional, tiltable telescopes, with nine actual examples for comparison. The Kitt Peak telescope is inside an open dome, thus passing the thermal limit set by sunshine and shadow. D is the telescope diameter in meters; λ is the shortest wavelength to be observed, in centimeters.

economical structure and is the same for steel and aluminum. It can be brought down to 5.3 cm with an uneconomically high total weight. Figure 1 shows that the gravitational limit is the essential one for large telescopes ($D \geqslant 100$ ft); some telescopes come very close to it, but not a single existing tiltable telescope passes it.

2. Homologous Deformations

There are several ways of passing this gravitational limit: (1) Fixed elevation transit telescopes (the 1000-ft dish in Arecibo does not move at all; the LFST* group has worked out three 600-ft telescopes moving 360° in azimuth); (2) motors in the structure or at the surface panels, correcting the deformations; (3) levers and counterweights as in large optical telescopes. The most natural way seems to be (4) designing a structure that deforms completely unhindered, but which deforms one paraboloid of revolution into another one, thus yielding a perfect mirror for any angle of tilt. Since this deformation transforms one member of a given family of surfaces into another member of the same family, we suggest calling it a "homologous deformation,"[1] and the permitted changes of focal length and axial direction "homology parameters."

* Largest Feasible Steerable Telescope.

Homologous deformations can be demanded for an arbitrary number N of equally spaced structural points, holding the surface or the panels, where N must be chosen so large that any deformation between neighboring points can be neglected ($\leqslant \lambda/16$). The minimum N then is proportional to λ^{-1}, and we must have $N = 2$ for $\lambda = \lambda_{gr}$ (two bearings of the conventional telescope). The design of the panels, however, is somewhat eased if we demand at least

$$N = 3\ \lambda_{gr}/\lambda. \tag{4}$$

Since small deformations can be superimposed, homology holds for all angles of tilt if it holds for two; a paraboloid of revolution is defined by six points, and since deformations parallel to the surface do not matter, we obtain homology if a set of $2(N - 6)$ conditions is fulfilled. On the other hand, a structure of N unconstrained points needs at least $3(N - 2)$ members just for stability, and even for a fixed geometry there are still $3(N - 2)$ degrees of freedom just for the bar areas. Since $3(N - 2) - 2(N - 6) = N + 6 > 0$, the problem is solvable and there is a family of solutions with at least $N + 6$ free

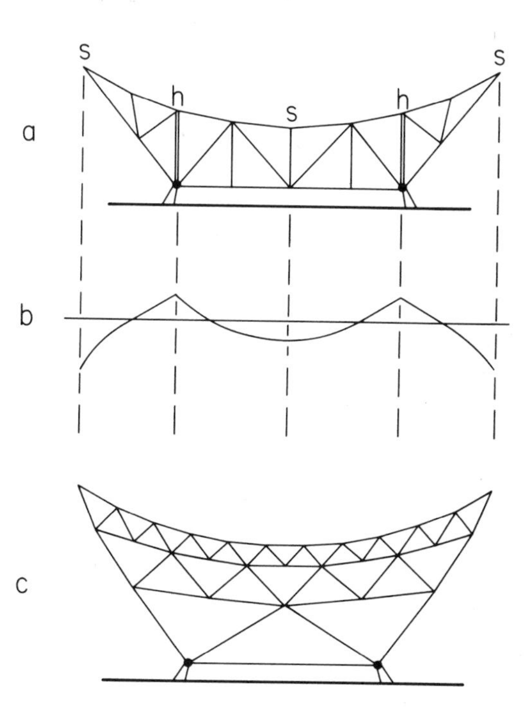

Figure 2. Equal-softness structures. (a) Conventional design, with hard (h) and soft (s) surface points. (b) Deformations of this telescope, looking at zenith; best-fit paraboloid is represented by a straight line. (c) Structure where all surface points have about equal softness.

parameters. This existence proof holds for "mathematical solutions," however, whereas a "physical solution" demands all bar areas to be positive and finite, and a "practical solution" must fulfill all specifications with a small total weight. There certainly are structures that have only mathematical solutions but no physical ones, for example, if we mount the two elevation bearings at surface points.

As a first approach to homology, the concept of an "equal softness structure" was introduced in Reference 1 (see Figure 2). A conventional structure usually has *hard* and *soft* surface points; this is clearly illustrated by the measured deformation patterns of several telescopes. Since we cannot make the soft points hard, we have to make the hard points soft. We obtain about equal softness when the structural paths at the nearest bearings are about equal for all surface points. A structure like Figure 2(c) already comes close to a homologous deformation, while a structure like Figure 2(a) usually has no practical solution at all.

3. The Homology Method

In 1965, a mathematical method was developed[2] for obtaining homology solutions on a computer. The homology problem would lead to a set of 2N highly nonlinear equations, but the method used is linearized and iterative. With the input data we give a complete structure its geometry (coordinates) as well as a "first guess" of all bar areas A_γ. The method then keeps the geometry unchanged, but changes all A_γ simultaneously in each iteration step, demanding a zero rms deviation ΔH between the surface points and a best-fit paraboloid of revolution. We define

$$\Delta H = \left\{ \frac{1}{2N} \left[\sum_{i=1}^{N} (\delta z_i)^2_{\text{zenith}} + \sum_{i=1}^{N} (\delta z_i)^2_{\text{horizon}} \right] \right\}^{1/2} \qquad (5)$$

as the "deviation from homology," where $\delta z_i = \Delta z_i - b(\Delta x_i, \Delta y_i)$ is the difference in z direction (parallel to the optical axis) between the deformed surface point i and the best-fit paraboloid of revolution; Δx_i, Δy_i, Δz_i are the deformations of point i. The task is to find a set of bar areas such that $\Delta H = 0$. This task is represented by a set of 2N linear equations, which we call "homology equations."

From all possible homology solutions (we have at least N+6 free parameters), the method selects that solution which is most similar to the first guess. We thus try to avoid impractical solutions. The first guess should be made such that the structure withstands the survival conditions and has only small wind deformations, both with a minimum total weight; the homology iterations should stay close to this condition. Finally, keeping all changes as small as possible gives the best hope for a good convergence. The method is described in full detail in Reference 2, we give only a brief outline here. The present method neglects the bending stiffness, regarding each joint as a pin point.

The method used is a generalization of Newton's method for finding the zero point of a function. If x is wanted such that y(x) = 0, Newton's method starts with some initial value x_0 (first guess), and iterates according to $x_{i+1} = x_i - y_i/(dy/dx)_i$. This is generalized to n variables, where n = m + 4 is the number of members plus the number of homology parameters. The quantity whose zero is wanted is ΔH from Equation 5; we now need $\partial \Delta H/\partial A_\gamma$, the derivatives of ΔH with respect to all bar areas. In Equation 5, ΔH goes back to the deformations Δz of the surface points, which are given as

$$\Delta z = K^{-1} F, \tag{6}$$

where K^{-1} is the inverse of the stiffness matrix K, and F is the force vector given by dead loads and surface weight. We then need the derivatives of all elements of K^{-1} with respect to all bar areas, $T_{ij\gamma} = K_{ij}^{-1}/\partial A_\gamma$ (a tensor of third order), and also all $\partial F_j / \partial A_\gamma$. Two facts make the method tractable. First, the derivatives of K^{-1} can be obtained from those of K with a formula derived in the appendix of Reference 2, which can be written as

$$T_{ijr} = - \sum_p \sum_q K_{ip}^{-1} \frac{\partial K_{pq}}{\partial A_\gamma} K_{qj}^{-1} \tag{7}$$

[This is merely the matrix equivalent to $(1/y)' = -y'/y^2$]. Second, the elements of K are always linear in the A_γ, and thus $\partial K_{pq}/\partial A_\gamma$ = const throughout all iterations for a given and unchanged geometry; the same is true for $\partial F_j / \partial A_\gamma$. In this way, the wanted change of ΔH (for obtaining $\Delta H = 0$) is finally traced back to the unknowns, the needed changes of the bar areas.

The combined task of (1) achieving homology, and (2) selecting that solution most similar to the first guess, is treated by the method of Lagrange multipliers, but we have also included the possibility of making the homology parameters as small as desired if they should turn out too large; we thus minimize

$$\sum_{\gamma=1}^{m} (dA_\gamma/A_\gamma)^2 + \omega \sum_{k=1}^{4} h_k^2 = \text{Min}, \tag{8}$$

where dA_γ is the needed change of bar area A_γ, and $h_1 \ldots h_4$ are the homology parameters; ω is a factor given with the input data which tells how important we consider small homology parameters. One iteration step then runs as follows. We build up the stiffness matrix, its inverse, and the force vectors. After several matrix operations we arrive at Equation 8 where a set of n linear equations must finally be solved, yielding the changes dA_γ, which are then added to the old values of A_γ. With these improved bar areas we again build up a *new* stiffness matrix and force vectors, and we repeat the procedure until a given number of iterations is finished.

The present method solves an optimum condition, Equation 8, together with a set of 2N constraint equations (homology equations) for obtaining $\Delta H = 0$; the external specifications (survival stress, wind deformation) are checked separately later on. This method was chosen for obtaining best convergence in case that convergence is a problem. But, from our present experience, convergence is no severe problem, and the best method would then be a different and more direct one. In addition to the homology equations, one should set up the external specifications in a linearized way. The demand on wind deformation would be represented by a set of w inequalities (if w different wind directions are specified), and the demand on maximum stress would yield a set of m inequalities. The optimum condition should then minimize the total weight. In the present method we have constraint equations only, whereas in the new method we would have inequalities as well as equalities. This task can be solved by a combination of Lagrange multipliers and Fritz John multipliers, as shown by Mangasarian and Fromovitz.[3] I would like to add that this method seems to be the best and most direct one for a large variety of optimization tasks in engineering. The only remaining problem is again the convergence; we found that the linearization of the homology equations gives fast convergence for a wide range of first guesses, but we cannot tell without trying whether this also holds for the linearization of the external specifications.

4. The Total Weight

Homology has nothing to do with the total weight. If we multiply all bar areas (and the surface weight) by a factor q, then the weight increases by a factor q, but so does the stiffness, and all gravitational deformations stay the same. Since the weight must be defined somehow, the present program keeps it constant. It also keeps constant the counterweight needed for balance. Usually we make our structures completely balanced (counterweight zero) before applying the homology iterations, but we can also choose any given amount of counterweight. The iterations then keep the counterweight constant, zero or not. This means we have two more constraint equations in addition to our 2N homology equations.

For an actual design, the total weight is defined by any one of three conditions:[1] (1) stable self-support under dead loads, for a telescope inside a radome; (2) stability under specified survival forces, for an exposed telescope and medium wavelengths; (3) specified wind deformations during observation, for an exposed telescope and very short wavelengths. In the present program, this final total weight is not obtained automatically. We start with a first guess fulfilling all conditions and then iterate until homology is reached; thereafter we check with a separate program whether all conditions still are fulfilled. If not, an improved first guess is made and the procedure tried again. (Sometimes, the geometry was also changed slightly.) Although this procedure does not look very elegant, it still seems to be the best one for gaining experience and understanding.

It turned out that this method is very easy for case (3), usually not too difficult for case (2), but a little troublesome for case (1) if low total weight is really desired. This difference is easily understood. If the wind deformation is too large by a factor q, we merely have to multiply each bar area and the surface weight by q, and we obtain an exact solution even without repeating any iteration. Survival stresses again go down (although by a smaller factor) if we multiply all bar areas by the same factor, but the dead load stresses stay the same. Thus in case (1) we have to try an essentially different first guess. It seems that one really should solve case (1) with the method suggested at the end of the last section.

The additional program calculates the stress in each bar in zenith position S_z and in horizon position S_h where the maximum stress for any elevation is then

$$S_m^2 = (S_z^2 + S_h^2)^{1/2}. \tag{9}$$

This is done either for dead loads in case of a radome, or for survival loads in case of an exposed telescope. The program then calculates the maximum allowed stress S_0 according to the l/r ratio of this member (more exactly, of its chords); each member is actually a built-up member according to Figure 3, where the influence of the lacings on weight and stiffness had been taken

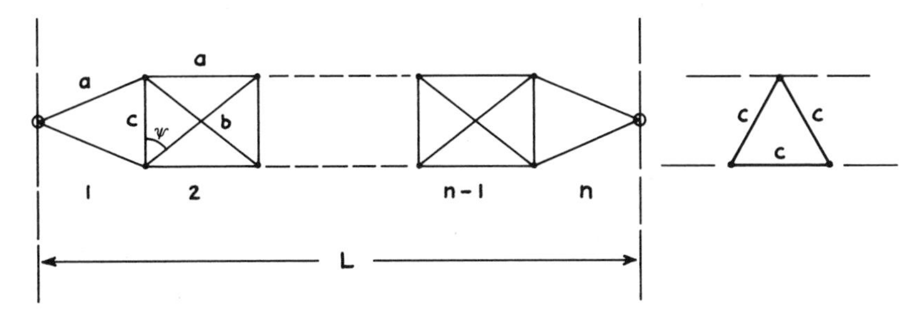

Figure 3. Built-up structural member. In principle all members could be different; but at present the same values are adopted for all members: $n = 10$, $\psi = 55^\circ$, $A_b = A_c = 0.3\ A_a$. This built-up member then is represented in our program by a single shape of area $A = 3.84\ A_a$, density $\rho = 1.19\ \rho_0$, and unchanged elasticity E. With respect to stability, we call $\Lambda = l/r$ ratio of the single chord, use standard pipes of the Steel Construction Manual, and obtain $\Lambda = L/(2.88\ A^{2/3})$.

into account. This stress factor

$$Q = S_m/S_o \tag{10}$$

is then printed for each member, and the stability condition is fulfilled if all $Q_\gamma \leqslant 1$. As to the numerical value of the survival condition, we have adopted a time of, say, 30 years after which a telescope becomes obsolete, and a chance of, say, 1 per cent for losing the telescope in a storm *before* it becomes obsolete. We then obtain from wind statistics at Green Bank [4] a survival wind of 110 mph at 200 ft height (90 mph as measured at 40 ft height, gusts included). Using this value, the telescope then is also stable against a snow load of 20 lb/ft^2, or 4 in. of solid ice. Moreover, since we should be able to dump the snow by tilting, we specify this load for any elevation angle according to Reference 5.

The wind deformation is, at present, calculated only for a wind face-on, assuming that this is the worst condition (different angles are planned for future checks). Since all gravitational deformations are omitted by homology, we omit the dead loads and regard the wind loads only. We calculate the rms surface deformation in z direction (no best fit this time) and call it $\Delta\zeta$. If the telescope is held and guided in an economical way (Section 6), the better part of $\Delta\zeta$ is just a parallel translation, which does not matter. The remaining part, which does matter, is due to (1) gusts of any size but faster than the servo loop of the drive, giving rise to pointing errors, and (2) gusts of any speed but smaller than the telescope radius, giving rise to surface deformations. An estimate of this remaining fraction of $\Delta\zeta$ is in preparation, but at present we adopted 0.5 $\Delta\zeta$. Furthermore, the calculations of Simpson, Gumpertz, and Hager (Section 7, 4) showed an rms difference in half the path length of only 0.777 $\Delta\zeta$, and both effects combined give 0.389 $\Delta\zeta$. On the other hand, the deformations of the towers holding the telescope should be added, where an estimate showed that about 60 per cent should be added (for an economical tower design), thus obtaining 0.632 $\Delta\zeta$. Finally, this value should be 1/16 λ; we arrive at a shortest wavelength λ for an rms deformation $\Delta\zeta$:

$$\lambda = 10\,\Delta\zeta. \tag{11}$$

How do we specify the highest wind during observation? Several observers were asked what fraction of their observing time they would be willing to lose at the shortest wavelength, due to high winds, in order to get the largest possible telescope for a given amount of money: the answer was "about one quarter." Our wind statistics,[5] show a wind speed of 22 mph at 300 ft height (17 mph as measured at 40 ft height, gusts included); the wind is higher than that for 15 per cent of all time during summer, 30 per cent during winter, and 24 per cent all year. This speed of 22 mph, together with Equation 11, then gives the specification for the deformation $\Delta\zeta$.

5. Miscellaneous

Sensitivity

The homology program delivers the final bar areas with six digits, but what accuracy is actually needed? What is the sensitivity of a homology solution to manufacturing deviations? These questions are answered by a small auxiliary program called "sensitivity." We start with perfect homology, and then assume that each bar area is changed according to $A_\gamma(1 + \epsilon_\gamma)$, where the ϵ_γ are uncorrelated random numbers with mean zero and variance $\epsilon^2 << 1$. The resulting deviation from homology ΔH can then be obtained analytically (without using actual random numbers); demanding $\Delta H \leqslant \lambda/16$ yields the maximum tolerated ϵ. Fortunately, it turned out that rms deviations of the bar areas of about 12 per cent can be tolerated for most practical purposes (see Table 4). This means we can use off-the-shelf shapes or pipes which give 10 per cent, if we always choose from the steel manual that size which comes closest to our A_γ from the computer output.

Thermal deformations

If the gravitational limit is passed by homology, the next natural limit is the thermal limit. The actual deformation of a given structure under a given temperature distribution is not calculated. Instead, we use with some confidence an estimate that assumes a temperature difference of $\Delta T = 5°C$ in unfavorable places; the estimate further assumes that all bars are made from steel and that the surface, if made from aluminum, is allowed to "float" on the backup structure of the panels. We used some experiments described in Reference 6 for the expected values of ΔT. For good protective white paint, we found $\Delta T = 5°C$ as the difference between sunshine and shadow on the average for clear, sunny summer days at noon. A second effect is given by the time lag of heavy members in the case of rapidly changing ambient air temperature (mostly around sunrise and sunset). For hollow members with white paint in winds below 5 mph, the time scale τ of thermal adaption was found by experiments as 1.73 hours per inch of wall thickness for steel, and 1.14 hours per inch of wall thickness for aluminum (length and diameter do not matter within wide ranges). The time scales are half these values for T and L shapes and solid rods; the time scales of unpainted aluminum and galvanized steel are 1.8 times longer. If the air changes by $\dot{T}$(°C/h), a member lags behind with $\Delta T = -\tau\dot{T}$.

At Green Bank, on one-quarter of all days, the measured maximum change of the day is $\dot{T} \geqslant 3.5°C/hr$. Thus, if $\Delta T \leqslant 5°C$ is demanded for three-quarters of all days, the heaviest members then must have a wall thickness below 0.83 in. for steel pipes, or below 1.66 in. for open shapes or rods. This, of course, can always be met by splitting up heavy members into several thinner ones; but then the wind resistance increases, and a good compromise is needed. One also could blow ambient air through hollow members at 15 - 20 mph, which would reduce thermal deformations by about a factor of 3.

I would like to add that the effect of thermal deformations (just as in case of wind deformations) will be reduced by at least a factor of 2, if the telescope pointing is done as suggested in Section 6.

Nonparabolic panels

It might be of advantage to make the surface panels of a shape that is different from a parabolic one, but easier to produce and to measure. For any given shape, a maximum size can be calculated. Formulas are given in Reference 7 for flat plates, spherical panels (to be measured and adjusted with a pendulum), and for toroidal panels (two-axis pendulum). These formulas are derived such that no deviation from the true paraboloid is more than $\lambda/16$, which means that the rms deviation from the best-fit paraboloid is about $\lambda/40$. In this way the values of Table 4 are calculated.

6. Telescope Pointing by Optical Means

Many of the older radio telescopes used a polar mount, which is the only way of driving a telescope without a computer (or analog coordinate converter). Since nowadays large telescopes always have an on-line computer, they usually use an altazimuth mount, which has many obvious structural advantages and even some observational advantages. In both types of mount, however, the pointing of the telescope is generally measured at the axes or drive rings (too far away from the telescope surface), and with respect to some structural elements or rails (stressed by heavy loads). The most logical way seems to be measuring the pointing where it matters (right at the apex), and with respect to something unstressed and unmovable (fixed points on the ground), which can be done by optical means. The JPL antenna at Goldstone, California, comes close to this demand, measuring the pointing at the apex and with respect to an internal unstressed pillar reaching close to the apex. But this internal pillar gives some structural (and financial) disadvantages. One should go one step further and use a sufficient number of light beacons right on the ground, as suggested in Reference 1.

Some satellites, rockets, and balloon telescopes already use optical pointing devices, "locked-in" to the bright rim of the earth or sun, or to some brighter stars. An investigation is planned into the availability, accuracy, and cost of such devices, and into their application to radio telescopes. The basic idea is to have a rotatable platform mounted behind the apex, with several small optical systems (theodolites) equipped with photocells, looking at as many light beacons (flashlights) mounted on concrete blocks on the ground; see Figure 4. A servo system keeps this platform "locked-in" to the beacons. At the joints between platform and telescope structure, we measure two angles and thus obtain the pointing direction of the telescope. Measuring these two angles and guiding the telescope into the desired direction with a second servo system can be done by normal techniques already in use; only the reference system defined by the platform, locked-in optically to the ground is new.

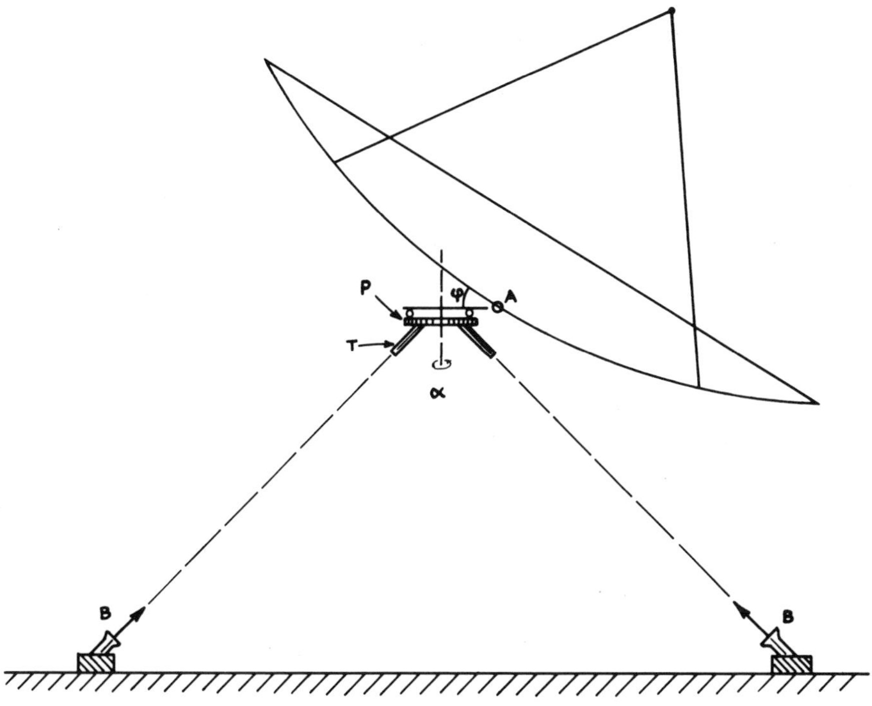

Figure 4. Position measurements by optical means. A small tiltable and rotatable platform P is mounted behind the apex A and looks with about six theodolites T to as many optical beacons B fixed at the ground. Three servo motors keep the platform "locked-in" to the beacons; elevation ϕ and azimuth α of the telescope then are measured between structure and platform. In this way, the position is measured where it matters and with respect to something unstressed and unmovable. No high accuracy is required for foundations, azimuth rails, and elevation ring; also, all deformations between apex and ground are omitted.

In principle, we need three beacons, and only two if the direction of gravity is measured independently by some kind of pendulum. Actually, we should have about twice as many beacons, because in some telescope positions one or another light path will be blocked by the surface or some structural members. In a first approach, the pointing direction of the telescope can be defined by the structural element where the platform is mounted. This is perhaps all that is needed in the future. In a more sophisticated version, a second platform, mounted in front of the apex, can look at the feed and three or more points on the dish surface, from which the computer can find the (best-fitting) pointing direction even of a slightly deformed dish.

The disadvantages of this method are, first, that it does not work in heavy fog or cloudburst (we cannot then observe at very short wavelengths anyway); since we do not need high accuracies for long wavelengths, the telescope could be equipped with an additional pointing system of conventional type for those cases. Second, there is the usual human inertia against any new method; large optical telescopes are *still* polar-mounted.

There are two major advantages. First, this method keeps the pointing accuracy completely independent of the accuracy of elevation rings and azimuth rails. As far as pointing is concerned, we could as well drive the telescope on a circular dirt road, and pull it into the right elevation by chains or ropes. Actually, one would use plain, normal railroad equipment for the azimuth ring, with $100,000 per mile for erection, $400 per mile and year for maintenance, and an accuracy of 1/4 to 1/2 in. vertical and lateral.[8] Second, with respect to thermal deformations, constant wind loads and all gusts slower than the servo loops, we completely omit all deformations occurring between the apex and the ground (telescope suspension, bearings, elevation ring, towers, rails, and foundation). This cuts down the remaining effective deformation by at least a factor of 2, even for the first approach, and the more sophisticated version should yield at least another factor of 2. It may cut down the costs of foundations and rails by almost a factor of 10.[8]

7. Numerical Results

The homology method was programmed at the Department of Civil Engineering of the University of Virginia, Charlottesville, Va. It was run with a Burroughs 5500 on several structures with good success. The results will be published elsewhere[9]; the following is a short summary.

In order to gain experience, we began with very simple structures and then proceeded to more complicated ones. The last one successfully run has $N = 13$ homologous surface points, a total of $p = 26$ points (pin joints), and $m = 112$ members. The next one, with $N = 21$, $p = 34$, $m = 128$, could not be run because of memory limitations (200,000 word disk; magnetic tape is too slow).

A second version of the program is now in preparation and half finished; it calculates only one *quadrant* of a symmetrical structure. We then hope to reach about N = 90, p = 190, m = 600. For a structure of this size, a full treatment three iterations, stress, and wind analysis) should take about 3- 4hr on our IBM 360/50.

Floating sphere telescope (structure 3a)

The first application to an actual telescope design was done for one of the LFST proposals: a complete sphere of 750-ft diameter, floating on water or pressurized air; one segment is cut off and replaced by a radome, and a stiffener ring at the opening holds the rim of a 565-ft parabolic reflector; see Figure 5. First, it was shown that a nondeforming ring can be obtained with no extra cost, just by a proper distribution of the stiffener ropes and counter-

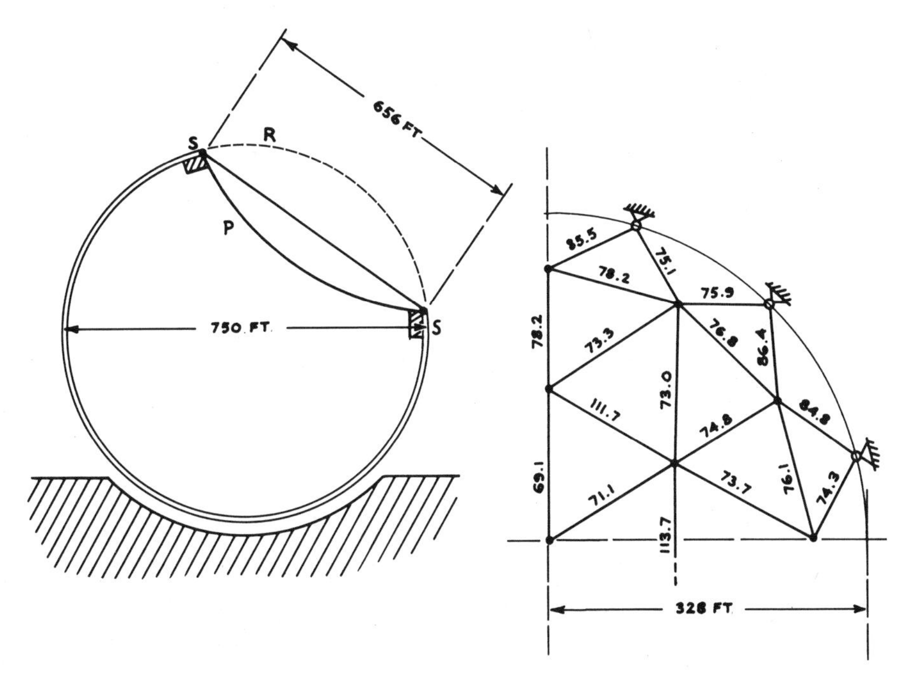

Figure 5. Parabolic reflector for floating-sphere telescope. The reflector structure P is two-dimensional network, suspended at the stiffener ring S inside the radome R. First, a nondeforming ring could be obtained. Second, all 66 members of the reflector were given the same area of 80 square inch for the first guess; after four iterations, the rms deviation between the deformed surface and a best-fit paraboloid was only $\Delta H = 7 \times 10^{-6}$ in. The final bar areas are written at each bar of one quadrant (all four quadrants are identical).

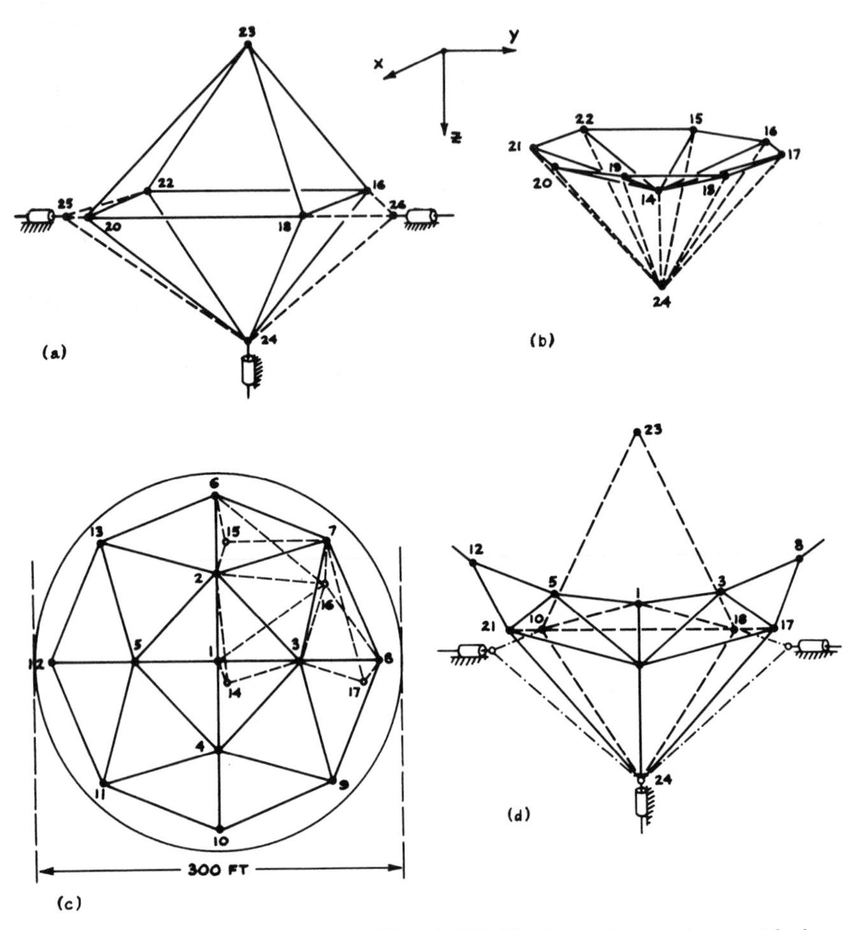

Figure 6. Geometry of structures 2e/4 to 2e/21. The basic structure is an octahedron, held by a suspension from two elevation bearings mounted on top of two towers. Bar areas are given in Table 2, final data in Table 3. This structure has 13 homologous surface points, and a total of 26 points (pin joints) and 112 members.

weight needed anyway. Second, supported at this ring, a parabolic surface structure was designed as a two-dimensional network of triangles, replacing a membrane by discrete members, with m = 66 and N = 19 (sufficient for $\lambda = 5$ cm). This structure ran successfully on first try; after only four iterations the calculating accuracy of the machine was reached, giving an rms deviation of $\Delta H = 7 \times 10^{-6}$ in. As to thefhomology parameters, the change of focal length was 3.9 in., and the change of axial direction only 0.72 min of arc. For the first guess, we took all bar areas equal; the largest change obtained in the final solution was 42 per cent, the average change 8 per cent. The result is not a shell that could be replaced by a membrane; it must be a framework since the radial members of the final solution are lighter than the ring members. Each iteration decreased ΔH by more than a factor of 10. This quick convergence showed that the floating sphere telescope could easily be supplied with a homologous mirror of any accuracy wanted. Furthermore, the convergence and the final value of $\Delta H = 7 \times 10^{-6}$ in. showed that the homology problem actually has *exact* solutions (as we have claimed with the existence proof), and that some structures also have practical solutions. Not only can we minimize ΔH, we can make it zero and thus omit gravity completely, in a natural and elegant way.

Octahedron and suspension (structures 2e)

Within the memory limit of the present program, we tried to obtain a sufficient number N of surface points, with a minimum of total points p and members m, for a telescope to be held at two elevation bearings on top of two towers. The best basic principle seemed to be an octahedron held with two suspensions [Figure 6(a)] which we called Structure 2. First, we experimented with N = 9 (Structures 2a, b, and c) in many variations, three-quarters of which converged to physical solutions. An attempt to obtain N = 20 with only p = 29 (Structure 2d) failed and was given up. We finally settled on Structure 2e as shown in Figure 6, with N = 13, p = 29, varying m from 102 to 116, and also slightly varying the geometry and the bearing restraints.

In all these experiments, we found that a good structure, close to equal softness and with a first guess well thought of, will generally give for the first guess a ΔH already small enough for practical purposes and then converge nicely (see Table 1); a wrong geometry or a bad first guess will generally give large ΔH and then converge to some negative bar areas. One of the major advantages of the homology program is that it tells you what is wrong it teaches its user after a few trials how to make a good first guess and how to choose a proper geometry. Furthermore, the numerical proof that ΔH can be made zero should encourage all designers who try to minimize ΔH by other methods, such as trial and error in the simplest case.

The best of our trial structures seemed to be Structure 2e/4 (see Table 1) with m = 112. Starting at two elevation bearings (points 25 and 26), two suspensions of three members each hold an octahedron, thus including the feed supports in the basic structure as suggested by Reference 1. The basic square of the octahedron then is used for obtaining an octagon. From the octagon and its center, 9 points, we reach the 13 surface points with a layer of 45 bars. The surface structure is represented by 28 surface bars, and the surface

Table 1. Convergence of the Homology Iterations. ΔH is the rms deviations of deformed surface from best-fit paraboloid; $\delta A/A$ is the rms relative change of bar areas between interations; df is the change of focal length between zenith and horizon; and i is the iteration (0 is the first guess).

Structure	i	ΔH (in.)	$\delta A/A$ (per cent)	df (in.)
3a	0	0.0856		− 2.99
	1	0.0149	15	− 3.71
	2	0.00074	3.2	− 3.96
	3	0.000016	0.4	− 3.92
	4	0.000007	0.3	− 3.87
2e/4	0	0.202		+ 0.26
	1	0.0240	21	− 0.46
	2	0.00089	4	− 0.52
	3	0.00024	3	− 0.49
2e/16	0	0.0248		+ 0.77
	1	0.0043	15	0.71
	2	0.0006	5	0.70
	3	0.0004	3	0.72
2e/18	0	0.0120		+ 0.80
	1	0.0045	7	0.84
	2	0.00006	1.4	0.85
2e/21	0	0.019		+ 0.98
	1	0.0015	6.7	1.00
	2	0.00011	2.4	1.03

itself by an additional load of 15,000 lb per surface point (2.76 lb/ft^2). The focus is at point 23. Each bar of this structure, is actually a built-up member as shown in Figure 3. The two bearings should be held on top of two towers moving on wheels on an azimuth ring. Since the telescope will have more stiffness in the x direction than the towers, we neglect their stiffness and let the bearings move freely along the x axis, making up for it by restraining point 24 in the x direction. Thus, the restraints of all three points 24, 25, and 26 are represented by gliding cylinder bearings. The actual tower stiffness will be introduced later on in the new program. Because of the present memory limit, we could not attach an elevation ring to the telescope.

After m and the geometry was settled, we tried to make a good first guess such that the bar areas would meet the survival condition by only a small margin, while the ratios of the areas would minimize the wind deformation. This is the Structure 2e/16 of Tables 1, 2, and 3. The diameter was chosen as D = 300 ft, for comparison with our 300-ft telescope at Green Bank. Structure 2e/16 started with ΔH = 0.025 in. and converged quickly (Table 1) to a physical solution. Table 2 shows the original bar areas and those after three iterations; the largest increase is a factor 1.75, and the largest decrease a factor 1.64, while

Table 2. Members of Structures 2e/16, 2e/18 and 2e/21. m is the number of identical members; A is the bar area, in inches squared; S_m is the survival stress according to Equation 9; Λ is the l/r ratio (of main chords of built up members); and S_o is the maximum allowed stress, for 33,000 yield steel and Λ.

Structure:		2 e/16		2 e/18				2 e/21			
Iteration:		0	3	2	2	2	2	2	2	2	2
Member	m	A_0	A_3	A_2	S_m	Λ	S_o	A_2	S_m	Λ	S_o
1 - 2	2	12	10.6	9.1	7.2	65	15.8	5.9	9.5	87	14.7
3	2	12	12.0	10.1	1.4	61	16.1	6.9	3.8	79	15.6
14	1	12	9.9	13.8	9.5	36	18.0	8.8	16.2	49	18.5
16	4	10	10.6	13.1	5.5	68	15.6	8.3	8.1	93	14.0
2 - 3	4	15	14.8	13.0	4.9	73	15.2	11.3	6.6	80	15.5
6	2	12	12.8	13.8	8.0	53	16.8	8.4	12.4	73	16.2
7	4	18	18.0	16.3	1.5	67	15.6	9.1	2.8	99	13.3
14	2	10	13.8	11.8	3.6	72	15.2	5.4	5.8	121	10.4
15	2	10	6.1	11.8	10.9	29	18.3	5.8	14.5	47	18.7
16	4	10	13.3	16.9	10.3	44	17.4	8.3	12.6	71	16.5
3 - 7	4	18	16.1	13.7	7.1	75	14.9	8.6	9.6	103	12.8
8	2	12	11.8	9.9	5.6	65	15.8	6.2	6.7	89	14.4
14	2	10	8.7	8.0	6.4	93	13.1	7.6	10.4	97	13.6
16	4	10	11.5	15.2	13.9	47	17.2	13.6	16.3	51	18.3
17	2	10	8.2	14.6	8.6	26	18.6	9.5	16.8	34	19.7
6 - 7	4	22	19.7	16.9	8.2	66	15.7	10.5	11.9	90	14.3
15	2	12	18.9	30.6	7.6	31	18.0	32.6	7.7	30	20.0
16	4	10	7.6	10.8	9.9	95	12.7	10.6	11.9	96	13.6
7 - 8	4	22	18.9	16.9	8.1	66	15.7	12.0	9.6	83	15.2
15	4	10	9.7	14.1	4.2	80	14.4	6.4	6.1	135	8.2
16	4	12	14.6	16.8	11.8	47	17.3	12.3	15.3	58	17.7
17	4	10	14.2	13.2	9.4	84	14.1	13.5	9.5	82	15.3
8 - 16	4	10	10.9	14.2	12.1	79	14.6	15.3	12.5	75	16.0
17	2	12	15.2	18.9	11.6	43	17.5	14.1	16.4	53	18.2
14 - 15	2	40	70.1	55.7	2.9	26	18.5	50.0	3.8	29	20.1
16	4	40	37.1	30.9	2.5	39	17.8	7.1	4.2	104	12.7
17	2	40	44.4	35.0	2.6	36	18.0	10.4	5.4	81	15.4
24	1	40	32.6	29.9	8.5	37	18.0	14.3	15.3	61	17.5
15 - 16	4	40	25.4	19.2	14.1	39	17.8	12.4	16.1	52	18.2
24	2	50	53.6	53.0	9.1	43	17.5	35.9	12.5	55	18.0
16 - 17	4	40	39.2	41.2	8.7	24	18.7	29.5	12.6	29	20.0
18	2	60	56.9	48.1	7.5	39	17.8	15.6	12.8	83	15.2
22	2	60	63.7	49.2	10.7	39	17.8	32.8	12.1	51	18.4
23	4	15	15.4	13.5	1.1	133	8.4	7.4	1.1	198	3.8
24	4	60	60.1	51.4	7.8	43	17.4	24.5	12.6	71	16.4
26	4	120	115.2	148.1	10.7	13	19.3	76.9	17.5	20	20.7
17 - 24	2	50	56.6	55.7	7.1	41	17.6	39.3	11.7	52	18.3
24 - 26	2	200	190.0	229.5	8.1	18	19.0	111.9	15.5	30	20.0

Table 3. Some Final Values for Structures 2e/16, 18, and 21.

		2 e/16	2 e/18	2 e/21
	Structure:	2 e/16	2 e/18	2 e/21
	Iterations:	3	2	2
rms	deviation from homology, ΔH (in.)	0.00040	0.00006	0.00011
	max surface deformation, ΔP_m (in.)	1.29	1.04	1.52
	change of focal length, df (in.)	0.72	0.85	1.03
	change of direction, $d\phi$ (min of arc)	4.20	2.37	3.06
	focal adjustment, rel. to point 23: ΔF_z (in.)	0.90	0.48	1.13
	focal adjustment, rel. to point 23: ΔF_x (in.)	1.89	0.84	0.89
	wind deformation, $\Delta\zeta$ (in.)	0.102	0.0835	0.172
	shortest wavelength, λ (cm)	2.59	2.12	4.40
	max stress factor, $Q = S_m/S_o$	1.27	0.83	0.90
	total weight on elev. axis (tons)	745	748	476

the average change is only 18 per cent. But we obtained survival stresses slightly larger than the allowed ones for three members ($Q = 1.27$ for member 2-15; 1.22 for 3-16; 1.12 for 8-16).

Improvement then was aimed in two directions; first, stability ($Q \leqslant 1$) with minimum weight just for survival, resulting in Structure 2e/21; second, stability and minimum weight for obtaining about $\lambda = 2$ cm for wind deformations, resulting in Structure 2e/18. Table 2 shows the final values of bar area A, survival stress S_m from Equation 9, the l/r ratio Λ, and the maximum allowed stress for this Λ and for steel of 33,000 psi yield. We see that $S_m < S_o$ for all members of both structures; thus both are stable in survival conditions. From Table 3 we see that $\lambda = 4.4$ cm, still a good wavelength for a 300-ft telescope, is achieved with only 476 tons total weight (the Green Bank 300-ft has 450 tons, the surface allows only $\lambda = 20$ cm, but the structure would allow $\lambda = 15$ cm). Structure 2e/18, with $\lambda = 2.12$ cm, is close enough to be called a final solution. Its weight is still fairly low with only 748 tons. Against survival, it is overdesigned by 21 per cent. For both structures, the change of focal length is about 1 in., and the direction changes by only 3 min of arc. The mechanical adjustment of the feed, with respect to point 23, is only about 1 in.

Weidlinger[5] has pointed out that the ideal minimum-weight structure should have the same stress S_m in all of its members; for any given structure with different S_m, the ratio of the smallest over the largest S_m is thus an estimate of how efficiently the material is used. However, this should be modified in two ways. First, with respect to members of different length, the stress factor Q is used instead of the stress S_m; second, the average, not whether one or the other member has $Q \approx 0$, is what is important. We thus define a weight efficiency η by

$$\eta = \frac{\text{maximum Q}}{\text{average Q}} \tag{12}$$

This is important for Structure 2e/18, where we just want to fulfill the survival condition. Figure 7 shows the distribution of the stress factors Q for 80 bars only, since 32 bars are defined by different criteria (28 surface bars that must be widely split-up and must resist bending forces, and 4 feed supports that should not go beyond $l/r = 200$). We see that 10 bars have rather low Q, but the remaining 70 bars form a dense group. For all 80 bars, we obtain $\eta = 0.786$, which we consider close enough to 1 to call it a final solution. The maximum is Q = 0.90, leaving a margin of 10 per cent for nuts and bolts and other things neglected.

Comparison with theoretical estimates

In 1965 several formulas were developed on purely theoretical grounds for estimating the weight of tiltable telescopes of a "near-to-ideal design," and it is interesting to compare these old theoretical estimates with the present actual designs. For a structure defined by survival, and for $\lambda \leqslant 5$ cm, formula 24 of Reference 1 reads (W measured in tons, D in 100 m)

$$W = 432\ D^3 + 160\ D^2, \tag{13}$$

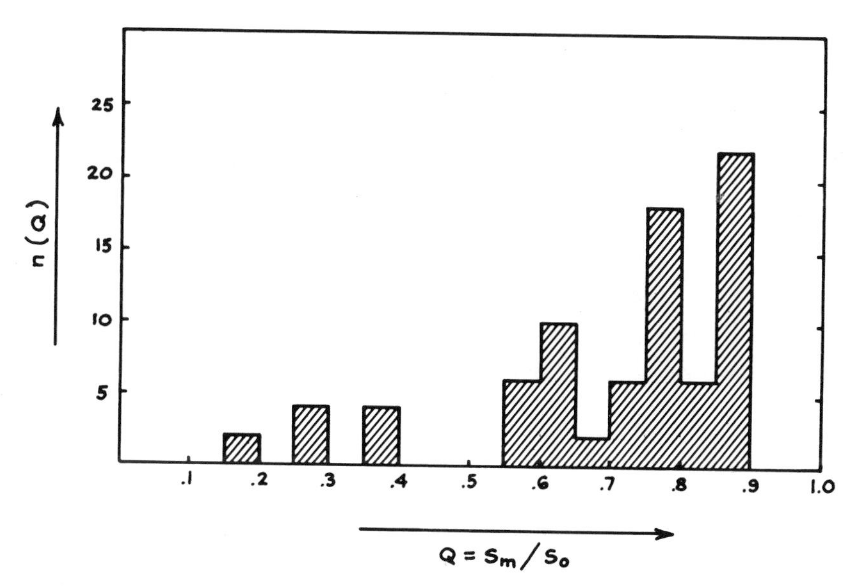

Figure 7. Distribution of stress factors Q from Equations 9 and 10, for Structure 2e/21.

which yields W = 467 tons as compared to W = 476 tons of Structure 2e/21. For a structure defined by wind deformations, Reference 1 used $\lambda = 16\ \Delta\zeta$ instead of $10\ \Delta\zeta$, used now in Equation 11 where we neglect parallel translations; if corrected for this difference, formula 30 of Reference 1, for $\lambda \leqslant 5$ cm, reads (W in tons, D in 100 m, λ in cm)

$$W = (2025\ D^4 + 181\ D^3) / \lambda + 122\ D^2 , \tag{14}$$

which yields W = 842 tons for $\lambda = 2.12$ cm, as compared to W = 748 tons of Structure 2e/18. This is a difference of 12 per cent, while the old estimates did not ask for more accuracy than, say, ± 30 per cent. Finally, if a structure is defined by survival, formula 18 of Reference 1 then gives the shortest wavelength from wind deformations, corrected for Equation 11, as (λ in cm, D in 100 m):

$$\lambda = 4.7\ D, \tag{15}$$

which yields $\lambda = 4.3$ cm as compared to $\lambda = 4.4$ cm of Structure 2e/21. All three results give a mutual confirmation of both estimate and design.

The fact that our homology solutions agree completely with the old estimates based on a near-to-ideal design, and that Structure 2e/21 gives the same weight as the 300-ft at Green Bank while inproving on its wavelength by more than a factor of 3, shows that we have omitted gravity without ill effect.

Check calculations

Usually, a computer program is checked before its application by some hand calculation of a simple case. This is impossible with our homology program. Since a paraboloid of revolution is defined by 6 points, we need at least 7 surface points to make the method work, which means at least a total of, say, 10 points and 30 members, far beyond the scope of a hand calculation. We thus have taken one of our final results, Structure 2e/18 after two iterations from Table 2, and have sent its coordinates and bar areas to Simpson, Gumpertz, and Heger (Cambridge, Mass.), asking for a complete stress and deformation analysis under all of our load conditions including survival. For dead loads only, a best-fit paraboloid of revolution and the deviations ΔH from it should be calculated for elevation angles 0°, 45°, and 90°, thus also checking at 45° our statement that homology holds in all angles if it holds in two.

The results are as good as can be expected for the finite calculating accuracy of the computers. All stresses and deformations agree with our results within five decimals. For the rms deviations ΔH from the best-fit paraboloid, Simpson, Gumpertz, and Heger obtain

Position	ΔH
horizon, 0°	0.000052 in.
45°	0.000051 in.
zenith, 90°	0.000049 in.

This agrees within the calculating accuracy with our value of $\Delta H = 6 \times 10^{-5}$ in. from Table 3 for the average of horizon and zenith according to Equation 5, and it also shows that our statement is correct at 45° elevation angle.

8. Further Possibilities and Plans

Radome

We also tried to work out results for a telescope of D = 300 ft to be enclosed in a radome (dead loads only). Since this turned out to be more difficult, as explained in Section 4, and since we think that exposed telescopes can be made less expensive than those in a radome, the solution was not pursued as extensively. A series of first guesses was run, called Structure 2f/1 to 2f/8, using about the same geometry as Figure 6, and varying m between 104 and 112.

A final result (all $Q \leqslant 1$ with low total weight) has not yet been obtained, but the best solution already comes close to it, still giving $Q > 1$ for 6 members but with a maximum of only Q = 1.24, similar to Structure 2e/16. This solution, Structure 2f/5, has m = 104 bars, and a total weight of W = 237 tons. From our previous experience we feel confident that a final solution with about W = 200 tons can be reached after some more trials, especially with the new program, and we have entered this expected value in Table 4 for comparison.

The new version of the program will be much more flexible than the present one, and it can be run in a mode that already comes close to the method suggested at the end of Section 3, which is better suited to this task.

Telescopes of other D and λ

Although we have calculated our structures only with D = 300 ft, we can predict a selection of other telescopes as shown in Table 4; N is found from Equations 4 and 3; ΔT from Equation 2; $\Delta A/A$ is based on the calculated sensitivity of Structures 2e/18 and 2e/21, and then is scaled according to $\Delta A/A \sim \lambda/D^2$; and formulas for n and l are given in Reference 7. The total weight for exposed telescopes is calculated from Equation 14, which gave good agreement with Structures 2e/18 and 2e/21. The weight of a telescope in a radome is based on the expected value for D = 300 ft, and then scaled according to $W \sim D^{2.5}$.

The first line of Table 4 is an "infrared telescope" and needs some explanation. Several colleagues have suggested building a small model of a homologous telescope, about 30 ft diameter. Its gravitational deformations ($\sim D^2$) then will be only 0.3 mm = 300 μ, and in order to show that they are homologous, they must be measured with an accuracy of only 10 μ; this can be done with an optical Michelson interferometer, as an experiment by J. Hungerbuhler of NRAO has shown. But a good model should do more than just demonstrate what a computer already has calculated; it should make itself useful as a telescope. The trouble then is that our atmosphere is opaque from 1 mm all the way down to 25 μ wavelength. A homologous telescope for that wavelength could be built, with N = 80 surface points, although it needs a

Table 4. Some Telescopes with Homologous Deformations.

D (ft)	λ (cm)	β (sec of arc)	N	ΔT (°C)	$\Delta A/A$ (per cent)	n	l (m)	W radome (tons)	W exposed (tons)
30	0.0025	0.7	80	0.06	1.7	770	0.01	3	–
85	0.1	10	17	0.8	9	165	0.11	25	131
300	2	54	10	4.5	14	60	0.96	200	888
400	3	61	12	5.2	12	56	1.35	410	1790
450	4	72	11	5.4	12	51	1.65	550	2150
500	5	81	11	6.7	12	47	1.95	750	2570
600	6	81	13	6.7	12	42	2.34	1300	4370

Free choice:
D is the telescope diameter
λ is the shortest wavelength } $\beta = 1.2\ \lambda/D$ is the half-power beamwidth.

Requirements:
N is the minimum number of homologous surface points (the present program is memory limited and gives N = 13; the planned version should reach N = 80 or more);
ΔT is the maximum tolerated temperature differences in the structure (good protective paint gives about 5°C in sunshine);
$\Delta A/A$ is the maximum tolerated rms deviation of bar areas from computed values (off-the-shelf structural pipes give 10 per cent).

Surface:
n is the minimum number of toroidal panels;
l is the maximum size of flat plates.

Weight (of elevation-moving structure: dish, surface, feed legs) = W
radome: minimum structure for stable self-support, inside radome;
exposed: a. wind deformation $\leqslant \lambda/10$ for 17 mph on ground (22 mph at 300-ft height);
b. survival = 20 lb/ft² of snow, or 4 in. solid ice, or 90 mph on ground (110 mph at 200-ft height);
comparison: the NRAO 300-ft telescope (λ = 15 cm) has W = 450 tons.

temperature stability of 0.06 °C and a structural accuracy of 1.6 per cent. The most severe problem, however, is how to obtain a large surface of almost optical quality with low costs, and this has not yet been solved. The second line of Table 4 is an 85-ft telescope for 1 mm wavelength which can certainly be built.

For the larger telescopes ($D \geqslant 300$ ft) we have always chosen the smallest λ such that N, ΔT, and $\Delta A/A$ can easily be met. We see that all large telescopes can be designed from Structure 2e/18 (all $N < 13$), they can be built from off-the-shelf pipes (all $\Delta A/A > 10$ per cent), and they can observe in sunshine (all $\Delta T \approx 5°C$).

Final designs

Our plan is, first, to finish the new program, allowing more complicated structures and more flexibility. Second, we will develop some good structures with 60 - 80 surface points; this is not necessary for observation, but it reduces the size of surface panels and eases the erection. Third, we will consider more details of survival stresses and wind deformations, such as various angles of the wind and the actual resistance and bending of the members. Fourth, a dynamical analysis of the more promising structures will be done somewhere else.

Finally, we plan to work out three complete designs (D = 85, 300, and 500 ft from Table 4) in all details, no matter what the financial hope for building them happens to be. They will be published, and anyone interested is welcome to use them.

References

1. S. von Hoerner, "Design of Large Steerable Antennas," *Astron. J.* **72**, 35 (1967). [First published as an LFST Report (June 1965). The LFST group, headed by Dr. Findlay of NRAO, investigates the possibilities for a Largest Feasible Steerable Telescope; its reports and summaries can be obtained from Green Bank on request.]

2. S. von Hoerner, "Homologous Deformations of Tiltable Telescopes," *J. Structural Div. ASCE* **93**, 461 (1967). [First published as LFST Report No. 4 (November 1965).]

3. O. L. Mangasarian and S. Fromovitz, "The Fritz John Necessary Optimality Conditions in the Presence of Equality and Inequality Constraints," *J. Math Analyt. Appl.* **17**, 37 (1967).

4. S. von Hoerner, "Statistics of Wind Velocities at Green Bank," LFST Report No. 16 (December 1966).

5. P. Weidlinger (private communication).

6. S. von Hoerner, "Thermal Deformations of Telescopes," LFST Report No. 17 (January 1967).

7. S. von Hoerner, "Non-Parabolid Panels, and Surface Adjustment," LFST Report No. 18 (June 1967).

8. S. von Hoerner, "Discussions with Railroad Engineers," LFST Report No. 14 (September 1966).

9. M. Biswas, R. Jennings, and S. von Hoerner (in preparation).

R. D'Amato
Lincoln Laboratory,
Massachusetts Institute of Technology
Lexington, Massachusetts

METAL SPACE FRAME RADOME DESIGN

1. Introduction

Other the past fifteen years the number of radomes in service throughout the world has steadily increased. As the advantages of radomes become more widely recognized, the rate at which radomes are added to those that are already in service will grow. In view of this situation, it is important that the design and performance of radomes be more thoroughly understood so that increasingly efficient radomes can be supplied to meet this growing need.

The purpose of this paper will be to review the state-of-the-art of radome design. The evolution of design methods will be discussed and progress will be noted along with areas where further work is needed. Where it is possible, various methods will be directly compared.

Radomes can be divided into the two broad classifications of air-supported and rigid. The air-supported radome consists of a thin, flexible envelope whose external shape is maintained by internal pressurization to provide its structural integrity. Although the air-supported radome has been made in various exterior shapes, the most common shape is a truncated sphere. A variation from the single thin envelope is the "air mat." The air mat consists of inner and outer walls connected with tension strands or ribbons. Pressurized air is introduced between the walls to provide its structural integrity.

The rigid or self-supporting radome consists of a relatively thin structural surface formed into a shell. The rigid radome derives its structural capability from the basic geometry of the shell which supports external loading primarily by direct tensile and compressive forces in the shell surface.

The primary types of surface structures that have been utilized for radome construction are: space frame, sandwich wall, and homogeneous wall. The space frame surface structure consists of an array of interconnected beams of metal or plastic covered by a thin dielectric membrane surface that transfers the external wind loads to the framework. The sandwich surface structure consists of an inner and outer thin wall of high-strength material separated by a low-density low-strength core material. The homogeneous wall is a single thickness of structural material of fiberglass or rigid foam plastic. Of all of the types of surface structure that have been used, the most common for large ground radomes (greater than 50 ft in diameter) has been the space frame. The reader is referred to References 1 and 2 for a more detailed description of radome types as well as a brief history of radome development.

2. Over-all Design Considerations

The three major considerations of radome design are: (1) electromagnetic performance, (2) structural integrity, and (3) cost.

For many specific installations, the radome designer may have to evaluate other special requirements, which may include such items as the mode of usage, the permanency of the installation, heating/air conditioning, and building interface. Since the special requirements are so dependent upon the specific installation, they will not be considered in this paper.

2.1. Electromagnetic considerations

The primary electromagnetic performance considerations that must be integrated into the structural design of a space frame radome are (1) frame blockage, (2) membrane loss, and (3) pointing errors. Many factors enter into the evaluation of space frame blockage. It is not the intent here to go greatly into detail regarding the calculation of blockage, since that has been considered in some detail by others.[3] It is important, however, that the structural designer have an appreciation of the major factors so that he makes his design more intelligently. Blockage is governed primarily by the wavelength of the radiation, the dimensions of the beams in the framework and ratio of the antenna diameter to radome diameter. In general, it is desirable to have the length of the beams in the framework long compared with the wavelength of the radiation. The cross-sectional dimensions of the beam are important in that the ratio of depth-to-width influences the scattering properties of the beams. When the width of the beam is as large or larger than the wavelength, this effect is minimized and optical shadowing of the framework on the antenna will be an excellent index of the blockage produced by the framework. A simple formula for the optical blockage of the radome frame is presented in Section 3. This formula, which will be discussed below, provides a reasonable index of the radome performance as long as the wavelength is small compared with the beam dimensions.

Another important contributor to the electromagnetic loss of the radome is the membrane that covers the space frame structure. This loss is governed primarily by the dielectric properties of the membrane and the ratio of membrane thickness to operating wavelength. For the smallest possible loss, it is important to make the membrane as thin as possible. For example, for a material with a dielectric constant of 4 and thickness to wavelength ratio of 1/40, the loss will be about 0.35 dB. It is of interest to note that with decreasing wavelength, the loss due to the membrane increases while the loss due to the radome frame remains substantially constant.

It has been found that the pointing errors introduced by space frame radomes are extremely small, usually less than 0.1 mrad. The primary parameter governing the pointing errors is the randomness of the space frame geometry. For a space frame to be random it must have no contiguous members that lie on the same great or minor circle. The effective beam dimensions and the distribution of the geometry in relationship to the pointing errors are affected in a minor way by the operating wavelength. There appear to be no simple procedures for making analytical estimates of pointing accuracies.

To sum up, the requirements for maximum electromagnetic performance dictate that the beam cross section be made as compact as possible and that the length of the beam be made as long as possible compared to the operating wavelength. The thickness of the membrane should be very small compared to the operating wavelength.

2.2. Structural considerations

The primary loads experienced by a radome are its own dead-weight loading and the pressure produced by the wind. The dead-weight loads on the structure are computed in a straightforward way and require no further elaboration. The wind forces, on the other hand, are influenced by a number of parameters and require a detailed exposition. It is to be noted that the wind forces produce the major loading on the radome, although the larger the radome becomes, the more significant will be the dead weight forces.

The size and the shape of the structure influence the distribution of pressures over the surface. The fact that the shape of the structure influences the pressure distribution is obvious; the influence of the size is more subtle. The size effect is important in that the winds to which a radome are exposed are not steady and uniform, but rather vary in intensity and duration with time and with spatial position. In general, for large structures one part of the structure will, at a given instant of time, be exposed to a different wind velocity than will another part that is some distance away. Thus, the wind loading on a large radome is an extremely complex problem. The way in which this comple problem is handled will be discussed further below.

In addition to its own size and shape, the wind pressures on a structure will also be strongly influenced by local terrain features. Local terrain features can provide either a sheltering effect or a wind-velocity amplification effect.

The over-all geographic location is important in that it governs the magnitude of the prevailing wind as well as influences the spatial distribution of the wind (velocity versus height and turbulence) by virtue of the roughness of the general terrain. For example, where the land is smooth and flat, a greater change of velocity with respect to height will occur over a smaller height than will be the case for a wooded area.

The useful life of the structure and the design risk will influence the extreme wind velocity to which a radome will be exposed at a particular location. The smaller the design risk for a given lifetime, the higher will be the wind velocity for which the structure must be designed.

Finally, the structural properties of the radome itself will affect the pressure distribution. In the case of the static structural behavior, the pressures induced by the wind velocity will change the shape of the structure thereby altering the local pressure distribution. In the case of the dynamic response of the structure, especially if the structural frequencies are close to the vortex shedding frequency, the structure effectively extracts energy from the air stream and thus may build up an excessive structural response. The first case is a static aeroelastic problem and the second case is a form of flutter. Neither of these cases have been found to be important for usual radome configurations.

The wind-induced pressure on the radome is resisted by the panels covering the frame, which produces tensile stress resultants T as shown in Figure 1. These tensile stress resultants produce a transverse loading $2T \sin \alpha$, which must be supported by the beam. The distribution of the lateral loading is approximately triangular. The transverse loading on the individual beams are, in turn, redistributed by "shell" action of the entire frame into primary axial loads P in the members. The primary axial loads in the beams are finally reacted at the base of the structure by the foundation. The dead-weight stresses are distributed in a similar fashion.

Several modes of failure are possible in the radome structure. For each beam it is possible for the stresses produced by the beam-column loads to exceed the allowable stress in the material and it is also possible for the beam to buckle as a column in the plane of the membrane. However, even if these two modes of failure are prevented, it is also possible for the level of loading to be sufficiently high for a particular structural configuration to cause collapse of a section of the radome consisting of several contiguous beams. Finally, in the

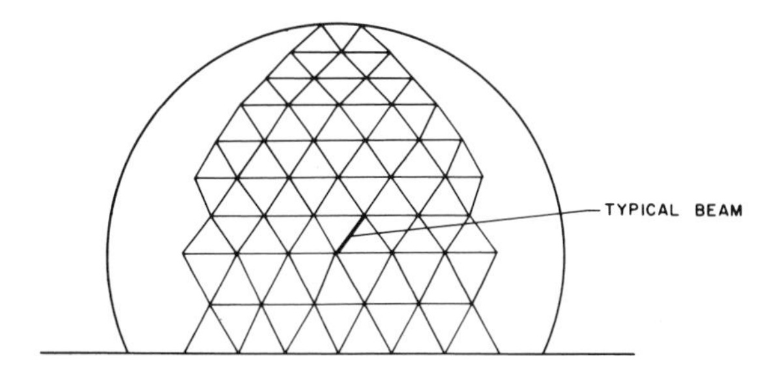

RADOME FRAME

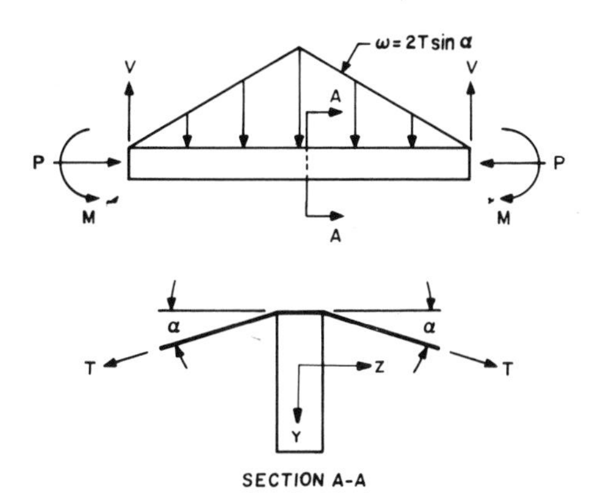

FORCES ACTING ON TYPICAL BEAM

Figure 1. Typical beam isolated from radome frame.

case of the membrane, failure can occur either by the stresses in the material tearing the material or by the tensile load being sufficiently high to fail the attachments of the membrane to the beam.

2.3. Interaction of electromagnetic and structural considerations

Given a particular loading condition, there are a large number of different combinations of beam dimensions which can satisfy the structural requirements outlined above. It remains to determine which combinations are the most suitable from the standpoint of electromagnetic performance and cost. For example, from a consideration of the electromagnetic performance it is desirable to have the smallest amount of material in front of the antenna. To achieve this, the designer would tend to specify long beams with compact cross sections. On the other hand, there are practical limits on the size of membrane panels that can be used without requiring a large thickness that could also contribute to radome losses. Furthermore, the large panels could increase the over-all cost of fabrication and erection of the structure. The minimum cross-sectional dimensions of the beam can be achieved by means of a solid cross section as compared with a hollow cross section. The solid cross section will, however, be heavier than the hollow cross section and the cost of the radome made with the solid cross section could be higher. Another method of reducing the blockage contributed by the frame is to "tailor" the cross section of each beam to its minimum size, which is a function of its location in the radome. A "tailoring" approach is one that is readily handled on a computer. However, having a large number of different sizes is certain to complicate fabrication and, hence, increase the cost. Even from this brief discussion it can be seen that the radome design process is one that involves an interaction of the contributing parameters.

3. Structural Design Methods

In Section 1 radome development was briefly outlined and in Section 2 an overview was given of the primary considerations involved in the design of space frame radomes. In this section, a more detailed consideration will be given of some of the tools and data for radome structural analysis.

The major areas in radome structural design are:

1. Wind loading.
2. Stress and stability analysis of beams.
3. Stress analysis of membranes.
4. Evaluation of space frame instability.
5. Optical blockage of space frames.

Following a discussion of these areas, a summary of simplified design equations will be outlined and set of recommended factors of safety presented.

3. 1. Wind loading

Since wind pressure is the governing loading condition for radome design, it is of great importance to specify the wind environment properly. The specification of the wind velocity should account for the particular site at which the radome is located, the number of years that the structure will be used, and the acceptable design risk. General data for peak wind velocity averaged over time periods of 5–20 min can be obtained by statistical analysis of existing weather records. The recorded wind velocities that are available have usually been measured with instruments that average the wind speed over some time period.

Over the past ten years there has been considerable amount of study to evaluate the effects of wind on buildings and structures and a considerable body of literature is being accumulated. Even so, there is still uncertainty in the wind environments. One of the major reasons for this is that the available wind speed measurements have often been influenced by obstructions and buildings in the immediate velocity of the wind-speed measuring instruments, and even by the towers on which these instruments have been mounted. This situation makes it imperative that the site at which the radome is to be located be thoroughly evaluated and that existing wind speed records be carefully examined.

After selecting the maximum wind speed that can be expected at a specific distance above the ground, the next step is to determine the variation of the wind speed with height. This is especially important for large radomes. The variation of wind speed with height will be a function of the roughness of the terrain in the vicinity of the site. A convenient way to express the velocity gradient is by means of the power law[4]

$$V = V_G(Z/Z_G)^{\alpha}, \tag{1}$$

where V_G is the velocity at the gradient height Z_G. It has been suggested[4] that Z_G and α be 900 ft and 0.17, respectively, for flat open areas and 1300 ft and 0.28, respectively, for rough wooded areas. The appropriate values of Z_G and α must be determined for the particular site.

Both the probability of occurrence and the variation of the design wind speed are shown graphically in Figure 2. Because the wind speed fluctuates considerably over the time averaging period, the wind velocity of structural significance to the radome will be in excess of the time average. The time variation of the fluctuating wind and its frequency spectrum are illustrated in Figure 3. The fluctuating wind velocity is physically experienced as "gustiness." Since structural damage can be produced by gusts that last only for a few seconds it is important that this gustiness be evaluated. The spectrum of the gustiness shown in Figure 3 is the amount of energy contained in the wind as a function of wind frequency, or inverse of the gust wavelength.

A rational way for utilizing the wind spectrum in the radome design cycle is shown in Figure 4. The procedure followed by this analysis is similar to that used in aircraft design[5]. The great difficulty in performing such an analysis on an actual radome is the lack of specific data concerning the aerodynamic and structural admittance. The analytical prediction of the stresses produced by fluctuating wind is not readily done for any structure. Most results obtained thus far have been obtained by an experimental approach with aeroelastic models in wind tunnels especially designed to produce a reasonable representation of the spectrum of wind turbulence. A good example of this approach is the analysis of a large cooling tower by Davenport and Isyumov in Reference 6. However, even the application of experimental methods to radome structures

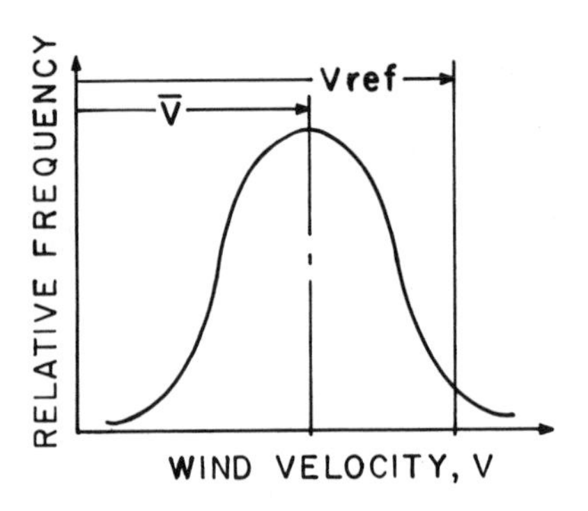

PROBABILITY OF OCCURRENCE OF AVERAGE WIND SPEED

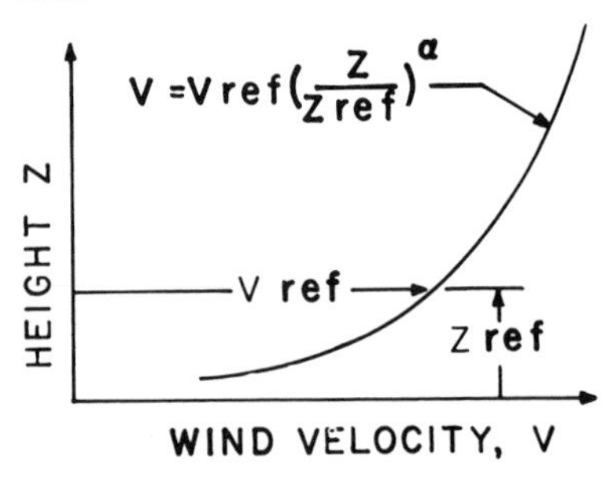

Figure 2. Probability of occurrence and variation with height of average wind speed.

is very difficult because of the density of the radome structural wall relative to the air density. This ratio for radomes makes the construction of small aeroelastic models very difficult. A less rational but more practical procedure for radomes is to multiply the reference wind speed by a gust factor and use the resulting wind speed in a static analysis. Some work has been done on this approach[7] and it has been found that the gust factor can be reasonable related to the size of the structure and the wind speed. Sherlock in Reference 7 suggests that the critical gust duration can be estimated by dividing a characteristic structural size by the wind speed. He suggests that the characteristic size be estimated as eight times the length of the structure in the wind direction. Some typical data for the variation of gust factor with gust duration are shown in Figure 5. For example, for a 300-ft-diameter radome exposed to a 100 mph wind, the gust duration will be $8(300)/147 \cong 16$ sec, which from Figure 5 will give a gust factor of about 1.25. Therefore, the design wind loading will be 125 mph at the particular height being considered.

After the environmental winds have been established, it is necessary to establish the distribution of wind pressures over the surface of the radome. At

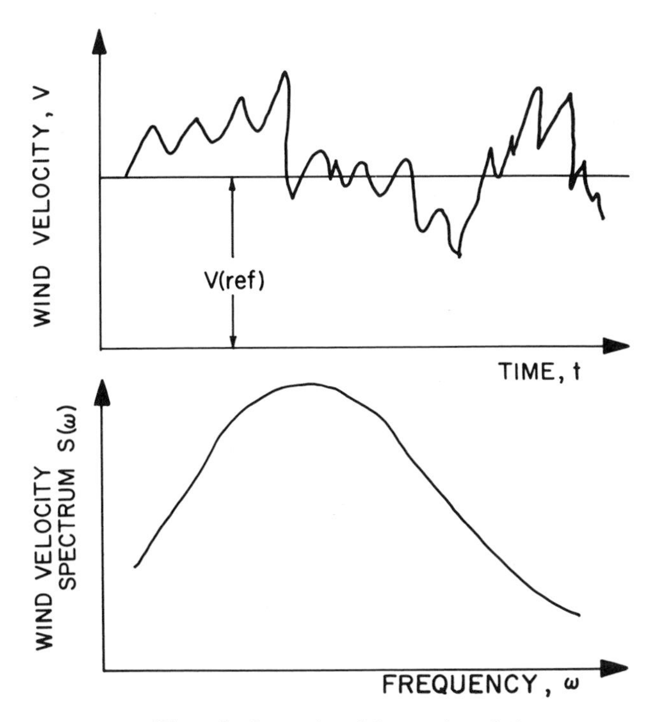

Figure 3. Properties of fluctuating winds.

the present time there are no adequate analytical procedures for computing the pressure distribution, and the only recourse is to use wind tunnel measurements. Up to the present time, all of the wind tunnel tests on radomes have been made under conditions of uniform flow and the measured pressure distributions have been used with a sufficiently high value of the wind speed to provide a conservative design. For radomes 150 ft in diameter or less this approach has been found reasonable. For very large radomes (500 ft in diameter), a more accurate evaluation of the wind pressure distribution should be carried out. A wind tunnel test program for radomes in a gradient flow is presently under way at the Wright Brothers Wind Tunnel at M. I. T. under the sponsorship of the CAMROC program. It is expected that the results from these tests will provide valuable data where none is presently available.

Until this gradient flow data become available, and for applications involving smaller size radomes, the uniform flow data will continue to be used with a conservative value of wind velocity specified. It is, therefore, worthwhile to look briefly at some of the data that are available so that a better appreciation may be had of its use in design.

Major considerations in the evaluation of the validity of aerodynamic test

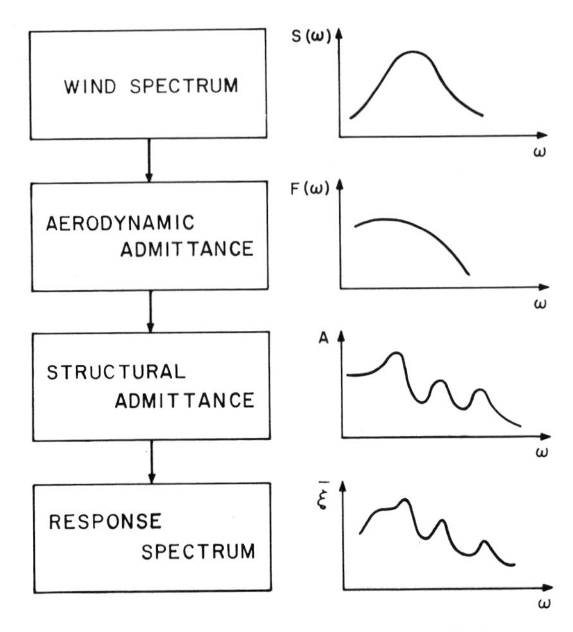

Figure 4. Procedure for dynamic analysis of radome.

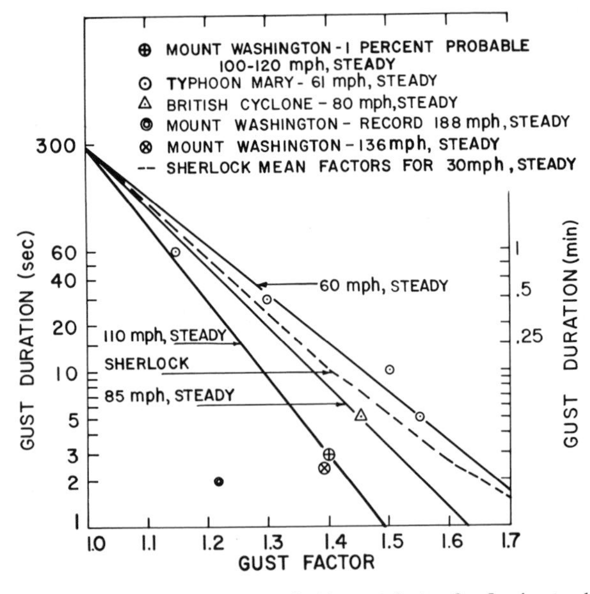

Figure 5. Estimated one per cent probable gust factor for 5-min steady wind.

data are the Reynolds number Re,

$$Re = VD/\nu \tag{2}$$

and the Mach number M.

$$M = V/C, \tag{3}$$

where V is the wind velocity, D is the radome diameter, ν is the kinematic viscosity, and C is the speed of sound. For exact similarity

(Re) Test = (Re) Full scale,

(M) Test = (M) Full scale.

It is not possible to fullfill these two requirements simultaneously. To avoid compressibility effects, it is absolutely essential that the test Mach number be less than about 0.3. The requirement of Reynolds number similarity can be somewhat relaxed by noting that above a critical Re there are no significant changes in the pressure distribution with increasing Re. For spheres, the critical Re is in the vicinity of 5×10^5.[8] It is important that the test Re be significantly above this value to make certain that the pressure distribution of the model be similar to the full scale.

Most of the wind tunnel measurements that have been made have utilized small rigid models having Reynolds numbers on the order of 2×10^6, which is significantly greater than the critical value for spheres. During the development of the Haystack radome, however, areodynamic tests were made on a 10-ft-diameter structural model mounted on a ground plane. The model was a truncated sphere with a height to diameter ratio of 0.786. For these tests the speeds ranged up to nearly 200 mph, giving test Reynolds numbers up to nearly 2×10^7.[9] It is of interest to compare the data obtained from this test with those from a smaller model which was a 27.5-in.-diameter 0.71 sphere mounted on a 12-in.-high twelve-sided tower,[10] The large model was a fairly accurate structural replica of the Haystack radome including such details as the network of individual beams having scaled cross-sectional properties, hubs, and a membrane covering. The 27.5-in. sphere was a solid wooden model that is common in wind tunnel testing. The geometric configurations of the two models are shown in Figure 6. The data from each test included an extensive survey of the pressures over the surface of the models. The pressure distribution was fitted with a four-term Fourier series in the azimuth angle θ at several horizontal cuts specified by the meridional angle ϕ measured from the top of the radome,

$$C_p(\phi,\theta) = C_{p_0}(\phi) + C_{p_1}(\phi)\cos\theta + C_{p_2}(\phi)\cos 2\theta + C_{p_3}(\phi)\cos 3\theta, \tag{4}$$

where the coordinates ϕ and θ are the usual shell coordinates. The pressure coefficient C_p in this equation is defined as the difference between the ambient

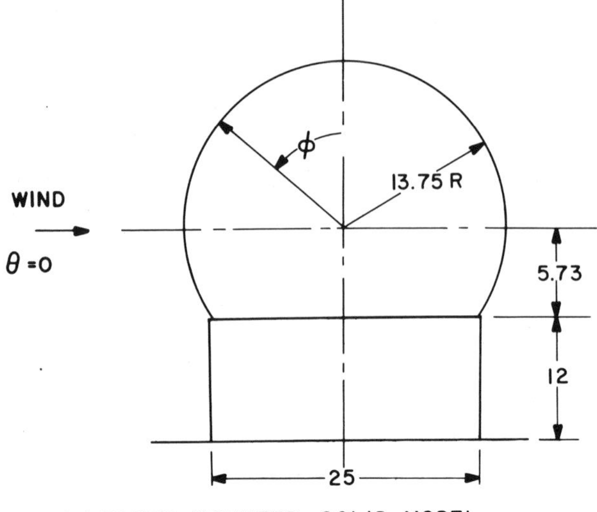

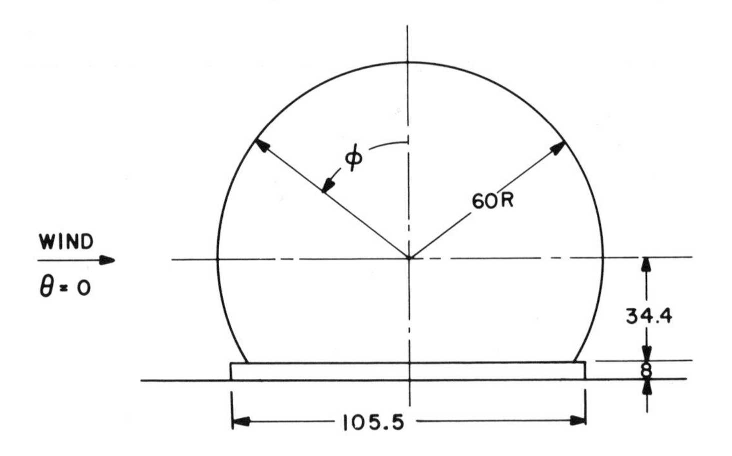

Figure 6. Wind tunnel test models.

and local pressures divided by the dynamic pressure,

$$q = \tfrac{1}{2}pv^2 \tag{5}$$

Figures 7, 8, and 9 show the pressure data at three typical meridional positions above, at, and below the equator (ϕ = 90°), compared with the four-term Fourier fit. It can be seen that the four-term Fourier series fits the data quite

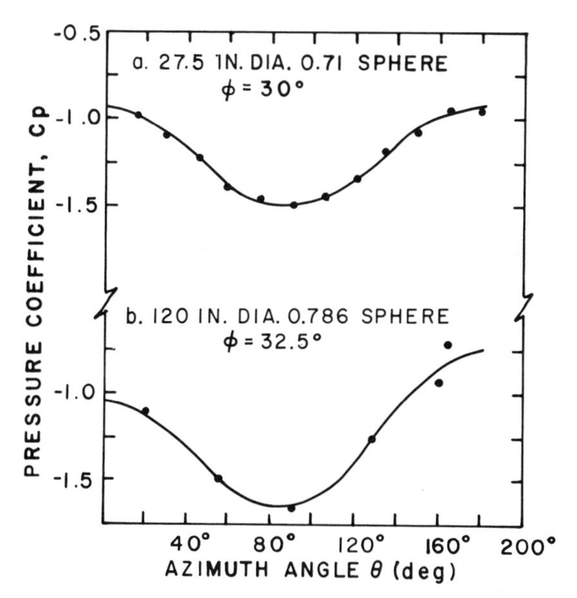

Figure 7. Comparison of measured pressures with four-term Fourier fit at typical position above equator.

well.

Figures 10, 11, and 12 compare the data for the two models at three typical meridional positions: above, at, and below the equator. Note that the peak negative pressure (at $\theta = 90^\circ$) for the larger model is consistently higher than for the smaller model and that the negative pressure at the base ($\theta = 180^\circ$) is higher for the smaller model. Figure 13 is a more comprehensive comparison of the pressure distribution of the two models in that it compares the four Fourier coefficients of the fitted distributions over the entire radome surface. Figure 13 shows that the poorest agreement occurs for the θ and 2θ terms. One possible explanation for the difference is that the mounting conditions for the two models were not the same. Another possible explanation for the difference is the fact that the 120-in. model was structurally flexible. Since the maximum shell deflection of the model was in excess of an inch, the possibility exists that the basic shell deflection and the local deflections of the beams and membranes could have contributed to the differences in the pressure distribution. Quantitative evaluation of these two effects is not possible without additional data. While the differences in the pressure distributions between the two models is significant, they would not be sufficiently large to cause a major change in design because:

1. The total lift on the radome for the two pressure distributions will be approximately the same since lift depends only upon the C_{p_0} term, which is

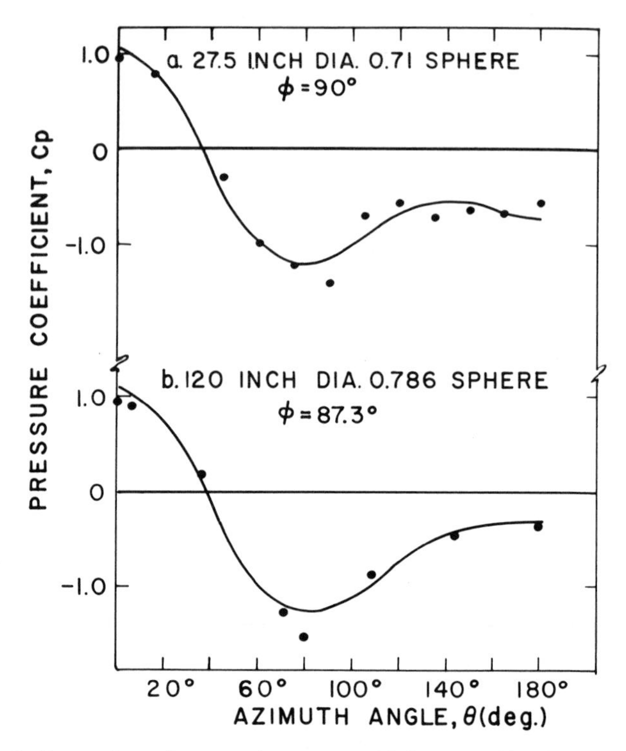

Figure 8. Comparison of measured pressures with four-term Fourier fit at equator.

nearly the same for each case.

2. The total drag on the radome will be smaller for the larger model since drag depends only upon the C_{p_1} term in the series. The difference, however, will be less than 10 per cent.

3. The difference in the C_{p_2} term will cause both the tensile and compressive loads to be higher for the larger model by something less than 10 per cent.

4. The difference in the maximum negative pressure coefficient will cause the stresses in the membrane to be about 10 per cent higher for the larger model.

3.2. The transverse load on the beam

The stress distribution in the membranes produced by the wind loading is computed using the large-deflection equations for triangular plates. Balmer, Witmer, and Loden[11] have shown that for an equilateral triangle the maximum stress occurs at the midpoint of the triangle leg. For an isotropic material the

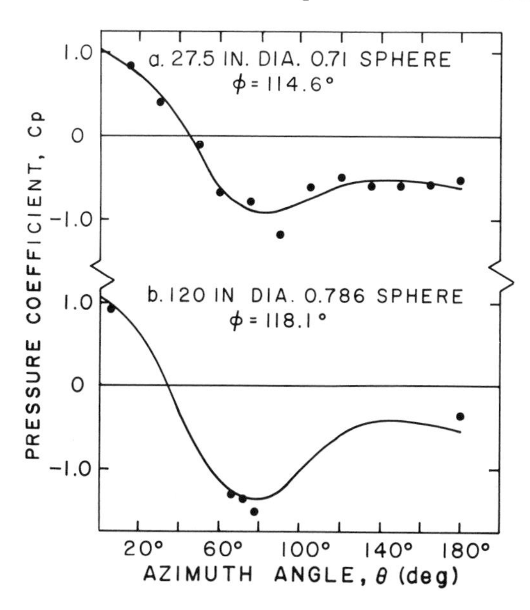

Figure 9. Comparison of measured pressures with four-term Fourier fit at typical position below equator.

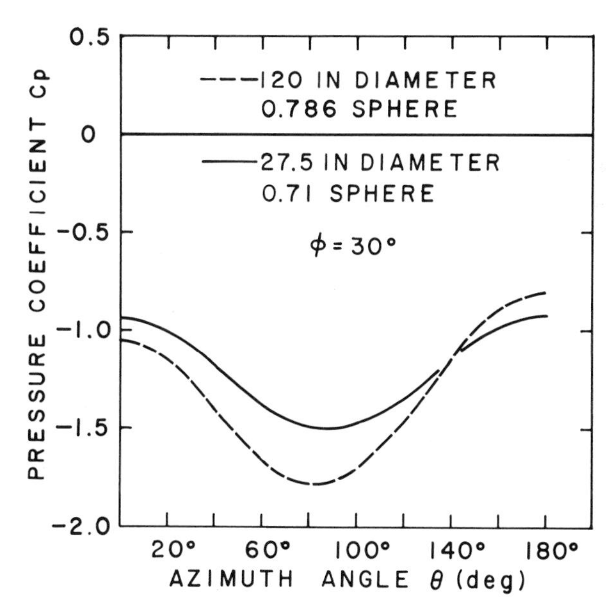

Figure 10. Comparison of pressure distributions for two models at typical position above equator.

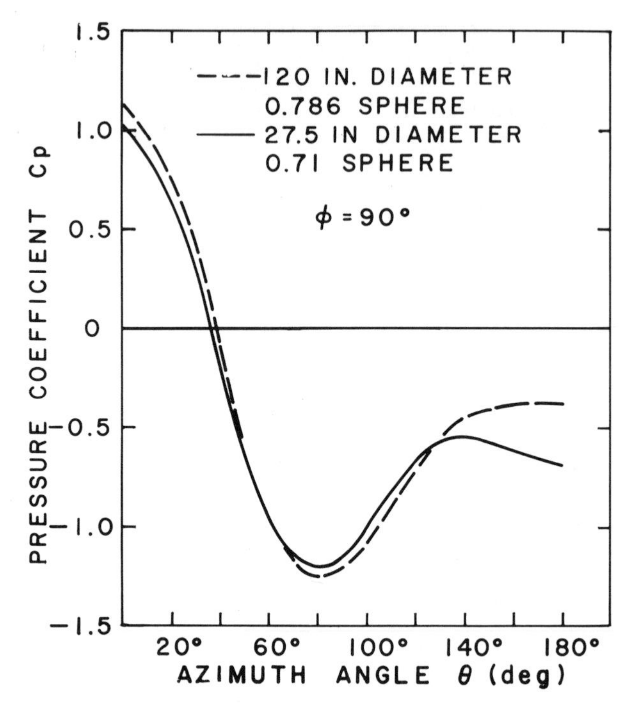

Figure 11. Comparison of two pressure distributions for two models at equator.

stress resultant N_m can be expressed as

$$N_m = 0.238\,(E_m\,t_m)^{1/3}(C_p qL)^{2/3}, \tag{6}$$

where t_m is the membrane thickness, C_p is the average pressure coefficient over the surface of the membrane, q is the reference dynamic pressure, and L is the length of the triangle side. The stress resultant is distributed approximately in a triangular manner along the beam with the maximum value N_m acting at the center and zero at the ends.

Consider now how the membrane load is transferred to the beam. Figure 14 is a cross section of the beam showing the position of the membrane before and after deformation. The transverse load per unit length at the center, S_m, (the Z component of N_m), is given as

$$S_m = 2N_m \sin(\alpha_1 + \alpha_2). \tag{7}$$

Since α_1 and α_2 will always be sufficiently small so that $\sin \alpha \approx \alpha$ and $\cos \alpha \approx 1$, this expression can be rewritten as

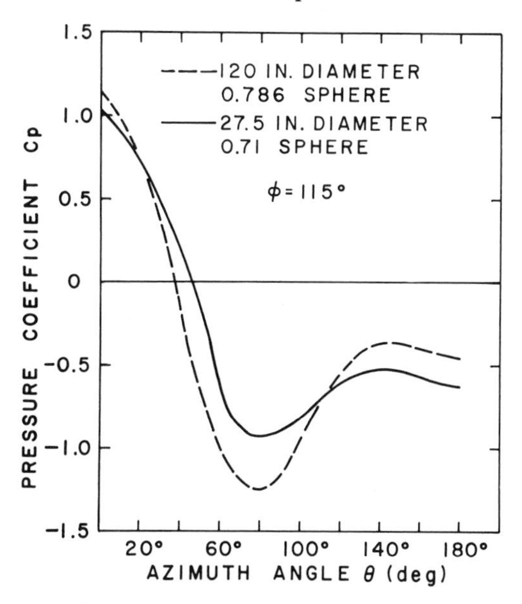

Figure 12. Comparison of pressure distributions for two radome models at typical position below equator.

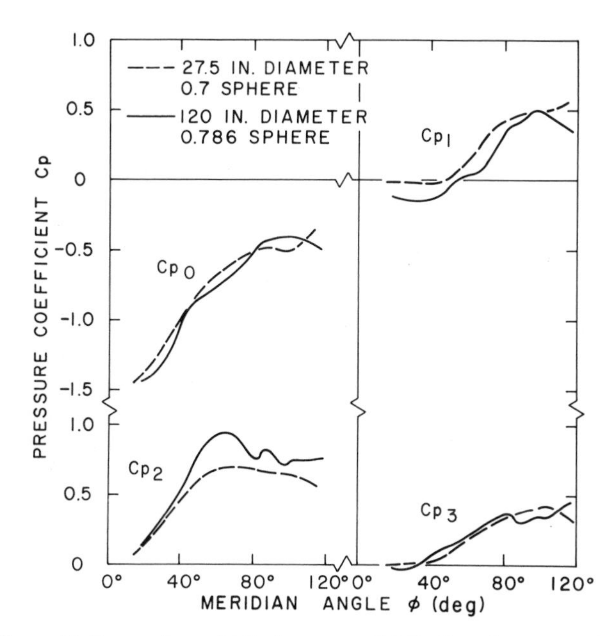

Figure 13. Comparison of fitted pressure distributions for two models.

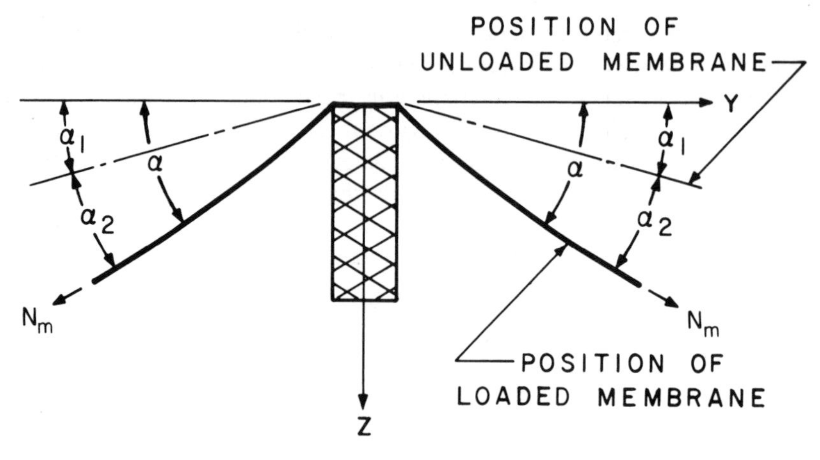

Figure 14. Deformation of membrane.

$$S_m = 2N_m(a_1 + a_2) \tag{8}$$

Assuming a triangular distribution of the membrane stress resultant along the beam and that $C_p q$ is the average pressure over the triangular panel formed by three beams, then the force equilibrium in the Z direction can be expressed approximately as

$$N_m a_2 = C_p qL/2\sqrt{3}\,. \tag{9}$$

From geometric considerations

$$a_1 = L/2R\sqrt{3}, \tag{10}$$

where R is the radius of the radome. Combining Equations 6 - 10 gives

$$S_m = \frac{qL}{\sqrt{3}}\left[C_p + 0.238\left(\frac{E_m t_m}{qR}\right)^{1/3}\left(\frac{L}{R}\right)^{2/3}\left(C_p{}^2\right)^{1/3}\right]. \tag{11}$$

The term in the brackets can be considered as the effective pressure resulting from the deformation of the membranes. It is to be noticed that when C_p is positive (wind pressure pushing membrane in towards the center of the radome), the effective pressure is larger than the actual pressure. When C_p is negative (wind pressure pulling membrane away from the center of the radome), the effective pressure is smaller than the actual pressure. The correction factor given by Equation 11 will be modified by the bending and axial deflection of the beams. The positive pressure correction will be reduced and the negative pressure correction will be increased by beam deformation.

Others[12] have noted the effect of membrane deflection but the corrections that have been suggested have not been in the explicit form of Equation 11. However, since Equation 11 represents only a simplified evaluation of the effects of membrane deflection, a more rigorous calculation is needed to check its validity. The calculation that is being planned under the sponsorship of the CAMROC program is a nonlinear analysis that utilizes finite elements to represent the structural properties of both the beam and the membrane. The calculation procedure will be similar to that used for the computer solution of frame work stability.[13]

3.3. The axial load in the beam

A straightforward procedure for stress analysis would be to utilize Equation 11, or some equivalent, to find a revised pressure distribution over the shell, which would then be used to compute the axial loads in the beams. This procedure has not been used in radome design up to the present time. One approach currently used is to increase or decrease the stresses computed with the uncorrected pressure distribution by a factor determined from an analysis similar to the one that led to Equation 11. Another approach is to apply a safety factor to the stresses computed from the uncorrected pressure distribution.

Two methods have been used to compute the axial loads in the beams of the radome frame. The first in historical sequence is the equivalent shell technique, which consists of first computing the membrane stress resultants in the shell and then transforming these stress resultants into beam axial loads by means of a frame-plate equivalence relationship. Most of the radomes now in service have been designed by this method. The second method in historical sequence is the discrete element method, which consists of representing the radome frame as a space truss with the applied wind loads acting at the joints. Since this method requires an extremely large capacity digital computer, it has not been as widely used in design as the equivalent shell technique.

In the application of the equivalent shell method, several procedures have been suggested for computing the stresses in truncated spherical shells, all of which are based upon approximating the shell as a membrane* (see References 15 and 16). For example, Foerster in Reference 12 develops the membrane stress equations for a three-term Fourier series representation of the pressure distribution. The three-term Fourier series he used ensured that the total lift and drag of the structure were maintained and that the stagnation point occurred on the most forward point of the radome facing the wind direction. The representation he used was not applicable to radomes, which were either tower mounted or larger than a hemisphere on the ground plane. The reason

* Recently, numerical methods have become available for solving the complete shell equations.[14]

for this is that the computed pressure on forward part of the radome below the equator will be lower than the actual pressure which exists on the radome. The lower pressure, in turn, causes the compressive forces to be underestimated. The four-term Fourier fit suggested in Reference 17 also suffered in the same way. In Reference 18, D'Amato indicated an approximate method for correcting this defect.

When the membrane stresses have been computed, the transformation to beam loads P_i can be carried out by using the following expression:

$$P_i = \frac{\sqrt{3}}{2}\left(\frac{4A_i}{\sqrt{3}}\right)^{1/2}\left(N_s - 1/3\, N_t\right), \tag{12}$$

where

$$N_s = N_\theta \cos^2\omega + N_\theta \sin^2\omega + N_{\phi\theta}\sin 2\omega, \tag{13}$$

$$N_t = N_\theta \sin^2\omega + N_\theta \cos^2\omega - N_{\phi\theta}\sin 2\omega, \tag{14}$$

N_θ, N_ϕ, $N_{\phi\theta}$ are the membrane stress resultants, ω is the angle that the beam makes with the θ coordinate direction, and A_i is the average of the areas of the two triangles adjoining beam. The transformation Equation 12 is based upon the Hrennikoff frame analogy[19] for a network of equilateral triangles. It is, of course, only approximate since radome frames are never a network of equilateral triangles.

Since the radome is composed of a space framework, the discrete element method is more rigorous than the equivalent shell method. During the past five years, much effort has been directed towards the development of computer programs for discrete element techniques. Among the programs that are capable of handling problems as large as a complete radome are STAIR,[20] FRAN,[21] STRESS,[22] and STRUDL.[23] There are other similar computer programs. The computation of the axial loads using these finite element programs requires (1) coordinates of all of the joints, (2) the connectivities of all of the members, (3) the loads at the joints for several orientations of the wind with respect to the frame, and (4) the structural properties of all of the members. The preparation of the data is generally so extensive that supplementary data handling programs must be employed.

In view of the magnitude of the effort required to perform a discrete element analysis of a space frame radome, it is natural to ask how valid are the results of an equivalent shell calculation, which is by far the easier to do. To answer this question, an analysis was carried out on a space frame radome using the STAIR program. The axial loads from the STAIR analysis were analyzed to determine the maximum loads in the beams as a function of elevation angle. In order to generalize the results, the beam loads P_i were normalized as

$$\eta_i = P_i/qRL_i, \tag{15}$$

where q is the reference dynamic pressure, R is the radius of the radome, and L_i is the length of the beam. The results of this analysis are plotted in Figures 15 and 16. Figure 15 presents the results for the horizontal or hoop members ($\omega = 0$). Also shown on this plot are the results of an equivalent shell analysis. As can be seen, the two methods are in quite good agreement. Figure 16 shows the results for diagonal members that range between beam angles ω of 30° and 60°. Shown on this curve are the results for the equivalent shell analysis. The band of results is not shown for the STAIR results since there was no distinction made of the beam angles other than that they were not hoop members. Again the results are in good agreement. For design purposes the envelope of the maximum values as a function of the meridional angle is the most significant parameter because cost considerations dictate that the number of different beam sizes be kept to a minimum.

Calculations using the discrete element method of analysis such as FRAN, STAIR, and STRUDL have indicated that the axial load distribution in the frame is not affected by the local bending stresses at the base attachment points. Therefore, in using the equivalent shell technique, membrane boundary conditions are used so that the membrane stress resultants are not influenced by the bending that would be produced by simply supported or clamped edges.

3.4. The stresses in the beam

When the distribution of axial and transverse loads have been obtained, the stresses in the beam can be computed. Using the beam column formula, the stresses at the center of the beam can be calculated as

$$\sigma_i = -\frac{P_i}{A_{S_i}} - \frac{M_i h_i}{2I_{S_i}} \frac{1}{1+(P_i/P_{E_i})} \tag{16}$$

where A_{S_i} is the area of the beam, h_i is the depth of the beam, I_{S_i} is the moment of the inertia of the beam, M_i is the bending moment at the center of the beam,

$$P_{E_i} = \pi^2 E_S I_{S_i}/\beta^2 L_i^2, \tag{17}$$

$$\beta_i = L_{e_i}/L_i, \tag{18}$$

L_i is the actual length of the beam, L_{e_i} is the effective length of the beam, and the quantity β_i is a function of the end fixity of the beam. The bending moment at the center of the beam can be expressed as

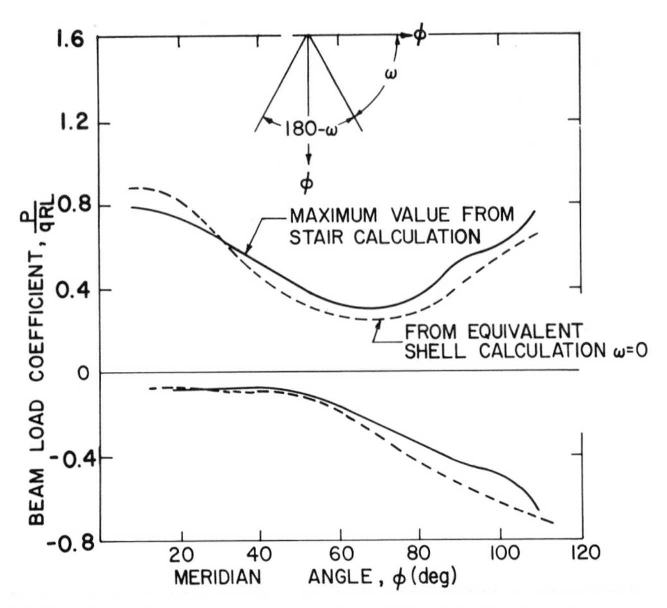

Figure 15. Envelope of maximum beam load coefficients in ¾ sphere for hoop members.

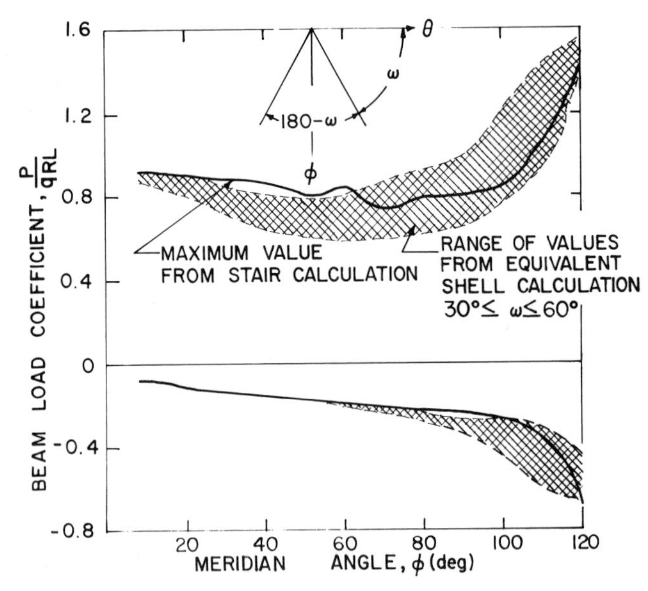

Figure 16. Envelope of maximum-beam load coefficients in ¾ sphere for diagonal members.

$$M_i = \frac{1}{24} L_i^2 \, S_{m_i} \beta_i^2 \, (3 - \beta_i) \tag{19}$$

for compressive axial loads, where S_{m_i} has been defined by Equation 11. For tensile axial loads the maximum stress will occur at the ends of the beam and, to be conservative, the full fixed end moment is used giving

$$\sigma_i = \frac{P_i}{A_{S_i}} + \frac{M_i h_i}{2 I_i}, \tag{20}$$

where $M_i = S_{m_i} L_i^2 / 19.2$ It is to be noticed that compressive axial loads are negative in sign and tensile axial loads are positive.

Using the definition for the axial load coefficient η_i, given in Equation 15, and setting the stress equal to the allowable stress σ_o, Equations 16 and 20 can be put into the following form:

$$\left(\frac{L}{R}\right)^2 = \frac{\left(\frac{\sigma_o}{q}\right)\left(\frac{h}{R}\right)^4 - \left(\frac{\eta}{K}\right)\left(\frac{L}{w}\right)\left(\frac{h}{R}\right)^3}{\left(\frac{C_p \beta^2 (3-\beta)}{4\sqrt{3}\, F_1}\right)\left(\frac{L}{w}\right)\left(\frac{h}{R}\right)^2 + \frac{12}{\pi^2}\left(\frac{\sigma_o}{E}\right)\frac{\eta\beta^2}{F_1}\left(\frac{L}{w}\right)\left(\frac{h}{R}\right) + \frac{12}{\pi^2}\left(\frac{q}{E}\right)\frac{\eta^2\beta^2}{KF_1}\left(\frac{L}{w}\right)^2}, \tag{21}$$

$$\left(\frac{L}{R}\right)^2 = \frac{\left(\frac{\sigma_o}{q}\right)\left(\frac{h}{R}\right)^2 - \left(\frac{\eta}{K}\right)\left(\frac{L}{w}\right)\left(\frac{h}{R}\right)}{0.180\left(\frac{C_p}{F_1}\right)\left(\frac{L}{w}\right)} \tag{22}$$

where w is the width of the beam and K, F_1, are dimensionless properties of the beam, given by

$$K = k_1 + k_2 - k_1 k_2, \tag{23}$$

$$F_1 = 1 - (1 - k_2)(1 - k_1)^3, \tag{24}$$

where k_1 is the amount of structural material in the depth direction of the beam and k_2 is the amount of material in the width direction of the beam. For solid beams $k_1 = k_2 = K = F_1 = 1$. Equations 21 and 22 define the required depth and width of the beam in terms of the beam length and load distribution in the radome. These two equations define the beam dimensions in terms of the maximum allowable stress criterion. It is to be noted that the absolute values of η and C_p are to be used in Equations 21 and 22.

3.5. Stability analysis

As previously mentioned, there are two modes of instability that must be protected against:

1. General instability or frame collapse in the radial direction.
2. Lateral instability in the plane of the membrane.

These two modes of failure are not independent. Frame collapse in the radial direction is influenced by the behavior of the beam in the plane of the membrane. It is important to notice that the membrane itself is considered as a structural element in restraining the beam against lateral buckling. The support provided by the membrane allows the width of the beam to be much smaller than the depth. In evaluating the general stability of the frame, it will be assumed that the beam is restrained in the plane of the membrane. The methods used to predict this mode of failure fall into three general classes: experimental measurements, the use of equivalent shell techniques with the coefficient of buckling determined by data obtained on uniform shells, and linear or nonlinear numerical analyses. The buckling of space frame shells is one that has received much attention in recent years, and other papers in this conference[13,24,25] deal with this subject in detail. For radome analysis, experimental evidence indicates the buckling pressure of space frame structures can be conservatively estimated from the relationship

$$p_{cr} = 0.3\, E_e\, (t_e/R)^2, \tag{29}$$

where E_e and t_e are the equivalent modulus and thickness of the space frame shell, respectively. With

$$t_e = \sqrt{12 I_s/A_s}, \tag{25}$$

$$E_e = E_s\, A_s/L\, t_e, \tag{26}$$

Equation 29 can be expressed as

$$p_{cr} = 1.2\, E_s \sqrt{I_s A_s}/L^2. \tag{27}$$

Setting p_{cr} equal to the dynamic pressure q, Equation 14 can be expressed as

$$\frac{h}{R} = \frac{(FS)^{1/2}}{(KF_1)^{1/4}} \left[\left(\frac{q}{E_s}\right)^{1/2} \left(\frac{L}{w}\right)\right]^{1/2} \tag{28}$$

where FS is a factor of safety applied to the dynamic pressure. It will be noted that the depth ratio for the radome beams is not sensitive to changes in the cross section solidity.

The buckling of the beam in the plane of the membrane, i.e., the lateral buckling, is complicated in that it involves an interaction between the axial load and transverse bending load. Furthermore, the buckling mode will be a combination of bending and twisting. An examination of the magnitudes of the various parameters involved indicates that the mode of failure is governed primarily by the action of the membrane that supports the beam in the manner of an elastic foundation. With this idealization, the lateral buckling load will be given by

$$P = 2\sqrt{K_m E_s I_y}, \tag{29}$$

where

$$I_y = \frac{1}{12} w^3 h F_2, \tag{30}$$

with

$$F_2 = [1 - (1 - k_1)(1 - k_2)^3], \tag{31}$$

and K_m is the membrane spring constant which can be estimated for an equilateral triangle to be

$$K_m = E_m t_m / \sqrt{\tfrac{3}{2}} L. \tag{32}$$

Using the definition of P from Equation 29 with a factor of safety there is obtained

$$\frac{h}{R} = 2.6 \left(\frac{\eta^2}{F2}\right) (FS)^2 \left(\frac{q}{E_s}\right) \left(\frac{Es}{E_m}\right) \left(\frac{R}{t_m}\right) \left(\frac{L}{w}\right)^3. \tag{33}$$

It can be seen that for lateral instability the h/R is strongly dependent upon all of the factors, especially the length to width ratio. It is also to be noticed that t_m becomes increasingly important for large radomes, since t_m is usually limited because of electromagnetic performance considerations.

Thus, when the load distribution (η and C_p) are known for a radome, Equations 6, 21, 22, 28, and 33 will determine the size of the membrane and the beam.

3.6. Radome blockage and weight

For completeness a simplified expression for optical blockage and weight will be developed. In developing the expression for the blockage it will be assumed that beams of length L are uniformly distributed over the surface of the sphere. It will be further assumed that half of the total length is distributed along great circles and the other half on minor circles. With the axis of the antenna through the point $\phi = 0$ on the radome, the blockage B_1 contributed by the beams along the great circle can be expressed as

$$B_1 = \int_0^{\phi_A} 2\sqrt{3}\left(\frac{R}{L}\right)^2 \frac{wL}{R_A^2} \sin\phi \cos\phi \, d\phi, \qquad (34)$$

where R_A is the antenna radius. Integrating and noting that $\sin \phi_A = R_A/R$, there is obtained

$$B_1 = \sqrt{3}\,(w/L). \qquad (35)$$

With the antenna in the same position the blockage B_2 contributed by the beams along the minor circles will be given as

$$B_2 = \sqrt{3}\,(w/L)\left[1 + \frac{h}{w}\,\frac{\sin^{-1}\eta_A\sqrt{1-\eta_A^2}}{\eta_A^2}\right], \qquad (36)$$

where $\eta_A = R_A/R$. The total blockage in percent will be

$$173\left(\frac{w}{L}\right)\left[2 + \frac{h}{w}\,G(\eta_A)\right], \qquad (37)$$

where

$$G(\eta_A) = (\sin^{-1}\eta_A - \eta_A\sqrt{1-\eta_A^2})/\eta_A^2$$

Equation 38 indicates that the blockage is most strongly affected by the width to length ratio w/L. For minimum blockage one should seek to keep w/L as small as possible. As shown in Figure 17, the function $G(\eta_A)$ increases sharply for values of η_A greater than 0.75.

With the assumption that the surface of the radome is covered by equilateral triangles of side length L, the total weight of the radome frame W can be expressed as

$$W = 4\sqrt{3}\,\pi\left(\frac{H}{R}\right)\left(\frac{R}{L}\right)^2 \rho\, A_s\, L, \qquad (38)$$

where H is the height of the radome, ρ is the weight density of the material, and A_s is the cross-sectional area of the beams.

3.7. Radome factors of safety

The basic reason for using safety factors is to account for uncertainties in the material properties, the applied loads, the analysis of stresses, and fabrication. If these items were known accurately, there would be no need to apply a safety factor. At the other extreme, if each of these areas is poorly known, large safety factors would be needed.

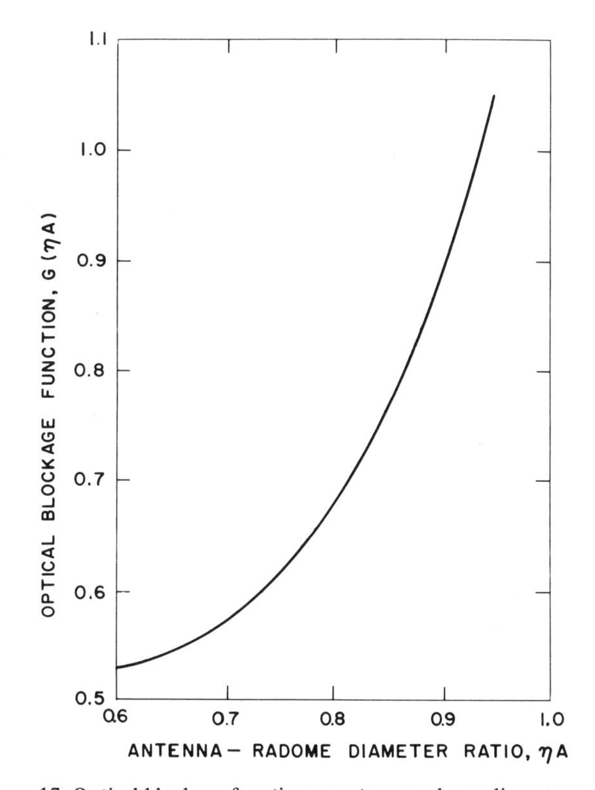

Figure 17. Optical blockage function vs antenna-radome diameter ratio.

Generally, the greatest uncertainties are associated with the loading. The least uncertainty is usually with the material properties. With the quality control of material that is current in the metal working industry, it is reasonable to expect that the basic minimum strength properties of the material are reliable. On the other hand, during fabrication the properties may be modified by welding or forming. There is, therefore, always some uncertainty of the material properties even though the uncertainty may be small.

In the case of wind loading on a radome, a maximum wind speed averaged over 5 to 20 min can usually be estimated with some specified confidence level by means of a statistical study of the data that are available. However, the gustiness and loading produced by the wind are less well known, although work is being done in this area. The wind velocity environment is used with pressure distribution data, which are usually from wind tunnel tests. With thorough wind tunnel measurements, which are carefully applied to the full scale structure, the uncertainty of loading can be reduced. But some uncertainty will remain since no full-scale test data are available.

In the case of radome design, the basic philosophical approach is as follows: First, an upper limit of the wind environment that can be expected during its useful lifetime is determined. Then, using this wind environment, the next step is to establish the pressure distribution over the surface of the radome. This pressure distribution with any other associated loads, such as those from the weight of the structure then constitutes the design load on the structure. Safety factors are applied to the design load to obtain ultimate loads. The structure should be so designed that it does not fail when subjected to this loading. It is at this point that the subjective nature of specifying safety factor enters. The development of safety factor levels depends on the confidence the designer has in the analysis he has performed as well as the service experience of similar structures. For large permanent radome installations having projected useful lives, greater than 20 years, the following factors of safety are currently recommended:

1. Frame collapse

$$\frac{\text{Collapse load}}{\text{Design load}} = 2.0.$$

2. Beam buckling

$$\frac{\text{Buckling load}}{\text{Design load}} = 2.0.$$

3. Maximum stress levels in beams or hubs

$$\text{Allowable stress} = \frac{\text{Yield stress}}{1.25}$$

4. Maximum stress levels in membranes

$$\text{Allowable stress} = \frac{\text{Material ultimate stress}}{2.0}$$

5. Maximum load in membrane attachments

$$\text{Allowable load} = \frac{\text{Ultimate attachment load}}{2.0}$$

For radomes which have projected lives of from 5 – 15 years, a greater risk may be acceptable and some reduction in the factors of safety may be allowable. In particular, the allowable stress in the material could be taken as the yield stress and the allowable buckling loads could be increased by about 1/3.

4. Applications

In Section 3 an overview has been given of the state of the art of methods for the structural design of radomes. In addition, a series of radome design formulas were presented. These formulas are useful in making parametric studies of radome configurations. The formulas are sufficiently comprehensive to allow the study of a great number of parameters and combinations. In this section, some simple examples will be given of the use of these formulas to illustrate their use.

Figure 18 is a plot of the required beam dimensions needed to prevent radial

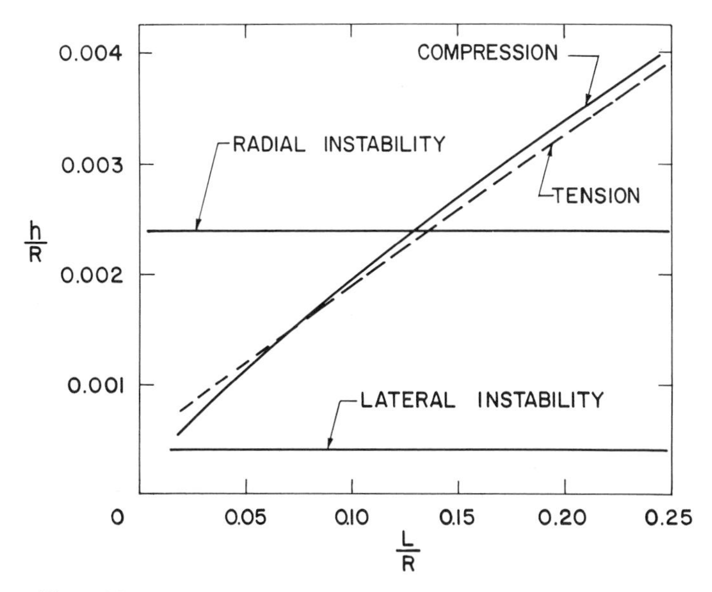

Figure 18. Beam depth requirements for steel radome with L/W = 100.

instability (frame collapse), lateral instability of the beam, and the attainment of maximum allowable tensile and compressive stresses. These conditions are covered by Equations 21, 22, 28 and 33, respectively. The following parameters have been used:

Material	E_s	$= 30 \times 10^6$ psi	(steel frame)
	E_m	$= 1.5 \times 10^6$ psi	(fiberglass membrane)
	R	= 3120 in.	(radius of radome)
	t_m	= 0.04 in.	(membrane thickness)
	q	= 0.3 psi	(wind dynamic pressure)
	β	= 0.8	(beam end fixity)
	k_1	$= k_2 = 1.0$	(solid beam)
	σ_0	= 40,000 psi	(tensile and compressive allowable stress)
	η	= 0.5	(compressive loads)
	C_p	= 1.0	
	η	= 1.0	(tensile loads)
	C_p	= 1.5	
	FS	= 2.0	(factor of safety)
	L/W	= 100	(length to width ratio)

These are values of the parameters suitable for a large radome. The values of the load coefficients are for all beams above the equator ($\phi = 90°$) as noted from Figures 15 and 16. For beams below the equator the load coefficients would be somewhat higher. It may be noted in Figure 18 that lateral instability of the beams is not important for the particular case illustrated. Notice also that there is very little difference between the tensile and compressive stress requirements. For values of L/R less than 0.13 the general instability is the governing requirement, while above a value of 0.13 the stress requirement is dominant. It is noted that a limiting value of L/R determined from membrane stress considerations could be plotted on Figure 18. For example, if a membrane allowable stress of 25,000 psi were used with a safety factor of 2.0, the value of L from Equation 6 would be 872 in. to give an L/R of 0.28. If the attachment allowable were 400 lb/in., the maximum L/R would be 0.20. Other considerations such as panel fabrication cost or panel handling may limit the largest length of beam that would be allowed. For example, if panel sizes above 35 ft became increasingly difficult to handle or expensive to fabricate, then the limiting L/R would drop to about 0.135. It turns out that this is approximately the intersection for stress and stability limitations in Figure 18. Carrying out this calculation for other values of L/w, and computing the optical blockage, a plot of blockage as a function of L/w can be determined as shown in Figure 19. It turns out that the intersection of the stability and stress limitations remains about constant when L/R varies between 75 and 150. It can be seen that as L/R decreases the optical blockage increases. Also shown

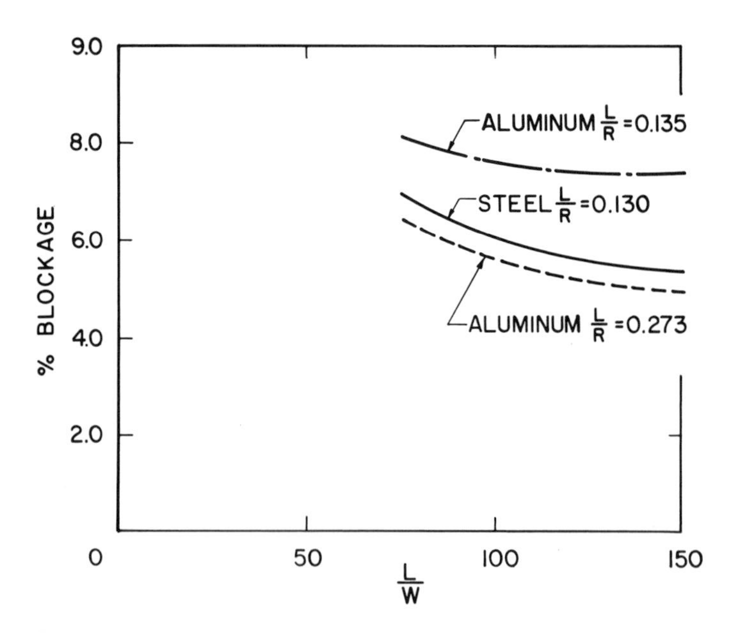

Figure 19. Comparison of optical blockage of aluminum and steel radomes.

in Figure 19 is a radome design using aluminum beams with an allowable stress of 30,000 psi. The intersection of the stress and instability failure modes remains constant at about L/R = 0.273. For this L/R the blockage is lower than for the steel frame. However, if the same L/R limitation used for the steel beam is applied to the aluminum beam, the blockage will be considerably higher.

Finally, it is of interest to consider the case of a hollow beam. A comparison between a hollow beam and a solid beam for an L/w of 150 is shown in Figure 20. The hollow beam has the same properties as the solid beam given in the

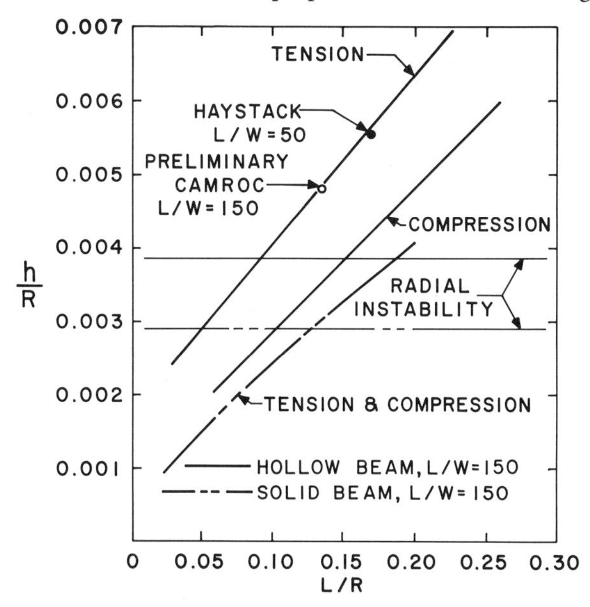

Figure 20. Comparison of solid and hollow beam sections.

table above except that $k_1 = {}^1/_3$ and $k_2 = {}^1/_6$. It is observed that the depth requirement for the hollow beam has increased over that for the solid beam for both the stress and the stability requirements. It may also be seen that there is a marked distinction between the tensile and compressive stress requirements for the hollow beam whereas there was none for the solid beam. For the hollow beam, the tensile stress limitation dominates the design for all L/R greater than 0.09. The basic reason for this behavior can be attributed to the large tensile loads in the radome. Also shown on this plot are points indicating the preliminary CAMROC beam design and the Haystack beam design.

These few examples give an idea of how the design equations may be used for evaluating the effect of various radome parameters on a general configuration. Utilization of these equations with the beam load distribution computed either by the equivalent shell method or the finite element method will allow rapid parameter studies.

References

1. R. B. Curtiss and J. Vaccaro, "Survey of Ground Radomes," RADC-TR-61-52 (May 1961).

2. J. A. Vitale, "Large Radomes," Chapter 5 of *Microwave Scanning Antennas, Volume I, Apertures,* R. C. Hansen, Ed. (Academic Press Inc., New York, 1964).

3. J. Ruze, "Electromagnetic Loss of Metal Space Frames, "International Symposium on Structures Technology for Large Radio and Radar Telescope Systems, M.I.T. (18-20 October, 1967).

4. A. G. Davenport, "Rationale for Determining Design Wind Velocities," *Journal of Structural Division,* American Society of Civil Engineers (May 1960).

5. R. L. Bisplinghoff, H. Ashley, and R. L. Halfman, *Aeroelasticity* (Addison-Wesley Publishing Company, Cambridge, Massachusetts, 1965).

6. A. G. Davenport and N. Isynmov, "The Dynamic and Static Action of Wind on Hyperbolic Cooling Towers I," University of Western Ontario Research Report BLWT-1-66.

7. R. H. Sherlock, "Variation of Wind Velocity and Gusts with Height," American Society of Civil Engineers Transactions, Paper No. 2553, published as Proceedings–Separate No. 126 (April 1952).

8. S. von Hoerner, *Fluid Dynamic Drag,* published by the author (1958).

9. W. H. Horten and B. H. McQuilkin, "Pressure Distribution on a 1/15 *Scale Model Radome,"* Data Report prepared for M.I.T. Lincoln Laboratory, Contract No. A-06085 (January 1964).

10. J. Bicknell and P. Davis, "Wind Tunnel Study of Spherical Tower Mounted Radomes," M.I.T. Lincoln Laboratory, Group Report No. 76-7 (November 1958).

11. H. A. Balmer, E. A. Witmer, and W. A. Loden, "A Theoretical Analysis and Experimental Study of the Behavior of Panels of Isotropic and Orthotropic Material under Static and Dynamic Loads," ASRL Technical Report 98-3, Lincoln Laboratory Report No. 71G-2 (February 1962).

12. A. F. Foerster, "Stress Distribution and Stability Criteria of Spherical Ground Radomes Subjected to Wind Loads," Proceedings of the OSU-WADC Radome Symposium, WADC Technical Report 58-272 (1958), Vol. I.

13. J. J. Connor, "Non-Linear Analysis Elastic Framed Structures," International Symposium on Structures Technology for Large Radio and Radar Telescope Systems, M.I.T. (18–20 October 1967).

14. J. H. Percy, *et al,* "Application of Matrix Displacement Method to Linear Elastic Analysis of Shells of Revolution," *AIAA J.* 2138 (1966).

15. S. Timoshenko, *Theory of Plates and Shells* (McGraw-Hill Book Company, Inc., New York, 1940).

16. W. Flugge, *Stresses in Shells* (Springer-Verlag, Berlin, 1960).

17. E. Sevin, "Analytical and Experimental Studies of Spherical Rigid Ground Radomes," RADC TR-60-261 (February 1961).

18. R. D'Amato, "Stress Analysis of Spherical Space Frame Shells," Electronic Space Structures Corporation TD 63-1 (December 1963).

19. A. Hrennikoff, "Solution of Problems in Elasticity by the Framework Method," *J. Appl. Mech.* 8 (1941).

20. "STAIR (Structural Analysis Interpretive Routine) Instruction Manual," Lincoln Manual No. 48, M.I.T. Lincoln Laboratory, Lexington, Massachusetts (March 1962).

21. "IBM 7090/7094 FRAN (Framed Structures Analysis Program) 7090-EC-OIX," International Business Machines Corporation, White Plains, New York (September 1964).

22. S. J. Fenves, R. D. Logcher, S. P. Mauch, and K. F. Reinschmidt, *STRESS:A Users Manual* (The M.I.T. Press, Cambridge, Mass., 1964); S. J. Fenves, R. D. Logcher, S. P. Mauch, *STRESS: A Reference Manual* (The M.I.T. Press, Cambridge, Mass., 1965).

23. J. M. Biggs and R. D. Logcher, "ICES STRUDL-I, General Description," Department of Civil Engineering, M.I.T. (September 1967); R. D. Logcher, B. B. Flachbart, E. J.s Hall, C. M. Power, and R. A. Wells, Jr., "ICES STRUD-I, Engineering User's Manual," Department of Civil Engineering, M.I.T. (September 1967).

24. D. T. Wright, "Instability in Reticulated Spheroids: Experimental Results and Effects of Nodal Imperfections," International Symposium on Structures Technology for Large Radio and Radar Telescope Systems, M.I.T. (18-20 October, 1967).

25. L. Berke and R. H. Mallett, "Automated Large Deflection and Stability Analysis of Three-Dimensional Bar Structures," International Symposium on Structures Technology for Large Radio and Radar Telescope Systems, M.I.T. (18-20 October, 1967).

D. T. Wright
Committee on University Affairs
Toronto, Ontario

INSTABILITY IN RETICULATED SPHEROIDS; EXPERIMENTAL RESULTS AND THE EFFECTS OF NODAL IMPERFECTIONS

1. Introduction

The development of radio and radar telescope systems involving reflectors of large physical size pose many structural problems both in the design of the reflectors and in the design of the radomes that may be required to protect them from the weather. Beyond sizes that are now regarded as rather small, metallic skeleton systems are required both for the reflectors and for their covers. There has been a tendency to develop space frameworks to meet the needs of the necessary physical form of such structures and their three-dimensional structural behavior.

While it is more or less trivial to acknowledge that in a three-dimensional world all real structure is three dimensional, it must be acknowledged that in most conventional structure the essentials of behavior, analysis, and design reflect two-dimensional considerations only. The structural behavior of three-dimensional systems is profoundly different from that of two-dimensional systems. Most importantly, three-dimensional structural systems offer potentialities of efficiency and economy far superior to those associated with two-dimensional behavior. Notwithstanding this circumstance it is still clear that many antenna structures are conceived and realized as assemblages of two-dimensional subcomponents with no effective three-dimensional action.

While this paper is primarily concerned with space-frame radome shell structural systems, it may be appropriate to note that related space-frame shell structures may also be exploited for antenna reflectors.

The space-frame radome is, in effect, an adaptation of the space-frame shell structural system, which has recently been developed for building construction. The significance of the space-frame shell in building structures may readily be seen by combining the benefits of skeleton building construction and surface or membrane structure. Starting in the late 19th century, the development of a light-weight metallic skeleton of beams and columns has permitted a tenfold increase in practicable building heights. Starting perhaps 30 years ago, the development of reinforced concrete monolithic shell structures has exploited the moment-free feature of curved surfaces in space. The combination of skeleton features with membrane or shell systems has evidently increased the practical shell span limit of about 50 m in the case of ordinary reinforced concrete shells by a factor of at least 10. For example, some design studies in Britain suggest that space-frame shell structures with clear spans of 500 m are quite practical and economical. (Costs as low as $6.00 per square foot, including cladding, have been proposed for such spans.)

The notion of the space-frame shell is that of a structure formed as a skeleton in space with the nodes or connections tracing out a smooth surface—or shell—in space. The space-frame shell, while a skeleton, is very different from the usual building skeleton. Reflecting the ability of a membrane structure to function without primary bending, members in a space-frame shell have only axial forces as primary loading.

The development of an effective technology for space-frame shell structures has been concluded only very recently. The pioneering work of such people as Buckminster Fuller has shown the practicability of the general structural forms, and has popularized the entire notion. But for the realization of effective and economic structures on a consistent basis more was needed than the development of notions of form. Perhaps the most difficult challenge has been the developing of practical and economical connection methods for members coming together from all directions to points in space. The other critical technological problem was the development of effective methods of analysis and design for the proportioning of such structures. Within the past decade a number of connection systems, including Triodetic, Mero, Octaplette, SDC, etc., have been shown to be effective. It is not within the scope of the present paper to deal in detail with these various methods of connection. Some attention will be given, however, to the influence on structural stability of joint characteristics. It is important only to acknowledge that effective connections are available, so that primary attention can be focused on such general questions as shape, geometry, and member selection. It is inevitable that in due time experience in construction and with costs will identify the most suitable connections.

Methods of Analysis and Design

The structural design of space-frame shells is challenging to conventional analysis by reason of the very great number of members involved. In the conventional multi-storey building, design considerations are usually restricted to plane frames, and a framework with 200 connections and 400 members is already considered to be large. Quite modest space-frame shells have over 1000 joints with 3000 members, and practical space frames may now be considered with up to 100,000 joints and perhaps 500,000 members. In a space frame there may be three to five times as many members as joints, and full three-dimensional behavior must be considered with three displacement unknowns per joint if bending is excluded, and six per joint if it is to be considered.

Three approaches appear to be available for the analysis of member forces in space-frame systems. The first, which some people have termed the "brute force" method, consists of dealing with the problem as a three-dimensional framework in much the same way as one would deal with a two-dimensional framework, and with the aid of sufficiently powerful computers determining forces and/or displacements under given external loadings. While forbidding at first sight, this approach has been greatly facilitated by the recent development of special computer programs and languages such as STRESS, FRAN, and STAIR. Even with the largest digital computers, however, some practical space-frame design problems are far beyond the range of capacity of such methods of analysis. Nevertheless, such methods may prove attractive for many ordinary problems in the structural analysis and design of space-frame shells.

A second method for the analysis of space-frame shells, only recently proposed,[1] provides for the analysis of the framework through the concepts of discrete field mechanics, in which an exact analysis may be obtained directly, analogous to the methods used to get field solutions for continuous shells. The significance of this method is that instead of solving partial differential equations as is ordinarily done in the case of continuous shell structures, one would deal with difference equations. While this method appears to be attractive, few solutions have yet been published, and it is evident that much work would be required to bring together a comprehensive collection of solutions parallel to those already available for ordinary shell structures.

The third approach available depends upon treating the framework shell as if it were a continuum.[2,3] To follow this procedure, one must first determine the effective elastic properties of an imaginary continuum that would behave in the same way as the framework shell system in question, and then analyze this analogous continuum according to the various loading and boundary conditions. Having determined the stress resultants, these must be transformed back to the framework to determine individual bar forces and moments. While such a procedure appears to be rather involved, it turns out to be fairly easy, and has the virtue that the designer can make direct use of the many published solutions for continuous shells of various form under various patterns of loading. For a great variety of practical shells, analysis is thus facilitated by using the analogous continuum approach. Perhaps a more important feature of the analogous continuum approach is that it provides a view of the behavior of the structure as an entity. This has particular significance for investigations of stability. Neither of the other two methods noted offer any direct access to questions of over-all behavior.

While in space-frame shell structures for buildings the choice of shell form is itself the most important matter of design, the spheroidal radome form restricts geometrical questions. There remain such important questions of form as: (1) the pattern of surface division, (2) the selection of an appropriate, typical, or average node spacing, and (3) the choice between single- and double-layer construction. The third question is most easily dealt with. The double-layer space-frame shell system is of course much more stiff and resistant to instability than the single-layer system, and for larger radii of curvature single-layer forms become uneconomic or even impossible. With ordinary tubular members, the maximum radius of curvature of a spheroid constructed with aluminum members is the order of 50 m, while the maximum radius of curvature with steel members is about 80 m. In double-layer forms spheroids with radii of curvature up to 500 m, at least, are quite practicable.

Surface divisions for space-frame shells may be developed either from notions of two-dimensional nets or, in the case of spheroids, may be related to the regular Archimedean solids. For roof structures shell surface divisions related to plane nets of equilateral triangles are often preferred because of regularity at the nodes and base and in construction. Electrical requirements in radomes suggest the use of forms derived from the Archimedean solids. These may be all subdivided to provide suitable member lengths; the continuum analog analytical procedures described in this paper are applicable to all such forms as long as elements of surface are triangular in form. Where there are less than three characteristic member directions, and triangles are not formed as the smallest surface units, it may be shown that membrane forces cannot be resisted without the introduction of primary bending moments and the simple continuum analog fails.

Continuum Analysis

Papers recently published have presented analyses establishing the properties. of analogous continua for reticulated single- and double-layer framework shells, and expressing transformation functions as between stress resultants and bar forces.[2,3] Accordingly, it is unnecessary here to reproduce these in detail. Simply, these methods provide for the determination of thickness, elastic moduli, and Poisson's ratio for an imaginary continuum that has the same extensional, flexural, shearing, and twisting resistances under external loadings as the framework system in question. Transformations between stress resultants and space-frame bar forces are based upon the conditions necessary for static equilibrium and deformational compatibility. Whereas the general case of a reticulated surface leads to an anisotropic equivalent continuum, the special case of the framework having an equilateral triangular pattern, with identical bar sizes, leads to an isotropic continuum. The solution in this case is so simple that it is worthwhile to reproduce the results here. For the single- and for the double-layer space-frame skeletons, with regular triangular faces, the analogous continuum is described as follows:

$$E' = AE/3\, r_g L, \tag{1}$$

$$t' = 2\sqrt{3}\, r_g, \tag{2}$$

$$\nu' = 1/3 = \nu'', \tag{3}$$

$$E'' = 4\, AE/3\, KL^2, \tag{4}$$

$$t'' = \sqrt{3}\, KL, \tag{5}$$

where a single prime refers to the single-layer system, and a double prime refers to the double-layer system, E represents elastic modulus, t thickness, ν Poisson's ratio, and A and r_g are the cross-sectional area and the radius of gyration of the skeleton member, with L the node spacing in the face. K is defined such that KL is the depth of the double-layer space-frame system. It has been shown that these analyses are applicable with only slight error in the case of triangulated systems which are not precisely equilateral in form.[4]

Stability

Instability in a framework shell may arise in any one of three different ways. Under the action of the membrane force system in the shell, axial loads in individual members may rise to levels at which such individual members become unstable. This case is readily treated since the load-carrying capacity of an individual bar in a space frame is that of the same member as a pin-ended strut.

Under the action of locally applied loads, with components normal to the shell surface, there may be a tendency to "snapping." This problem has been studied by Wright and Lind for the reticulated shell.[2,5] Analyses show that resistance to local point loads is very low in the reticulated single-layer system if the joints are pinned. With perfectly stiff joints these analyses have shown that snapping will *not* occur under a point load if

$$L^2/ar_g \leqslant 9 . \tag{6}$$

where a is the radius of curvature of the shell surface.

It has been proposed that the question of over-all stability of a framework shell may be dealt with approximately by noting that for a homogeneous isotropic continuous shell of spheroidal form, the critical uniform radial pressure may be expressed approximately as

$$q_{cr} = C\ Et^2/a^2 , \tag{7}$$

in which C is a constant, E the elastic modulus of the shell material, and t the thickness of the shell. Estimates of C have been made theoretically,[6] but the value is clearly affected by imperfections, residual stresses, etc.

For the reticulated single-layer framework shell the critical radial pressure may be written, then, as

$$q'_{cr} = 4\ C(A\ E\ r_g/L\ a^2). \tag{8}$$

from Equation 7 and using values for effective elastic modulus and thickness from Equations 1 and 2. For the double-layer framework shell the critical radial pressure may similarly be written as

$$q''_{cr} = 4\ c(A\ E\ K/a^2). \tag{9}$$

At the time of publication of the original analysis, containing Equation 8 above,[2] the only evidence beyond logic and intuition to support the general notion of the continuum analysis and its extension to embrace instability lay in a study of the failure of the National Economy Exhibition Pavilion space-frame shell roof in Bucharest in 1963.

That structure, of clear span 93.5 m, with a radius of curvature of 65.25 m, had been constructed in 1961 and had collapsed in 1963 when it was loaded by about a meter of fresh snow. While initial reports were meager, a subsequent paper by Beles and Soare[7] has described the failure, which clearly was a case of loss of stability, and which resulted in the structure literally turning itself inside out. Noting the dead load of 11 lb/ft^2, and a probable value of the snow load as 20 lb/ft^2, Equation 8 above suggests an effective value of C of 0.49 when the appropriate dimensions for the structure are substituted in the formula. This compares remarkably well with the von Karman – Tsien value of 0.366.[6] It must be acknowledged, as Beles and Soare have pointed out, that the loading in fact was quite irregular and this evidently good agreement between failure and analysis may be partly fortuitous.

Since 1965, when this analysis was first published,[2] model tests have been carried out at the University of Waterloo to provide a better basis for validation. Some data have also become available from Czechoslovakia on related model tests.

The tests at Waterloo were undertaken in relation to the development of a design for a large-span aluminum space-frame dome in Britain. The model had a radius of curvature of 180 in. with a base diameter (span) of 259 in. These dimensions were one-eighth of the proposed prototype. Two models were, in fact, constructed and tested.

In the first, an attempt was made to use relatively small members for the sake of better modeling, and the sections chosen were aluminum tubes of wall thickness 0.022 in. with diameters of 3/8 in., 7/16 in., and 1/2 in. in different regions. Figure 1 shows the model, which had a regular geometry with 80

Figure 1. Model of aluminum space frame dome.

divisions at the base. With such geometry, the characteristic member length diminishes with increasing latitude so that the analogous continuum is, in fact, not homogeneous. The use of members of different diameter in different regions tends to offset this variation and the resulting value of the quantity $E't'^2/a^2$ varied only over the range 2.18 to 2.84. The model was loaded by evacuating a plastic bag that had been placed over the dome and sealed to the floor. At a pressure of 0.639 psi there was a sudden loss of stability with the structure actually imploding (Figure 2). This corresponded to a critical buckling coef-

Figure 2. Implosion stability failure of space frame dome.

ficient C of 0.29, somewhat lass than the von Karman – Tsien value. Examination of the members after the failure showed that the connections had been inadequate, with the members bending out under a moment hardly 10 per cent of their own flexural strength.

Subsequently a second test involving a model of the same base diameter and curvature was performed using more effective Triodetic joints and slightly heavier members (7/16-in. diameter by 0.065-in. wall thickness) with 48 divisions at the base. Following the same testing procedure, this dome resisted a load of 0.96 psi, corresponding to a value of C of 0.48, and instead of an implosion, the structure became dimpled (Figure 3) over an area $\gg L^2$

While nonuniform loadings are important for design, it is, of course, very difficult to test models of such character except under uniform load in the fashion indicated. An attempt was made, however, to determine the influence of asymmetric loading by testing the second dome, initially, under an asymmetric gravity load. A mixture of sand and gravel, with enough water to induce bulking, was used and applied on one-half of the structure alone. The test was not carried to failure because it was impossible to accumulate enough material on the model without it sliding off. A maximum load of intensity of 100 lb/ft^2 was, however, reached. This load was about 75 per cent of the buckling strength under a uniform radially symmetric load. While it may be improper to make much of the results of such a single test, it appears reasonable to think that asymmetric loadings are probably not more critical than uniform

Figure 3. Dimpled stability failure of space frame dome.

loadings having an intensity equal to the maximum level or the nonuniform load. If this were accepted, design for nonuniform loadings could become quite easy!

Following the failure of the Bucharest dome, Professor F. Lederer of the Brno Technical University conducted a number of tests to study the stability of reticulated domes. These tests were reported in a discussion[8] to Reference 2. In that presentation, the test results were not directly compared with the continuum analysis proposed. Table 1 here gives these results of the five tests

Table 1. Reticulated Dome Stability Tests at Technical University Brno.

Model number	(1)	(2)	(3)	(4)	(5)	Units
Radius of curvature, a	892.6	633	632.7	624.2	643.4	mm
Base diameter	906	885	887	887	885	mm
Max node spacing, L	31.6	30.8	30.9	30.9	30.8	mm
Member section, A	0.502	1.13	1.54	3.14	0.502	mm^2
Failure pressure, q_{cr}	0.014	0.058	0.092	0.350	0.0225	kg/cm^2
Buckling coefficient, C	0.420	0.251	0.253	0.320	0.340	–

completed at Brno in a form related to the continuum approach. It will be seen that these tests gave values to the buckling coefficient ranging between 0.25 and 0.42. It must be noted that these tests were of relatively small diameter models in which the members were continuous bent wires, soldered together at the connections. With such a style of fabrication, it is inevitable that there would have been both significant residual stresses and some significant geometrical departures from the intended sperical shape.

It is, of course, well known that experimental values of the buckling coefficient C for spherical shells are strongly influenced by residual stresses and initial deformations. Reported values of C, determined by experiment, range from a low of under[9] 0.20 to a high of the order of 0.80.[10] Given that a prefabricated space-frame skeleton will have little or no residual stress and will fit the intended shape very precisely, the second Waterloo test result may seem most appropriate of the lot reported. In any event, it seems quite proper to use for design at least the value 0.36 suggested initially by von Karman and Tsien from a theoretical basis.

Connection Imperfections

The influence of connection imperfections can be seen from the general dependence of the stability of reticulated spheroidal shells on stiffness with respect to both axial and flexural deformations. While it is not possible to produce convenient algebraic expressions indicating the influences of connection imperfections on stability, effects can be indicated quantitatively using Equation 7 with values of effective thickness and effective elastic modulus modified from the values given in Equations 1 - 5 to reflect joint flexibility.

The significance of joint flexibility is best indicated by an example. An interesting example, perhaps, is the BMEWS antenna.[11] An analysis indicating the critical load of such a structure, but with perfect connections, was given in Reference 2. Now, assume joints with 50 per cent of the axial and flexural rigidity of the main members over the same length as the total diameter of the connection (about 10 per cent of the node spacing). It may be shown that the indicated loss in the product $E't'^2$, which is the measure of critical load, is about 30 per cent. The indicated initial load falls from approximately 175 to about 120 lb/ft^2.

Given practical detailing it is, of course, difficult to estimate actual joint performance without experimental evidence. But it is important to realize that simple tests on a single joint may be used in the manner shown to indicate joint effects on the buckling load in a large and complex structure.

The example cited is relatively favorable, since many practical connections are bound to be even more flexible. It is clear, then, that joint flexibility may have a very important influence on over-all stability of practical reticulated spheroids.

References

1. D. L. Dean and C. P. Ugarte, in "Membrane Forces and Buckling in Reticulated Shells," *Proc. Am. Soc. Civil Engrs.* **91** (No.1 ST5), 378 (October 1965).

2. D. T. Wright, "Membrane Forces and Buckling in Reticulated Shells," *Proc. Am. Soc. Civil Engrs.*, Structural Division **91** (No. ST1), 173 (February 1965).

3. D. T. Wright, "A Continuum Analysis for Double Layer Space Frame Shells," *Mem. Intern. Assoc. Bridge Structural Eng.* **26**, 593 (1966).

4. J. H. Lane, "Bar Forces in Latticed Spherical Domes by Means of the Membrane Analogy for Shells," Ph. D. dissertation, North Carolina State University, Department of Civil Engineering (1965).

5. N. C. Lind, "Stability Analysis of Symmetric Dome Frameworks," *Proc. Am. Soc. Civil Engrs.* (to be published).

6. T. von Karman, and H. S. Tsien, "The Buckling of Spherical Shells by External Pressure," *J. Aerospace Sci.* **7** (No. 2) (1939).

7. A. A. Beles, and M. Soare, "Some Observations of the Failure of a Dome of Great Span," Proceedings, International Conference on Space Structures, London, 1966 (to be published).

8. F. Lederer, in "Membrane Forces and Buckling in Riticulated Shells," *Proc. Am. Soc. Civil Engrs.* **91** (No. STS), 385 (October 1965).

9. K. Kloppel and O. Jungbluth, *Beitrag zum Durchschlagproblem Dunnwandiger Kugelschalen,* (Der Stahlbau, Wilhelm Ernst & Sohn, Berlin-Wilmersdorf, 1953), Vol. 22, No. 6, pp. 121-130.

10. M. A. Krenzke, "Tests of Machined Deep Spherical Shells under External Hydrostatic Pressure," Report No. 1601, Structural Mechanics Laboratory, David Taylor Model Basin, U.S. Department of the Navy, Washington, D. C. (May 1962).

11. D. Smollet, "Nuts and Bolts Zip up a Radome," *Eng. News-Record* **168** (No. 23), 29 (7 June 1962).

Laszlo Berke
Air Force Flight Dynamics Laboratory,
Wright-Patterson Air Force Base, Ohio
Robert H. Mallett
Bell Aerosystems, A Textron Company,
Buffalo, New York

AUTOMATED LARGE DEFLECTION AND STABILITY ANALYSIS OF THREE-DIMENSIONAL BAR STRUCTURES

1. Introduction

The importance of elastic stability phenomena in the design of flight hardware and other light-weight structures, such as radomes, has provided strong motivation for generalizing the powerful finite element methods of structural analysis to accommodate geometric nonlinearities that are fundamental to stability analyses. Material nonlinearities are also contributory to some complex stability phenomena; however, linear elastic material behavior is generally sufficient when considering the stability of the important class of light-weight structures.

There exists a hierarchy of nonlinearities which may be retained in generalizing linearized methods to treat elastic stability. Associated with each level in this hierarchy is a level of complexity in formulation and subsequent solution, and a class of problems for which the solution furnishes a realistic measure of actual behavior. As additional nonlinearities are considered, the pertinent problem class is expanded and complexity of formulation and solution is increased.

For the most part, stability analyses have been transformed into eigenvalue problems that yield buckling load estimates. This approach is well suited to problems in which the internal force distribution varies only in intensity in consequence of the buildup of applied loading. While an eigenvector buckling mode shape is predicted, no knowledge of the post-buckling behavior can be obtained.

The more ambitious goal of the analysis of nonlinear pre- and post-buckling behavior is pursued in the finite element method presented here. Internal force distributions are related nonlinearly to applied load levels. Internal force redistributions occasioned by local instabilities are also accounted for. General instability can arise from a sequential accumulation of local instabilities. This yields a realistic estimate of the maximum load that can be sustained. The capability to predict post-buckling behavior depends upon the type of nonlinear analysis method employed.

Two fundamental types of nonlinear analysis methods exist: iterative and incremental. The iterative approach is a direct attack on the nonlinear problem. The primary advantage of this approach is the fact that it furnishes solutions directly to the posed nonlinear problem, and at any load level, without step-by-step buildup. Solutions for post-buckling behavior arise naturally. A disadvantage of this direct approach is the need to solve sets of nonlinear algebraic equations. In addition, total rotation displacement variables are necessarily manipulated as vectors. This can be restrictive although the utilization of linearized curvatures tends to be a more restrictive assumption.

The restrictions imposed by linearized curvatures defy elimination; however, a linearized incremental method can circumvent the direct solution of nonlinear equations and the vector manipulation of total rotation variables. An incremental method is generally more efficient where applicable. Unfortunately, it breaks down with the first occurrence of an instability. For this reason, a direct iterative approach was essential in the subject investigation of pre- and post-buckling phenomena. Accepting the necessity of a direct attack on the nonlinear problem, attention was given to obtaining a formulation of simple expression and amenable to calculation.

The displacement methods of finite element formulation have emerged as predominant in the structural analyses technology based upon finite element idealization. Automated linearized tools for analysis and design, evolved within this framework, have revolutionized practices in the aerospace industry. This widespread utilization of the linear displacement method has been somewhat fortuitous from the viewpoint of stability analysis since geometric nonlinearities are most readily incorporated within a displacement formulation.

The simplicities afforded by displacement formulations and finite element idealizations are preserved in the subject nonlinear analysis method.

The concepts of finite element idealization allow formulative effort to involve only typical finite elements. In accordance with the well-known procedure for deriving linear finite element representations for displacement methods, element strain–displacement relations and displacement mode shapes are selected. Development of element models for stability analyses is distinguished by the retention of nonlinear terms in the strain–displacement relations. Specifically, second-order terms in the normal rotations are retained. Special care is given herein to the selection of element displacement modes. In selecting these modes an attempt is made to satisfy the differential equations obtained from the variational statement insofar as is possible. Extended comment is given to this point in Section 2.

As the next step, an algebraic element potential energy function is developed from the selected strain-displacement relations and displacement modes. In a departure from the conventional practice of deriving a stiffness matrix, this element potential energy function is employed as the element mathematical model. Such models for the individual elements of a structure are then readily combined to obtain an expression for the total potential energy of an assembled structure. At this point recourse is taken to the powerful numerical methods of mathematical programming to seek out the minima of the total potential energy function which are known to characterize the stable equilibrium displacement states. Direct minimization of the total potential energy as the basis for a general-purpose finite-element structural-analysis method was originally reported in Reference 1. Reference 2 includes an application of this method to shell buckling. References 1 and 3 include numerous small-scale truss and frame applications. Direct minimization has been used previously as a numerical solution scheme only for specific structural problems.

Further applications of the potential energy minimization method of analysis are presented herein with emphasis on the general radome design problem. Presentation of these applications is prefaced with a review of the specific element mathematical models employed and a description of the numerical search procedure. The paper is concluded with identification of some fruitful areas for further study.

2. Description of the Numerical Method

The computational procedure presented here utilizes directly the principle of the minimum of the total potential, which states that the total potential is a minimum when an elastic body is in a state of stable equilibrium. This statement is valid without regard to whether the structure reached this stable equilibrium position through linear or nonlinear response to loading. Moreover this response need not be a continuous function of a monotonically increasing or decreasing (quasi-static) loading. Stable equilibrium positions can be found, using this principle, both in the pre-buckling and post-buckling range even if they are separated by sudden jumps of the structure from one equilibrium position into another. These potentialities can be realized, however, only if a computational procedure is used which establishes a solution for a given load independently from a (nearby) solution obtained previously for another (nearby) load. The direct minimization of the total potential with respect to the displacement variables is such a procedure provided appropriate nonlinear strain-displacement relationships and appropriate admissible displacement patterns are used in developing the total potential expression.

The above statements are valid for any elastic body; however, in developing the present computer program to study the feasibility of direct minimization as a general purpose, automated numerical solution scheme, attention was limited to the simplest of structural discrete elements, namely, bars and beams. A bar or truss element is defined here as a slender prismatic element, straight when unloaded, which carries axial loads only but can buckle at its Euler buckling load. A beam or frame element is also defined as a slender prismatic element, straight when unloaded, which carries not only axial loads but also bending and torsional moments, and can buckle into bending and torsional modes. In this paper attention is focused on the bar element and structures made up of bar elements. Only a few simple examples are given to show the behavior of the frame element.

The deformations of the bar and frame elements are described in terms of nodal point translations, nodal point rotations, and buckling mode amplitudes. The total potential is obtained as the sum of the strain energy contributions of the individual elements minus the sum of the potentials of the applied loads all written in terms of the above displacement variables. The vector $\{x\}$ of these displacement variables for which the total potential attains a local minimum describes an equilibrium state according to the minimum principle of potential energy. The basic ingredients of the numerical method described here then are an expression for the strain energy of a typical element in terms of the discrete displacement variables and a method for finding the unconstrained local minimum of the total potential as a function of the displacement variables of the assembled structure. These two ingredients are described next.

Strain energy expression

The derivation of the strain energy expressions will be given here in detail only for the algebraically simpler case of the truss element. The derivation proceeds along the same lines for the frame element but with considerably more algebra.[1]

The truss element is assumed to be straight and free of stress in the undeformed structure. The undeformed length L of a jm discrete element is defined by the initial positions of the j and m nodes which it joins (Figure 1).

$$\bar{L} = \bar{r}^{(m)} - \bar{r}^{(j)}. \tag{1}$$

Given that the initial position vectors to all nodes are prescribed in a common reference coordinate system $(\bar{i}_{ok})$, the undeformed length of a jm element is

$$L = \left\{ \sum_{k=1}^{3} [x_{ok}^{(m)} - x_{ok}^{(j)}]^2 \right\}^{1/2}. \tag{2}$$

Superscripts denoting the jm element are omitted without ambiguity since attention is focused on a single arbitrary truss discrete element throughout this development. The nodal distance spanned by the truss element in the loaded structure S is the deformed length.

$$\overline{S} = (\overline{r} + \overline{u})^{(m)} - (\overline{r} + \overline{u})^{(j)}. \tag{3}$$

Given that all nodal displacements are prescribed in the reference coordinate system, the deformed length is

$$S = \left\{ \sum_{k=1}^{3} [(x_{ok} + u_{ok})^{(m)} - (x_{ok} + u_{ok})^{(j)}]^2 \right\}^{1/2}. \tag{4}$$

Primary deformation of a truss element is manifested as a change in length imposed by the displacements of the j and m nodes. Secondary deformation, exhibited as transverse bending, is admitted in the truss element representation in order to provide for buckling of individual elements within a structural system (local buckling). Accordingly, the strain response that arises within a truss discrete element due to an imposed change in length is expressed in terms of an axial deformation mode u and a transverse bending mode v.

$$\epsilon(x, y) = u_x(x) - yv_{xx}(x) + \tfrac{1}{2} v^2{}_x(x). \tag{5}$$

The x, y coordinate system is defined by the displaced positions of the j and m nodes and the plane of bending of the element. Bending is assumed to take place in a plane of symmetry of the cross section. Characterization of the strain response by Equation 5 implies that the deformations are small though finite.

Integration of the strain energy density definition

$$dU = \int_0^{\epsilon} \sigma d\epsilon \tag{6}$$

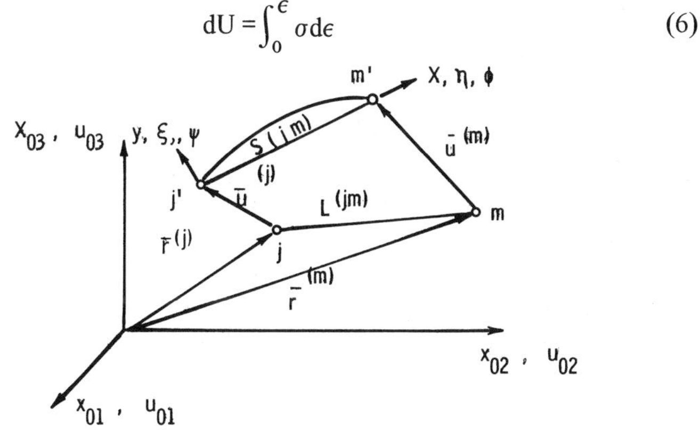

Figure 1. Truss element deformation.

under the assumption of ideal linear elastic material behavior

$$\sigma = E\epsilon \tag{7}$$

results in the following expression for the strain energy in terms of the strain:

$$U = \int_V \tfrac{1}{2} E\epsilon^2 \, dV. \tag{8}$$

Substitution for the strain from Equation 5 and integration over the cross section yields the strain energy functional.

$$\Phi = \tfrac{1}{2}\int_0^{(S/L)} [(\phi_\eta + \tfrac{1}{2}\,\psi_\eta^2)^2 + (\frac{I}{AL^2})\,\psi_{\eta\eta}^2]\, d\eta. \tag{9}$$

The strain energy functional is presented here in nondimensional form for convenience. The transformation to nondimensional variables is straightforward.

$$x = L\eta, \tag{10}$$

$$u(x) = L\phi(\eta),$$
$$v(x) = L\psi(\eta), \tag{11}$$

$$U = AEL\ \Phi. \tag{12}$$

The total potential energy is the same as the strain energy since it is assumed that no nonzero forces are prescribed to act on the truss element. Energy is stored in the element through the change in length imposed by the nodal displacements.

The governing differential equations for the truss discrete element are derived by executing the variation of the total potential energy in order to obtain additional information for use in the construction of deformation modes.

$$\begin{aligned}
\delta\Phi = &-\int_0^{(S/L)} \left\langle \frac{d}{d\eta}[\phi_\eta + \tfrac{1}{2}\psi_\eta^2]\right\rangle \delta\phi\, d\eta \\
&+\int_0^{(S/L)} \left\langle \left(\frac{I}{AL^2}\right)\psi_{\eta\eta\eta\eta} - \frac{d}{d\eta}[\psi_\eta(\phi_\eta + \tfrac{1}{2}\psi_\eta^2)]\right\rangle \delta\psi\, d\eta \\
&+ \left\langle (\phi_\eta + \tfrac{1}{2}\psi_\eta^2)\right\rangle \delta\phi\, \Big|_0^{(S/L)} \\
&- \left\langle \left(\frac{I}{AL^2}\right)\psi_{\eta\eta} - \psi_\eta(\phi_\eta + \tfrac{1}{2}\psi_\eta^2)\right\rangle \delta\psi\, \Big|_0^{(S/L)} \\
&+ \left\langle \left(\frac{I}{AL^2}\right)\psi_{\eta\eta}\right\rangle \delta\psi_\eta\, \Big|_0^{(S/L)}.
\end{aligned}$$

It is known from the potential energy stationary (minimum) principle that the actual state of displacement is one for which the total variation is zero. It is clear that the total variation is zero if each contribution is itself zero. It follows that the expressions in the brackets <> must be zero in contributions (a) and (b) since the variations ($\delta\phi$, $\delta\psi$) are arbitrary. The following equations are obtained for equilibrium in the longitudinal and in the lateral directions, respectively:

$$\phi_\eta + \tfrac{1}{2}\psi_\eta^2 = k_t, \tag{14}$$

$$(I/AL^2)\psi_{\eta\eta\eta\eta} - \frac{d}{d\eta}[\psi_\eta(\phi_\eta + \tfrac{1}{2}\psi_\eta^2)] = 0\,. \tag{15}$$

The rotations of the ends of the truss element are not imposed since moment free connections have been assumed; therefore, the variational quantity ($\delta\psi_\eta$) in contribution (e) is arbitrary on the boundaries and the expression in the brackets must be zero. The resulting natural boundary conditions are

$$\psi_{\eta\eta}(\eta)|_{\eta=0} = 0, \tag{16}$$

$$\psi_{\eta\eta}(\eta)|_{\eta=(S/L)} = 0. \tag{17}$$

The variational quantities ($\delta\phi$, $\delta\psi$) in contributions (c) and (d) are prescribed zero on the boundaries since the translational displacements of the ends of the truss element are not subject to variation but rather are imposed by the nodal displacements. The imposed boundary conditions stem from the problem description (Figure 1).

$$\phi(\eta)|_{\eta=0} = 0, \tag{18}$$

$$\phi(\eta)|_{\eta=(S/L)} = [(S/L) - 1], \tag{19}$$

$$\psi(\eta)|_{\eta=0} = 0, \tag{20}$$

$$\psi(\eta)|_{\eta=(S/L)} = 0. \tag{21}$$

A polynomial with five undermined coefficients

$$\psi(\eta) = c_0 + c_1\eta + c_2\eta^2 + c_3\eta^3 + c_4\eta^4 \tag{22}$$

is arbitrarily assumed to approximate the buckled shape of the element with sufficient precision. The imposed boundary conditions 20, 21 are admissibility requirements that determine two of the arbitrary coefficients. The natural boundary conditions 16, 17 are invoked to determine two further coefficients. The final form of the transverse deformation includes the remaining degree of freedom.

$$\psi(\eta) = \frac{16}{5}\left[\left(\frac{L}{S}\right)\eta - 2\left(\frac{L}{S}\right)^3\eta^3 + \left(\frac{L}{S}\right)^4\eta^4\right]C. \tag{23}$$

The coefficient C, which describes the midspan amplitude of the bending mode, is retained in the formulation as a generalized displacement variable.

Substitution of the preselected bending deformation mode into the equation governing axial equilibrium allows calculation of an appropriate axial deformation mode.

$$\int_0^{(S/L)} k_t\, d\eta = \int_0^{(S/L)} \phi_\eta d\eta + \tfrac{1}{2} \int_0^{(S/L)} \psi_\eta^2\, d\eta. \tag{24}$$

The constant K_t is obtained explicitly by integrating and introducing the imposed boundary conditions 18, 19.

$$K_t = [1 - (L/S)] + 2.4868571\,(L/S)^2\, C^2. \tag{25}$$

This determination of K_t implies an axial deformation mode that is compatible with the preselected transverse deformation mode in the sense that axial equilibrium is satisfied over the entire span of the truss discrete element.

Introduction of the assumed deformation modes into the potential energy functional (Equation 9) and subsequent integration over the deformed length yields the potential energy formulation for a truss discrete element.

$$\Phi = \tfrac{1}{2}(S/L)K_t^2 + 24.576\,(I/AL^2)(L/S)^3 C^2. \tag{26}$$

Function minimization

The formulation of the structural analysis problem has been cast here into a function of the displacement behavior variables $F(\{x\})$, defined as the total potential. The formulation is such that the prediction of the behavior corresponds to the selection of the behavior variables so as to minimize the function. The minimization of a function $F(\{x\})$ by choice of the vector $\{x\}$ is the problem of mathematical programming. The particular type of mathematical programming problem encountered herein is one in which the variables x_i are not restricted to certain intervals and is referred to as unconstrained minimization.

The functions considered here are both continuous and differentiable for all $\{x\}$. A point $\{x\}^*$ is said to be a local minimum if the inequality

$$F(\{x\}^*) \leqslant F(\{x\}) \tag{27}$$

holds for all $\{x\}$ in some neighborhood of $\{x\}^*$. A necessary condition for the occurrence of an extremum is that the gradient vanish.

$$g_i \equiv \partial F/\partial x_i = 0 \qquad (i = 1, \ldots, n). \tag{28}$$

One approach to the finding of a local minimum is to attack Equation 28 directly. However, this task is a formidable one for nonlinear systems and may yield extrema of the function which are not local minima. These considerations have led to the development of efficient alternative procedures for the minimization of functions.

Not all search techniques that have been developed are sequential in the sense that improved points are sought sequentially by modification of the best available point.[2-4] However, sequential search techniques are appropriate for the analytic functions of many variables considered herein. The sequential search techniques seek to move from a given point $\{x\}_q$ a distance t_q along a direction $\{\phi\}_q$

$$\{x\}_{q+1} = \{x\}_q + t_q \{\phi\}_q \tag{29}$$

such that the function value is reduced:

$$F_{q+1} < F_q.$$

With reference to Equation 29, the unconstrained minimization problem is reduced to determining which way to go ($\{\phi\}_q$) and how far to go (t_q) in the modification of the current point ($\{x\}_q$) so as to obtain a solution ($\{x\}^*$) with minimum effort. The selection of the optimum minimization technique is dependent upon the nature of the function.[5]

The negative gradient direction was originally proposed by Cauchy in 1847 as the answer to the "which way to go" question.[8] This approach has come to be known as the method of steepest descents or the method of optimum gradients depending upon the criterion employed in determining "how far to go." In the method of optimum gradients the step length is determined so as to minimize the function along the gradient direction, while in the method of steepest descents it is required only that some improvement in the function be achieved. Both methods have been found inefficient to the point of being useless for solving large nonlinear problems. The inefficiency is characterized geometrically by zigzag behavior caused by the eccentricity of the hyperspheres representing function contours. A number of gradient minimization techniques have been developed which substantially overcome the zigzag problem. A variable metric gradient technique described by Fletcher and Powell[7] and based on Davidon's work[8] seems to be the best of these. It is applicable if analytic gradients are available and the storage requirement for the nxn metric matrix (n is the number of variables) is not prohibitive. The requirement for analytic gradient has been eliminated recently in a modified version[9] but the storage requirement remains.

Relaxation minimization techniques result from the use of univariate modifications as an alternative answer to "which way to go."[10,11] Recent innovations have served to improve the effectiveness of this well-known approach to minimization.[12] The use of these methods is particularly well suited to problems in which a univariate modification allows a simple correction of the function value or when an analytic gradient is not available. The best gradient methods are likely to be more efficient when the formulation and programming are extended to include analytic expressions for the gradient components.

Composite technique

The search technique developed here for use in the analysis of structural systems of one-dimensional elements embodies three interdependent techniques of seeking an improved point which are coordinated to obtain maximum benefit from each. A similar technique has been reported by Hooke and Jeeves.[12]

The basic modification technique is univariate. Each variable is considered in turn. The gradient component is estimated by finite difference, from function observations made to determine whether to increment or decrement, and the single variable is modified so as to minimize the function value.

If the total modification achieved by the consideration of all n variables is small, a second modification technique is initiated. The gradient components of the univariate mode are taken as an approximate gradient at the current point and the variables are modified so as to minimize the function along the gradient direction. This mode of modification serves to perturb the relaxation, thereby improving its effectiveness, while at the same time contributing significantly to the minimization of the function.

A third modification technique is initiated after two consecutive cycles of the search modes previously described. An extrapolation is made along the line defined by the current point and a point stored upon the previous initiation of this mode. The variables are then modified so as to minimize the values of the function along this line. This mode, which has been used with remarkable success in overcoming the zigzag tendency of the optimum gradient method,[13] is used to advantage here even though the interim steps are not governed by an analytic gradient.

A local minimum is identified when no small perturbations yield an improved function value in the univariate mode and the subsequent gradient modification is unsuccessful.

The minimization procedure just described evolved from the study and numerical testing of techniques reported in the substantial literature that exists on unconstrained minimization. The use of analytic gradients were precluded at the outset in the development of this minimization technique to minimize the effort required to formulate a problem. The minimization is conducted entirely by interrogation of the function value. This feature allows application to problems where the analytic gradients are considered intractable. The relatively small computer storage requirement serves as a further recommendation of the method.

Computer program

The strain energy expressions for the truss and the frame elements and the composite minimum search technique were incorporated into a pilot computer program suitable for numerical experimentation with structural configurations exhibiting various types of geometric nonlinearities in their response to loads. These nonlinearities might be an increase or decrease in stiffness during loading, snap-through or bifurcation type of general instability, local buckling of elements, or general instability triggered by local buckling of elements. Imperfection sensitivity is also correctly detected. Equilibrium configurations of the structures can be found both in the pre- and post-buckling range and without the knowledge of previous solutions.

To facilitate numerical experimentation both nodal forces and nodal displacements may be prescribed either separately or in proper combinations. Because in the case of nonlinearity one is usually interested in tracing some curve characteristic of the structural behavior, an initial loading condition and various increments of the same or some other loading condition may be prescribed. This load incrementation is a convenience for the user, since the method itself does not rely on incrementation.

The program was originally envisioned to be used in connection with small structures with interesting behavior rather than for actual large design problems. Although it is written to operate completely in core, accuracy rather than efficiency was the goal. Running times become prohibitive for the present program for structures larger than the largest problem discussed here. Efficiency could be increased especially as a tradeoff for accuracy simply by using more liberal convergence criteria for accepting a point as a minimum. However, there are irreducible difficulties inherent in a nonlinearized nonincremental large deflection analysis method, and the analyst will have to trade the power and range of applicability of the nonincremental methods for the size of the problem that can be analyzed in acceptable computer times. The details of the computer program are given in Reference 14.

3. Sample Problems

The first few problems are intended to introduce the reader to the basic capabilities of the truss and frame elements. A shallow A frame and a well-known simple imperfection sensitive model are used for this purpose (Figures 2, 7, 8).

The second group of problems is derived from the shallow geodesic dome shown on Figure 9. The particular geometry of this dome was chosen because of the simplicity of the input data. The structure need not be shallow; the computer program is applicable to any three-dimensional assembly of truss and frame elements. Increasing portions of this dome were isolated and analyzed for a few loading conditions. Only truss elements are considered for reasons of economy. The intent is to show some possible uses of the method with a few examples rather than to attempt to analyze the multitude of problems that could be posed for this shallow geodesic dome.

Problem 1

The shallow A frame shown in Figure 2 was analyzed using truss and then frame elements for the two members. The truss elements represent the case when all nodes are hinged, and the frame elements when the middle node is framed and the nodes at the support are either fixed or hinged. The behavior of this structure is usually discussed referring to its force-deflection curve, the force P applied at the middle node point plotted against the vertical deflection of the same node point. The curves of Figure 3 are such force-deflection curves. For bars such curves are always symmetric about the point $\Delta = h$ on the Δ axis. This family of curves was generated using truss elements ($A = 0.1$ in.^{2}, $E = 10^7$ psi) and various values of the rise h of the midpoint of the frame. The buckling of the truss elements was suppressed for these examples. This can be done either by using oversized moments of inertia or by the option provided in the computer program to suppress member buckling degrees of freedom. In order to trace the complete curves, increments of nodal displacements rather than forces had to be prescribed since the reversed slope portion of the curves corresponds to equilibrium configurations that are not stable for prescribed

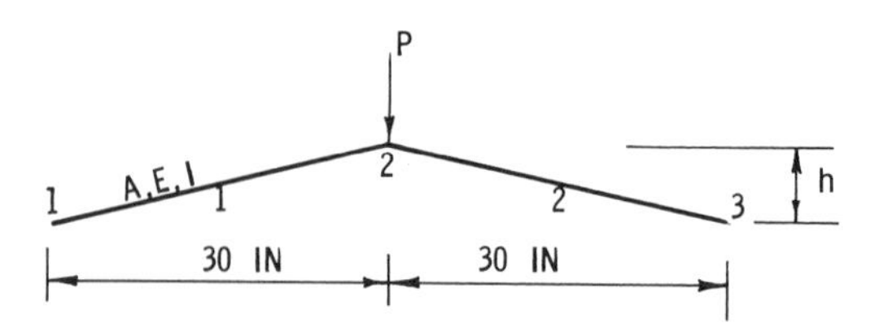

Figure 2. Shallow A frame snap-through structure.

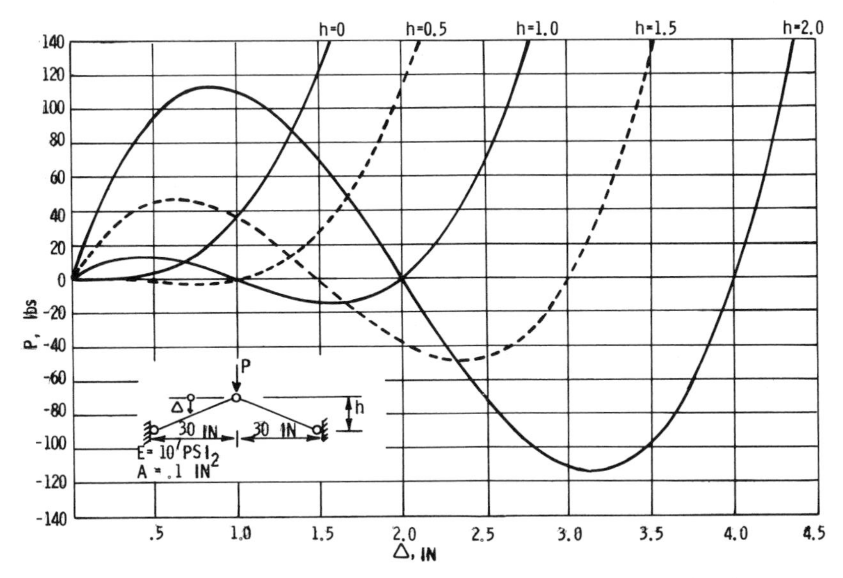

Figure 3. Load-deflection curves for A frames with hinged middle joint.

nodal forces. One could remark here that the case of $h = 0$ would pose problems for a linearized incremental scheme. This structure is a mechanism in the undeformed state resulting in a singular stiffness matrix for the first increment.

The curves of Figure 4 were generated using $h = 1.771$ in. With this value of h the A frame is the basic building block of the geodesic dome of Figure 9. (For the middle node it is exactly and for the other nodes approximately.) The curves marked P_1 and S_1 are the nodal load and member force, respectively, for the case of suppressed member buckling. Both are plotted against the deflection Δ. If increments of the load P are prescribed a jump from the unbuckled stable maximum load configuration at point A will occur to the buckled stable configuration at B. The maximum for the member force S_1 is reached when $\Delta = h$. If the Euler buckling load of the bar is less than this maximum then it can not be reached, since the bar will buckle at its Euler load at some Δ less than h. If this occurs before the nodal load would reach its maximum at point A, the whole structure will snap-through prematurely due to member buckling. For example, the truss element of this particular A frame and with $I = 0.0075$ in.^{4} will buckle at 819.6 lb. If this buckling is not suppressed the path C-D marked P_2 will be obtained while the member force S_2 will stay nearly constant. The computer value of S_2 actually increases and then decreases, similar to S_1, but varying only between 823.1 and 825.1 lb. This is significant because it shows that the actual column action is properly approximated, and from the stiff side, for large ranges of the end-shortening. The truss element incorporated

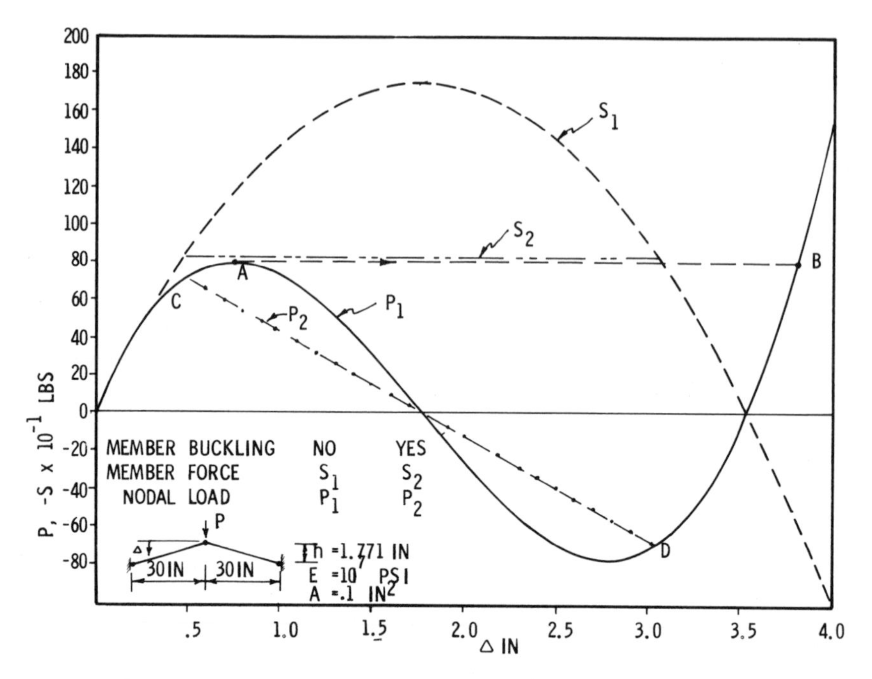

Figure 4. Load deflection and member force curves for hinged A frames.

in the present computer program therefore appears to function realistically under large deflections and both in the pre- and post-buckled state.

The frame element capability can be shown in a similar manner and Figures 5 and 6 were generated for this purpose. Figure 5 is again a family of curves,

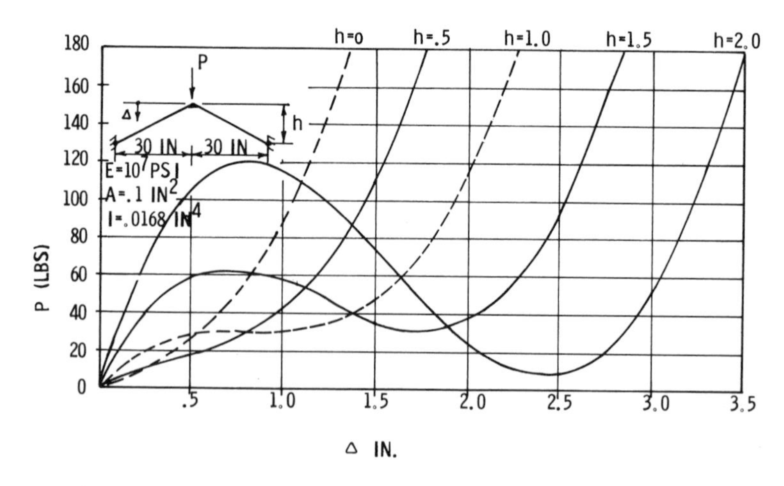

Figure 5. Load-deflection curves for A frames with framed middle joint.

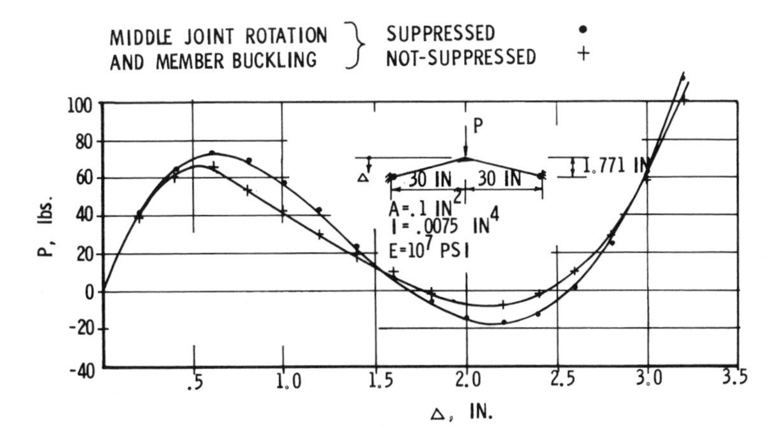

Figure 6. Load-deflection curves for A frame with framed middle joint.

similar to that of Figure 3, and were generated for the same geometries. The frame elements have the same properties as the truss elements did ($A = 0.1$ in.^{2}, $E = 10^7$ psi) but in this case a moment of inertia also had to be assigned. $I = 0.0169$ in.^{4} was chosen, which prevents premature buckling. Comparing Figure 3 and Figure 5, the difference in the behavior is apparent. The well-known transition from the snap-through to the nonlinear beam behavior for decreasing values of h is shown to be correctly reproduced using the frame element incorporated in this computer program.

Figure 6 is to be compared with Figure 4 discussed earlier. It was generated using frame elements with properties identical to that of the bar elements used to generate Figure 4. The difference actually is that the middle node is considered hinged in the case of the bar elements and framed in case of the frame elements. The other two nodes were considered pinned in both cases. Again two force-displacement curves were generated, one with member buckling suppressed and one with member buckling allowed. Comparing the corresponding curves of the two figures one might observe that in this particular case considering the middle node framed instead of hinged, did not improve the structure.

The above sample problems, even if somewhat primitive, attempt to serve the purpose of exposing the basic behavior of both the truss and the frame element. The problems also represent the simplest cases of the snap-through type instability, one of the collapse modes of geodesic domes.

Problem 2

Imperfection sensitivity for buckling is an important consideration in judging the usefulness of a structural configuration for practical engineering applications. Imperfection sensitivity necessitates either large safety factors in the design formulas or extreme requirements on fabrication tolerances, as is the case for cylindrical shells in uniform axial compression and spherical shells under uniform external pressure, the two most celebrated cases. It is of importance therefore for a method to be able to correctly predict the buckling of imperfection sensitive structures for any given actual imperfection. A well-known simple imperfection sensitive model is used to show the capability of the method in this respect. The model is essentially a beam with a nonlinear elastic support following a cubic law at its midspan. The beam might be replaced with two bars hinged at midspan. The cubic spring behavior is supplied by a shallow A frame. The structure shown in Figure 7 is sensitive in its buckl-

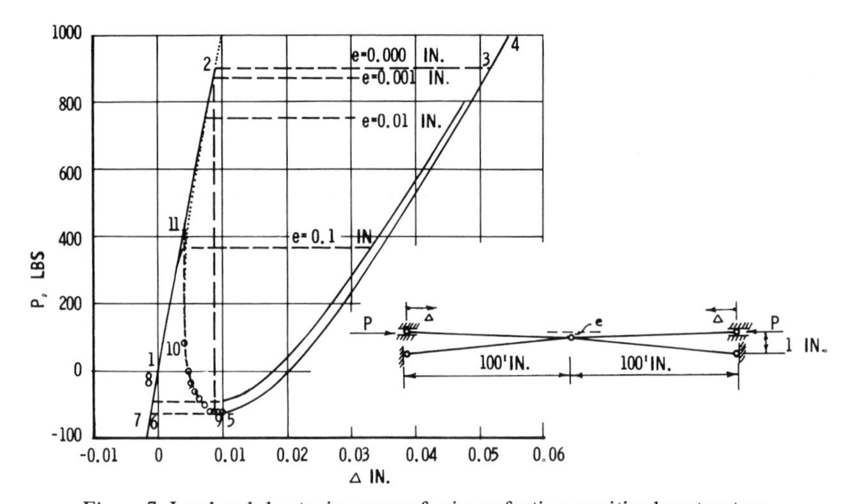

Figure 7. Load-end shortening curves for imperfection-sensitive bar structure.

ing behavior to the degree of the "imperfection," that is, the value of "out of straightness e." The curves of Figure 7 were obtained by assigning various values to this out of straightness e in the unloaded state. The buckling loads are shown for e = 0.0 in., e = 0.001 in., e = 0.01 in., and e = 0.1 in. Such eccentricities are barely perceptible for the 100-in.-long bars, yet the reductions in buckling loads are quite noticeable.

Using the load incrementation option to prescribe the proper sequence for the nodal forces P, the path 1-2-3-4-3-5-6-7-8 can be traced. The jump 2-3 represents the snap buckling and the jump 5-6 the return to the straight configuration. This path is for the "perfect" structure with e = 0.0 in. A similar path is traced out for the structure with the largest imperfection investigated, one with e = 0.1 in. Observe that for this larger value of imperfection the buckling portion of the curve also departs from the straight line for loads close to the snap buckling load. Enforcing proper sequences of prescribed nodal displacements Δ the path 1-2-9-4-9-10-11-1 may be traced since the equilibrium configurations represented by circles on the segment 5-10 are also stable for the case of prescribed nodal displacements. The dotted line segment 10-2 that represents unstable equilibrium configurations both for prescribed nodal forces and nodal displacements was traced using a special computer program. For the case of enforced nodal displacements the jumps 2-9 and 10-11 indicate the sudden changes in the equilibrium configuration.

The snap-through buckling load for the "perfect" structure, that is, the load level at point 2, is 903 lb. A simple calculation shows that the classical buckling load is 1000 lb. In the range of 800–1000 lb the stationary points of the total potential representing stable and unstable equilibrium are very close, as is signified by the "unstable" dotted line very closely approaching the "stable" straight line. The search procedure in effect perturbs the configuration of the structure, causing it to jump prematurely through the near unstable to the far stable configuration represented by point 3. This premature jump could be postponed somewhat by refining the present step sizes but the theoretical 1000 lb could not be reached for this highly imperfection-sensitive structure with the present method of analysis. Similar behavior is observed, for example, in connection with the classical problem of buckling of isotropic circular cylindrical shells under uniform axial load. By exercising extreme care in the fabrication and testing of a cylindrical shell, the classical buckling load can be approached very closely. Values as high as 90 per cent have been attained by skilled and patient experimenters, but to attain 100 per cent is impossible in practice due to the always-present slight disturbances that are analogous to the perturbing step sizes, which can be made small but have to be nonzero in the search procedure.

The structure shown in Figure 8 is essentially the same as the structure of the previous problem. The two horizontal bar members are replaced by two frame members. Instead of the geometric imperfection e of the previous problem, a small disturbing force Q of various magnitude was used to generate the family of curves shown in Figure 8. With the lateral load Q kept at a constant magnitude, equal and opposite displacements were prescribed for the nodes at the ends of the frame members through two bars representing an elastic, displacement enforcing test machine.

For Q = 0 lb loading and unloading produced the path 1-2-3-4-3-5-6-1 with point 2 representing the buckling load. A similar path 1-7-8-9-8-10-11-1 was produced for Q = 1 lb. A loading value of Q = 5 lb creates a large enough disturbance to take the sudden snap out of the structure for the case of displacement enforcing stiff loading fixture so that every point on the load–displacement curve represents a stable configuration. One might again conclude that the intricate behavior of imperfection sensitive structural systems is realistically predicted by the numerical method presented here. The direct minimum

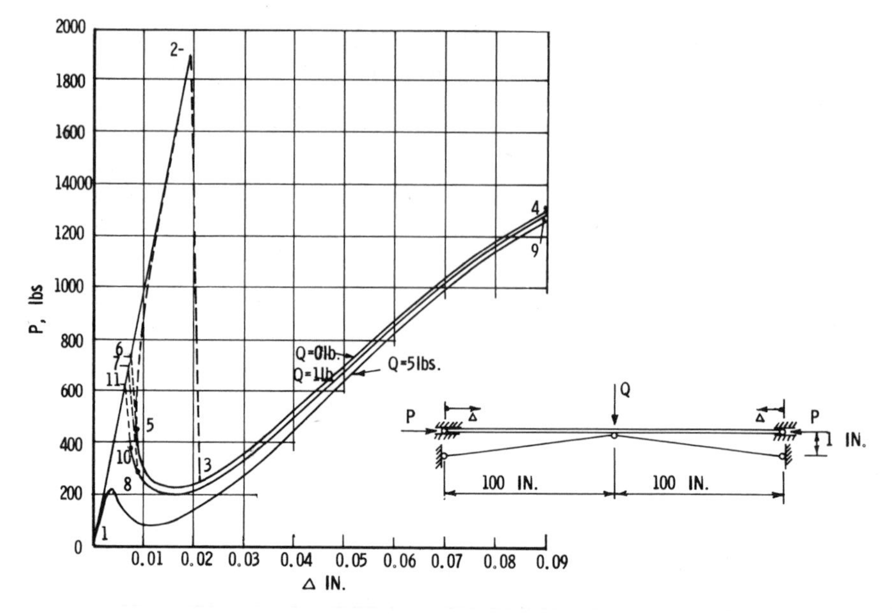

Figure 8. Load-end shortening curves for imperfection-sensitive beam structure.

search actually predicts the behavior of a structure as it would be observed in the test machine, force or displacement enforcing, or elastic, since neither the actual structure nor the direct minimum search can find an unstable equilibrium position.

Problem 3

Depending on the chosen boundary conditions the structure shown in Figure 9 may be taken as a shallow geodesic dome or a shallow cap isolated from a deep geodesic dome. The plan view was chosen to be regular as shown. The z coordinates were calculated to place the nodes on a sphere with a radius of 255 in. A shallow dome was chosen to keep the lengths of the bars close to the nominal 30 in. The above geometry helped to simplify the generation of the input data. For all members $E = 10^7$ psi and $A = 0.1$ in.^{2}. The truss elements also must have a value for the moment of inertia to define their Euler buckling load. The choice of this value limits the maximum axial load the individual truss elements can carry. Reaching this load will cause the element or elements to buckle. The resulting load redistribution might precipitate the collapse of the entire structure. With the present method member buckling can be artifically suppressed, as discussed previously. Such an analysis will give displacements and member forces that could be achieved if buckling of the bars could be also avoided in practice. Even if the results might be illusory it is desirable to analyze the structure with member buckling suppressed. Such analyses give information about the response of the total structure as a configuration which is an upper bound on load carrying capacity. In the analyses that follow member buckling was suppressed unless otherwise noted.

Various portions of the structure shown on Figure 9 were isolated and analyzed for a few load conditions. The intent of these examples is to present the capability and behavior of this particular numerical method through a few representative examples, rather than the much more ambitious goal of thorough analysis of the complex pre- and post-buckling behavior of geodesic domes. Increasingly larger and larger portions of the shallow dome, designated STR-I, -II, -III, -IV, were analyzed for decreasingly less and less number of loading conditions for obvious reasons of economy. The results of some of these analyses are presented next.

STR-I

The structure designated STR-I, shown in Figure 10, is isolated from the geodesic dome as the first central hexagon. It is made up of the six radial and the six circumferential elements, and its behavior is not much different from the A frames of Problem 1. If all joints are considered hinged, and member buckling is suppressed, an upper bound for snap-through load is obtained if nodes 2 to 7 are not allowed to move. A lower bound is obtained if nodes 2 to 7 are allowed to move in the x-y plane. The load-displacement curves for these

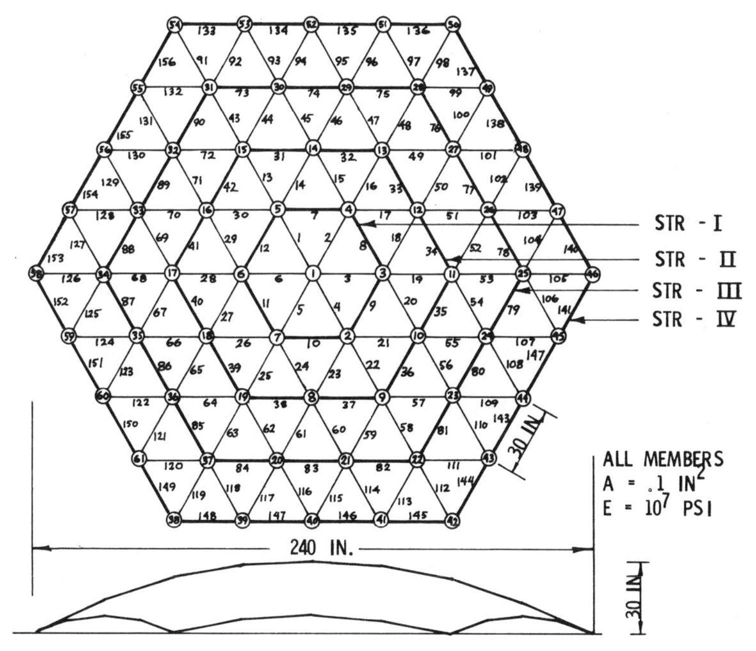

Figure 9. Shallow geodesic dome.

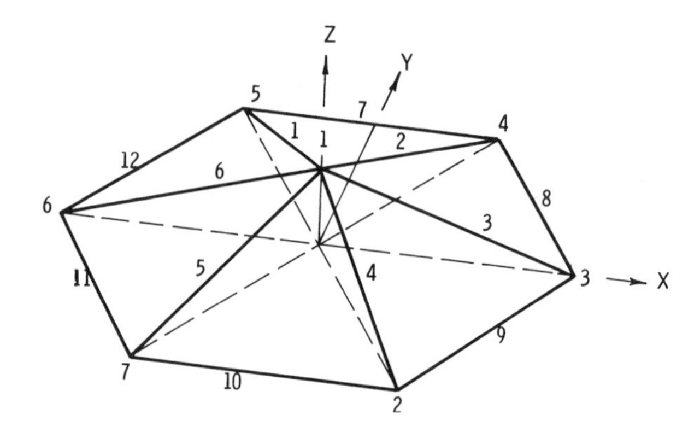

Figure 10. Six bar snap-through structure (STR-I).

two cases are shown on Figure 11. The load P, acting on node 1, is plotted against the deflection Δ, both taken positive in the negative z direction. It is also of some interest to know what the snap-through load of node 1 is when it is the central node of larger portions of the sample geodesic dome. STR-II was analyzed with displacements of nodes 8 to 19 suppressed. As seen in Figure 11 the resulting snap-through load is much closer to the lower than to the upper bound curve. STR-III was analyzed next with displacements of nodes 20 to 37 suppressed. The snap-through load of node 1 was almost identical with that of STR-II. Consequently, STR-IV was not analyzed for this condition. For all of the above cases displacements of node 1 were prescribed in the negative z direction, and member buckling was suppressed.

STR-I was also analyzed considering members 1 to 6 to be frame members. Only a general description of the results is given here. Depending on the geometry assumed for the cross section, load deflection curves of the types shown on Figure 5 for the A frame can be obtained for this structure. This time the bending stiffnesses of the frame elements were increased in steps. This has the same general effect as decreasing the rise h. Above a certain bending stiffness the structure will not snap buckle. Depending on the values of the moments of inertia of the cross section, node 1 might not only translate but will tilt, and also rotate about the z axis resulting in a spiral-like collapse mode. This was found to be particularly pronounced for very slender frame members.

STR-II

The structure outlined by truss members 31 to 42 is designated STR-II. All 42 members will be assumed to be truss elements; in other words, all the nodes are assumed to be hinged in all directions.

Four loading conditions were considered. The first case, snap-through loading of node 1, has been discussed previously in connection with upper and lower bounds for the snap-through load of the center node. As the next case, nodes 1, 2, 3, 4, 7 were loaded in 50-lb increments in the negative z direction; that is, only one-half of the structure was loaded with a gravity type load, say snow. Figure 12 (a) shows the load distribution and the resulting buckled state. Such buckled states, and they are not unique, are possible only if member buckling does not occur. The first snap-through occurred between 250 and 300 lb per node. This is higher than the upper bound, 234 lb, for the case when only a single node is loaded (Figure 11). Tables 1 and 2 are the nodal deflections and member forces prior to and after buckling.

If all nodes are loaded the buckled state shown on Figure 12(b) results. The first buckling in this case occurred between 350 and 400 lb. The nodal deflections and member forces prior to and after buckling are given in Tables 3 and 4. This analysis was repeated with member buckling not suppressed. As expected, the structure completely snapped through since the high compressive forces (Table 4) of the buckled state of Figure 12(b) could not be maintained. Even if compressive member forces are maintained to the point of general buckling, most of the buckled equilibrium states require very high compressive member forces which in practice cause member buckling and the final result is total collapse.

In order to assess the initial imperfection sensitivity of spherical geodesic domes, STR-II was loaded with radial loads on nodes 1 to 7 representing uniform external pressure and at the same time corresponding radial displacements of nodes 8 to 19 were prescribed. These prescribed sets of combined forces

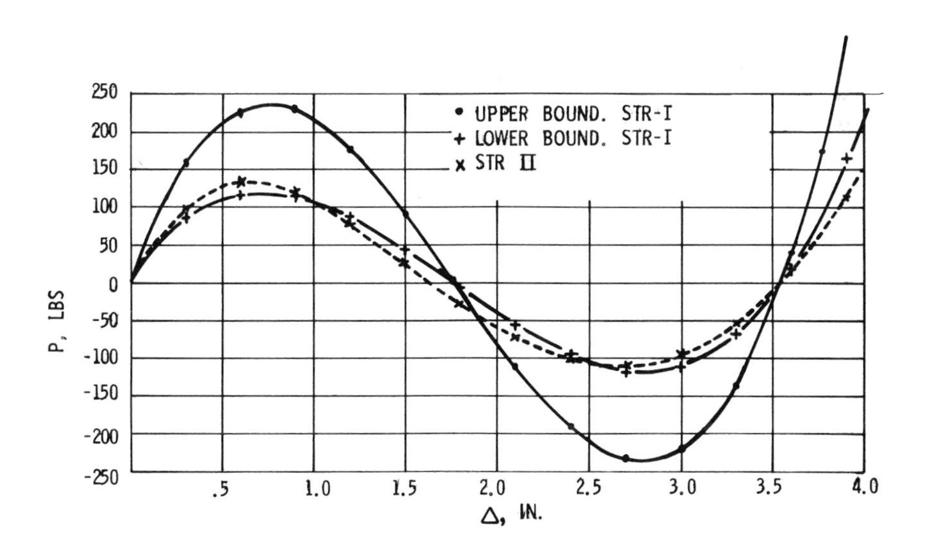

Figure 11. Load-deflection curves for center node of geodesic dome.

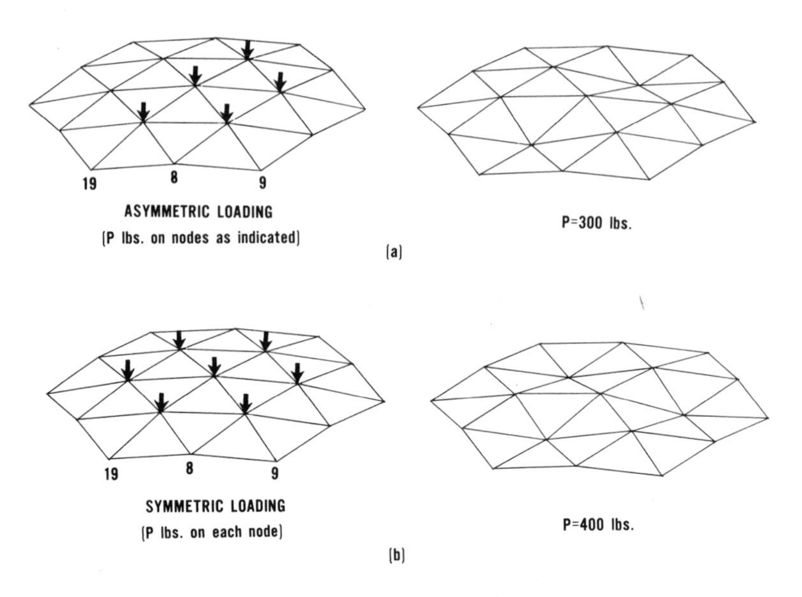

Figure 12. Buckled equilibrium mode shapes of STR-II.

and displacements were calculated to simulate uniform compression of a complete spherical geodesic dome and considering STR-II to be an isolated segment of it. The test of an "ideal" uniformly compressed sphere is that it is in a purely uniform hydrostatic membrane stress state. In the case of our geodesic dome with its regular geometry the uniformity of the member forces is the requirement. Tables 5 and 6 are the nodal displacements and member forces, respectively, for the last unbuckled and the first buckled load condition. Even though the member forces are not exactly uniform, the "perfect membrane" state appears to be very closely approximated. The buckling load is between 551 and 620 lb per node, much higher than for any of the previous loading conditions. It is probably a very close approximation to the classical buckling load for the uniform radial loading of the geodesic dome under consideration. It is characteristic of the extreme initial imperfection sensitivity of complete spherical geodesic domes under uniform external compressive load that for a slightly "imperfect" radial loading, a 1½ per cent difference due to an error in its calculation, the structure buckled between loads 477 - 542 lb instead of 551 - 620 lb. With this in mind the degradation in the buckling loads is understandable as one goes from the "perfect" uniform compression case to the uniform vertical load on the total structure, vertical load on one-half of the structure and finally vertical load only on one node. Such loading conditions might be thought of as highly imperfect cases of the uniform external pressure. Starting then from the above "ideal" case, a systematic study of the

Table 1. Nodal Displacements of STR-II. Asymmetric Loading (Fig. 12a).

	PRE-BUCKLING P = 250LBS.				POST-BUCKLING P = 300LBS.		
	X	Y	Z		X	Y	Z
1	-0.2509E-01	0.1440E-01	-0.3965E-00	1	-0.1009E-01	0.5805E-02	-0.7477E 00
2	-0.2058E-01	0.2998E-01	-0.3448E-00	2	-0.4587E-01	0.4254E-01	-0.5822E 00
3	-0.3644E-01	0.2752E-02	-0.3462E-00	3	-0.5971E-01	0.1848E-01	-0.5817E 00
4	-0.3435E-01	-0.2706E-01	-0.4049E-00	4	-0.2451E-00	-0.3878E-00	-0.4527E 01
5	-0.2934E-01	0.2836E-01	0.1793E-00	5	0.2679E-01	-0.1234E-01	-0.2616E-00
6	-0.3926E-01	0.1119E-01	0.1793E-00	6	0.2415E-01	-0.1702E-01	-0.2623E-00
7	0.6170E-02	0.4306E-01	-0.4032E-00	7	0.2132E-00	0.4062E-00	-0.4528E 01

Table 2. Member Forces in STR-II. Asymmetric Loading (Fig. 12a).

PRE-BUCKLING P = 250LBS.				POST-BUCKLING P = 300LBS.			
1	-0.4737E 03	16	-0.9498E 03	1	-0.1963E 04	16	-0.3118E 03
2	-0.1332E 04	17	-0.3682E 03	2	0.2126E 03	17	0.1641E 04
3	-0.4749E 03	18	-0.7717E 03	3	-0.1964E 04	18	-0.1643E 04
4	-0.4732E 03	19	-0.7669E 03	4	-0.1966E 04	19	-0.1264E 04
5	-0.1333E 04	20	-0.6146E 03	5	0.2141E 03	20	-0.5868E 03
6	-0.4752E 03	21	-0.6107E 03	6	-0.1965E 04	21	-0.5884E 03
7	0.2424E 02	22	-0.7660E 03	7	0.1163E 04	22	-0.1265E 04
8	-0.8947E 03	23	-0.7726E 03	8	0.1238E 03	23	-0.1643E 04
9	-0.1052E 04	24	-0.3671E 03	9	-0.9262E 03	24	0.1642E 04
10	-0.8910E 03	25	-0.9482E 03	10	0.1234E 03	25	-0.3120E 03
11	0.2731E 02	26	-0.1291E 04	11	0.1161E 04	26	0.5995E 03
12	0.6604E 03	27	0.3947E 03	12	0.1790E 03	27	-0.1080E 04
13	-0.2088E 03	28	-0.2096E 03	13	-0.7036E 03	28	-0.7050E 03
14	0.3961E 03	29	-0.2421E 03	14	-0.1079E 04	29	-0.1091E 03
15	-0.1294E 04	30	-0.2428E 03	15	0.5992E 03	30	-0.1075E 03

Table 3. Nodal Displacements of STR-II. All Nodes Loaded (Fig. 12b).

	PRE-BUCKLING P = 350LBS.				POST-BUCKLING P = 400LBS.		
	X	Y	Z		X	Y	Z
1	0.2211E-04	-0.1790E-04	-0.2365E-00	1	0.5797E-05	-0.1581E-05	-0.1740E 01
2	-0.1947E-01	0.3379E-01	-0.4458E-00	2	-0.6296E-01	0.1090E-00	-0.1315E 01
3	-0.4108E-01	-0.2242E-04	-0.4613E-00	3	-0.4347E-00	0.1859E-04	-0.5440E 01
4	-0.1958E-01	-0.3394E-01	-0.4471E-00	4	-0.6295E-01	-0.1090E-00	-0.1315E 01
5	0.2060E-01	-0.3571E-01	-0.4622E-00	5	0.2174E-00	-0.3765E-00	-0.5440E 01
6	0.3897E-01	-0.2699E-06	-0.4454E-00	6	0.1259E-00	0.3162E-05	-0.1315E 01
7	0.2066E-01	0.3580E-01	-0.4633E-00	7	0.2173E-00	0.3765E-00	-0.5440E 01

Table 4. Member Forces in STR-II. All Nodes Loaded (Fig. 12b).

PRE-BUCKLING P = 350LBS.				POST-BUCKLING P = 400LBS.			
1	-0.8985E 03	16	-0.1222E 04	1	0.4937E 03	16	-0.2636E 04
2	-0.8644E 03	17	-0.1001E 04	2	-0.4956E 04	17	-0.2141E 04
3	-0.8973E 03	18	-0.1017E 04	3	0.4933E 03	18	0.2117E 04
4	-0.8626E 03	19	-0.1236E 04	4	-0.4956E 04	19	-0.1488E 04
5	-0.9008E 03	20	-0.1019E 04	5	0.4948E 03	20	0.2118E 04
6	-0.8605E 03	21	-0.1000E 04	6	-0.4957E 04	21	-0.2141E 04
7	-0.1341E 04	22	-0.1221E 04	7	0.1902E 03	22	-0.2637E 04
8	-0.1340E 04	23	-0.9985E 03	8	0.1895E 03	23	-0.2141E 04
9	-0.1339E 04	24	-0.1020E 04	9	0.1887E 03	24	0.2117E 04
10	-0.1340E 04	25	-0.1239E 04	10	0.1899E 03	25	-0.1489E 04
11	-0.1341E 04	26	-0.1021E 04	11	0.1898E 03	26	0.2117E 04
12	-0.1339E 04	27	-0.9989E 03	12	0.1898E 03	27	-0.2141E 04
13	-0.1236E 04	28	-0.1220E 04	13	-0.1488E 04	28	-0.2636E 04
14	-0.1018E 04	29	-0.9989E 03	14	0.2117E 04	29	-0.2141E 04
15	-0.1000E 04	30	-0.1019E 04	15	-0.2140E 04	30	0.2118E 04

Table 5. Nodal Displacements of STR-II. "Ideal" Loading.

	PRE-BUCKLING P = 551LBS.				POST-BUCKLING P = 620LBS.		
	X	Y	Z		X	Y	Z
1	0.6172E-05	0.6796E-06	-0.3852E-00	1	0.4021E-02	0.9228E-05	-0.6409E 01
2	-0.2354E-01	0.4079E-01	-0.3937E-00	2	-0.7132E-01	0.1248E-00	-0.9767E 00
3	-0.4765E-01	-0.3730E-05	-0.3977E-00	3	-0.1427E-00	-0.8773E-05	-0.9699E 00
4	-0.2356E-01	-0.4082E-01	-0.3940E-00	4	-0.7130E-01	-0.1248E-00	-0.9769E 00
5	0.2383E-01	-0.4129E-01	-0.3978E-00	5	0.7349E-01	-0.1260E-00	-0.9930E 00
6	0.4709E-01	0.3986E-05	-0.3937E-00	6	0.1469E-00	-0.6130E-05	-0.1001E 01
7	0.2385E-01	0.4132E-01	-0.3981E-00	7	0.7346E-01	0.1260E-00	-0.9931E 00
8	0.	0.8160E-01	-0.3924E-00	8	0.	0.9180E-01	-0.4414E-00
9	-0.4700E-01	0.8140E-01	-0.3892E-00	9	-0.5287E-01	0.9157E-01	-0.4378E-00
10	-0.7066E-01	0.4080E-01	-0.3924E-00	10	-0.7949E-01	0.4590E-01	-0.4414E-00
11	-0.9400E-01	0.	-0.3892E-00	11	-0.1057E-00	0.	-0.4378E-00
12	-0.7066E-01	-0.4080E-01	-0.3924E-00	12	-0.7949E-01	-0.4590E-01	-0.4414E-00
13	-0.4700E-01	-0.8140E-01	-0.3892E-00	13	-0.5287E-01	-0.9157E-01	-0.4378E-00
14	0.	-0.8160E-01	-0.3924E-00	14	0.	-0.9180E-01	-0.4414E-00
15	0.4700E-01	-0.8140E-01	-0.3892E-00	15	0.5287E-01	-0.9157E-01	-0.4378E-00
16	0.7066E-01	-0.4080E-01	-0.3924E-00	16	0.7949E-01	-0.4590E-01	-0.4414E-00
17	0.9400E-01	0.	-0.3892E-00	17	0.1057E-00	0.	-0.4378E-00
18	0.7066E-01	0.4080E-01	-0.3924E-00	18	0.7949E-01	0.4590E-01	-0.4414E-00
19	0.4700E-01	0.8140E-01	-0.3892E-00	19	0.5287E-01	0.9157E-01	-0.4378E-00

Table 6. Member Forces in STR-II. "Ideal" Loading.

PRE-BUCKLING P = 551LBS.				POST-BUCKLING P = 620LBS.			
1	-0.1562E 04	22	-0.1544E 04	1	0.8544E 03	22	-0.1747E 04
2	-0.1552E 04	23	-0.1557E 04	2	0.8568E 03	23	-0.2180E 04
3	-0.1562E 04	24	-0.1564E 04	3	0.8514E 03	24	-0.2235E 04
4	-0.1551E 04	25	-0.1549E 04	4	0.8579E 03	25	-0.1763E 04
5	-0.1563E 04	26	-0.1564E 04	5	0.8542E 03	26	-0.2201E 04
6	-0.1550E 04	27	-0.1557E 04	6	0.8546E 03	27	-0.2227E 04
7	-0.1583E 04	28	-0.1544E 04	7	-0.4861E 04	28	-0.1771E 04
8	-0.1583E 04	29	-0.1557E 04	8	-0.4829E 04	29	-0.2227E 04
9	-0.1583E 04	30	-0.1564E 04	9	-0.4828E 04	30	-0.2199E 04
10	-0.1583E 04	31	-0.1571E 04	10	-0.4861E 04	31	-0.1767E 04
11	-0.1583E 04	32	-0.1571E 04	11	-0.4898E 04	32	-0.1767E 04
12	-0.1583E 04	33	-0.1570E 04	12	-0.4897E 04	33	-0.1767E 04
13	-0.1549E 04	34	-0.1571E 04	13	-0.1763E 04	34	-0.1768E 04
14	-0.1563E 04	35	-0.1571E 04	14	-0.2236E 04	35	-0.1768E 04
15	-0.1558E 04	36	-0.1570E 04	15	-0.2180E 04	36	-0.1767E 04
16	-0.1544E 04	37	-0.1571E 04	16	-0.1748E 04	37	-0.1767E 04
17	-0.1558E 04	38	-0.1571E 04	17	-0.2218E 04	38	-0.1767E 04
18	-0.1563E 04	39	-0.1570E 04	18	-0.2191E 04	39	-0.1767E 04
19	-0.1549E 04	40	-0.1571E 04	19	-0.1743E 04	40	-0.1768E 04
20	-0.1563E 04	41	-0.1571E 04	20	-0.2192E 04	41	-0.1768E 04
21	-0.1557E 04	42	-0.1570E 04	21	-0.2217E 04	42	-0.1767E 04

imperfection sensitivity of this structure could be made similar to Problem 2 discussed previously. One such analysis was performed, a + 0.5 in. "assembly error" was made in the z coordinate of node 2. Buckling occurred between loads 346 and 415 lb, a substantial drop from the 551 - 620 lb of the "ideal" case. The curves of Figure 13 tend to indicate the degree of nonlinearity of the

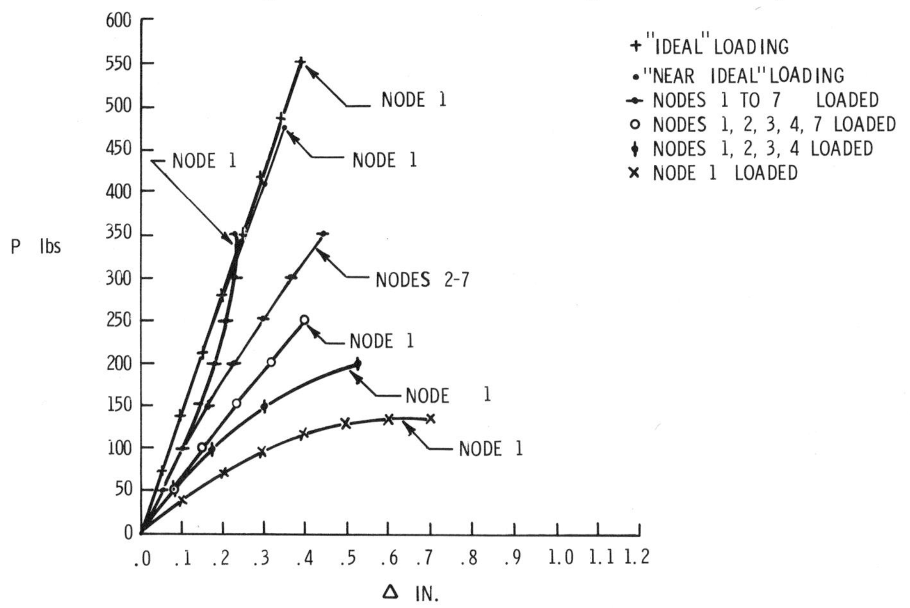

Figure 13. Load-deflection curves for various loading conditions of STR II. For each curve the next load increase caused buckling.

pre-buckling behavior for the various loading conditions discussed above. Selected nodal displacements are plotted against nodal load levels. The closer the loading is to the "ideal," the closer the pre-buckling behavior is to the linear, and the higher the buckling load.

STR-III

Three loading conditions were considered for this structure. First the snap-through load of node 1 was determined. It was found to be 128 lb, only slightly less than the 130 lb for node 1 of STR-II. The nodes on one half of the structure [Figure 14(a)] were loaded next in 50-lb increments. Buckling occurred between 250 and 300 lb, the same as for STR-II. Two consecutive buckled shapes are shown in Figure 14(a). Tables 7 and 8 give the nodal deflections and member forces, respectively, for the two load levels prior to and after buckling. Finally, all nodes were loaded in 50-lb increments. Buckling occurred between 350 and 400 lb, again the same as for STR-II. Tables 9 and 10 give the nodal deflections and member forces, respectively, for both load levels.

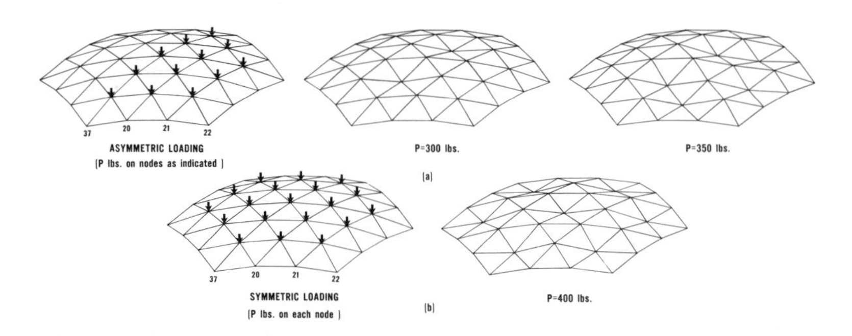

Figure 14. Buckled equilibrium shapes of STR-III.

All of the above analyses were made with member buckling suppressed. The buckled stable configuration is, of course, again dependent on whether the high member loads can or cannot be sustained in practice by the truss elements. It is of some interest that the buckled shape of Figure 14(b) did not change by further collapse up to 950 lb per node, the highest load considered, thus exhibiting high post-buckling strength. This post buckling strength, however, is associated with a network of very high compressive bar forces making its utilization difficult, at least for slender bars. The curves of Figure 15, like the

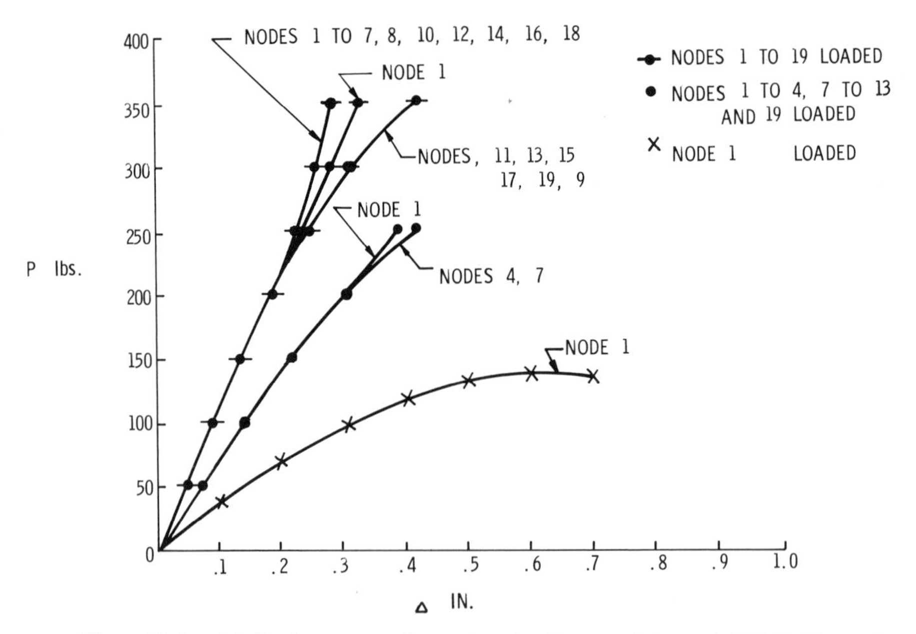

Figure 15. Load-deflection curves for various loading conditions of STR-II. For each curve the next load increase caused buckling.

Table 7. Nodal Displacements of STR-III. Asymmetric Loading (Fig. 14a).

	PRE-BUCKLING P = 250LBS.				POST-BUCKLING P = 300LBS.		
	X	Y	Z		X	Y	Z
1	-0.4614E-01	0.2661E-01	-0.3932E-00	1	-0.6300E-01	0.3640E-01	-0.4721E-00
2	-0.3775E-01	0.4013E-01	-0.3100E-00	2	-0.9830E-01	0.8638E-01	-0.6777E 00
3	-0.5350E-01	0.1259E-01	-0.3095E-00	3	-0.1239E-00	0.4201E-01	-0.6776E 00
4	-0.6301E-01	-0.1545E-01	-0.4216E-00	4	-0.4127E-00	-0.4517E-00	-0.4935E 01
5	-0.5149E-01	0.4214E-01	0.1787E-00	5	-0.3228E-01	0.2448E-01	-0.4229E-01
6	-0.6217E-01	0.2353E-01	0.1783E-00	6	-0.3729E-01	0.1574E-01	-0.4251E-01
7	-0.1828E-01	0.6187E-01	-0.4168E-00	7	0.1849E-00	0.5834E 00	-0.4936E 01
8	-0.1381E-01	0.5901E-01	-0.3000E-00	8	-0.2437E-01	0.2020E-00	-0.9637E 00
9	-0.3875E-01	0.5907E-01	-0.3161E-00	9	-0.2494E-01	0.3745E-01	-0.2234E-00
10	-0.5305E-01	0.3066E-01	-0.2973E-00	10	-0.1146E-00	0.6619E-01	-0.6043E 00
11	-0.7005E-01	0.4021E-02	-0.3142E-00	11	-0.4501E-01	0.2902E-02	-0.2239E-00
12	-0.5832E-01	-0.1766E-01	-0.3020E-00	12	-0.1870E-00	-0.7981E-01	-0.9633E 00
13	-0.5349E-01	-0.4793E-01	-0.3223E-00	13	-0.8899E-01	-0.9053E-01	-0.4967E-00
14	-0.2858E-01	0.5309E-01	0.1800E-00	14	-0.2608E-01	-0.2608E-02	-0.2054E-01
15	-0.2499E-01	0.2829E-01	0.1032E-00	15	-0.3816E-01	0.4501E-01	0.1639E-00
16	-0.3884E-01	0.2244E-01	0.1288E-00	16	-0.4610E-01	0.2661E-01	0.1500E-00
17	-0.3709E-01	0.7507E-02	0.1035E-00	17	-0.5818E-01	0.1054E-01	0.1643E-00
18	-0.6013E-01	-0.1600E-02	0.1795E-00	18	-0.1068E-01	0.2397E-01	-0.2098E-01
19	0.1558E-01	0.7164E-01	-0.3282E-00	19	0.3401E-01	0.1224E-00	-0.4973E-00

Table 8. Member Forces in STR-III. Asymmetric Loading (Fig. 14a).

PRE-BUCKLING P = 250LBS.				POST-BUCKLING P = 300LBS.			
1	-0.4050E 03	37	-0.7968E 03	1	-0.1598E 04	37	-0.1185E 04
2	-0.1437E 04	38	-0.9204E 03	2	0.1290E 03	38	-0.2762E 04
3	-0.4048E 03	39	0.3162E 03	3	-0.1600E 04	39	-0.1006E 04
4	-0.4090E 03	40	0.3526E 02	4	-0.1600E 04	40	0.5177E 02
5	-0.1433E 04	41	0.4514E 03	5	0.1298E 03	41	0.6334E 03
6	-0.4077E 03	42	0.4495E 03	6	-0.1598E 04	42	0.6328E 03
7	-0.1820E 03	43	-0.1513E 03	7	0.8252E 03	43	-0.2135E 03
8	-0.6445E 03	44	0.2305E 03	8	0.8125E 03	44	0.3449E 03
9	-0.1059E 04	45	-0.4958E 03	9	-0.1712E 04	45	-0.4967E 03
10	-0.6430E 03	46	0.4034E 03	10	0.8127E 03	46	0.3240E 03
11	-0.1766E 03	47	-0.1377E 04	11	0.8252E 03	47	-0.1712E 04
12	0.7142E 03	48	-0.8944E 03	12	0.3356E 03	48	-0.7894E 03
13	-0.3739E 03	49	-0.1249E 03	13	-0.5044E 03	49	0.6991E 02
14	0.6823E 03	50	-0.8704E 03	14	-0.7550E 03	50	-0.1772E 04
15	-0.7739E 03	51	-0.4332E 03	15	0.5800E 03	51	-0.1000E 04
16	-0.1327E 04	52	-0.7891E 03	16	0.3128E 03	52	-0.6540E 03
17	-0.3070E 03	53	-0.7662E 03	17	0.5851E 03	53	-0.7064E 03
18	-0.9715E 03	54	-0.5639E 03	18	-0.3352E 04	54	-0.4915E 03
19	-0.5073E 03	55	-0.5650E 03	19	0.3135E 02	55	-0.8314E 03
20	-0.5551E 03	56	-0.5641E 03	20	-0.8214E 03	56	-0.8313E 03
21	-0.5531E 03	57	-0.5662E 03	21	-0.8226E 03	57	-0.4913E 03
22	-0.5110E 03	58	-0.7681E 03	22	0.3350E 02	58	-0.7054E 03
23	-0.9713E 03	59	-0.7916E 03	23	-0.3352E 04	59	-0.6524E 03
24	-0.2956E 03	60	-0.4285E 03	24	0.5857E 03	60	-0.1000E 04
25	-0.1332E 04	61	-0.8642E 03	25	0.3126E 03	61	-0.1772E 04
26	-0.7662E 03	62	-0.1333E 03	26	0.5806E 03	62	0.6888E 02
27	0.6766E 03	63	-0.8998E 03	27	-0.7563E 03	63	-0.7892E 03
28	-0.3727E 03	64	-0.1384E 04	28	-0.5022E 03	64	-0.1711E 04
29	-0.2191E 03	65	0.4065E 03	29	-0.2805E 03	65	0.3245E 03
30	-0.2186E 03	66	-0.4959E 03	30	-0.2796E 03	66	-0.4971E 03
31	0.3866E 02	67	0.2307E 03	31	0.5215E 02	67	0.3455E 03
32	0.3242E 03	68	-0.1515E 03	32	-0.1005E 04	68	-0.2133E 03
33	-0.9106E 03	69	-0.1884E 03	33	-0.2761E 04	69	-0.2435E 03
34	-0.7949E 03	70	-0.2235E 03	34	-0.1185E 04	70	-0.2863E 03
35	-0.1016E 04	71	-0.2239E 03	35	-0.1344E 04	71	-0.2860E 03
36	-0.1018E 04	72	-0.1886E 03	36	-0.1344E 04	72	-0.2441E 03

Table 9. Nodal Displacements of STR-III. All Nodes Loaded (Fig. 14b).

	PRE-BUCKLING P = 350LBS.				POST-BUCKLING P = 400LBS.		
	X	Y	Z		X	Y	Z
1	0.1796E-04	0.2810E-04	-0.3246E-00	1	0.3069E-04	0.1176E-02	-0.5853E 01
2	-0.1426E-01	0.2469E-01	-0.2946E-00	2	-0.4320E-01	0.7521E-01	-0.1009E 01
3	-0.2847E-01	0.1800E-04	-0.2948E-00	3	-0.8600E-01	0.3406E-03	-0.1004E 01
4	-0.1419E-01	-0.2459E-01	-0.2942E-00	4	-0.4283E-01	-0.7389E-01	-0.1002E 01
5	0.1428E-01	-0.2467E-01	-0.2950E-00	5	0.4289E-01	-0.7391E-01	-0.1002E 01
6	0.2842E-01	0.2051E-04	-0.2940E-00	6	0.8606E-01	0.3669E-03	-0.1004E 01
7	0.1425E-01	0.2473E-01	-0.2948E-00	7	0.4322E-01	0.7524E-01	-0.1008E 01
8	0.1022E-04	0.4702E-01	-0.2905E-00	8	-0.1788E-04	0.1341E-00	-0.9014E 00
9	-0.4260E-01	0.7376E-01	-0.4202E-00	9	-0.5383E 00	0.9326E 00	-0.5031E 01
10	-0.4095E-01	0.2371E-01	-0.2921E-00	10	-0.1158E-00	0.6698E-01	-0.8999E 00
11	-0.8437E-01	0.6120E-05	-0.4171E-00	11	-0.1076E 01	0.1543E-03	-0.5029E 01
12	-0.4113E-01	-0.2376E-01	-0.2931E-00	12	-0.1155E-00	-0.6664E-01	-0.8985E 00
13	-0.4225E-01	-0.7315E-01	-0.4175E-00	13	-0.5383E 00	-0.9324E 00	-0.5028E 01
14	0.1086E-04	-0.4741E-01	-0.2926E-00	14	0.1187E-04	-0.1333E-00	-0.8977E 00
15	0.4225E-01	-0.7314E-01	-0.4175E-00	15	0.5383E 00	-0.9323E 00	-0.5027E 01
16	0.4103E-01	-0.2364E-01	-0.2923E-00	16	0.1155E-00	-0.6669E-01	-0.8980E 00
17	0.8501E-01	0.3459E-04	-0.4194E-00	17	0.1076E 01	0.1282E-03	-0.5028E 01
18	0.4075E-01	0.2358E-01	-0.2909E-00	18	0.1158E-00	0.6686E-01	-0.8996E 00
19	0.4271E-01	0.7402E-01	-0.4214E-00	19	0.5383E 00	0.9327E 00	-0.5031E 01

Table 10. Member Forces in STR-III. All Nodes Loaded (Fig. 14b).

PRE-BUCKLING P = 350LBS.				POST-BUCKLING P = 400LBS.			
1	-0.1006E 04	37	-0.1147E 04	1	0.6484E 03	37	0.3453E 03
2	-0.1004E 04	38	-0.1147E 04	2	0.6489E 03	38	0.3451E 03
3	-0.1005E 04	39	-0.1148E 04	3	0.6535E 03	39	0.3477E 03
4	-0.1006E 04	40	-0.1146E 04	4	0.6486E 03	40	0.3448E 03
5	-0.1006E 04	41	-0.1149E 04	5	0.6488E 03	41	0.3484E 03
6	-0.1004E 04	42	-0.1145E 04	6	0.6528E 03	42	0.3457E 03
7	-0.9504E 03	43	-0.1258E 04	7	-0.2869E 04	43	-0.1010E 04
8	-0.9498E 03	44	-0.1015E 04	8	-0.2874E 04	44	0.1792E 04
9	-0.9505E 03	45	-0.9095E 03	9	-0.2887E 04	45	-0.2841E 04
10	-0.9518E 03	46	-0.9098E 03	10	-0.2893E 04	46	-0.2841E 04
11	-0.9510E 03	47	-0.1015E 04	11	-0.2888E 04	47	0.1793E 04
12	-0.9499E 03	48	-0.1259E 04	12	-0.2876E 04	48	-0.1010E 04
13	-0.1089E 04	49	-0.1014E 04	13	0.5911E 03	49	0.1794E 04
14	-0.8920E 03	50	-0.9100E 03	14	-0.2811E 04	50	-0.2845E 04
15	-0.8900E 03	51	-0.9103E 03	15	-0.2810E 04	51	-0.2842E 04
16	-0.1088E 04	52	-0.1014E 04	16	0.5915E 03	52	0.1790E 04
17	-0.8912E 03	53	-0.1258E 04	17	-0.2803E 04	53	-0.1012E 04
18	-0.8928E 03	54	-0.1013E 04	18	-0.2815E 04	54	0.1798E 04
19	-0.1088E 04	55	-0.9088E 03	19	0.5924E 03	55	-0.2844E 04
20	-0.8912E 03	56	-0.9070E 03	20	-0.2804E 04	56	-0.2840E 04
21	-0.8883E 03	57	-0.1017E 04	21	-0.2821E 04	57	0.1792E 04
22	-0.1093E 04	58	-0.1261E 04	22	0.5885E 03	58	-0.1017E 04
23	-0.8873E 03	59	-0.1018E 04	23	-0.2814E 04	59	0.1798E 04
24	-0.8864E 03	60	-0.9046E 03	24	-0.2814E 04	60	-0.2843E 04
25	-0.1095E 04	61	-0.9043E 03	25	0.5890E 03	61	-0.2844E 04
26	-0.8875E 03	62	-0.1019E 04	26	-0.2821E 04	62	0.1801E 04
27	-0.8868E 03	63	-0.1263E 04	27	-0.2801E 04	63	-0.1015E 04
28	-0.1092E 04	64	-0.1020E 04	28	0.5912E 03	64	0.1792E 04
29	-0.8887E 03	65	-0.9049E 03	29	-0.2818E 04	65	-0.2842E 04
30	-0.8908E 03	66	-0.9063E 03	30	-0.2801E 04	66	-0.2841E 04
31	-0.1145E 04	67	-0.1015E 04	31	0.3472E 03	67	0.1797E 04
32	-0.1146E 04	68	-0.1259E 04	32	0.3479E 03	68	-0.1012E 04
33	-0.1145E 04	69	-0.1017E 04	33	0.3416E 03	69	0.1790E 04
34	-0.1145E 04	70	-0.9082E 03	34	0.3472E 03	70	-0.2840E 04
35	-0.1144E 04	71	-0.9094E 03	35	0.3410E 03	71	-0.2840E 04
36	-0.1148E 04	72	-0.1013E 04	36	0.3522E 03	72	0.1793E 04

curves of Figure 13, show the degree of nonlinearity of the pre-buckling behavior for the various loading conditions.

STR-IV

For reasons of economy this structure was analyzed only for one loading condition, vertical loads on one-half of the structure as shown on Figure 16. Buckling occurred between 250 and 300 lb, the same as for STR-II and STR-III under this loading condition. The 50-lb load increment is admittedly coarse

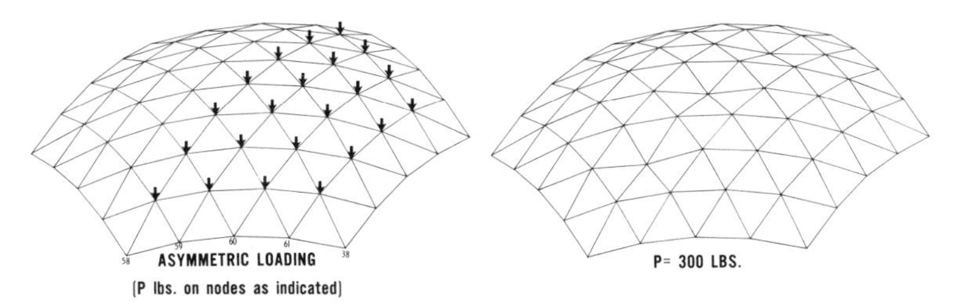

Figure 16. Buckled equilibrium shapes of STR-IV.

in determing the buckling load but again it was dictated by economy of computer running times. The size effect, if any, is apparently less than the 50-lb increment. The buckled equilibrium mode shape for 300-lb load level is shown in Figure 16. Again, as for STR-II and STR-III, nodes along the loaded diameter snapped through first. Figure 17 contains some explanation for this behavior. The angle changes of the bars connecting to node 1 is the largest, as shown by the plots on Figure 17(b), leading up to the snap-through of node 1. There are three curves plotted on Figure 17(a) for the 250-lb load level. The labeling 03, 04, 05 refers to convergence criteria in terms of function exploration step sizes in the order of lessening requirements. It is interesting to note that as the accuracy requirement is relaxed, the resulting solution starts to resemble the buckled mode shape. For very small exploration steps the neighborhood of the minimum point on the potential surface is very smooth and shows no sign of buckling. At larger distances the decrease of the potential along certain displacement degrees of freedom starts to develop and this is seen by the "fuzzy ball" solutions of the less stringent convergence criteria labeled 04 and 05. The lack of symmetry of the displacements is a sign of the deterioration of the solutions. Computer times are, however, considerably reduced for relaxed convergence criteria. This is particularly significant if post-buckled equilibrium configurations are sought since "buckling" takes four to eight times more iterations than a nonlinear pre-buckling solution. Tables 11 and 12 have been generated with convergence criteria relaxed to 04. Member buckling was suppressed again in order to obtain information on the general buckling of the configuration.

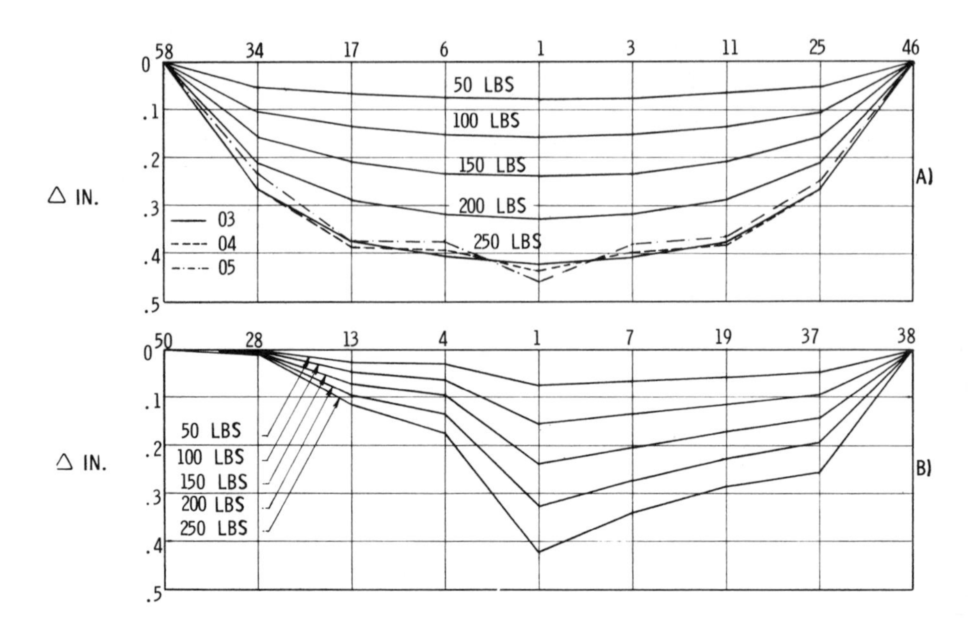

Figure 17. Vertical displacements of the indicated nodes of STR-IV.

4. Conclusions and Recommendations

This study has served to demonstrate the applicability of a new methodology of stability analysis to radome problems. While the applications presented were necessarily relatively simple, the finite element basis of idealization clearly enables generalization to more practical models.

One of the important analytical concepts employed herein was that of generalized element degrees of freedom. Such degrees of freedom take the form of amplification factors on assumed buckling mode shapes. This is closely analogous to the use of supplementary vibration mode shapes in formulating element representations for dynamic analyses.[15] These generalized mode shapes serve to reduce the total number of degrees of freedom required to obtain a given level of accuracy.

A formalization of notation has been evolved in Reference 16 which can be expected to facilitate generalization of the subject nonlinear analysis capability to accommodate more complex types of finite elements. Therein a systematic matrix notation amenable to data processing is achieved. This matrix notation also clarifies the interrelation among the direct energy minimization, direct matrix iterative and matrix incremental approaches to stability analysis.

Table 11. Nodal Displacements of STR-IV: Asymmetric Loading (Fig. 16a).

	PRE-BUCKLING P = 250LBS.				POST-BUCKLING P = 300LBS.		
	X	Y	Z		X	Y	Z
1	-0.8055E-04	0.7804E-01	-0.4352E-00	1	0.1794E-03	0.1579E-00	-0.4890E 01
2	-0.1671E-01	0.7727E-01	-0.3403E-00	2	-0.4722E-01	0.1546E-00	-0.7298E 00
3	-0.4146E-01	0.7275E-01	-0.3965E-00	3	-0.5739E-01	0.9263E-01	-0.4878E-00
4	0.1148E-01	0.9046E-01	0.1714E-00	4	-0.1133E-01	0.6748E-01	-0.3636E-01
5	-0.1158E-01	0.9045E-01	0.1714E-00	5	0.1151E-01	0.6742E-01	-0.3668E-01
6	0.4107E-01	0.7275E-01	-0.3937E-00	6	0.5762E-01	0.9269E-01	-0.4889E-00
7	0.1667E-01	0.7736E-01	-0.3408E-00	7	0.4741E-01	0.1546E-00	-0.7302E 00
8	-0.1109E-04	0.8144E-01	-0.3033E-00	8	0.7003E-04	0.8997E-01	-0.2938E-00
9	-0.2818E-01	0.7499E-01	-0.2836E-00	9	-0.4223E-01	0.1211E-00	-0.4305E-00
10	-0.4689E-01	0.6651E-01	-0.3091E-00	10	-0.8914E-01	0.1396E-00	-0.6255E 00
11	-0.8235E-01	0.5722E-01	-0.3812E-00	11	-0.1163E 01	0.1313E-00	-0.4815E 01
12	0.3338E-01	0.8029E-01	0.1712E-00	12	-0.9057E-02	0.4959E-01	-0.1036E-00
13	0.1448E-01	0.6337E-01	0.1165E-00	13	0.1442E-01	0.7173E-01	0.1228E-00
14	-0.4693E-04	0.6798E-01	0.1194E-00	14	0.2688E-04	0.9345E-01	0.1841E-00
15	-0.1461E-01	0.6352E-01	0.1171E-00	15	-0.1441E-01	0.7179E-01	0.1231E-00
16	-0.3342E-01	0.8036E-01	0.1714E-00	16	0.9235E-02	0.4962E-01	-0.1043E-00
17	0.8347E-01	0.5732E-01	-0.3864E-00	17	0.1164E 01	0.1314E-00	-0.4817E 01
18	0.4673E-01	0.6644E-01	-0.3082E-00	18	0.8927E-01	0.1396E-00	-0.6257E 00
19	0.2811E-01	0.7486E-01	-0.2826E-00	19	0.4245E-01	0.1215E-00	-0.4327E-00
20	0.1389E-01	0.8297E-01	-0.2655E-00	20	0.1961E-01	0.1215E-00	-0.3736E-00
21	-0.1391E-01	0.8344E-01	-0.2671E-00	21	-0.1962E-01	0.1215E-00	-0.3733E-00
22	-0.3860E-01	0.7565E-01	-0.2474E-00	22	-0.4926E-01	0.1012E-00	-0.3221E-00
23	-0.5298E-01	0.6245E-01	-0.2600E-00	23	-0.4566E-01	0.5950E-01	-0.2381E-00
24	-0.7202E-01	0.4694E-01	-0.2775E-00	24	-0.2304E-00	0.1199E-00	-0.7971E 00
25	-0.8639E-01	0.2595E-01	-0.2591E-00	25	-0.1284E-00	0.3429E-01	-0.3458E-00
26	0.5542E-01	0.5776E-01	0.1777E-00	26	-0.6643E-02	0.2842E-01	0.7153E-02
27	0.2561E-01	0.4732E-01	0.1071E-00	27	0.4010E-01	0.6612E-01	0.1592E-00
28	0.1460E-01	0.3829E-01	0.8003E-01	28	0.1694E-01	0.4573E-01	0.9435E-01
29	0.6616E-02	0.5224E-01	0.1096E-00	29	0.8250E-02	0.6415E-01	0.1351E-00
30	-0.6683E-02	0.5238E-01	0.1101E-00	30	-0.8179E-02	0.6422E-01	0.1351E-00
31	-0.1452E-01	0.3817E-01	0.7960E-01	31	-0.1729E-01	0.4576E-01	0.9547E-01
32	-0.2572E-01	0.4735E-01	0.1074E-00	32	-0.4005E-01	0.6658E-01	0.1597E-00
33	-0.5579E-01	0.5784E-01	0.1784E-00	33	0.6772E-02	0.2807E-01	0.6606E-02
34	0.8457E-01	0.2601E-01	-0.2548E-00	34	0.1287E-00	0.3469E-01	-0.3464E-00
35	0.7194E-01	0.4699E-01	-0.2776E-00	35	0.2313E-00	0.1198E-00	-0.7995E 00
36	0.5305E-01	0.6270E-01	-0.2603E-00	36	0.4514E-01	0.5930E-01	-0.2360E-00
37	0.3941E-01	0.7708E-01	-0.2515E-00	37	0.4920E-01	0.1006E-00	-0.3210E-00

Table 12. Member Forces in STR-IV. Asymmetric Loading (Fig. 16a).

PRE-BUCKLING P = 250LBS.				POST-BUCKLING P = 300LBS.			
1	-0.4377E 03	67	-0.1634E 03	1	0.7353E 03	67	0.7188E 03
2	-0.4367E 03	68	-0.1262E 04	2	0.7350E 03	68	0.2417E 03
3	-0.1452E 04	69	-0.9044E 03	3	0.1890E 03	69	0.4461E 03
4	-0.4352E 03	70	0.6493E 03	4	0.7325E 03	70	-0.7681E 03
5	-0.4389E 03	71	-0.5279E 03	5	0.7342E 03	71	-0.7520E 03
6	-0.1450E 04	72	0.4151E 03	6	0.1897E 03	72	0.6101E 03
7	0.7680E 03	73	0.3825E 03	7	-0.7625E 03	73	0.4624E 03
8	-0.1902E 03	74	0.4430E 03	8	-0.1382E 04	74	0.5471E 03
9	-0.5415E 03	75	0.3840E 03	9	-0.1932E 04	75	0.4531E 03
10	-0.1114E 04	76	0.3560E 02	10	-0.3169E 04	76	0.6912E 02
11	-0.5386E 03	77	0.1987E 03	11	-0.1931E 04	77	0.3238E 03
12	-0.1874E 03	78	0.4811E 03	12	-0.1386E 04	78	-0.6455E 03
13	-0.3875E 03	79	-0.9089E 03	13	-0.3726E 03	79	-0.2499E 04
14	-0.2443E 03	80	-0.7657E 03	14	-0.2868E 03	80	-0.1143E 04
15	-0.2447E 03	81	-0.6633E 03	15	-0.2850E 03	81	-0.7798E 03
16	-0.3888E 03	82	-0.8917E 03	16	-0.3749E 03	82	-0.1183E 04
17	0.7196E 03	83	-0.9282E 03	17	0.3411E 03	83	-0.1310E 04
18	-0.6031E 03	84	-0.8962E 03	18	-0.1861E 04	84	-0.1187E 04
19	-0.1412E 04	85	-0.6693E 03	19	0.1182E 03	85	-0.7563E 03
20	-0.2504E 03	86	-0.7689E 03	20	-0.1311E 04	86	-0.1159E 04
21	-0.1115E 04	87	-0.8978E 03	21	-0.1786E 04	87	-0.2483E 04
22	-0.4490E 03	88	0.4891E 03	22	-0.6699E 03	88	-0.6687E 03
23	-0.5381E 03	89	0.2016E 03	23	-0.5392E 03	89	0.3460E 03
24	-0.5372E 03	90	0.3771E 02	24	-0.5410E 03	90	0.4767E 02
25	-0.4512E 03	91	-0.1142E 03	25	-0.6727E 03	91	-0.1339E 03
26	-0.1117E 04	92	-0.1665E 03	26	-0.1784E 04	92	-0.2001E 03
27	-0.2447E 03	93	-0.1915E 03	27	-0.1312E 04	93	-0.2308E 03
28	-0.1414E 04	94	-0.2127E 03	28	0.1166E 03	94	-0.2574E 03
29	-0.5978E 03	95	-0.2142E 03	29	-0.1862E 04	95	-0.2548E 03
30	0.7171E 03	96	-0.1918E 03	30	0.3424E 03	96	-0.2305E 03

31	0.4881E 03	97	-0.1649E 03	31	0.6035E 03	97	-0.2139E 03
32	0.4881E 03	98	-0.1128E 03	32	0.6026E 03	98	-0.1425E 03
33	-0.6118E 02	99	0.1871E 03	33	-0.1778E 03	99	0.2301E 03
34	0.2469E 02	100	-0.3729E 03	34	0.9430E 03	100	-0.4731E 03
35	-0.7090E 03	101	0.2558E 03	35	0.6656E 03	101	0.3236E 03
36	-0.6058E 03	102	-0.6375E 03	36	-0.6150E 03	102	-0.5772E 03
37	-0.9762E 03	103	0.3724E 03	37	-0.1121E 04	103	0.2729E 03
38	-0.9767E 03	104	-0.1388E 04	38	-0.1119E 04	104	-0.1618E 04
39	-0.6028E 03	105	-0.8240E 03	39	-0.6224E 03	105	-0.7168E 03
40	-0.7126E 03	106	0.2168E 02	40	0.6732E 03	106	0.2453E 03
41	0.1904E 02	107	-0.9377E 03	41	0.9437E 03	107	-0.1782E 04
42	-0.6134E 02	108	-0.2894E 03	42	-0.1815E 03	108	-0.6308E 03
43	-0.3165E 03	109	-0.8461E 03	43	-0.3819E 03	109	-0.8643E 03
44	-0.1416E 03	110	-0.4982E 03	44	-0.1817E 03	110	-0.4420E 03
45	-0.2490E 03	111	-0.7366E 03	45	-0.2901E 03	111	-0.9762E 03
46	-0.2491E 03	112	-0.7241E 03	46	-0.2915E 03	112	-0.8778E 03
47	-0.1431E 03	113	-0.4977E 03	47	-0.1846E 03	113	-0.5433E 03
48	-0.3156E 03	114	-0.6142E 03	48	-0.3788E 03	114	-0.7091E 03
49	0.4145E 03	115	-0.5372E 03	49	0.6129E 03	115	-0.6005E 03
50	-0.5261E 03	116	-0.5345E 03	50	-0.7568E 03	116	-0.6030E 03
51	0.6439E 03	117	-0.6093E 03	51	-0.7687E 03	117	-0.7122E 03
52	-0.8964E 03	118	-0.5032E 03	52	0.4525E 03	118	-0.5506E 03
53	-0.1263E 04	119	-0.7286E 03	53	0.2388E 03	119	-0.8794E 03
54	-0.1540E 03	120	-0.7425E 03	54	0.7172E 03	120	-0.9691E 03
55	-0.1034E 04	121	-0.4946E 03	55	-0.3123E 04	121	-0.4325E 03
56	-0.3611E 03	122	-0.8471E 03	56	-0.2217E 02	122	-0.8610E 03
57	-0.9406E 03	123	-0.2900E 03	57	-0.1231E 04	123	-0.6407E 03
58	-0.5156E 03	124	-0.9409E 03	58	-0.5951E 03	124	-0.1782E 04
59	-0.5644E 03	125	0.2792E 02	59	-0.7152E 03	125	0.2555E 03
60	-0.5507E 03	126	-0.8217E 03	60	-0.5574E 03	126	-0.7173E 03
61	-0.5500E 03	127	-0.1385E 04	61	-0.5529E 03	127	-0.1630E 04
62	-0.5573E 03	128	0.3703E 03	62	-0.7175E 03	128	0.2706E 03
63	-0.5222E 03	129	-0.6378E 03	63	-0.5934E 03	129	-0.5714E 03
64	-0.9368E 03	130	0.2560E 03	64	-0.1231E 04	130	0.3303E 03
65	-0.3638E 03	131	-0.3714E 03	65	-0.2255E 02	131	-0.4781E 03
66	-0.1029E 04	132	0.1863E 03	66	-0.3128E 04	132	0.2282E 03

The solution of nonlinear problems necessarily involves a great deal more computation than the solution of linear problems. However, the increasing attention being given to algorithms for solving nonlinear problems can be expected to improve computational efficiencies significantly. In the area of minimization procedures alone, a number of advances have been reported recently.[17,18] The utilization of these advanced procedures will improve the efficiency of present day analyses and will enable the analysis of large scale structures not possible with existing capabilities. Progress in this regard can be expected to benefit from further advances in computer hardware and software as well.

The example problems considered in this study were designed to exhibit behavior representative of the general radome problem. Thus, the results obtained identify relevant stability phenomena and establish feasible goals regarding behavior prediction in actual design situations. Further insight can be gained by consideration of additional examples. For example, buckling associated with joint rotation about an axis normal to the shell surface warrants examination. The computer program employed in the subject study is suitable for such investigations. Utilization in actual design studies is thought to be an ambitious, though feasible, goal. New advancements in minimization procedures and a method for automatic rescaling of the variables to improve conditioning of the total potential expression are needed before application to structural systems with large number of behavior variables can become routine.

References

1. F. K. Bogner, R. H. Mallett, M. D. Minich, and L. A. Schmit, Jr., "Development and Evaluation of Energy Search Methods of Nonlinear Structural Analysis," AFFDL-TR-65-113 (October 1965).

2. R. L. Anderson, "Recent Advances in Finding Best Operating Conditions," Institute of Statistics, North Carolina State College.

3. G. E. P. Box, "The Exploration and Exploitation of Response Surfaces: Some General Considerations and Examples," *Biometrics* (March 1954).

4. S. H. Brooks, "A Discussion of Random Methods for Seeking Maxima," *Operations Res.* **6**, 244 (1958).

5. N. S. Bromberg, "Maximization and Minimization of Complicated Multivariable Functions," *Commun. Electronics* **58**, 725 (1962).

6. A. Cauchy, "Methode Generale Pour la Resolution des Systems d'Equations Simultanees," *Compt. Rend.* **25**, 536 (1847).

7. R. Fletcher and M. J. D. Powell, "A Rapidly Convergent Descent Method for Minimization," *Computer J.* **6** (No. 2), 163 (July 1963).

8. W. C. Davidon, "Variable Metric Method for Minimization," AEC Research and Development Report, ANL-5990 (1959).

9. G. W. Stewart, III, "A Modification of Davidon's Minimization Method to Accept Difference Approximations of Derivatives," *J. Assoc. Comp. Machinery* **14** (No.1), 72 (January 1967).

10. J. L. Synge, "A Geometrical Interpretation of the Relaxation Method," *Quart. J. Appl. Math.* **1,** 87 (1943).

11. G. Temple, "The General Theory of Relaxation Methods Applied to Linear Systems," *Proc. Roy. Soc. (London).* **A169,** 476 (1939).

12. R. Hookes and T. A. Jeeves, "Direct Search Solution of Numerical and Statistical Problems," Westinghouse Research Laboratories, Pittsburgh, Pennsylvania (1960).

13. R. L. Fox, "An Integrated Approach to Engineering Synthesis and Analysis." Ph. D. thesis, Case Institute of Technology (1965).

14. R. H. Mallett and L. Berke, "Automated Method for the Large Deflection and Instability Analysis of Three-Dimensional Truss and Frame Assemblies," AFFDL-TR-66-102, AD 810 499.

15. R. H. Mallett, S. Jordan, and J. R. Batt, "Manageable Finite Element Models for Large Scale Structural Dynamics Analyses," Bell Aerosystems Technical Paper No. 2500-941020 (September 1967).

16. R. H. Mallett and P. V. Marcal, "Consistent Matrices and Computational Procedures for Nonlinear Pre- and Post-Buckling Analyses," Bell Aerosystems Technical Paper No. 2500-941019 (August, 1967).

17. R. H. Mallett and L. A. Schmit, Jr., "Nonlinear Structural Analysis by Energy Search," *ASCE J. Structures Div.* **93** (No. ST3) (June 1967).

18. R. L. Fox and E. L. Stanton, "Developments in Structural Analysis by Direct Energy Minimization," Case Western Reserve University, Cleveland, Ohio, Technical Paper (August 1967).

J. Connor, Jr.
Department of Civil Engineering,
Massachusetts Institute of Technology,
Cambridge, Massachusetts

ANALYSIS OF GEOMETRICALLY NONLINEAR FRAMED STRUCTURES

1. Introduction

The analysis of geometrically nonlinear framed structures has received considerable attention during the last decade, due primarily to the ability to solve such problems on large-scale digital computers. Livesley,[1] in 1956, developed a program for analyzing plane frames comprised of prismatic linear elastic members and subjected to in-plane loads applied only at the joints, In this study, he neglected the nonlinear effect of flexure and chord rotation on the member end shortening but did include the nonlinear effect of chord rotation on the transverse end actions. His applications were restricted to bifurcation buckling, i.e., where the displacements defining the prebuckled configuration are negligible. Renton[2] extended Livesley's formulation for the three-dimensional case and included torsion-flexure coupling. His applications were also restricted to bifurcation buckling.

Saafan[3] presented a nonlinear formulation for the planar case which includes flexural end shortening and does not place any restriction on the magnitude of the chord rotation. Allowing for an arbitrary chord rotation complicates the formulation considerably and is debatable since the "small" rotation theory is valid for a chord rotation less than approximately 15°. His solution procedure involves iteration on the displacements but is described from a physical point of view so that it is difficult to correlate it with standard iterative procedures. Saafan's examples do illustrate an important point, namely, that the end shortening due to flexure may be of the same order of magnitude as the linear extensional term. This has also been shown by Williams[4] and by Merchant and Brotton.[5]

Argyris[6] employed an incremental technique to determine the approximate nonlinear load-deflection behavior of rigid-jointed space frames. He assumes the behavior is linear within an increment and uses the tangent stiffness matrix evaluated at the initial position, i.e., the position prior to the application of the load increment, to determine the incremental displacements. In forming the tangent stiffness matrix, he neglects the effect of axial force on the member flexibility. This assumption corresponds to neglecting the nonlinear flexural rotation effect and is reasonable only when the axial force is small in comparison to the Euler load.[3]

Livesley[7] described the application of the Newton - Raphson procedure to nonlinear structures. The discussion is general and no equations are presented for framed structures. However, he does illustrate the analysis of a guyed tower.

Millar *et al.*,[8] in a recent paper, described a nonlinear formulation for plane frames which is essentially the same as Saafan's except that their solution procedure is more straightforward. Basically, it involves applying functional iteration to the equilibrium equations with the nonlinear end shortening terms combined with the applied external loads. They suggest three-point interpolation formulas to accelerate convergence but do not present any comparison studies.

In another recent paper, Johnson and Brotton[9] discuss the analysis of rigid-jointed space frames. They neglect the effect of axial force on the member flexibility, i.e., they work with linear force-displacement relations referred to the deformed chord. This assumption is also introduced by Argyris.[6] Their solution procedure is basically the Newton - Raphson method. However, in forming the tangent stiffness matrix, they neglect terms due to the incremental chord rotation. What effect this has on the convergence rate remains to be studied. One immediate consequence is that the formulation cannot be specialized for bifurcation buckling.

In this paper, we present a consistent nonlinear formulation for a rigid-jointed space frame comprised of prismatic linear elastic members. The derivation is restricted to the "small" rotation case, i.e., where the squares of the rotation angles are negligible with respect to unity. Torsion-flexure coupling is neglected to simplify the formulation. The appropriate equations for solution by successive substitution and Newton - Raphson iteration are developed. A linearized incremental solution technique is also described. Finally, comparison studies that illustrate the influence of load increment and convergence criteria on the total computational effort and examples of solutions for a reticulated radome-usage type dome are presented.

2. Member Force-displacement Relations

2.1. Planar deformation

Figure 1 shows the initial and deformed positions of the member. The undeformed centroidal axis coincides with the X_1 direction and X_2 is a principal inertia direction. We work with displacement quantities (u_1, u_2, ω_3), distributed external force (b_2), and end actions $(\bar{F}_1, \bar{F}_2, \bar{M}_3)$ referred to the initial $(X_1 - X_2 - X_3)$ member frame. The rotation of the chord is denoted by ρ_3 and is related to the end displacements by

$$\rho_3 = (1/L)(u_{B2} - u_{A2}). \tag{1}$$

Figure 2 shows the initial and deformed positions of a differential element. The deformed unit vectors are ν_1, ν_2 and y_2 is the normal coordinate. We neglect transverse shear deformation and transverse extensional strain, i.e., we take $y_2 = x_2$ and ν_1, ν_2 to be orthogonal.

We let u_1^*, u_2^* represent the translations of a point at x_1, x_2 and u_1, u_2 the

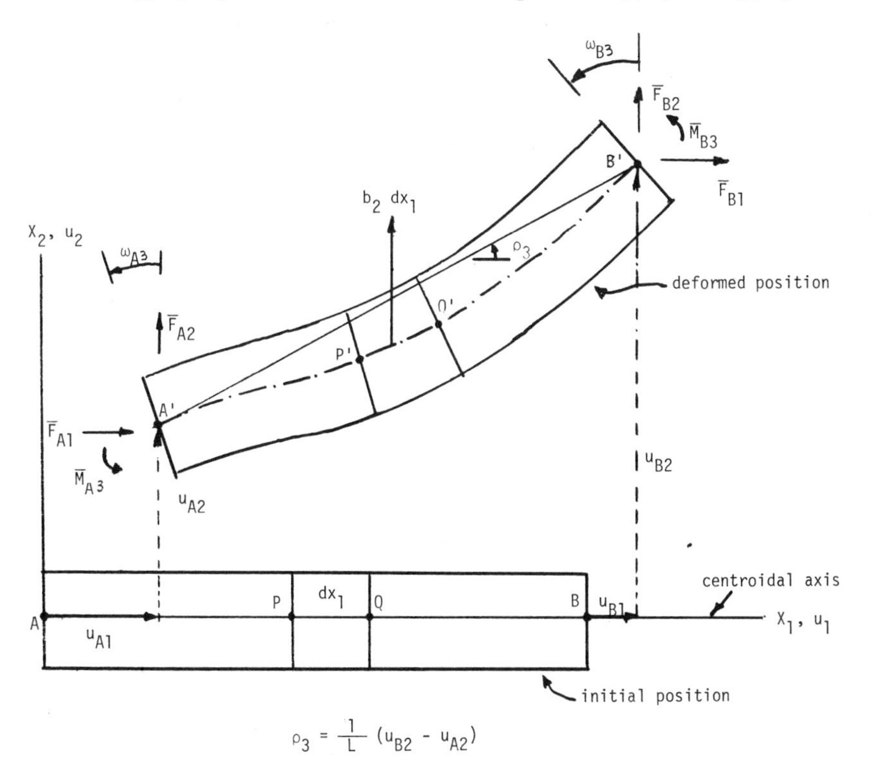

Figure 1.

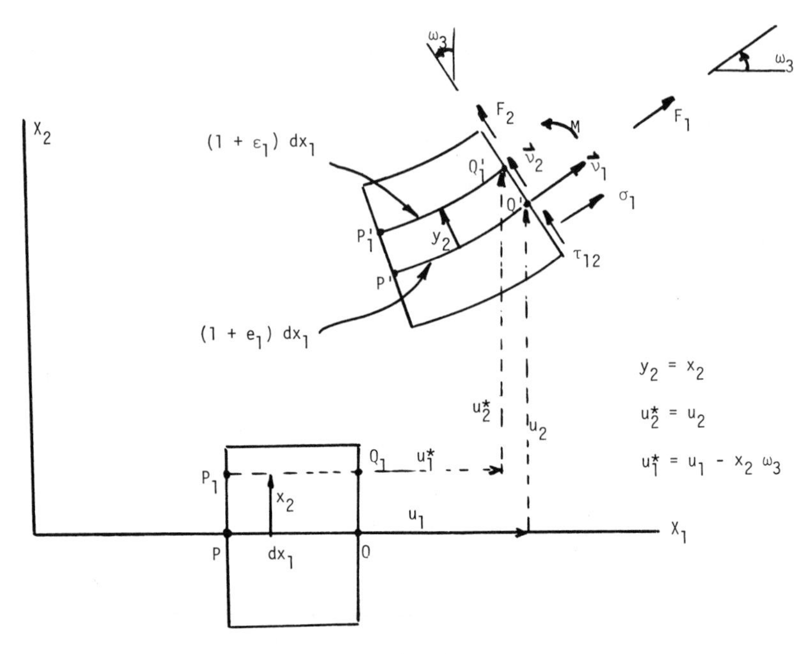

Figure 2.

translations of the centroid. For small rotation ($\omega^2 \ll 1$), they are related by

$$u_1^* \approx u_1 - x_2\omega_3,$$
$$u_2^* \approx u_2. \tag{2}$$

Also, the expression for the extensional strain ϵ_1 reduces to

$$\epsilon_1 = e_1 - x_2 k_3, \tag{3}$$

Where e_1, k_3 are one-dimensional stretching and bending measures defined by

$$e_1 = (du_1/dx_1) + \tfrac{1}{2}\,\omega_3^2,$$
$$k_3 = d^2u_2/dx_1^2, \tag{4}$$
$$\omega_3 = du_2/dx_1.$$

We consider the material to be linear elastic and the initial strain to vary linearly over the cross section. The tangential stress resultant and moment are related to the displacements by

$$F_1 = AE\,(e_1 - e_1^0), \tag{5}$$

$$M_3 = EI_3\,(k_3 - k_3^0), \tag{6}$$

where A is the cross-sectional area, E is Young's modulus, I_3 is the cross sectional moment of inertia about the X_3 axis and e_1^0, k_3^0 are the initial deformation measures.

The equilibrium equations consistent with the geometric approximation, $\omega_3^2 \ll 1$, are obtained by applying the principle of virtual displacements. We omit the details since the procedure is straightforward and list only the final equations:

$0 \leqslant x_1 \leqslant L$:

$$dF_1/dx_1 = 0, \tag{7}$$

$$\frac{d^2 M_3}{dx_1^2} - \frac{d}{dx_1}(F_1\,\omega_3) = b_2; \tag{8}$$

$x_1 = 0$:

$$\begin{aligned} \bar{F}_{A1} &= -\,F_1, \\ \bar{F}_{A2} &= -\,F_1\,\omega_3 + (dM_3/dx_1), \\ \bar{M}_{A3} &= -\,M_3; \end{aligned} \tag{9}$$

$x_1 = L$:

$$\begin{aligned} \bar{F}_{B1} &= +\,F_1, \\ \bar{F}_{B2} &= F_1\,\omega_3 - (dM_3/dx_1), \\ \bar{M}_{B3} &= M_3. \end{aligned} \tag{10}$$

Note that the term $F_1\,\omega_3$ is due to the nonlinear rotation term in the expression for e_1.

The rotation ω_3 is the sum of the rotation due to flexure and the rigid-body chord rotation ρ_3. We shall show later that it is reasonable to neglect nonlinear terms due to flexural rotation when the axial force is small in comparison to the Euler load. This assumption is introduced by taking $\omega_3^2 = \rho_3^2$ in the expression for e_1. The consistent equilibrium equations are obtained by setting $\omega_3 = \rho_3$ = const in Equations 8, 9, 10. Note that for this case, the transverse equilibrium equation has the *same* form as for the geometrically linear case.

Solving Equation 7 we have

$$F_1 = \text{const} = \bar{F}_{B1} = -\,\bar{F}_{A1}. \tag{11}$$

The relative axial displacement is obtained by integrating Equation 5. We write the result as

$$\bar{F}_{B1} = (AE/L)(u_{B1} - u_{A1}) + AE(\delta_2 - e_1^0), \tag{12}$$

where

$$\delta_2 = \frac{1}{2L}\int_0^L \omega_3^2 \, dx_1. \tag{13}$$

Note that $-L\delta_2$ is the relative axial displacement due to rotation. Also, δ_2 reduces to $\frac{1}{2}\rho_3^2$ when the nonlinear flexural rotation effect is neglected.

The governing equation for u_2 is obtained by substituting for M_3 in Equation 8. We assume the initial deformation is constant. Then,

$$\frac{d^4 u_2}{dx_1^{\ 4}} + \frac{\phi_2^2}{L^2}\frac{d^2 u_2}{dx_1^{\ 2}} = \frac{b_2}{EI_3}, \tag{14}$$

where

$$\phi_2 = \left\{\frac{-F_{B1}L^2}{EI_3}\right\}^{1/2} = \text{axial force parameter}. \tag{15}$$

The term involving ϕ_2 is due to nonlinear flexural rotation. We neglect this effect by setting $\phi_2 = 0$. Solving Equation 14 leads to the following expressions for the end actions,*

$$\begin{aligned}
\bar{F}_{B2} &= \bar{F}_{B2,i} + \rho_3 \bar{F}_{B1} + \frac{EI_3 k_{23}}{L^2}\left[-\omega_{B3} - \omega_{A3} + \frac{2}{L}(u_{B2} - u_{A2})\right], \\
\bar{F}_{A2} &= \bar{F}_{A2,i} - \rho_3 \bar{F}_{B1} + \frac{EI_3 k_{23}}{L^2}\left[\omega_{B3} + \omega_{A3} - \frac{2}{L}(u_{B2} - u_{A2})\right], \\
\bar{M}_{B3} &= \bar{M}_{B3,i} + \frac{EI_3}{L}\left[k_{21}\omega_{B3} + k_{22}\omega_{A3} - \frac{k_{23}}{L}(u_{B2} - u_{A2})\right], \\
\bar{M}_{A3} &= \bar{M}_{A3,i} + \frac{EI_3}{L}\left[k_{22}\omega_{B3} + k_{21}\omega_{A3} - \frac{k_{23}}{L}(u_{B2} - u_{A2})\right],
\end{aligned} \tag{16}$$

where the initial end actions and stiffness coefficients are nonlinear functions of the axial force parameter. Figure 3 shows the variation of the stiffness factors. For $\phi_2 = 0$,

$$\begin{aligned}
k_{21} &= 4, \\
k_{22} &= 2, \\
k_{23} &= 6.
\end{aligned} \tag{17}$$

* The general solution for u_2 and expressions for the coefficients are listed in the Appendix.

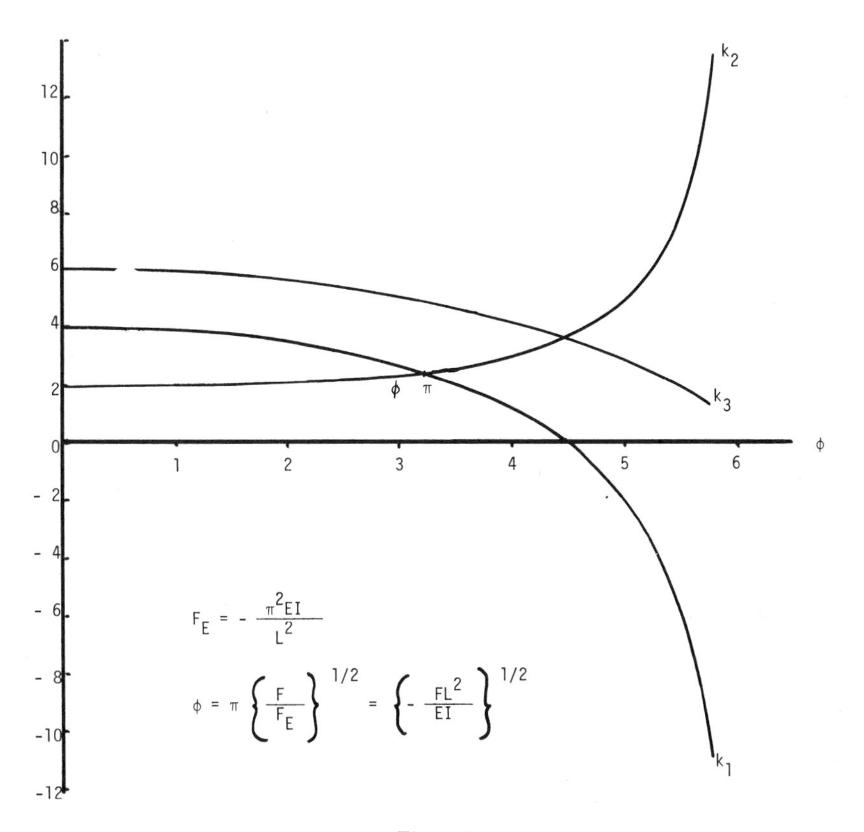

Figure 3.

As $\phi_2 \to 2\pi$,

$$k_{21} \to -\infty,$$

$$k_{22} \to +\infty, \tag{18}$$

$$k_{23} \to 0.$$

These curves show that it is reasonable to neglect the nonlinear flexural rotation effect when the axial force is small in comparison to the pin-ended Euler buckling load, say $F/F_E < 1/4$.

It should be noted that the nonlinear term $\rho_3 \bar{F}_{B1}$ is due to chord rotation. Also, it is inconsistent to neglect δ_2 and retain $\rho_3 \bar{F}_{B1}$ as Livesley[1] has done. Argyris[6] neglects δ_2 and the nonlinear flexural effect, i.e., he takes $\phi_2 = 0$. This is also inconsistent, since $\delta_2 \approx \frac{1}{2}\rho_3^2$ for $\phi_2 \approx 0$. In this study, we use the exact expressions for K_{2j} and δ_2.

At this point, it is convenient to express the end action - end displacement relations in matrix form, We let

$$\bar{\mathbf{F}} = \left\{ \bar{F}_1, \bar{F}_2, \bar{M}_3 \right\} = \text{end action matrix},$$
$$\mathbf{U} = \left\{ u_1, u_2, \omega_3 \right\} = \text{end displacement matrix}, \tag{19}$$

and write the complete set of relations as

$$\bar{\mathbf{F}}_B = \bar{\mathbf{F}}_{B,i} + \mathbf{k}_{BB}\,\mathbf{U}_B + \mathbf{k}_{BA}\,\mathbf{U}_A + \mathbf{F}_g,$$
$$\bar{\mathbf{F}}_A = \bar{\mathbf{F}}_{A,i} + \mathbf{k}_{BA}^T\,\mathbf{U}_B + \mathbf{k}_{AA}\,\mathbf{U}_A - \mathbf{F}_g, \tag{20}$$

where $\mathbf{F}_g$ contains the nonlinear terms due to rotation,

$$\mathbf{F}_g = \left\{ AE\delta_2, \rho_3\,\bar{F}_{B1}, 0 \right\}. \tag{21}$$

Note that the initial end action and stiffness matrices reduce to the corresponding linear forms when the nonlinear flexural effect is neglected.

Later we shall need expressions for the incremental end actions. Now we neglect the *incremental* nonlinear flexural rotation terms; i.e., we assume ϕ_2 is constant when forming the differentials. Then,

$$d\bar{\mathbf{F}}_B = d\bar{\mathbf{F}}_{B,i} + \mathbf{k}_{BB}\,\Delta\mathbf{U}_B + \mathbf{k}_{BA}\,\Delta\mathbf{U}_A + d\mathbf{F}_g,$$
$$d\bar{\mathbf{F}}_A = d\bar{\mathbf{F}}_{A,i} + \mathbf{k}_{BA}^T\,\Delta\mathbf{U}_B + \mathbf{k}_{AA}\,\Delta\mathbf{U}_A - d\mathbf{F}_g, \tag{22}$$

where $d\bar{\mathbf{F}}_{B,i}$, $d\bar{\mathbf{F}}_{A,i}$ are the linearized incremental initial end actions and

$$d\mathbf{F}_g = \begin{Bmatrix} AE\,d\delta_2 \\ d(\rho_3 F_{B1}) \\ 0 \end{Bmatrix}. \tag{23}$$

To be consistent with our assumption that the incremental nonlinear flexural effect is negligible, we must take $d\omega_3 = d\rho_3$ when forming $d\delta_2$.

$$d\delta_2 = \frac{1}{L}\int_0^L \omega_3\, d\omega_3\, dx_1$$
$$\approx \rho_3\, d\rho_3 = (\rho_3/L)(\Delta u_{B2} - \Delta u_{A2}). \tag{24}$$

Also,

$$d\bar{F}_{B1} = (AE/L)(\Delta u_{B1} - \Delta u_{A1}) + AE\,d\delta_2. \tag{25}$$

Substituting Equations 24, 25 in Equation 23 leads to

$$\mathbf{dF_g} = \mathbf{k_g}\,(\Delta \mathbf{U_B} - \Delta \mathbf{U_A}),$$

$$\mathbf{k_g} = \frac{AE}{L} \left[\begin{array}{c|c|c} 0 & \rho_3 & 0 \\ \hline \rho_3 & \dfrac{\bar{F}_{B1}}{AE} + \rho_3^2 & 0 \\ \hline 0 & 0 & 0 \end{array}\right] \tag{26}$$

Finally, the approximate incremental expressions are

$$\begin{aligned} \mathbf{d\bar{F}_B} &= \mathbf{d\bar{F}_{B,i}} + (\mathbf{k_{BB}} + \mathbf{k_g})\,\Delta \mathbf{U_B} + (\mathbf{k_{BA}} - \mathbf{k_g})\,\Delta \mathbf{U_A}, \\ \mathbf{d\bar{F}_A} &= \mathbf{d\bar{F}_{A,i}} + (\mathbf{k_{BA}} - \mathbf{k_g})^T\,\Delta \mathbf{U_B} + (\mathbf{k_{AA}} + \mathbf{k_g})\,\Delta \mathbf{U_A}. \end{aligned} \tag{27}$$

One can interpret $\mathbf{k_g}$ as the *geometric* tangent stiffness matrix.

If the displacements defining the equilibrium position are such that the chord rotation is negligible, $\mathbf{k_g}$ reduces to

$$\mathbf{k_g} = \frac{\bar{F}_{B1}}{L} \begin{bmatrix} 0 & 0 & 0 \\ 0 & 1 & 0 \\ 0 & 0 & 0 \end{bmatrix} . \tag{28}$$

We shall utilize this form in linearized stability analysis.

2.2. Three-dimensional equations

The positive sense for the end actions and displacements for the three-dimensional case coincides with the corresponding direction of the undeformed member frame (e.g., M_2 points in the $+X_2$ direction). In this study, torsion-flexure coupling and warping restraint are neglected. The force-displacement relations are obtained by combining the expressions associated with flexure in the $X_1 - X_2$ plane, flexure in the $X_1 - X_3$ plane, and unrestrained torsion.

In the formulation for flexure in the $X_1 - X_2$ plane, a subscript 2 was used. Now, to obtain the corresponding equations for flexure in the $X_1 - X_3$ plane, one has only to change the first subscript according to

$$\begin{aligned} 2 &\rightarrow 3, \\ 3 &\rightarrow -2, \end{aligned} \tag{29}$$

and replace I_3 by I_2. The only quantities with a subscript 3 as the first subscript are the moments M and rotations ω, ρ. Then, 29 amounts to interchanging the first subscript on all terms and reversing the sign on M, ω, ρ. For example,

$$\rho_3 = (u_{B2} - u_{A2})/L \tag{30a}$$

becomes

$$-\rho_2 = (u_{B3} - u_{A3})/L. \tag{30b}$$

The three-dimensional end action and end displacement matrices are

$$\begin{aligned} \bar{\mathbf{F}} &= \left\{ \bar{F}_1, \bar{F}_2, \bar{F}_3, \bar{M}_1, \bar{M}_2, \bar{M}_3 \right\}, \\ \mathbf{U} &= \left\{ u_1, u_2, u_3, \omega_1, \omega_2, \omega_3 \right\}. \end{aligned} \tag{31}$$

We express the three-dimensional relations in the same form as for the planar case which are given by Equations 20 and 27. The expanded form of the various stiffness matrices and generalized definitions for the coefficients are listed in the Appendix.

3. System Equations

The complete set of joint force equilibrium equations referred to the global reference frame are written as

$$\mathbf{P}_e = \mathbf{P}_m = \mathbf{P}_i + \mathbf{P}_g + \mathbf{KU}, \tag{32}$$

where $\mathbf{P}_e$ contains the external joint forces, $\mathbf{U}$ contains the joint displacements, $-\mathbf{P}_m$ contains the joint forces due to the member end actions, $\mathbf{P}_i$ is due to the external member loads, and $\mathbf{P}_g$ is due to the nonlinear rotation term $\mathbf{F}_g$. The most convenient procedure for assembling the system equation is the "direct stiffness method," which involves superimposing the contributions of the members in $\mathbf{P}_i$, $\mathbf{P}_g$, and $\mathbf{K}$. We suppose the boundary conditions have been introduced, i.e., we consider all the elements of $\mathbf{U}$ to be variables in Equation 32. The introduction of displacement boundary conditions is also quite straightforward.

We obtain the incremental system joint force matrix by applying the direct stiffness method to Equation 27. This leads to

$$d\mathbf{P}_m = d\mathbf{P}_i + \mathbf{K}_t \, \Delta\mathbf{U}. \tag{33}$$

The tangent stiffness matrix can also be expressed as

$$\mathbf{K}_t = \mathbf{K} + \mathbf{K}_g, \tag{34}$$

where $\mathbf{K}_g$ is due to the geometric stiffness term $\mathbf{k}_g$. This form is utilized in linearized stability analysis.

The nonlinear terms in $\mathbf{P}_i$ and $\mathbf{K}$ are due to ϕ, i.e., to the nonlinear flexural rotation effect. When this effect is neglected, $\mathbf{P}_i$ and $\mathbf{K}$ reduce to their *linear* forms. Note, however, that $\mathbf{P}_g$ and $\mathbf{K}_g$ are *still nonlinear* since they involve the chord rotation, ρ.

4. Solution Procedures

In this section, we discuss the application of two well-known iterative techniques for solving a set of nonlinear algebraic equations, namely, successive substitution and Newton - Raphson iteration. We also describe a linearized procedure that can be interpreted as an abbreviated form of Newton-Raphson. Finally, we discuss the question of stability.

4. 1 Successive substitution.

In successive substitution, we first rewrite Equation 32 in the following form:

$$\mathbf{K}\mathbf{U} = \mathbf{P}_e - \mathbf{P}_i - \mathbf{P}_g. \tag{35}$$

Let $\mathbf{U}^{(n)}$ denote the nth estimate for the solution corresponding to $\mathbf{P}_e$. The next estimate is determined from

$$\mathbf{K}^{(n)}\mathbf{U}^{(n+1)} = \mathbf{P}_e - \mathbf{P}_i^{(n)} - \mathbf{P}_g^{(n)}, \tag{36}$$

where

$$\mathbf{K}^{(n)} = \mathbf{K}|_{\mathbf{U}=\mathbf{U}^{(n)}}$$

etc.

A convenient displacement measure is the Euclidean norm,

$$N^{(n)} = [(\mathbf{U}^{(n)})^T (\mathbf{U}^{(n)})]^{½}. \tag{38}$$

Also, one can take as a convergence criterion the relative change in N,

$$(N^{(n+1)} - N^{(n)})/N^{(n)} \leqslant \epsilon, \tag{39}$$

where ϵ is the desired accuracy.

Successive substitution has only first-order convergence, i.e., the error for the (n+1)th step is a linear function of the error for the nth step. However, it is the most convenient solution procedure to program. Also, when the non-linear flexural rotation effect is neglected, the recurrence relation takes a very simple form,

$$\mathbf{K}\mathbf{U}^{(n+1)} = \mathbf{P}_e - \mathbf{P}_i - \mathbf{P}_g^{(n)}, \tag{40}$$

where $\mathbf{K}$ and $\mathbf{P}_i$ are the *linear* stiffness and end action matrices.

4.2. Newton – Raphson iteration

To apply the Newton–Raphson method, we first write the governing equation in the following form:

$$\psi = \mathbf{P}_e - \mathbf{P}_m = \mathbf{0}. \tag{41}$$

Let $\mathbf{U}^*$ be the exact solution and $\mathbf{U}^{(n)}$ the nth estimate. Expanding $\psi(\mathbf{U}^*)$ about $\mathbf{U}^{(n)}$ and retaining only the first differential leads to

$$\psi(\mathbf{U}^*) = \mathbf{P}_e - \mathbf{P}_m^{(n)} - d\mathbf{P}_m^{(n)} = \mathbf{0}. \tag{42}$$

Finally, we substitute for $d\mathbf{P}_m$ using Equation 33. Note that we are considering only displacement increments and therefore $d\mathbf{P}_i = \mathbf{0}$. The recurrence relation is

$$\begin{aligned} \mathbf{K}_t^{(n)} \Delta\mathbf{U}^{(n)} &= \mathbf{P}_e - \mathbf{P}_m^{(n)}, \\ \mathbf{U}^{(n+1)} &= \mathbf{U}^{(n)} + \Delta\mathbf{U}^{(n)}. \end{aligned} \tag{43}$$

Newton–Raphson iteration has second-order convergence and is generally more efficient than successive substitution. However, it is somewhat more difficult to program. Also, there is no simplification in the recurrence relation when the nonlinear flexural effect is neglected since $\mathbf{K}_t$ depends on the chord rotation.

4.3. Linearized incremental procedure

Suppose we have obtained the solution $\mathbf{U}$ for a particular load $\mathbf{P}_e$ and we now apply incremental joint and member loads. The incremental equilibrium equations follow from Equation 32.

$$\Delta \mathbf{P}_e = \Delta \mathbf{P}_m. \tag{44}$$

In the linearized procedure, we take

$$\Delta \mathbf{P}_m \approx d\mathbf{P}_m = d\mathbf{P}_i + \mathbf{K}_t \, \Delta \mathbf{U}, \tag{45}$$

where $\mathbf{K}_t$ is evaluated at $\mathbf{U}$, i.e., at the position prior to the application of the load increment. The incremental displacement due to $\Delta \mathbf{P}_e$, $d\mathbf{P}_i$ is determined from

$$\mathbf{K}_t \, \Delta \mathbf{U} = \Delta \mathbf{P}_e - d\mathbf{P}_i. \tag{46}$$

We evaluate $\mathbf{K}_t$ at $\mathbf{U} + \Delta \mathbf{U}$ and then apply the next increment. This procedure corresponds to truncating the Newton-Raphson method after the first cycle.

Since the iterative methods converge to the exact solution, it is not necessary to obtain the exact solution at each load. One can determine the approximate solution for a limited number of load steps using the linearized procedure and then correct with an iterative method.

4.4. Stability analysis

We consider next the question of stability. In general, an equilibrium position is stable when the corresponding value of the total potential energy Π_p is a relative minimum, i.e., when the second differential $d^2 \Pi_p$ is positive definite. The form of $d^2 \Pi_p$ for a member system is

$$d^2 \Pi_p = \Delta \mathbf{U}^T d\mathbf{P}_m = \Delta \mathbf{U}^T \mathbf{K}_t \Delta \mathbf{U} \tag{47}$$

and it follows that $\mathbf{K}_t$ must be positive definite for the equilibrium position to be stable. Now, a matrix is positive definite when all its discriminants* are positive. If a discriminant is zero, there exists a nontrivial solution of

$$\mathbf{K}_t \, \Delta \mathbf{U} = 0 \tag{48}$$

and the equilibrium position corresponds to a bifurcation. If a discriminant is negative, the position is unstable.

The signs (and values) of the discriminants can be obtained by triangularizing $\mathbf{K}_t$ using elementary row operations. The nth discriminant is the product of the first n diagonal elements in the triangular form. Initially, the discriminants are

* The nth discriminant is the determinant of the array contained in the first n rows and columns.

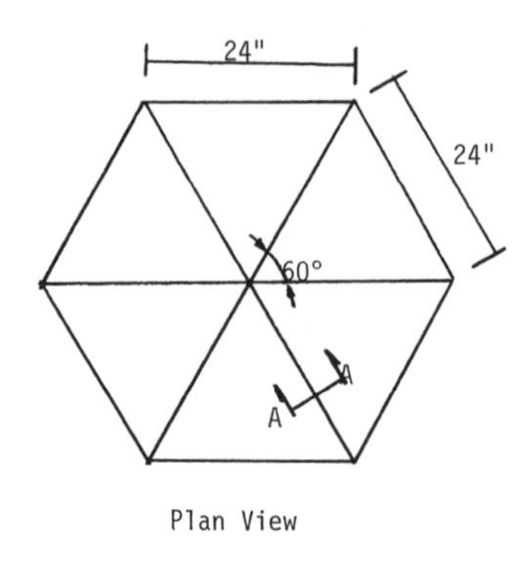

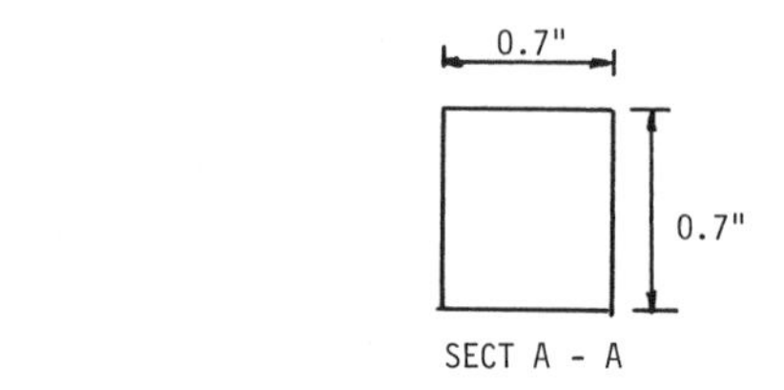

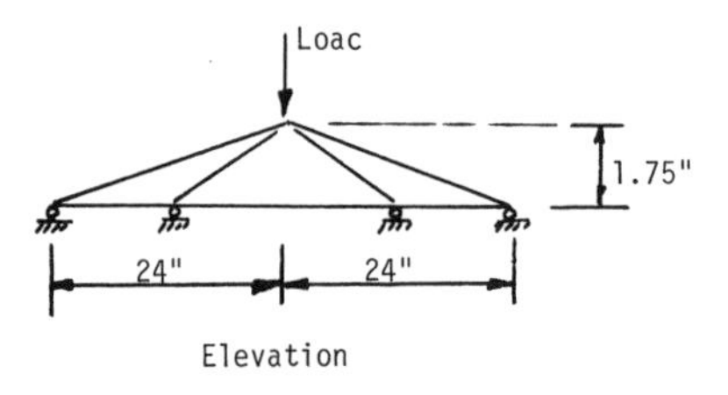

Figure 4. Hexagon geometry. Notes: All members have the same cross section; E=460,000 psi (average value).

all positive since $|\mathbf{K}_t| = \mathbf{K}_L$, the linear stiffness matrix. As the loading is increased, the nonlinear terms in $\mathbf{K}_t$ become more significant and the diagonal elements may decrease, depending on the structure and loading. If $|\mathbf{K}_t|$ decreases with increasing load, one should anticipate a possible instability. If only one diagonal element changes sign (positive to negative), the determinant $|\mathbf{K}_t|$ also changes sign. Usually $|\mathbf{K}_t|$ is plotted versus load and the load at which $|\mathbf{K}_t| = 0$ is determined by interpolation. However, one must be careful since two diagonal elements may change sign at essentially the same load and $|\mathbf{K}_t|$ will still be positive. Finally, we make one additional comment. The magnitude of $|\mathbf{K}_t|$ is quite large and to avoid exceeding the computer capacity we determine log $|\mathbf{K}_t|$, which is just equal to the sum of the logarithms of the diagonal elements.

5. Applications

The analytical procedure was first applied to the shallow hexagonal frame shown in Figure 4. The relative dimensions (except for the cross-sectional width) are typical of a large radome. Figure 5 shows the load-deflection curves for various cycling schemes. It appears that the linearized procedure (curve 1) considerably overestimates the critical load even for a reasonably small load increment. Also, there is some difference, as regards computational effort, between correcting after each step or after every third step (curves 2, 3). Curve 3 required less computation, particularly for the initial portion. However, this is offset by the less accurate estimate of the limiting load. We believe that each nonlinear problem must be treated individually, so we make no attempt to recommend a cycling shceme. We do feel, however, that one should first apply the linearized step method in order to obtain some indication of the behavior.

We applied the procedure next to a radome model constructed in the M.I.T. Aeroelastic and Structures Research Laboratory[10] for CAMROC. A typical sector is shown in Figure 6. These are 121 unrestrained joints and 440 members. The members are prismatic, of rectangular cross section, except for the

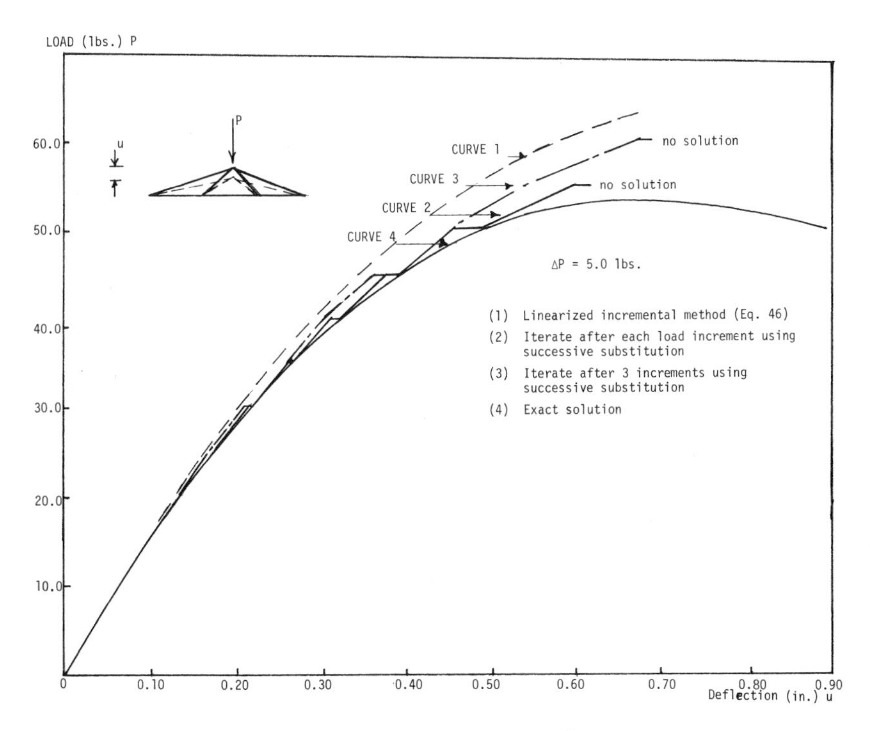

Figure 5. Load-deflection curves (hexagon).

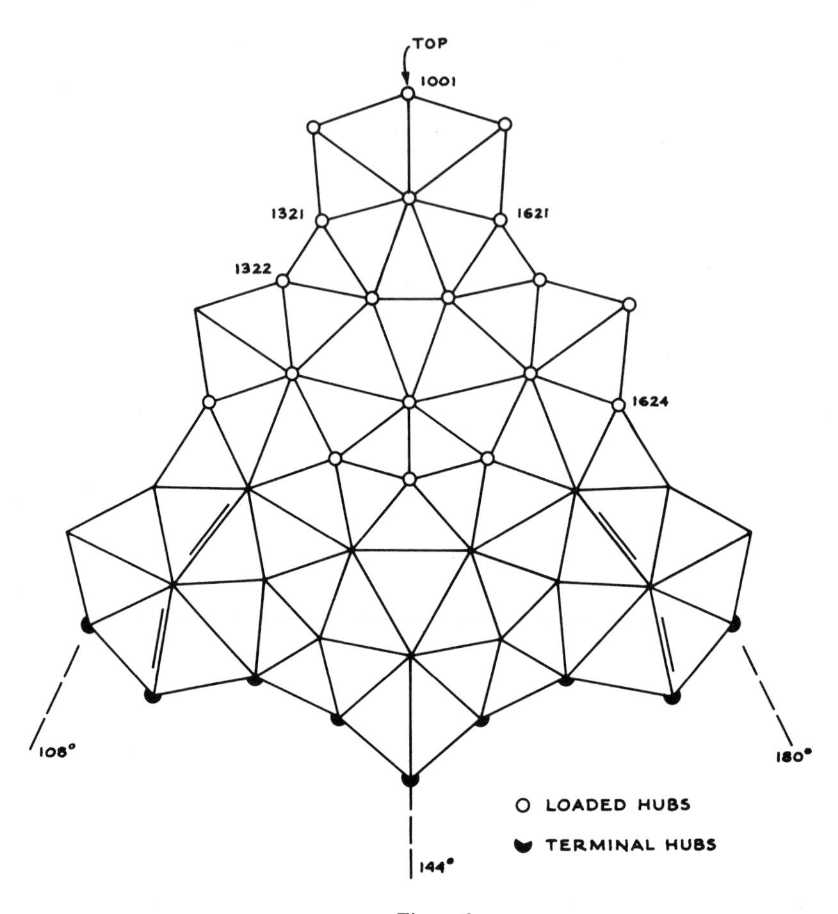

Figure 6.

four bars indicated with double lines. These bars were reinforced with tee sections to prevent Euler buckling. Equal radial loads were applied to 61 joints.

Figures 7 and 8 show the variation in radial displacement and meridional rotation for joint 1624.* The numbers in parentheses are the cycles of successive substitution required for 5 per cent accuracy (relative difference in successive norms = 0.05). The linearized step procedure indicated that $|\mathbf{K}_t|$ was negative for a radial load/joint of 75 lb. When successive substitution was applied at 75 lb, it indicated that the values of the axial force parameters for certain bars were approaching the limiting value of 2π. A check had been

* These computations were carried out by personnel of Simpson, Gumpertz, and Heger, Cambridge, Massachusetts, who implemented the nonlinear analysis program on Lincoln Laboratory's IBM 360 system.

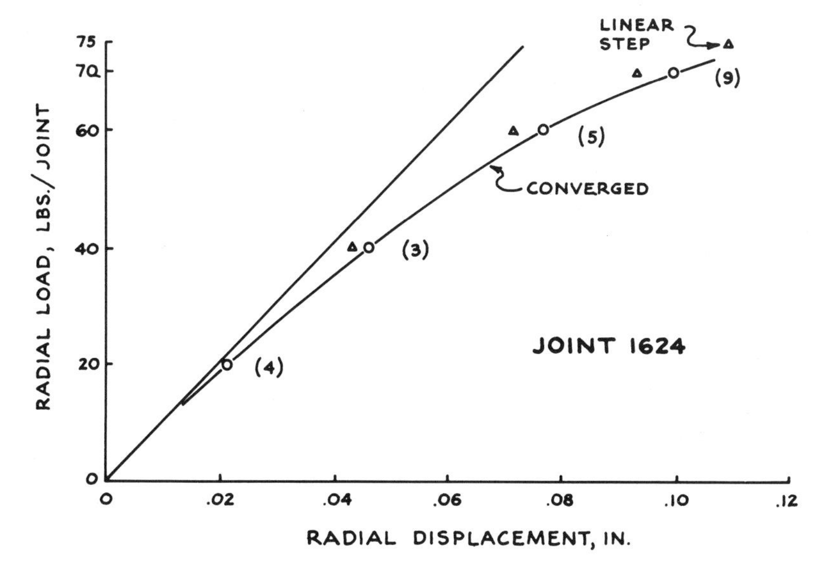

Figure 7.

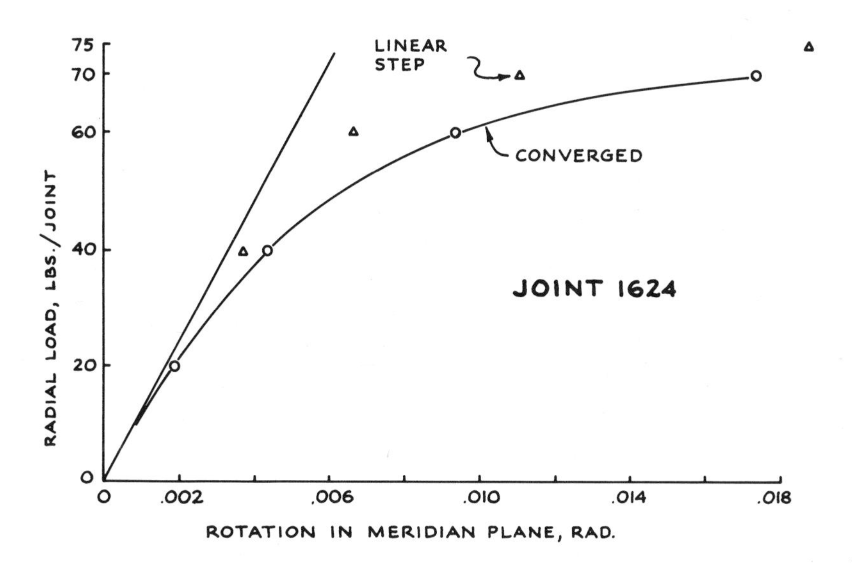

Figure 8.

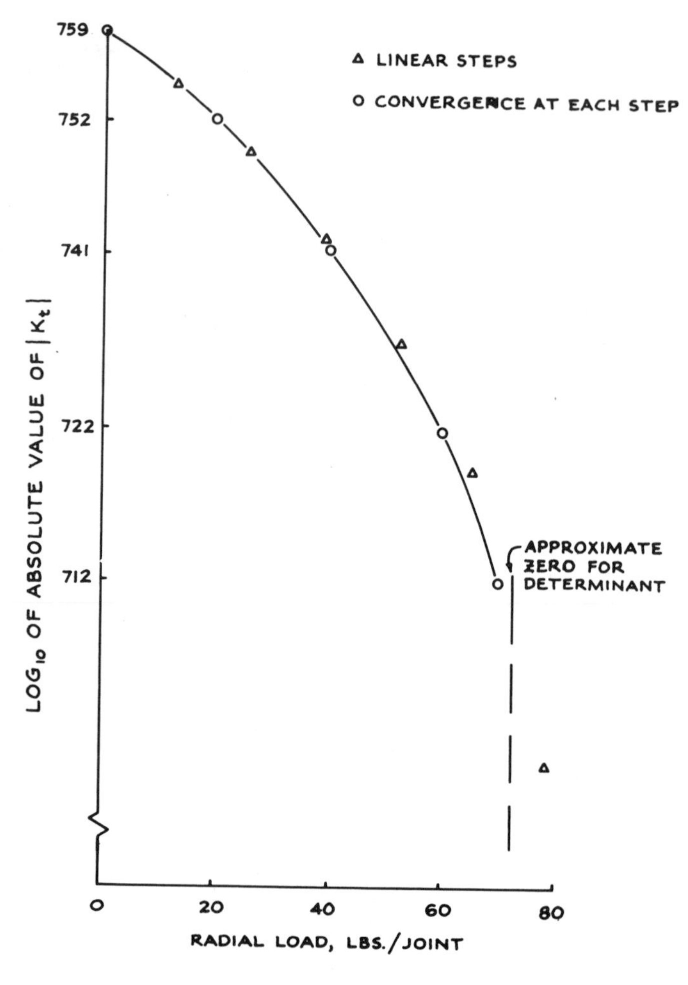

Figure 9.

included in the program to detect member buckling and it automatically stopped the iteration after five cycles.

Figure 9 shows the variation of $|\mathbf{K}_t|$ with radial load. It is interesting to note that the points obtained with the linearized procedure fall on the "exact" curve. This implies that cycling on the displacements at a given load has essentially no effect on the magnitude of the determinant of $\mathbf{K}_t$.

Table 1 contains a summary of the computed and experimental values of radial displacement for joint 1621. The maximum load was 95.1 lb. At this load, joint 1321 was touching its stop. When the load was reduced to zero, the radial positions of the joints were, at most, only 0.003 in. from the initial

Table 1. Radial displacement for node 1621.

Radial load lb/joint	Computed (in.)	Experimental (in.)
20	0.0238	0.019
40	0.0492	0.044
60	0.0780	0.069
70	0.0963	
80		0.096
84.5		0.101
89.5		0.116
94.5		0.131
95.1		0.182
94.5		0.169

zero position. This indicates that the instability was elastic. The difference between the theoretical and experimental values can be attributed to the joint connection detail employed in the model. Two circular plates approximately 2 in. in diameter were used to fasten the members at each joint. This detail reduced the effective length of the members by 2 in. Now, the average member length (distance between joints) is 10 in. so that the member buckling load is increased by about 50 per cent. Although the instability was not due to *member* buckling of an individual member, the increase of 36 per cent (70 vs 95) appears to be reasonable.

Acknowledgments

This work has been supported by Lincoln Laboratory, M.I.T., under Grant GP5832 from the National Science Foundation for Project CAMROC.

References

1. R. D. Livesley, "The Application of An Electronic Digital Computer to Some Problems of Structural Analysis," *Struct. Engr.* **34** (No. 1), 1.

2. J. D. Renton, "Stability of Space Frames By Computer Analysis," *J. Struct. Div. ASCE* **88** (No. ST4), 81 (August 1962).

3. S. A. Saafan, "Nonlinear Behavior of Structural Plane Frames," *J.Struct. Div. ASCE* **89** (No. ST4), 451 (August 1963).

4. F. W. Williams, "An Approach to the Nonlinear Behavior of the Members of a Rigidly Jointed Plane Framework with Finite Deflections," *Quart. J. Mech. Appl. Math.* **17**, 451 (1964).

5. W. Merchant and P. M. Brotton, "A Generalized Method of Analysis of Elastic Plane Frames," IABSE Symposium, Rio de Janeiro (1964).

6. J. H. Argyris, *Recent Advances in Matrix Methods of Structural Analysis* (Pergamon Press, New York, 1964), pp. 115-145.

7. R. K. Livesley, *Matrix Methods of Structural Analysis* (Pergamon Press, New York, 1964), pp. 241-252.
8. M. A. Millar, D. M. Brotton and W. Merchant, "A Computer Method For the Analysis of Nonlinear Elastic Plane Frameworks," International Symposium on Use of Computers in Structural Engineering, Department of Civil Engineering, University of Newcastle (1966).
9. D. Johnson and D. M. Brotton, "A Finite Deflection Analysis for Space Structures," International Conference on Space Structures, Department of Civil Engineering, University of Surry, (1966).
10. S. D. Lewis, "Progress Report on Camroc Radome Model Program," M. I. T. Aeroelastic and Structures Research Laboratory (June 1967).

Appendix

General definitions for the terms employed in the member force-displacement relations and the expanded forms of the various matrices are listed below:

$$j \neq k,$$
$$j, k = 2, 3,$$
$$\delta_{2j} = \begin{cases} 1 \text{ for } j = 2, \\ 0 \text{ for } j = 3. \end{cases}$$

$$\phi_j = \left\{ - \bar{F}_{B1} L^2 / EI_k \right\}^{1/2},$$

$$C_j = \cos \phi_j,$$
$$S_j = \sin \phi_j,$$

$$k_{j1} = \frac{\phi_j (S_j - C_j \phi_j)}{2(1 - C_j) - S_j \phi_j},$$

$$k_{j2} = \frac{\phi_j (\phi_j - S_j)}{2(1 - C_j) - S_j \phi_j},$$

$$k_{j3} = k_{j1} + k_{j2}.$$

$$\rho_j = \frac{1 - 2\delta_{2j}}{L} (u_{BK} - u_{AK}),$$

$$2\delta_j = \frac{1}{L} \int_0^L (\omega_K)^2 \, dx_1 = \frac{1}{L} \int_0^L (u_{j,1})^2 \, dx_1$$

$$= D_{j3}^2 + \frac{2D_{j3}}{\phi_j^2} [- S_j D_{j1} + (1 - C_j) D_{j2}]$$

$$+ D_{j1} D_{j2}\left(-\frac{S_j^2}{\phi_j^3}\right) + \frac{D_{j1}^2}{2\phi_j^2}\left(1 + \frac{S_j C_j}{\phi_j}\right) + \frac{D_{j2}^2}{2\phi_j^2}\left(1 - \frac{S_j C_j}{\phi_j}\right),$$

$$D_{j2} = (2\,\delta_{2j} - 1)(-\,k_{j1}\,\omega_{Ak} - k_{j2}\,\omega_{Bk} + k_{j3}\,\rho_k),$$

$$D_{j1} = (2\,\delta_{2j} - 1)\left(\frac{\phi_j}{1 - C_j}\right)(\omega_{BK} - \omega_{AK}) - \left(\frac{S_j}{1 - C_j}\right)D_{j2},$$

$$D_{j3} = (2\,\delta_{2j} - 1)\,\omega_{Ak} + \frac{1}{\phi_j} D_{j1},$$

$$D_{j4} = \frac{u_{Aj}}{L} + \frac{1}{\phi_j^2}\,D_{j2}.$$

$$e_1^0 = \frac{1}{A}\int_A \epsilon_1^0\, d\,A,$$

$$k_j^0 = \frac{(2\,\delta_{2j} - 1)}{I_j}\int_A x_k\,\epsilon_1^0\, d\,A\,.$$

$$k_{BB} = \begin{bmatrix} \frac{AE}{L} & & & & & \\ & 2k_{23}\frac{EI_3}{L^3} & & & & -k_{23}\frac{EI_3}{L^2} \\ & & 2k_{33}\frac{EI_2}{L^3} & & k_{33}\frac{EI_2}{L^2} & \\ & & & \frac{GJ}{L} & & \\ & \text{Sym} & & & k_{31}\frac{EI_2}{L} & \\ & & & & & k_{21}\frac{EI_3}{L} \end{bmatrix},$$

$$
\mathbf{k}_{AA} = \begin{bmatrix}
\frac{AE}{L} & & & & & \\
& 2k_{23}\frac{EI_3}{L^3} & & & & k_{23}\frac{EI_3}{L^2} \\
& & 2k_{33}\frac{EI_2}{L^3} & & -k_{33}\frac{EI_2}{L^2} & \\
& & & \frac{GJ}{L} & & \\
& \text{Sym} & & & k_{31}\frac{EI_2}{L} & \\
& & & & & k_{21}\frac{EI_3}{L}
\end{bmatrix},
$$

$$
\mathbf{k}_{BA} = \begin{bmatrix}
\frac{-AE}{L} & & & & & \\
& -2k_{23}\frac{EI_3}{L^3} & & & & -k_{23}\frac{EI_3}{L^2} \\
& & -2k_{33}\frac{EI_2}{L^3} & & k_{33}\frac{EI_2}{L^2} & \\
& & & \frac{-GJ}{L} & & \\
& & -k_{33}\frac{EI_2}{L^2} & & k_{32}\frac{EI_2}{L} & \\
& k_{23}\frac{EI_3}{L^2} & & & & k_{22}\frac{EI_3}{L}
\end{bmatrix},
$$

$$\mathbf{F}_g = \left\{ AE(\delta_2 + \delta_3);\ \rho_3 \bar{F}_{B1};\ -\rho_3 \bar{F}_{B1};\ 0;\ 0;\ 0 \right\},$$

$$\mathbf{k}_g = \left[\begin{array}{c|c} \mathbf{k}_{g,1} & \underset{\sim}{0} \\ \hline \underset{\sim}{0} & \underset{\sim}{0} \end{array} \right],$$

$$\mathbf{k}_{g,1} = \frac{AE}{L} \left[\begin{array}{c|c|c} 0 & \rho_3 & -\rho_2 \\ \hline \rho_3 & \dfrac{\bar{F}_{B1}}{AE} + \rho_3^2 & -\rho_2\rho_3 \\ \hline -\rho_2 & -\rho_2\rho_3 & \dfrac{\bar{F}_{B1}}{AE} + \rho_2^2 \end{array} \right].$$

S. Dean Lewis and E. A. Witmer
Aeroelastic and Structures Research Laboratory
Massachusetts Institute of Technology
Cambridge, Massachusetts

BUCKLING TESTS ON SPACE FRAME RADOME

1. Introduction

In recent years two computer analyses have become available to aid in the detail design of space frames. The first is STAIR, which assumes pin-ended beams (columns) and the second is FRAN, which considers the beams rigidly attached to the hubs. In either of these programs, all of the loads imposed on the structure may be analyzed to determine loads on individual members. However, neither of these programs considers failure by buckling, either in individual members, i.e., Euler column buckling, or of all the members, i.e., a general instability of the radome.

In order that the state of the art be advanced in the design of radomes, it was desired that a new analysis be created which would be able to contend with nonlinear effects. Construction of a model radome was undertaken to provide test data on a well-defined model to check the results of the new nonlinear program, particularly to discover if it is possible to predict the maximum load which the radome can carry.

2. Model Design

2.1. General

The design for the space-frame radome model (see Fig. 1) was based upon providing a small space frame that can yield experimental measurements to verify analytical techniques for load-deflection and maximum load predictions.

Figure 1. Over-all view of the CAMROC space frame model.

Five items of special concern, based on previous radome model testing in this laboratory, were

1. To seek an initial condition of zero stress in the model.
2. To seek geometric continuity of the members at the joints.
3. To seek a construction procedure to position joints on the desired spherical surface at the design coordinates.
4. To provide a method to measure model geometry as constructed.
5. To employ an adequate and feasible model size.

The first two items listed involved design of the hubs (the fitting that connects the members together at the joints). Of several possible construction methods available, the best seemed to involve a pair of flat disks with the beam ends clamped in between. With beams curved to lie on the spherical surface, each would come tangentially into a hub and it would have been a simple matter to fasten every beam end with two bolts in double shear. In fact, some previous model work had been done with curved beams.

However, it was decided that in spite of increased model construction complexity and cost, the model should be built with straight beams, which are representative of usual full-scale radome construction techniques. While this decision slightly simplified beam construction, it greatly increased the complexity of design and construction of the hubs.

Item 3 was influenced basically by model construction techniques and by care in manufacture. The technique chosen required the use of a rigid mandrel (which will be more fully described later).

Item 4 involved a master template that was fitted to the construction mandrel which was useful in construction in checking the free-standing radome geometry, and ultimately in measuring hub radial deflections under load.

The last item 5 above, model size, was limited by the space available in the ASRL laboratory and by cost. The model chosen represents a part of a 14-ft-diameter sphere and, in fact, covers the spherical surface within a cone of half-angle 42.7° to the lowest row of unsupported hubs, or 52.0° to the lowest row of base hubs. This sector was judged to be adequate to study any type of instability that the complete sphere is capable of exhibiting, uninfluenced by the boundaries.

2.2. Detail design

A random geometry configuration was chosen for the model as representative of a typical space frame. Lincoln Laboratory supplied the complete geometry for this space frame to model scale, including hub positions in angular and rectangular coordinates, beam lengths and angles between beams at the hubs. From these data, 121 free hubs and 40 base hubs were chosen symmetrically located about the "north pole."

It was found that 14 kinds of hub designs were required. Each hub was made in two parts, an upper and a lower hub disk. Each hub required a milled surface in both parts to receive each beam end. These milled surfaces were cut to a bevel angle determined by the beam length and held within ± 10 min. of arc to minimize initial bending in the beams. The two parts of a complete hub appear in Figure 2.

The beams were thus rigidly clamped between the hub disks about one axis. About the other axis the beams were locked to the hubs by two bolts in double shear, pressed into drilled and reamed holes. The hubs were manufactured with these holes drilled slightly under size, and precisely located by means of a jig.

2.3. Material selection

In selecting a beam material to study elastic buckling, it was desired to obtain a high ratio of the yield strength to the modulus of elasticity. The most promising materials considered are listed as follows:

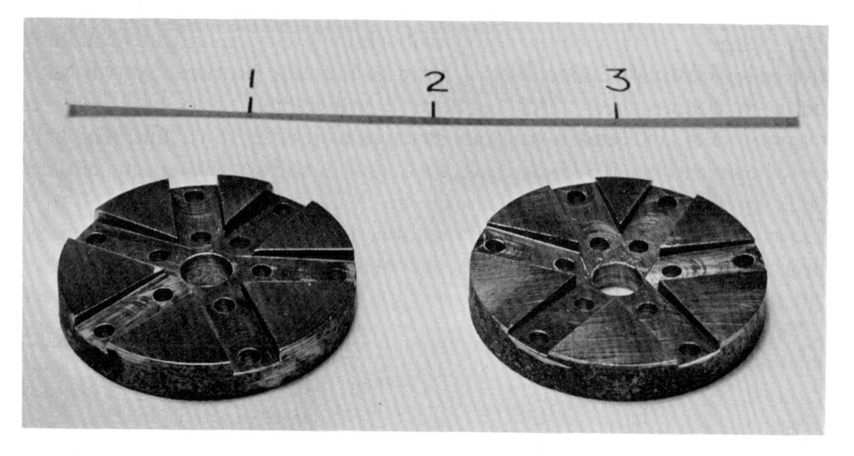

Figure 2. Photograph of inner faces of a pair of two parts of a complete hub disk.

Item	Material	f_{yp}	E	f_{yp}/E
1	7178-T6, T651	78,000. psi	10.4×10^6 psi	7.5×10^{-3}
2	7075-T6, T651	69,000. psi	10.4×10^6 psi	6.6×10^{-3}
3	4340	162,000. psi	29.0×10^6 psi	5.6×10^{-3}
4	4150	135,000. psi	29.0×10^6 psi	4.6×10^{-3}

Of these materials, item 1 was not readily available commercially. Therefore, item 2 was selected. The aluminum alloys have a further advantage over steel in easy machineability even when hardened.

2.4. Beam-column design

Having selected aluminum alloy for the beam material, the cross sections were determined from equivalent shell calculations. While a full-scale radome may be designed for lower pressures, practical considerations of model construction indicated that a buckling pressure close to 1 psi was advisable. The most appropriate rectangular dimensions for the beam cross section were found to be 0.20 in. in width by 0.30 in. in depth. Consideration of a typical hub-panel geometry indicated that the 1-psi design-pressure loading was equivalent to concentrated loads of about 100 lb per hub for the subject model. Since 64 hubs were to be so loaded, a total force of over 6000 lb would be required to cause buckling.

Euler buckling of individual beams in this model structure, based on a rigid-end attachment at the hubs, was calculated to occur at the following loads:

Beam length	Euler buckling axial load
7.91 in. (shortest)	1312 lb
13.15 in. (longest)	475 lb

This compared with a compressive nonbuckling yield strength of 4260 lb for the beam material. Connection of beam to hub was provided by two No. 4 bolts in double shear. This connection was computed to be capable of carrying 2080 lb, limited by beam material at the first bolt hole. The bolts were separated by 0.375 in. in order to give adequate end rigidity to delay buckling about the more flexible axis. For hub construction AlSl 4150 steel, heat-treated to Rockwell C 33, was selected. This material exceeded the bearing strength requirements but was useful in the construction procedure wherein the beams were drilled in place through the hubs. Also, it was anticipated that beams could be replaced after a model has been buckled, and the hubs re-used.

2.5. Proof testing

Before ordering beam and hub materials, sample hub-beam models of both longest and shortest beams were tested under compressive loading. While it is not difficult to calculate the strength of such a joint, it was uncertain what end fixity would be obtained. These models were mounted in the testing machine with the hubs rigidly fixed and the beam material initially under no bending load. The buckling load for a rigid-ended beam is

$$P_B = \frac{4\pi^2 EI}{l^2} = \frac{4\pi^2 Eh^2 A}{12 l^2}, \tag{1}$$

where E is Young's modulus of elasticity, h is the beamwidth, A is the beam cross-sectional area, and l is the beam length (to *outer* bolts). For the short beam, measured buckling loads were 1290 and 1305 lb, while for the long beam, measured buckling loads were 470 and 435 lb. The last-mentioned load was disregarded due to an initial beam curvature incurred during manufacture.

Defining the ratio of the measured to the theoretical buckling load as

$$\epsilon = \frac{P_{B\ \text{Measured}}}{P_{B\ \text{Calculated}}}, \tag{2}$$

values of ϵ were found to be as follows:

Beam	ϵ
Long beam	0.992
Short beam	0.988

Aside from achieving a close approximation to the theoretical buckling load, it was also observed that no end failure occurred, even when the beam center had been buckled laterally by about 1/2 in. Therefore, it was concluded that the hub-to-beam connection design was adequate. Note particularly that the above correlation requires that beam length must be measured to the outer bolt holes.

3. Model Construction

3.1. Mandrel

A large steel frame called a mandrel (see Figure 3) was essential to the build-

Figure 3. Photograph of the completed mandrel.

ing and testing of the model. It served as a rigid frame for the following requirements:

1. Attachment point for base hubs.
2. Attachment point for supports for "free" hubs during model construction.
3. A precise circular track for the "template" (described later) for measuring latitude, longitude, and radius on the space frame model.
4. Attachment point for the screw jack to load the whiffletree apex. This structure carries internally the reaction from the 40 base hubs back to the screw jack.

5. Support for pulleys that direct the loading cables radially to the hubs and vertically to the whiffletree.
6. Attachment point for supports under free hubs to prevent excessive damage of model when buckling load is reached.

The central part of the mandrel was a spherical shell with a 76-in. radius of curvature and a 3/8-in. thickness. It was located 8 in. inside the nominal radome surface and obtained rigidity from its spherical shape. Around the perimeter of the dished head were two rolled 4-in. channels bolted to a machined flat surface on a 1½-in.-thick plate, to which the base hub supports were bolted. The outer rim of this plate was machined to an accurate circle. A rim was bolted to this surface, level on top, on which template cam followers roll. Under this plate, a rolled-up 15-in. channel added stiffness to the mandrel. This structure, weighing about 6400 lb, had to be designed in small pieces for delivery into the laboratory for final assembly.

3.2. Template

For stiffness, the template was a closed box beam in cross section, formed of rolled-up 6-in. aluminum channels top and bottom and 5/16-in.-thick aluminum sides or webs. The box beam was formed as an arch, with one of the wide webs particularly flat and with inner and outer edges cut precisely as circular arcs, 86,100 and 102.500 in. in radius, respectively. This template was mounted on the mandrel as shown in Figure 4 with the plane surface as a great circle of both the model and the mandrel, and the inner arc 2.1 in. outside the model nominal spherical surface. Two cam followers on either end maintained the template in this position, riding on the outer rim of the mandrel.

Longitude of the template position was measured on a steel tape attached to the outside of the mandrel rim using a vernier scale mounted to the template. Latitude was read from a steel scale attached to the large radius arc of the great circle web of the template, using a radially directed straight edge clamped to the flat template surface, with a similar vernier for improved accuracy. Distances below the template lower arc, which were read on a dial micrometer, allowed precise determination of radius anywhere on the model.

The template was first used to locate each hub on a rigid support as accurately as possible. Latitude, longitude, radius from spherical center and axis along the radial direction were all important in this process. With hubs in position, beams were cut to length, clamped between the hub disks and drilled, reamed, and bolted. This process strated at the north pole and was continued until the lowest row of beams terminated at the base hubs.

Figure 4. Photograph of the template mounted on the mandrel.

3.3. Model instrumentation

Lincoln Laboratory supplied linear analysis results for this model from both the STAIR and FRAN programs. These analyses were carried out for cases of 64, 61, and 3 loaded hubs for a radial loading of 100 lb per hub.

These analyses were an aid in the instrumentation of the model. Measurements were desired of radial deflections of all 121 free hubs at each loading condition. With approximately 0.10 in. deflection predicted at a loading of 100 lb per hub, it was evident that 1 per cent accuracy required measurements to 0.001 in. A dial indicator mounted on the template did in fact allow repeatable readings to ± 0.001 in. It was also desired to measure stresses in some of the members. Thirteen beams were selected where compressive stresses were predicted to be relatively high in terms of their Euler buckling stresses. Two beams in each of the five sectors were chosen plus three additional beams in one sector. In each of these, the beam midspan location received a strain gauge on all four faces of the beam. One of these beams, 1008-1135, had additional sets of strain gauges close to either hub.

3.4. Whiffletree loading system

After completion of the space frame model, all of the free hubs were released from the supports, and the supports were dropped back about 1/4 in. A whiffletree loading structure, designed to apply equal radial loads to 64 hubs symmetrically located about the vertical axis, was then installed. Twenty small load cells were used to monitor individual loads applied to selected hubs to verify uniformity of loading. At the whiffletree apex, a large load cell was used to measure the total load on the model applied by a worm-drive screw jack attached to the mandrel base. The entire loading system was conservatively designed to minimize deflections and preclude the possibility of loading system failure before model failure.

3.5. Actual model geometry

The three space coordinates of all hubs were measured before release of the "free" hubs and again after completion of the whiffletree loading mechanism. Very small changes could be observed in some hub locations. In particular, radius measurements decreased a few thousandths. This is understandable since the whiffletree adds a radial force of 2.6 lb per loaded hub and the space frame itself weighs roughly 0.4 lb per hub. It is interesting to note from the radius measurements in the self-supporting condition that all values fall within the limits

$$R = 84.010 {}^{+\,0.029}_{-\,0.018} \text{ in.}$$

and that the rms error is 0.008 in. from a true sphere. Figure 1 shows the completed model and the template mounted on the mandrel, the apex of the whiffletree, and also some instrumentation associated with strain measurements.

4. Radome Static Test: Phase I

4. 1 Preliminary tests.

Several tests were conducted on the model in order to learn something about the model itself, the rigidity of the supporting structure, the loading system, and the instrumentation. Based on the "equivalent shell" prediction of buckling at 100 lb per hub, and to avoid damage to the model, loads of 0 (whiffletree only) 10, 20, and 30 lb per hub were applied. Radial locations of all 121 free hubs were measured at each load level and again at zero load after each test. Similarly, all strain gauges located on the beams were read for zero load and each of the above loads. Also, each of the 20 small load cells were read at each loading condition. The major load cell was, of conrse, used to determine and monitor the load applied at the base of the whiffletree.

Additional dial gauge readings were made at selected points on the steel mandrel to determine its deflection relative to the floor as the several loads were applied.

After considering the results of preliminary tests, the following conclusions were reached:

1. These tests did not exceed the elastic limit of the model, since radial hub measurements at zero load after this series repeated within ±0.002 in. of those measured earlier.
2. The steel mandrel deflections due to carrying the reactions from loads applied to the model were negligible.
3. Measurement results on 20 small load cells indicate that the whiffletree had applied equal loads to the 64 loaded hubs.
4. No modification was required in the model, the loading rig, or the instrumentation.
5. A decision was made to continue with this testing procedure, reading radial location of all 121 free hubs at each load level and all strain gauges until buckling was detected.

4.2. Test procedure

After making these preliminary decisions, testing of the model was continued as before, except that measurements of mandrel deflections were discontinued. After reaching 40 lb per hub, testing continued at intervals of 5 lb per hub. Finally, at a load of 80 lb per hub, one member, connecting hubs 1326-1517, was observed to have buckled as an Euler column, without any noticeable hub rotations.

It was observed at this loading condition that no general instability existed in the model and that the model was still self-supporting and capable of carrying increasing total load. This buckle had evidently resulted from a single member's having reached its Euler buckling stress, based on a rigid end constraint, before the onset of general instability.

After taking data at the 80 lb per hub loading condition, the applied load was removed and all hub radial positions were measured again. There was substantially no change from previous measurements at zero load. Also, the buckled member was nearly straight again.

4.3. Test results

Typical results from hub radial deflection measurements are presented in Figure 5. For hubs 1326 and 1517 an abrupt change of slope, indicative of

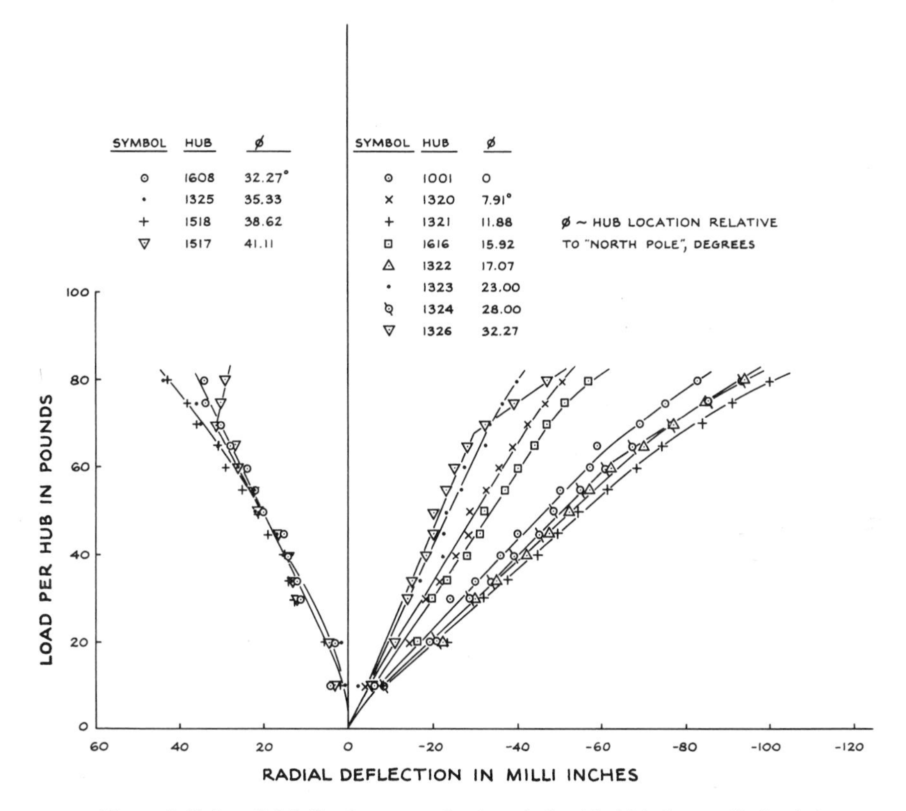

Figure 5. Hub radial deflection versus load per hub with 64 hubs equally loaded.

incipient buckling, is observed at about 68.5 lb per hub, which would correct to about 71 lb per hub considering the initial loading from the whiffletree. It is interesting to compare the test result with the FRAN-computed results for this model with 64 loaded hubs, although some incorrect base hubs were used in that calculation. At a loading of 100 lb per hub, member 1326-1517 is predicted to have a compressive stress level of 8695 psi. If proportionality between loading, stress, and deflection is assumed, which is the basis for FRAN, this member would have reached its computed Euler buckling stress of 8050 psi (based on a column length to the outer bolts justified by individual beam testing) at (8050/8695) 100 lb per hub, that is, 92.6 lb per hub.

4.4. Six-component measurements

At the conclusion of this phase of the testing, hub 1926 was selected to try out a system, previously bench tested, designed to measure three linear and three angular displacements during loading of the model. This system required considerable time for installation, but worked very well and presented the information conveniently on six dial micrometers. A computer program was written to allow these measurements to be converted to linear and angular displacements of the hub center. Drawbacks to measurement of six-component data in general included the large number of dial micrometers required and the space required around each hub so measured. Thus, on the entire radome only a few such hubs could have been instrumented and the usual radial measurements became almost impossible due to restrictions on the template travel. The results were interesting, and though not included here, they were in fair agreement with both FRAN and the new nonlinear program preliminary calculations.

5. Radome Static Test, Phase II

5.1. Modified loading system

As described in Section 4, the first phase of testing did not yield the desired general instability. The question then arose as to the best way to complete the investigations.

A FRAN analysis of this model run early in 1967 had considered both 64 and 61 loaded hubs. Comparisons of the actual beam stresses measured on the model with the FRAN analysis, both with 64 loaded hubs, indicated good correlation. Having thus gained confidence in the computed results, the FRAN-predicted loading of the model with 61 hubs was investigated. These results indicated that no member would reach its Euler load until the loading reached 141 lb/hub. To effect this 61-hub loading application, the load heretofore carried by the three hubs involved was carried instead by individual columns supported by the floor, allowing the whiffletree to remain otherwise unchanged. Also, this 61-hub loading provided a more symmetrical loading distribution than that involving the 64 loaded hubs.

5.2. Selective beam stiffening

However, an additional margin was desired, lest once again a member should buckle before general instability was encountered. First, an investigation was made to find the most critical beams. For each beam in the radome, the ratio of the fixed-ended-column Euler buckling stress (for lateral buckling) to the stress indicated in the FRAN analysis for 61 loading hubs was multiplied by 100 lb per hub. The result was assumed to indicate the hub loading at which that member will buckle individually. A listing of all the beams in order of decreasing susceptibility to buckling could then be made.

If the ten most critical beams on the list were suitably stiffened against lateral buckling, this approximate analysis indicated an increse from 141 lb per hub to 168 lb per hub before the onset of member buckling. This was believed to be an adequate margin above the "equivalent-shell predicted" general instability load of about 100 lb per hub.

As these ten members were viewed on the model, it appeared that an additional ten members should also be stiffened to preserve strict model symmetry; thus, a total of 20 members was selected to be stiffened.

5.3. Beam stiffening method

While several methods were considered to delay lateral buckling of these selected beams, it was decided to cement 7075-T6 stiffening material to the lateral faces with epoxy adhesive. This method of model modification was readily accomplished and can be precisely expressed in the model description required in any computer analysis.

Standard beams with stiffeners having several cross-sectional geometries were buckled in the testing machine. The stiffening material in the shape of a T section, which provided a generous glue surface area to prevent a bond shear failure was chosen. Samples of "long" beams tested with these stiffeners (with ends clamped) buckled at about 1475 lb about the major axis of the standard beam compared with 475 lb about the minor axis for the standard beam. This accomplished the intention to prevent any chance of buckling laterally in the modified beams.

5.4. Instrumentation

In this series of tests, measurements were made as before at each loading condition of radial positions of all free hubs, all strain gauges, and all load cells. Due to having added stiffeners to some of the beams, most of the side-side strain gauges had to be removed. They were originally instrumented because of being critically loaded, and were consequently the first beams to need stiffening.

5.5. Test procedure

This series of tests was conducted in the same fashion as Phase 1. All readings were taken at zero load. Then at each new applied load, all load cells and strain gauge readings were recorded initially.

Next, all measurements of hub radial positions were made, which requires about 1½ hours for each load level. At the end of each run, all strain gauges and load cells were again read just before going to the next load. After setting the next load, all strain gauge measurements were again recorded. By reading before and after each change, drift of the strain gauge instrumentation could be taken into account. Totals are then deduced from the sum of the increments. In general, loads were not removed between runs.

Having the results of the first test, more confidence in the model allowed the loading program to proceed with 20 lb per hub increments on up to 80 lb per hub and thence to loads of 90, 92.5, and 95 lb per hub. At 95 lb per hub, a number of beams started to buckle and hub 1921 was observed to have touched its stop. Unfortunately, as the data were plotted it was evident that this hub had been against the stop starting at 60 lb per hub.

After taking data at 95 lb per hub, the loading was removed and all hub radial positions were again measured. Fortunately, no permanent set was observed from the hub position data and the beams were essentially straight again. At this time sketches and measurements were made of a number of beams to find the actual out-of-straightness. After lowering a number of stops, testing was resumed. New runs at 40, 60, 80, 85, 90, and 96 lb per hub were conducted.

As the load per hub was being increased gradually above 95 lb per hub, it was noticed that simultaneously the load measured at the main load cell dropped and hub 1321 moved down rapidly to touch its stop. The maximum load recorded from the main load cell was 95.9 lb per hub. After this event, the load was reduced to 95 lb per hub and all readings were repeated.

After lowering the stop under hub 1321, loading was again increased very slowly. This time incipient snap-through occurred at 95.8 lb per hub. A complete set of readings was made at this time, with hub 1321 touching its stop. Then, photographs were taken, a number of sketches were made to show the mechanism of general instability, and curvature measurements were made of a number of beams.

After completing the taking of data at this loading, the load was removed and all zeros were read. It was interesting to discover that buckling had been almost completely elastic. At zero load, the hub radial positions were *at most* 0.003 in. from the previous measurements at zero load, and many hub locations duplicated exactly their previous zero-load positions.

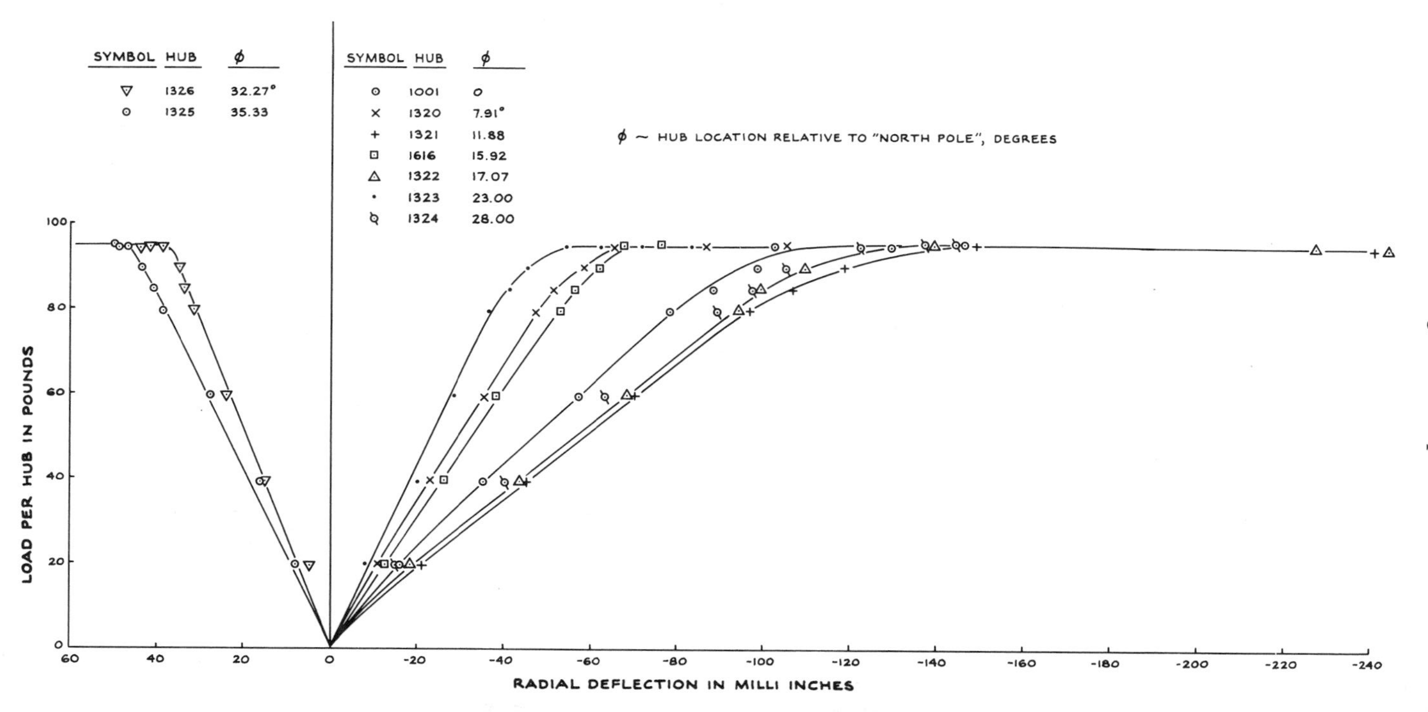

Figure 6. Hub radial deflection versus load per hub with 61 hubs equally loaded.

5.6. Test results

Test results include variation of radial deflection at each hub with load, strain gauge measurements of stress in certain beams, measurements of unit loads at 20 of the 61 loaded hubs and main load cell reading indicative of total load at the whiffletree apex. Also sketches, measurements and photographs were made to indicate the mode of buckling. Figure 6 presents typical results of the hub radial deflections, plotted against the load per hub from the latest main-load cell calibration data. This indicated that hubs 1321 and 1322 have become unstable, with deflections increasing without a corresponding change in load. Once again, the small load cells fulfilled their purpose by indicating that the whiffletree has uniformly distributed the applied load to each loaded hub.

Figure 7 presents qualitative information in the region adjacent to hub 1321 where the actual snap-through occurred. Hub 1001 is on the axis of symmetry and has been referred to as the "north pole." The arrows about the hubs indicate the direction of hub rotation about the radial axis due to the applied load. Curvature caused in the beams in the tangent plane is indicated in the sketch. In the normal direction, the plus and minus signs indicated the sense of the deflection, positive outward. It is quite obvious that, in contrast to Phase I, a very complicated mode of failure is represented here, with rotation of the hubs playing a major role. The beams were found to have deflected about both axes, and to have twisted. It was also apparent from observation that the hubs rotated about all three axes. Of interest is the fact that hub 1001, which is common to each of the five equal sectors, has also rotated.

As noted earlier, the model was designed with a geometry which repeats five times in identical sectors having a common point at hub 1001. Also, the loading in Phase II is entirely symmetrical. Therefore, probably some small abnormality in one sector has triggered the snap-through. The results in the immediate region of the snap-through point and also their counterparts in the other four sectors are presented in Figure 8. Information presented includes the direction of hub rotations about the radial axis and the deformed appearance of the beams in the tangent plane. Additional information regarding quantitative bending of the beams was measured using a curvature gauge. The numbers presented as fractions, such as 48/6, indicate a curvature of 48 thousandths of inches in a gauge length of 6 in., measured in the tangent plane. Similar numbers in parentheses indicate like information in the normal direction. It appears that all five sectors have distorted in a similar manner but in the actual snap-through location, all the distortions are greater.

The photograph in Figure 9 shows the visible distortions at the region of greatest interest near hubs 1321 and 1322. At this time, the maximum load was still applied and hub 1321 is resting on its stop. All other hubs are still self-supporting.

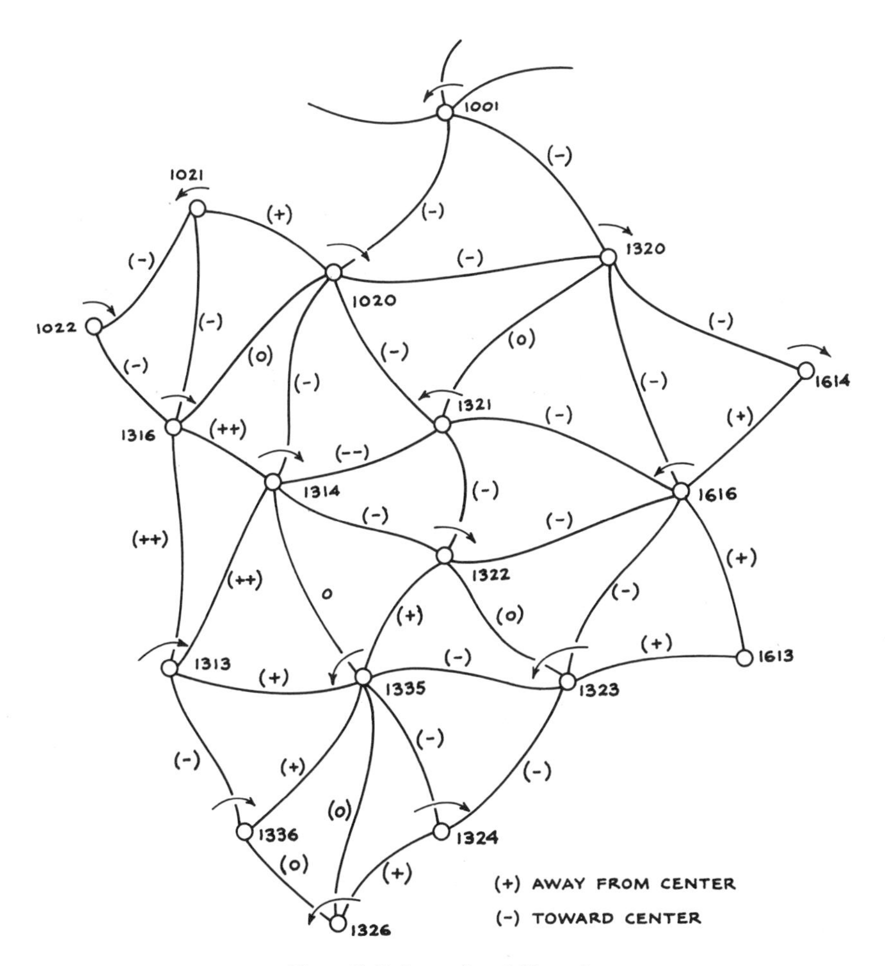

Figure 7. Pattern of unstable region.

5.7. Comments on equivalent-shell-theory comparisons

The maximum screw-jack-applied load carried by this model was 95.8 lb per hub, which caused a general instability in the model and would have caused hub 1321 to snap through except for the presence of the stop. If 2.6 lb per hub is added to allow for the dead weight of the whiffletree, the maximum load carried was about 98.4 lb per hub. An analysis of this model mentioned earlier, based on equivalent shell theory, predicted general instability to occur under uniform pressure loading of 1 psi. While no exact equivalence between uniform radial hub loading and constant pressure loading exists for a random

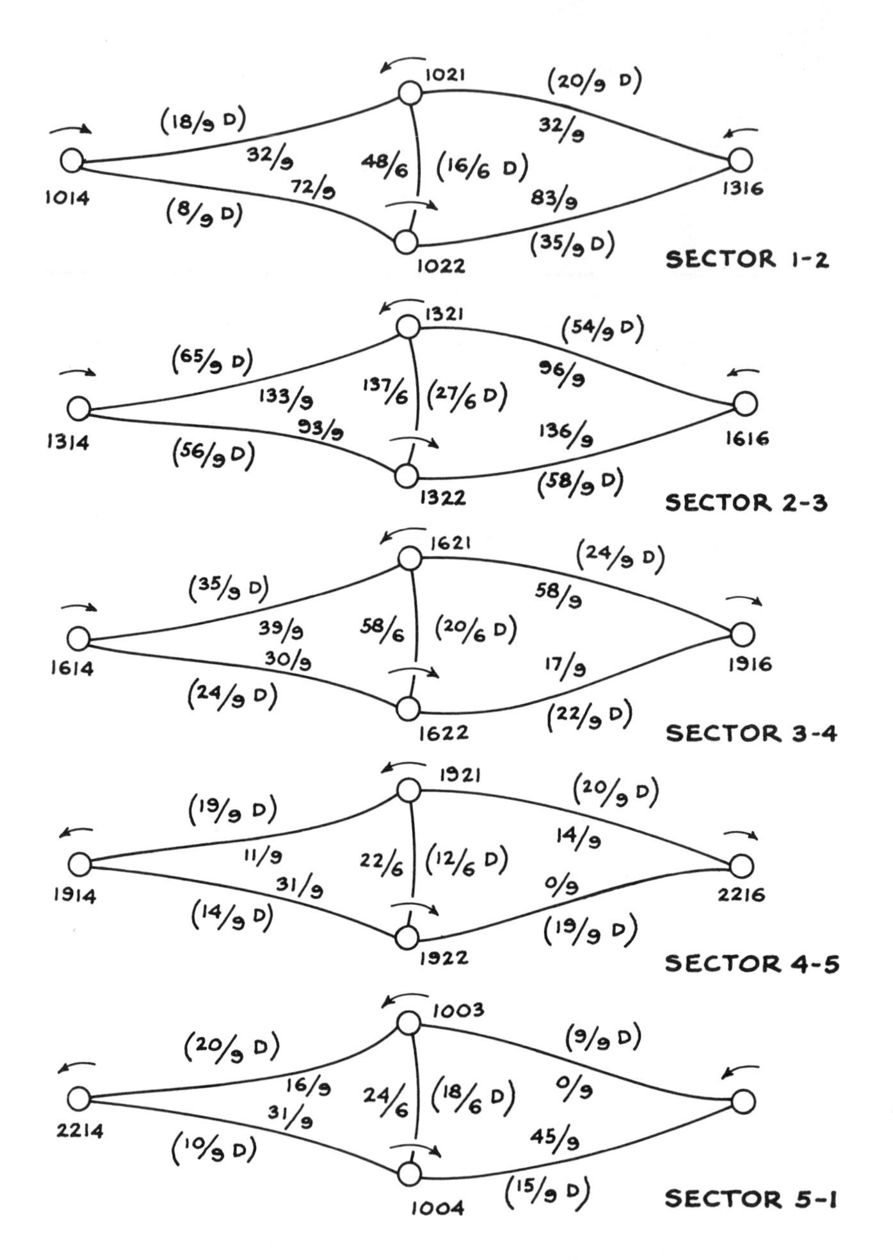

Figure 8. Beam curvature measurements in critical regions of the five sectors.

Figure 9. Photograph of unstable region.

radome, yet by a crude approximation, it is found that 61 loading hubs are distributed over about 6204 in.^{2}, or that

$$\text{load/hub equivalent to 1 psi} = \frac{6204}{61} = 101.7 \text{ lb/hub.}$$

Thus in spite of using several approximations, the comparison of the model results with equivalent shell theory predictions of general instability type of failure are as follows:

Experiment	98.4 lb/hub,
Theory	101.7 lb/hub.

While this is perhaps interesting, a more rational analysis and comparison is needed.

6. Summary Comments

The model space frame, which has been designed, constructed, and tested, has reached its objectives. For a model of known geometry and materials, two kinds of accurately applied static loads have provided several types of test data as the loading was progressively increased. These two kinds of loads have resulted in two kinds of buckling failure: (1) the first was a rather simple local buckling of an individual beam which did not result in a collapse and (2) the second was of a more complex nature, called a general instability, wherein many beams and hubs had deformed more or less about all axes and ultimately one hub required support to prevent a snap-through and a catastrophic collapse of the entire structure.

Acknowledgments

This work has been supported by Lincoln Laboratory, M.I.T., under Grant GP5832 from the National Science Foundation for Project CAMROC.

B. E. Greene, E. D. Herness, M. W. Ice,
W. L. Salus, and D. R. Strome
Aerospace Group,
The Boeing Company
Seattle, Washington

ASTRA – BOEING'S ADVANCED STRUCTURAL ANALYZER

Introduction

The Boeing Company has been developing the direct stiffness method of structural analysis since the early 1950's. This formulation of the matrix displacement method is based on the powerful finite element technique. Presently the Boeing Company's Aerospace Group is developing an advanced digital structural analysis system designed to meet the major structural analysis requirements of the Company during the next ten years. When completed, this system will be capable of obtaining numerical solutions within the requirements of engineering accuracy to practical problems in all fields of structural mechanics. The salient feature of the system is the ASTRA (Advanced Structural Analyzer) program which is the central theme of this paper.

ASTRA is based on the displacement, or stiffness, approach to the finite element method. From the description of the idealized structure as an assemblage of small, interconnected structural elements, ASTRA will generate a mathematical model of the structure consisting of a stiffness matrix, a mass matrix, and a damping matrix. Load vectors are also generated to represent the structural environment. From these basic generated quantities, ASTRA will produce solutions for static stresses and deflections due to loads and thermal effects, for natural vibration modes and frequencies, for critical buckling loads and modes, and for dynamic response to harmonic forcing functions. In addition to these solutions. ASTRA will produce generalized mass and stiffness matrices and modal damping matrices that will allow it to couple with other programs in the system for solving problems in transient dynamic response, random vibrations, structural control feedback, and flutter.

Inevitably, the development of a structural analysis system of this scope is the result of dedicated effort on the part of many people. The authors wish to acknowledge the work of many others in both structures and digital computing staff organizations throughout the company for contributions, either directly or indirectly, to ASTRA and to other programs that will be linked to ASTRA to provide all the capabilities mentioned above. Specific acknowledgement is due Richard B. Kaylor for his development of sophisticated sparse matrix techniques, which will allow the program to work with very large order matrices in an economical manner.

The bibliography dealing with theoretical development of the matrix displacement, or stiffness, method over the past decade is voluminous and the works of many contributors are hereby acknowledged without attempting to reference them individually. Only those that bear directly on the work carried on at Boeing have been referenced in this paper.

While active development of the ASTRA program has been under way for about 2½ years, the work that has made it possible began at The Boeing Company in 1953. With the advent of swept wings, low aspect ratio wings, and other structural configurations associated with high-speed high-performance aircraft, conventional methods of analysis previously employed in aircraft design were no longer adequate. Under the direction of M. J. Turner, a new method of analysis was sought which would solve the structural dynamics problems associated with such high-speed aircraft. The results of this original work were reported in References 1, 2, and 3. Later work[4] was directed toward exploiting the power of the stiffness method in the field of stress and deflection analysis of complex structures.

During this early period, small digital computer programs were used to verify the concepts of the direct stiffness method. Some of these results are reported in References 1, 2, 3, and 5. The comparison between digital and theoretical results were very encouraging and therefore, with the availability of the IBM 704 computer at Boeing, the development of a general purpose computer program to implement the theory was undertaken in 1958. Out of this effort emerged the COSMOS program. COSMOS (from Comprehensive Option Stiffness Matrix Organizational System) was one of the first fully automated digital computer programs capable of performing a complete stress and deflection analysis on complex structures of arbitrary configuration. COSMOS became operational in late 1960 and since that time it has been continually maintained and improved by adding new features as they became available. Most notable of the additions are: a method for treating large deflections and geometrically nonlinear structural behavior,[6,7] the development of peripheral programs for extending its capabilities in specific areas, the addition of an eigenvalue solution routine for obtaining natural vibration modes and frequencies, and a conversion of the program to the IBM 7094 computer in 1962. The success of the COSMOS program is attested by the fact that today, nearly ten years after its initial inception, it is still used 400 computer hours a month in filling structural analysis requirements throughout The Boeing Company.

With the structural analysis capability provided by COSMOS already operational it is necessary to discuss briefly the need for ASTRA. The fundamental fact behind this need is that more sophisticated space-age configurations with increasing demands on structural efficiency and ever more stringent design requirements require continual improvement of our analytical tools. This improvement takes place in several ways. First, advances in the finite element theory of structural analysis are continually being made. These often take the form of new or improved element types for representing or idealizing the structure. The stiffness matrices derived for these elements are the basic building blocks of the method and our ability to represent any structural configuration is limited only by the element types available. Typical of recent developments in finite element types are those for curved shell surfaces and for representing three-dimensional elastic continua. Other means by which the theory is advanced have to do with improved capability in specific fields of analysis such as plasticity and nonlinear material behavior. To take advantage of these advances in theory the computer program must be easily modified so that new developments may be tested and added permanently after they have been proved. The COSMOS program does not meet this requirement.

The second way in which our analytical capability is improved is through advances in digital computer technology. Larger memory and greater storage capacity with rapid access have made today's computers capable of handling much larger problems economically than was possible a decade ago. Similarly during the next decade many significant advances are anticipated which will greatly enhance our potential problem solving capability. Some of these may be (1) the use of central processors with multiple processing of jobs through time sharing, (2) direct communication of the analyst with the computer through remote consoles with graphic display of input and output, (3) faster processing through the use of optical switching. To be able to realize the advantages that such technological improvements will make possible, the program must be as machine independent as possible so that conversion can be made efficiently. COSMOS, written in machine language, requires a complete rewrite for conversion to a new computer configuration.

There is also a third way in which analytical capability is improved. That is by making the program more convenient and more efficient from the user's standpoint. Much of the voluminous input data required to describe the finite element idealization of a complicated structure can be generated automatically. This removes much of the burdensome, time-consuming, repetitive work, where errors are most likely to occur, from the hands of the engineer. Better organization and display of output data can remove much of the confusion and delay caused by sorting through countless pages of printed output to locate and interpret desired information.

The means by which ASTRA meets these needs are discussed in the description of the ASTRA system, its specifications and specific design features.

The stiffness method has been chosen as the basis for the ASTRA system because (1) the degree of redundancy of the structure is immaterial; (2) there is a wide choice of elemental structural types available; (3) modification of the structure to accommodate design changes requires a minimum of changes in the basic input data defining the structural model; (4) it is highly versatile in the analysis of structures of a completely arbitrary configuration; (5) it is convenient for representing dynamic behavior of structures; (6) it is adaptable to extension into the fields of nonlinear geometry and material behavior; (7) its use allows any degree of refinement required in the mathematical representation of the structure. An indication of the versatility of the method is its use in analysis problems ranging from stress and vibration analysis of airframes to stress concentration factors surrounding cavities in rock due to blast overpressure loads. Figures 1, 2, and 3 show some typical structures to which ASTRA is adapted. Of particular interest to this meeting is ASTRA's ability to analyze (1) three-dimensional space frameworks and truss structures typical of large, steerable reflectors, as shown in Figure 1, (2) shell-like structures, both monocoque and semimonocoque, typical of smaller reflectors and radomes, and (3) thick shells and bodies of revolution typical of radomes in the nose cones of missiles and re-entry vehicles. Several examples are given at the end of this paper to show how ASTRA might be applied to the structural analysis and design of some typical radar antenna configurations.

The ASTRA System

The method used in ASTRA for determining static deflections and stresses as the basis for all other analytical capabilities is that of the direct stiffness method. This method determines the structural characteristics of a complex structure as the composite of the structural characteristics of many individual finite elements. The method of combining the elastic characteristics of individual elements by direct superposition to form a gross stiffness matrix for the structures has given the method its name. Clearly, the concept also lends itself to the superposition of inertial and dissipational characteristics to form mass and damping matrices to complete the mathematical model of the gross structure for dynamic purposes.

The individual element characteristics are derived from relatively simple displacement fileds assumed over the element. These displacement fields are

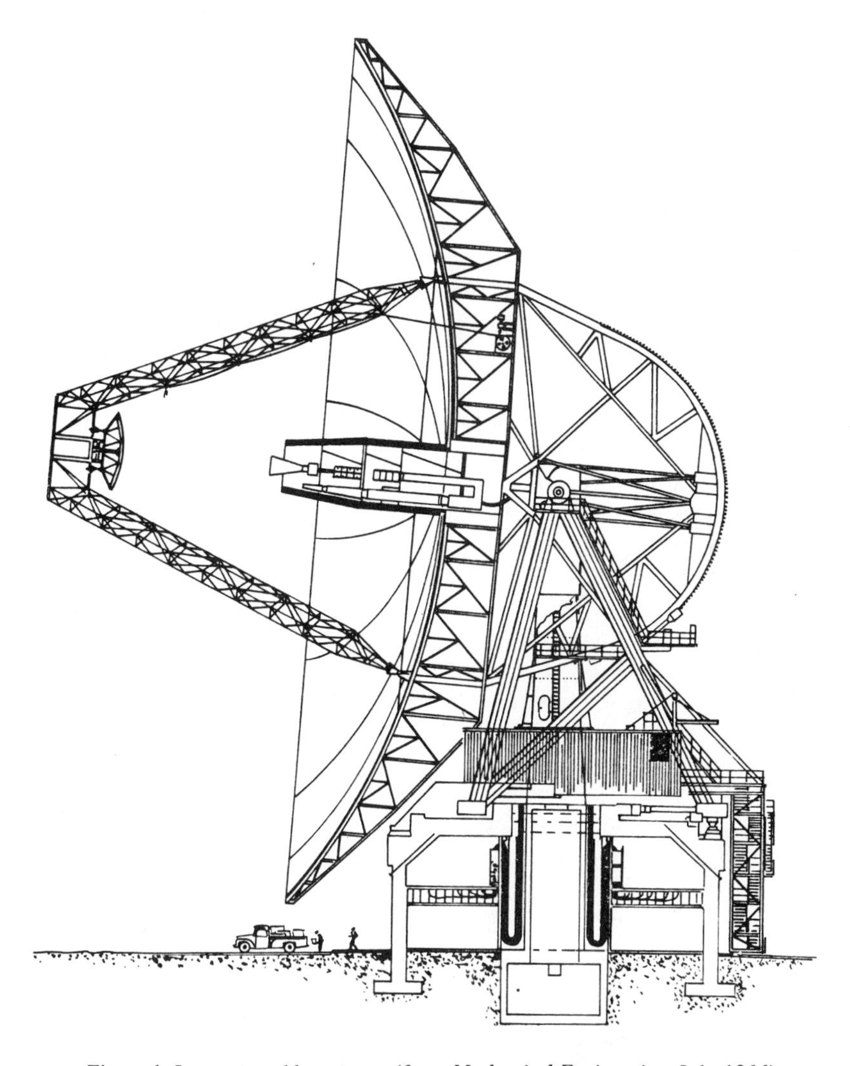

Figure 1. Large steerable antenna (from *Mechanical Engineering*, July 1966).

defined by displacements of specific points on the element boundary, which will hereafter be called nodes. The complete displacement field over the structure is the superposition of the elemental displacement fields obtained by requiring continuity of displacements between adjacent elements at common nodes. Hence, the deformation of the entire structure is completely defined by the nodal displacements.

The direct stiffness method is actually an extension of the Ritz method in

Figure 2. S-IC stage Saturn V launch vehicle.

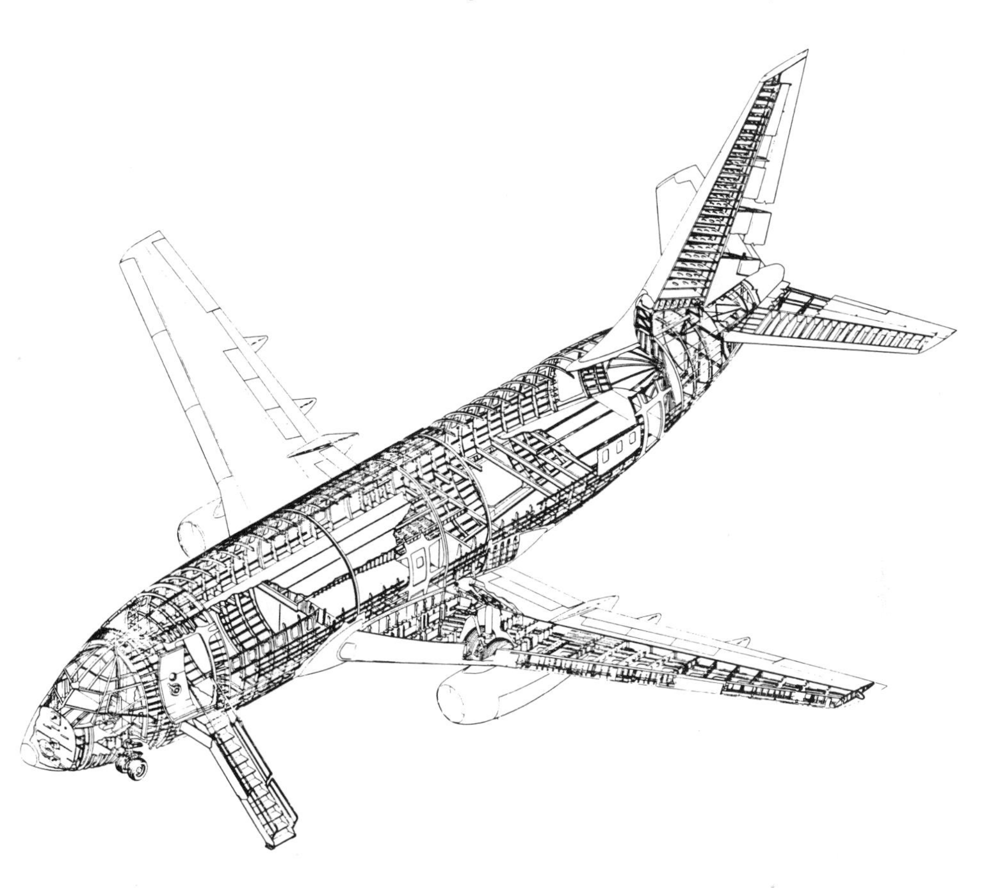

Figure 3. Typical airplane structure–Boeing 737.

which the assumed displacement states are defined over individual elements rather than continuously over the entire structure. In order to preserve the continuity of the entire displacement field, which is essential to the Ritz method, elemental displacement fields are chosen so that no cuts or hinges occur in the structure along element boundaries.

One of the unique features of the ASTRA system which makes it economically feasible to solve very large problems involving thousands of nodal displacements as degrees of freedom is the merge and reduce procedure. This is a procedure by which generation and solution of the system equations are carried on simultaneously so that a complete stiffness matrix need not be formed. This is made possible by the sparse, banded character of the stiffness matrix. When all elements connecting to a given node have been brought into the stiffness matrix no further coupling with the degrees of freedom at that node is possible and those freedoms may be eliminated by Gaussian reduction.

Thus, by proper ordering of the structural elements, a stiffness matrix of several thousandth order can be continually reduced as it is formed and never allowed to grow beyond a few hundredth order. Of course, a full stiffness matrix may be generated if desired.

General specifications

In order to meet the requirements of an advanced structural analysis system, as outlined in the Introduction, ASTRA has been designed to the following specifications.

Generality. The program is general enough to accept all structural configurations that can be adapted to a finite element type of idealization. Thus, it is limited in this respect only by the catalogue of element types available. Most structures, such as those shown in Figures 1, 2, and 3, are idealized by one- and two-dimensional elements, i.e., beams and plates. Very thick plates and shells and other massive bodies require the use of three-dimensional elements. A full complement of elements in all three categories is available for use in ASTRA.

For the elements in the ASTRA program, the analyst has the option to relax nodal freedoms at the elemental level. With proper usage of this feature, pins, slots, ball and socket joints, hinges, etc., can be simulated.

Specifically, the elements immediately available are:

1. Straight beams of uniform section resisting stretching, torsion, and bending about two axes. The effect of shear deformation on bending stiffness may be included if desired. The node points defining the ends of the element may be offset from the elastic axis in order to properly represent the behavior of a stiffener attached to one side of a plate or shell. The element may be subjected to a linearly varying distributed lateral load between node points and a linear temperature distribution.

2. Flat, orthotropic triangular or quadrilateral panels of uniform thickness, having membrane and bending properties and subject to linearly varying distributed lateral load and a linear temperature distribution.

With these elements, structural configurations similar to each of those shown in Figures 1, 2, and 3, have been successfully analyzed. The idealization used in the analysis of the airframe of Figure 3 is shown in Figure 4. Members having variable section properties are adequately represented by a stepped approximation, and curves are approximated by polygons. Other elements that have been derived and proven reliable, and which will be added to ASTRA include:

3. Straight and curved beams resisting stretching, torsion and bending, and having arbitrarily variable section properties, loading, and thermal distribution.[8]

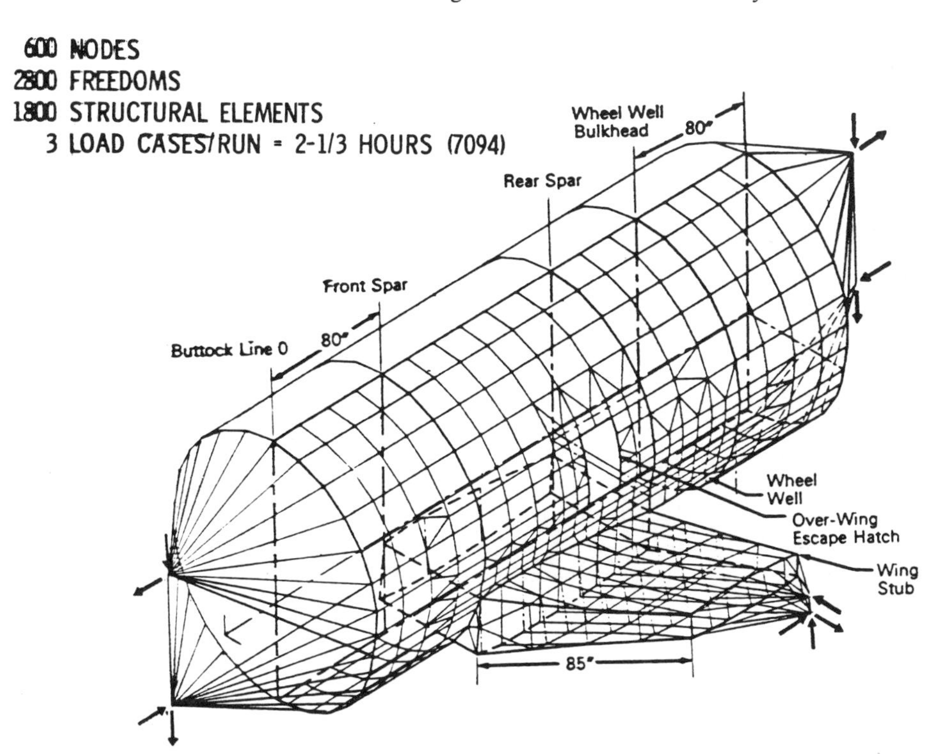

Figure 4. Finite element idealization–Boeing 737 wing-body junction.

4. Conical and doubly curved axisymmetric shell elements with symmetric or unsymmetric loading.[9-11]

5. Circular ring element of triangular cross section for axisymmetric solid bodies subjected to symmetric or unsymmetric loading.[12-14]

6. Tetrahedron element for building massive solid bodies of any shape, and for three-dimensional elastic continuum problems.[15-17]

Machine Independence. ASTRA is written entirely in FORTRAN IV except for a few library subroutines. Thus, ASTRA can be made operational on any computing facility having a FORTRAN IV compiler, subject to certain minimum configuration requirements, with a minimum amount of rewrite. Current conversions are being made for the IBM 360/67 and the CDC 6600 computers.

Ease of Modification. The program is written in modular form. That is, each computational step is performed by an independent unit, or module. Data transmittal between modules and peripheral storage is accomplished through a Data Retrieval System (DRS), which allows any module to be modified, or a new module to be added to provide additional capability, without affecting the rest of the program. The only requirements are that each module must have input and output formats compatible with those of other modules to which it links and with the DRS.

Size of Problem. There is no program limitation on the maximum number of variables in a static stress or deflection analysis. However, practical limitations may be imposed by the configuration of the computing facility with respect to memory size and amount of peripheral storage available. Practical limitations may also be imposed by the impact of running time on economy and on reliability, and by the effects of roundoff error on solution accuracy. In the initial version of ASTRA (UNIVAC 1108) it is estimated that the practical problem size limit will be on the order of 12,000 degrees of freedom for an ordinary static stress and deflection analysis. The eigenvalue solution is limited to 2000 degrees of freedom, although it is estimated that most problems will be of smaller size.

Input-Output. Input to the ASTRA program is via two user-oriented languages, one for describing the structure and its environment, the other for defining the analysis to be performed. These languages contain advanced features that will greatly enhance efficiency and flexibility in the use of ASTRA by making it possible for the program to generate much of the repetitive geometric and descriptive data automatically, and by allowing the user complete control over solution paths. These input languages are described in detail in the following section of this paper.

Graphic plots are used for checking input data and will be used for displaying output as selected by the engineer. The engineer will also have control over the volume and format of printed output to prevent the generation of useless reams of paper.

Special Features. 1. Troubleshooting – Graphic displays of both input and output will be a great help in detecting input data errors. An over-all equilibrium check is made to indicate solution accuracy for static problems. Input format errors and machine or tape errors will be detected whenever possible and suitable diagnostic comments will be printed out. When an error is detected the program will not exit, but will continue processing the data so that as many errors as possible will be located on the first run.

2. Re-entry – The program uses a save and re-entry feature that allows the user to interrupt the problem solution at any time. The program may be re-entered later using the data generated and saved from the previous run along

with other input data to complete the analysis. This is particularly useful where the user wishes to examine intermediate results and make data changes accordingly before proceeding. The save and re-entry feature includes an automatic error recovery procedure. If a processing error, such as a tape read error, is detected by the program, it will save all usable data generated. The analysis may then be re-entered at the point where the failure occurred and carried to completion.

3. Constraints – A completely general use of constraints[18] will be incorporated into the program to aid in properly representing boundary conditions, rigid inclusions in a flexible structure, and other kinematic restrictions on structural response. Some of the problems that can be solved through the use of this feature are: the inextensional analysis of arches, rings, and frames; insertion of pins or hinges in otherwise rigidly connected frames; rigid rings and fittings in shell structures; free-free vibrations; and the static analysis of unsupported structures.

4. Assemblies – To assist both the engineer and the program in dealing efficiently with large, complex structures, the structure may be decomposed into major subdivisions called assemblies. Each assembly may be considered as a separate structure in describing input data and in the generation and solution parts of the program until the final solution step in which all assemblies are united. Output data are organized by assembly and printed output may be limited to any assembly as desired.

5. Coordinate systems – A choice of coordinate systems is available to aid in representing different structural geometries most efficiently. Missile structures and bodies of revolution, for example, are most conveniently described in cylindrical coordinates, while airframes, buildings, and irregular structures are usually described in a Cartesian coordinate system.

The specific coordinate systems now offered are rectangular, spherical, and cylindrical. Other orthogonal, curvilinear coordinate systems that may be defined by parametric equations can be added easily at any time. Furthermore, the engineer may supply his own coordinate system by supplying the appropriate transformation matrices.

Different coordinate systems may be used for different assemblies within the same structure. This will be particularly useful in describing large antenna structures where the reflector, a paraboloid of revolution, is supported on a tower structure that is best described in rectangular coordinates. Joining of two assemblies having different coordinate systems is easily done in the final solution with the use of the constraints feature mentioned above.

6. Mass and loads data generation – Structural mass matrices and load vectors to represent distributed loads and thermal effects are derived according to the procedures described in References 19, 20, and 21, and are generated automatically. In addition, lumped masses and point loads specified by the

engineer at nodes are included by being added directly into the generated mass matrix and load vectors. Since the dynamic problem size limitation is considerably less than for static problems, reduction of the mass matrix is necessary. This is done as described in Reference 22.

7. Eigenvalue solution – The principal eigenvalue solution module will use Householder's method. The advantages of this method are: (1) that all eigenvalues are obtained, (2) solutions are obtained for multiple roots, (3) the stiffness matrix need not be inverted, (4) matrices up to 2000th order are accepted. A backup routine using power iteration will also be available. This will be limited to smaller problems, and may prove useful where only a few of the lower-order eigenvalues and eigenvectors are desired.

8. Small problem solution – For small problems that can be contained entirely in core there is a very efficient solution package called BABY ASTRA. This will produce static stress and deflection solutions and will be of particular value in checking out new elemental stiffness matrix derivations where many small problems are solved.

Flexibility. The program is capable of operating in two general modes. In the first mode of operation, the program will select the solution path from a minimum of control data supplied by the analyst. This results in a highly automatic solution for most normal structural analysis requirements. In the second mode of operation the analyst will specify the solution path by detailed control input, thus obtaining maximum flexibility in solving his problem. The ultimate in flexibility will be provided eventually by the incorporation of a matrix abstraction routine that will be able to perform any desired algebraic operations on the matrices generated within ASTRA or input by the use. It will be able to communicate directly with any of the computational modules through the DRS subject only to the restriction that proper compatibility and identification requirements be observed.

Documentation. The efficient use of a sophisticated computer program requires concise and informative documentation. Documents published with respect to the ASTRA system will include an introduction, basic theory, element deviations, idealization techniques, user manuals, and programmers documents.[23–28]

Input languages

The engineer will communicate with the ASTRA system via two input languages. The Structural Analyzer Input Language (SAIL) is used to describe the physical and geometrical properties of the structure. The Structural

Analyzer User Control Language (SAUL) is used to define the appropriate analysis path through the various computational modules of ASTRA. These languages are discussed in greater detail below.

SAIL – Structural Analyzer Input Language. The description of a complex structure required by a general purpose structural analysis program such as ASTRA comprises a tremendous volume of input data. These data consist essentially of (1) identification number and coordinates for all nodes, (2) description of all structural elements as to type, location, and their material and sectional properties, (3) structural environment consisting of temperature and load distribution specified by node and/or by element. To illustrate what this means in terms of a typical application, consider the wing body juncture of the airframe structure shown in Figure 3 and idealized into a finite element assemblage as shown in Figure 4. This idealization contains 600 nodes, totaling 2800 deg of freedom, and 1800 structural elements. The nodal data alone for this structure consisted of approximately 6000 words while the structural data contained over 14,000 words. For this particular problem only three static load cases of simple configuration were analyzed, but if distributed loads or temperatures had been included this could easily have added another 1200 words per load case. Thus, it is seen that only a moderately complex structure, in terms of the capability offered in ASTRA, may require over 20,000 words of data for a complete description to the program.

The performance of the burdensome and repetitive task of generating, tabulating, and verifying this seemingly endless quantity of data by the engineer has been a consistent source of input data errors that are difficult to find and result in failure of the analysis, or worse yet, in undetected inaccuracies in the results. The Structural Analyzer Input Language (SAIL) is designed to relieve the engineer of this burden by simplifying the input to ASTRA and automating the generation of data wherever practical.

The advanced features of SAIL represent a radical departure from the normal methods of entering numerical data into a structural analysis program of this type. To use them effectively will require imagination, study, and experience on the part of the engineer. Thus, they offer a challenge which may prevent even the most prosaic of structural analysis problems from becoming boring.

In addition to the advanced features, which will be discussed in detail, SAIL also allows simple tabular input similar to that used in the past so that the engineer may begin using ASTRA immediately, before he has become familiar with the use of the advanced features. Even here he has some labor-saving devices available. One of these is an autoduplication feature which eliminates the need for entering the value of a specific input parameter each time it is repeated; e.g., the moment of inertia of a beam element need only be entered at the beginning of the tabulation of beam structural data, and whenever its value changes. The other is the use of tables which allow the engineer to specify a complete list of structural parameters for an element by giving a table reference, a single number, corresponding to a previously compiled table of data.

The bulk of the advanced capability in SAIL for describing the structure and automatically generating input data is contained in three features which are now discussed in detail.

1. Defining and manipulating substructures – The ASTRA user has the ability to group together some of his input data to form a unit. This unit (i.e., the collection of data) is called a substructure and can be manipulated to form other units (i.e., other collections of data). This operation can be interpreted as defining a substructure, and then, by appropriately translating and rotating this substructure, forming other substructures. Therefore, when the user wishes to transform a substructure that he has defined, he gives a translation vector v_0 and a rotation matrix A. The new substructure is automatically generated from the original substructure and the transformation data as follows:

A. The coordinates of each node of the new substructure,

$$\begin{Bmatrix} x_2 \\ y_2 \\ z_2 \end{Bmatrix},$$

are formed from coordinates of the original substructure,

$$\begin{Bmatrix} x_1 \\ y_1 \\ z_1 \end{Bmatrix},$$

by

$$\begin{Bmatrix} x_2 \\ y_2 \\ z_2 \end{Bmatrix} = \begin{Bmatrix} v_0 \end{Bmatrix} + \begin{bmatrix} A \end{bmatrix} \begin{Bmatrix} x_1 \\ y_1 \\ z_1 \end{Bmatrix}.$$

B. A set of concentrated loads associated with each node of the new substructure is formed from concentrated loads of the original substructure by

$$\begin{Bmatrix} P_{x_2} \\ P_{y_2} \\ P_{z_2} \\ \hdashline M_{x_2} \\ M_{y_2} \\ M_{z_2} \end{Bmatrix} = \left[\begin{array}{c:c} A & O \\ \hdashline O & A \end{array}\right] \begin{Bmatrix} P_{x_1} \\ P_{y_1} \\ P_{z_1} \\ \hdashline M_{x_1} \\ M_{y_1} \\ M_{z_1} \end{Bmatrix} .$$

C. A structural element in the new substructure is formed from each element of the original substructure by connecting corresponding transformed nodes with a similar element.

As an example of the use of substructure transformation, one-half of a structure (symmetric about the x-z plane) could be input, then transformed to get its mirror image by the following transformation:

$$\begin{Bmatrix} v_0 \end{Bmatrix} = \begin{Bmatrix} 0 \\ 0 \\ 0 \end{Bmatrix} , \begin{bmatrix} A \end{bmatrix} = \begin{bmatrix} 1 & 0 & 0 \\ 0 & -1 & 0 \\ 0 & 0 & 1 \end{bmatrix} .$$

The unit, or collection, of data comprising a substructure is a completely general set. It may contain nodal, structural, or loads data, or any combination thereof. For example, a pattern of nodes with no connecting structural elements or loads may be manipulated as a substructure if desired.

2. Computational loops – Perhaps the most powerful SAIL feature the ASTRA user has available in the ability to establish computational loops. Before a computational loop is described, consider the pattern of nodes shown in Figure 5. Notice that this pattern of nodes is formed by six successive columns of four points each. Notice also that any node can be completely defined by specifying the column in which it appears j and the position within the column i. That is, the label or identification of the node is given by 4(j-1) + i, the x coordinate is given by 3(j-1), and the y coordinate is given by 2(i-1). Therefore this entire pattern of nodes could be generated by the descriptive statement, "establish a node having 4(j-1)+i as its label, 3(j-1) as its x coordinate, and 2(i-1) as its y coordinate: let j assume the values 1, 2, 3, 4, 5, and 6, and for each value of j, let i assume the values 1, 2, 3, and 4." The generation of this pattern of nodes is a simple application of the computational loop feature in SAIL.

A computational loop is established by enclosing a set of input within two statements indicating the beginning and end of the loop. There is an index associated with the loop which is given an initial value v_i, a final value v_f, and a increment Δv. The increment is added to the index at the end of each pass through the loop so that, as the loop is executed, the loop index assumes the values v_i, $v_i + \Delta v$, $v_i + 2\,\Delta v, \ldots, v_f$. The looping procedure halts and normal sequential processing continues after the loop executes the final value of the loop index v_f.

The computational loops which might be used to form the nodal pattern shown in Figure 5 are stated below.

A. Begin a loop using j as the loop index. The initial value of j is 1, the final value is 6, and the increment is 1.

B. Begin a loop using i as the loop index. The initial value of i is 1, the final value is 4, and the increment is 1.

C. Establish node having 4(j−1)+i as a label, 3(j−1) as its x coordinate, and 2(j−1) as its y coordinate.

D. End loop using i as the loop index.

E. End loop using j as the loop index.

Notice that the loop within a loop has precisely the effect: "let j assume the values 1, 2, 3, 4, 5, and 6, and for each value of j, let i assume the values 1, 2, 3, and 4." In reality, each of the five steps given above can be specified by a simple SAIL statement, allowing the ASTRA user to input the entire set of 24 node points with only five statements. It should be noted that by merely changing the final values for the loop indices, these five SAIL statements will generate any similar rectangular pattern of nodes. Let us now consider the slightly more complex pattern of nodes shown in Figure 6 and indicate how that pattern might be handled in SAIL.

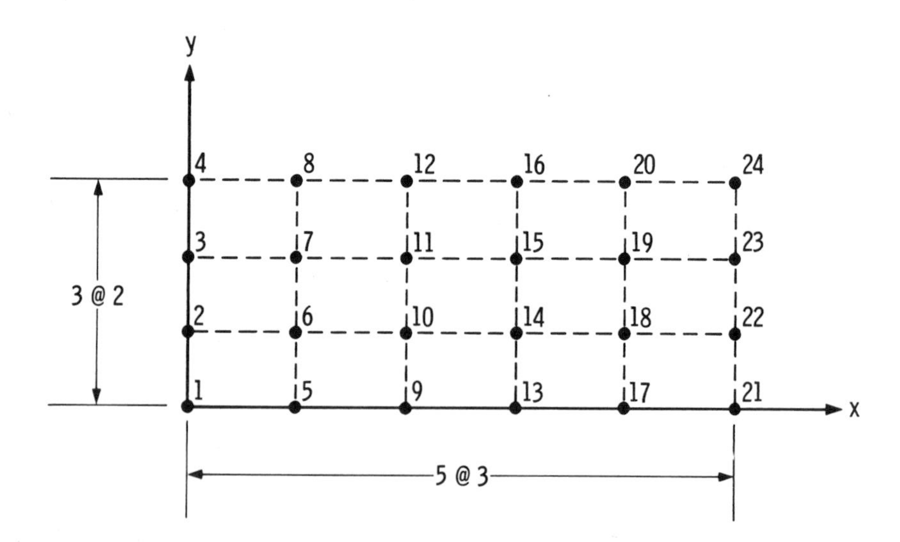

Figure 5. Rectangular array of nodes.

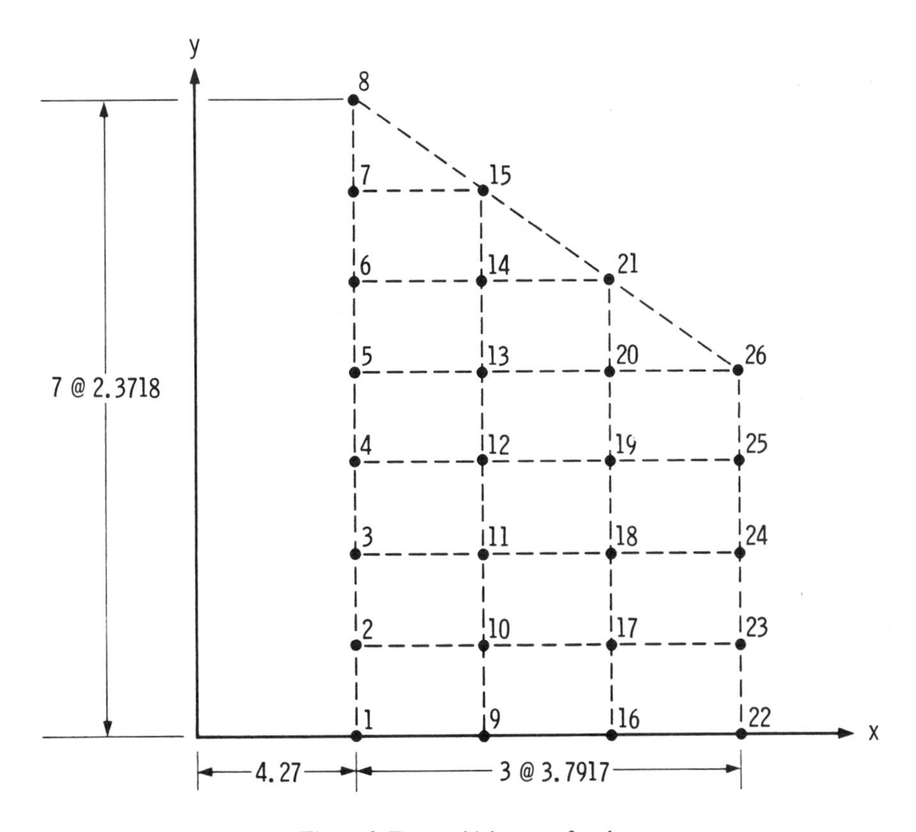

Figure 6. Trapezoidal array of nodes.

A. Initialize an index k to be zero.
B. Begin a loop using j as the loop index. The initial value of j is 1, the final value is 4, and the increment is 1.
C. Initialize and index jj to be 9–j.
D. Begin a loop using i as the loop index. The initial value of i is 1, the final value is jj, and the increment is 1.
E. Increment the index k by 1.
F. Establish a node having k as a label, 3.7917(j– 1)+4.27 as the x coordinate and 2.3718 (i– 1) as the y coordinate.
G. End loop using i as the loop index.
H. End loop using j as the loop index.

Again these eight steps represent eight simple SAIL statements. Naturally, structural elements, loads, and temperatures can also be specified within the computational loops as functions of the loop indices.

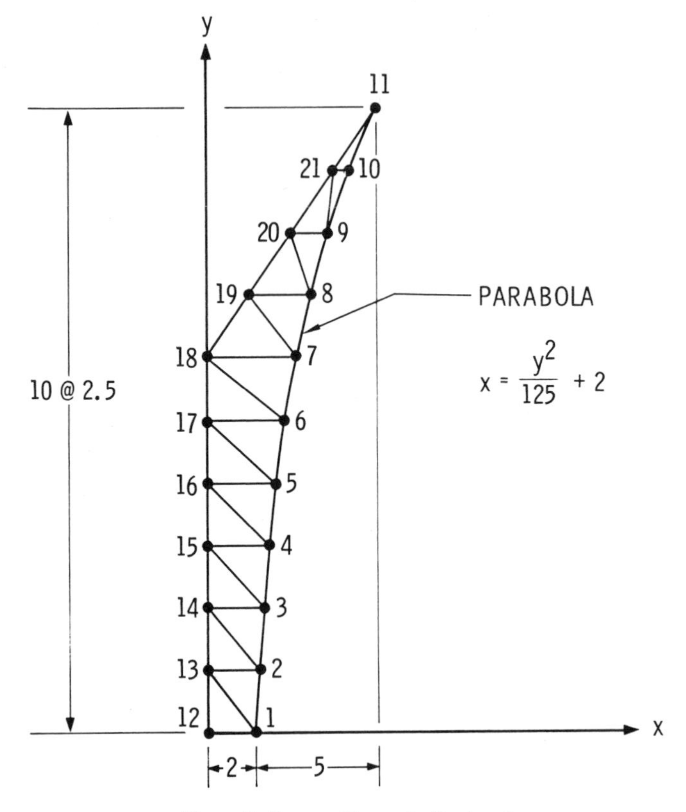

Figure 7. Truss with parabolic chord.

As a final example of how the computational loop can be used to automate ASTRA input, steps will be given to indicate how the truss shown in Figure 7 might be input.

A. Begin a loop using i as the loop index. The initial value of i is 1, the final value is 11, and the increment is 1.

B. Establish a node having i as a label, $((2.5(i-1))^2/125)+2$ as the x coordinate, and $2.5(i-1)$ as the y coordinate.

C. End loop using i as the loop index.

D. Begin a loop using j as the loop index. The initial value of j = 1, the final value is 7, and the increment is 1.

E. Establish a node having j+11 as a label, 0 as the x coordinate, and $2.5(j-1)$ as the y coordinate.

F. End loop using j as the loop index.

G. Begin a loop using k as the loop index. The initial value of k is 1, the final value is 3, and the increment is 1.

H. Establish a node having k+18 as a label, (7/4) (k) as the x coordinate, and 15+(2.5) (k) as the y coordinate.
I. End loop using k as the loop index.
J. Establish an element type.
K. Establish a table of element properties.
L. Begin a loop using *l* as the loop index. The initial value of *l* is 1, the final value is 9, and the increment is 1.
M. Establish element connecting nodes *l*, *l*+1.
N. Establish element connecting nodes *l*, *l*+11.
O. Establish element connecting nodes *l*, *l*+12.
P. Establish element connecting nodes, *l*+11, *l*+12.
Q. End loop using *l* as the loop index.
R. Establish element connecting nodes 10, 11.
S. Establish element connecting nodes 10, 21.
T. Establish element connecting nodes 21, 11.

Again each of the 20 steps shown here can be specified by one simple SAIL statement. Therefore, with only 20 SAIL statements, the input data describing the 21 nodes and the 39 structural elements making up the truss, has been generated. This example illustrates another advantage that SAIL offers. In addition to greatly reducing the amount of input required, it facilitates the description of mathematical curves and surfaces. Notice that the coordinates of the nodes lying on the parabola in Figure 7 are specified mathematically. SAIL computes the coordinates automatically at the time of input, and the user does not have to scale them from a drawing nor calculate them by hand. This feature is invaluable when analyzing antenna reflectors or other structures containing mathematical curves and surfaces.

The full potential of the computational loop feature in SAIL is brought forth when the nodal grid is a repetitive geometric pattern. Such patterns are often found in practical structures such as the first stage of the Saturn V shown in Figure 2. Previous forms of input required tedious hours of engineering time generating and preparing these data in the proper format. SAIL, through a few simple statements, transfers the generation of this data to the computer, thus making more efficient use of the engineer's time while improving the accuracy of the input.

The geometric pattern of the cylindrical shell and supporting structure is recognized as a series of concentric circles. The generation of the nodal coordinates becomes trivial in ASTRA when cylindrical coordinates are used. One nodal circle is generated when r and z are constant and θ is incremented around the circumference of the circle. A second nodal circle has the same θ and z coordinates as the first, only the r coordinate is changed by a constant increment. Furthermore, these two nodal circles are repeated along the axis of the cylinder when r and θ are held constant, but z is changed by a constant increment. The description above is that of three SAIL computational loops, which will generate all the nodal data of one of the cylindrical sections. The others are equally simple.

One problem remains, that of generating the structural elements. In Figure 2, there is a circular annulus that is repeated along the axis and a cylindrical shell that is also repeated between nodal circules along the axis. Thus if a repetitive pattern of node numbers was chosen along the cylinder axis, the structural elements may be generated using two tables and two computational loops. Similar reasoning can be used for the associated trusswork.

3. External data generator – The last significant feature of SAIL to be discussed is the external data generator. An external data generator is a group of SAIL statements which generates a set of ASTRA input having a variable configuration controlled by one or more parameters. At the time the external data generator is input, no ASTRA data are actually produced. Rather, the SAIL processor "remembers" the algorithm defined by the statements in the external data generator. Thereafter, the engineer may reference the external data generator, each time supplying the desired values for the control parameters, to obtain the corresponding set of ASTRA data. Normally an external data generator is only temporarily "remembered" by the SAIL processor (i.e., for only the current computer run). However, it is a simple matter to permanently incorporate any external data generator into the ASTRA system whenever desired.

As an example, an external data generator can be set up to produce the truss shown in Figure 8(A). Let the control parameters for such an external data generator be i, a, b, and n, as indicated in Figure 8(A). With such an external data generator available any of the trusses shown in Figures 8(B) through 8(E) can be generated with one reference to the external data generator giving the appropriate values of i, a, b, and n.

External data generators can be set up to automatically produce the input for any class of structure. Examples include flat plates idealized by a regular array of elements, stiffened plates and cylinders, frames for airplane fuselages and wings, etc.

The use of an external data generator for an antenna structure is illustrated in the application section of this paper. The external data generator allows the engineer to make very rapid trade and parameter studies of various structural configurations and therefore should prove very useful in preliminary design.

SAUL - Structural Analyzer User Control Language. The ASTRA system contains many different computational boxes, each of which performs some specific function in the processing of the input data to produce a solution.

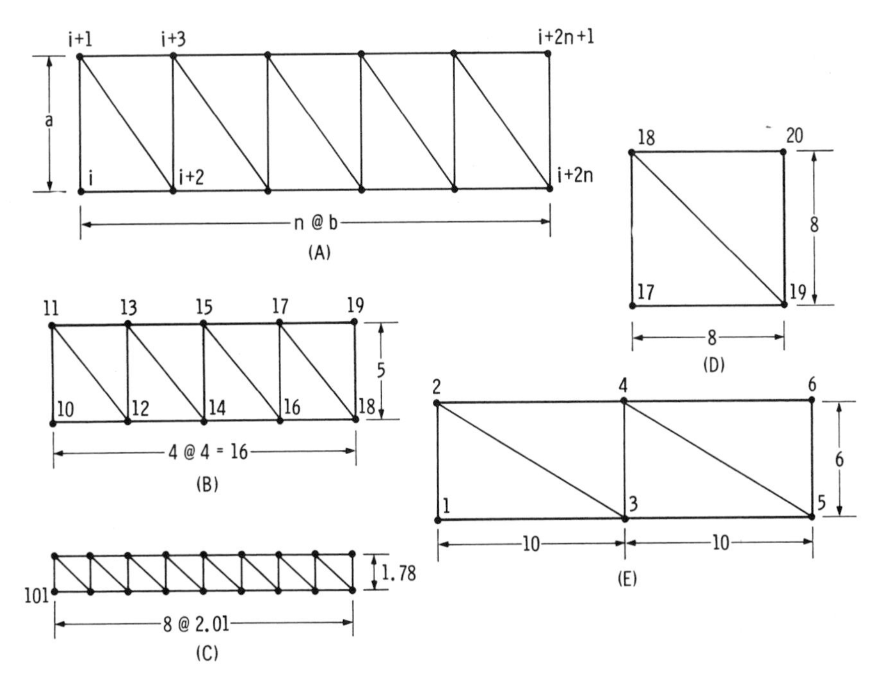

Figure 8. Application of external data generator to a truss structure.

For the engineer to call each box in the correct order to perform his analysis would require an intimate knowledge of the internal workings of ASTRA. Therefore, a Structural Analyzer User Control Language, SAUL, has been developed which will generate the control required to call the necessary computational boxes in the correct sequence to perform the desired analysis. At the same time, SAUL will allow the advanced user to describe his own processing sequence if so desired.

SAUL consists of problem-oriented language statements that allow the engineer to conveniently describe the desired solution of his problem. One or more SAUL statements describes the solution procedure he wishes ASTRA to perform on his input data. The SAUL statements are transformed by the SAUL processor to form the executive control for the current solution path through ASTRA.

The SAUL language has been designed to be very simple for the standard solution paths through ASTRA but flexible enough so that the engineer can choose nonstandard paths that will give him greater flexibility in performing his analysis. A single SAUL statement is sufficient to perform any one of the eighteen standard ASTRA analyses. SAUL is designed such that additional standard solution procedures can be readily added.

Each specific analysis capability presently contained in ASTRA corresponds to one or more of the standard solution paths available. The statement PERFORM STRESS ANALYSIS, for example, implies a standard solution for a stress analysis known as the Gross Merge procedure. A flow diagram for this standard solution procedure is shown in Figure 9. In this procedure a single

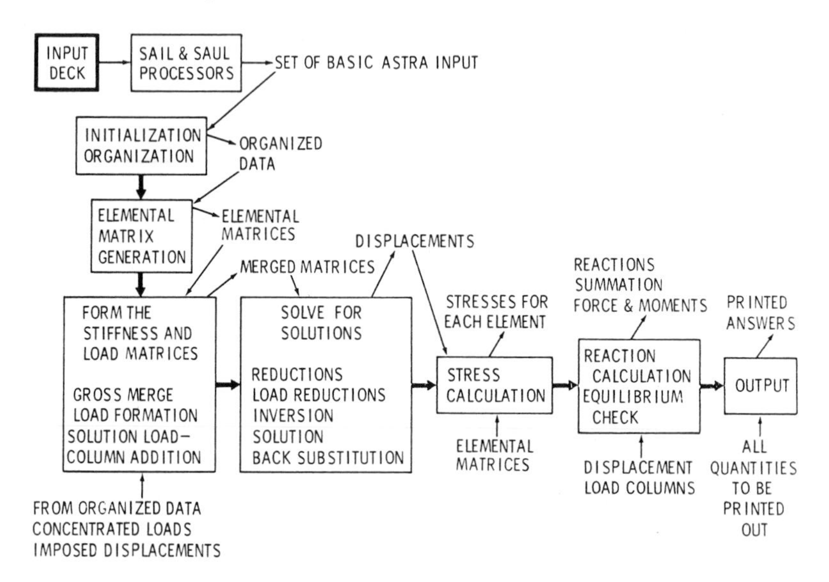

Figure 9. Flow diagram for a typical stress analysis in ASTRA.

Gross Stiffness Matrix is formed before any reductions are performed. This standard solution procedure has the advantage of having no programming limitations as to the size or type of problem that can be solved. In this procedure there are practically no limits on the number of imposed displacements, nor does it require the stiffness matrix to be banded. The only limitations on the size of problem that can be solved are practical limitations on the size of the available memory, the amount of peripheral storage available, and the computer run time.

The statement PERFORM STRESS ANALYSIS, MACRO MERGE PROCEDURE is a SAUL statement that implies another standard stress analysis solution path through ASTRA. In this procedure the engineer divides his structure into convenient parts called assemblies. For each assembly, the elemental stiffness matrices are then merged into an assembly stiffness matrix which is reduced as it is formed. The reduced assembly stiffness matrices are merged into a stiffness matrix for the structure. This technique of merging assembly stiffness matrices into the stiffness matrix for the entire structure is termed the macro merge procedure. Although an assembly may contain up to 900 nodes, a reduced assembly stiffness matrix, as it is formed and reduced, must be an in-core matrix (i.e., something less than 200 freedoms for the current computer core size). There is no programming limitation on the number of assemblies into which a structure may be divided.

ASTRA Programming Design Features

The major part of the ASTRA system is made up of a group of some 500 subroutines that perform the various computations required in the solution of structural problems. Most of these subroutines are grouped according to their function to form independent modules known as computational boxes. The rest of the subroutines make up the support modules such as the Data Retrieval System (DRS), sparse matrix routines, plus a set of miscellaneous routines. The solution part of the ASTRA system performs the various solution procedures as directed by a main control program generated by SAUL.

Computational Boxes. The computational boxes have been defined such that each one performs a logical task of the solution. Each computational box is independent of the other computational boxes. Control information for the various options within a computational box is obtained from the data generated by SAUL and passed to the computational box in a control array. Data generated by one computational box is stored in peripheral storage and obtained by means of the data retrieval system. The order in which the boxes are entered is arbitrary as long as the data required as input to each box exist in the DRS.

All computational boxes are written in FORTRAN IV and contain no input or output statements except those that are made by calls to the DRS. The communication of a computational box with the DRS is done by subroutine calls to the DRS subroutines. The FORTRAN language, plus isolation of input-output through the use of the DRS, make each box completely machine independent.

Data Retrieval System. In developing a general purpose computer program, such as ASTRA, a large part of the effort must be concerned with transmittal and storage of data. The vast quantity of data generated during the solution of a structural problem requires a sophisticated data storage procedure. For a specific solution procedure and problem size data could be assigned to specific units. However, for a general purpose program of modular design, preassigning of data to given units is not feasible. The DRS is designed to perform this function in ASTRA, and to handle all communication of data between the computational boxes.

The DRS is written almost entirely in FORTRAN IV. It is designed to be operational on any type of digital computer having certain minimum capabilities and peripheral storage configurations. It has the ability to use several different types of peripheral storage devices simultaneously. Moreover, it has tables that describe the various properties of all tapes, disks, and drums, and uses only those units needed for a given analysis.

All data in ASTRA follows a single data format consisting of two records known as a data entry. The first record is an identification record that describes the length and content of the second record, known as the data record. The identification record also has three numbers that are used by DRS to identify the data. The identification of data in this manner is a powerful feature of DRS. All that is needed to locate any data entry in the ASTRA system is to give these integer numbers to DRS. The computational boxes never need to know on what physical unit a particular piece of data is stored.

The DRS allows the computational boxes to be selective, to a degree, in allocating storage of output data. For example, if a computational box requires input which is stored on units A and B it can instruct the DRS to store its output on unit C, which is different from A or B. This is particularly important for sequential operation, where a number of similar data entries are input, processed, and then output. The ability to simulate all peripheral storage as one continuous unit and still maintain uniqueness of units is characteristic of DRS. A structural problem is never too big for the ASTRA system unless the storage required exceeds that available.

The DRS relieves the computational boxes from the responsibility of determining the configuration in peripheral storage that the data will take; that is, the data will automatically be placed into areas of storage that DRS considers most efficient. Having an automatic selection of data distribution in storage is especially important in view of the vast and varied quantities of data produced by the numerous paths through ASTRA. The ability to select a configuration of units and the ability to delete and re-use peripheral storage greatly increases the efficiency of the program.

Sparse Matrix Notation. In designing a computer capability to solve the structural problems of the space age, a program must be capable of handling large matrices. A problem with 2000 nodes may have a stiffness matrix of 12,000 freedoms, If all of the terms of this matrix were to be stored on magnetic tape, it would fill several dozen reels of tape. Furthermore, to write that many terms onto tapes would take a prohibitive amount of time with the present generation of computers. Obviously, to solve a problem of this magnitude, Sparse Matrix techniques must be used to minimize the amount of data that is generated during the solution.

When looking at a stiffness matrix formed by the direct stiffness method, it is apparent that most of the terms are zero. Depending upon the structural coupling, most of the rows of the matrix will have less than 100 nonzero terms in them. Sparse schemes, which minimize the handling of zero terms in a matrix, may take several forms. For example, a matrix partitioning scheme is often used where only those partitions are kept which contain nonzero terms. This scheme is attractive for analyzing multistory buildings, bridge trusses, and other "linear" configurations, for which the stiffness matrix can be represented as a partitioned tridiagonal matrix, with densely populated nonzero partitions. However, for arbitrarily shaped structures in which the partitions have many more zero terms that nonzero terms, the scheme becomes needlessly inefficient.

The Sparse Matrix scheme in ASTRA considers only the nonzero terms of the matrix. Matrices are stored by row; only the nonzero terms for a row of a matrix plus a few words of sparse data are used in the matrix operations. Using this sparse notation for a problem whose stiffness matrix has 10,000 freedoms, it is common to have several hundred rows of the matrix in core at one time during a matrix operation.

Sparse matrix notation lends itself to very efficient computational operations for several reasons. First, only the nonzero terms of a row and a few words of sparse index data are used, transmitted, and stored during the matrix operations. Secondly, since the notation is so compact a larger portion of the matrix is operated upon at one time thus reducing the amount of data needed to be temporarily stored and transmitted. Finally, since the addition and multiplication of sparse notation data for rows of a matrix is a logical operation a negligible amount of machine time is used in generating sparse notation data.

Plans for future expansion

In the last decade drastic changes have taken place in the field of structural analysis, which have dramatically affected the ability of the structural engineer to accurately analyze complicated structures. This was primarily due to the simultaneous introduction of (1) equations of structural mechanics written in matrix form, (2) finite element development, and (3) the development of the digital computers and computing sciences. It is difficult to predict where the state of the art will be in another decade; therefore, the following discussion is confined to the development of ASTRA in the next two or three years.

Additions to ASTRA. The nonlinear features of the ASTRA program are designed to analyze the geometrically nonlinear problems associated with large deflections of structures. These problems include those in which large thermal deflections occur as well as those in which large deflections are caused by applied forces. The nonlinear feature, as presently contained in ASTRA, does not account for physical nonlinearities associated with the variation of material properties as functions of stress, temperature, or time. For certain structures and environments, a rational structural analysis must take into account nonlinear material behavior such as creep, plasticity, and viscoelasticity.

The state of the art in digital structural analysis can be improved significantly by developing better idealization techniques. One obvious way to do this is to develop a variety of basic structural elements with more sophisticated geometry or more sophisticated permissible strain patterns. There are available in the literature, and from Boeing research, derivations for many types of elements, several of which, are to be coded in ASTRA. Some of these elements have been discussed under General Specifications to the ASTRA system.

A program, recently developed at The Boeing Company, for solving the equations of motion for structural dynamics problems is titled, "Computer Language for Structural Dynamics Problems" (LSD), Reference 29. The equations of motion are derived automatically in the program using Lagrange equations. The program also generates a FORTRAN program and numerically integrates the equations. The time history of the solution may be either obtained on printed output or plotted on an Orthomat plotter.

Applications

To illustrate the method of analysis as applied to representative problems, several features of SAIL will be examined and some typical results of an analysis of a radar antenna will be shown.

Input generation examples

The following discussion concerns the application of SAIL to two types of problem: (1) the generation of data for an analysis of a particular problem with specified geometry, structure, and loads, (2) the application of SAIL to preliminary design analysis through the use of an external data generator.

Applications of SAIL to a Typical Antenna Configuration. Figure 10 shows an example of a radar antenna where the geometry of the surface is a paraboloid of revolution with a projected contour of an ellipse. The focal length of the paraboloid is 60 in. and the semimajor and semiminor axes of the ellipse are 84 and 30 in., respectively. The reflecting surface is 0.032 in. aluminum. Tapered beams, also of aluminum, comprise the supporting structure. The load is a uniform pressure of 0.229 lb/in.^{2} corresponding to a 100 mph wind. Since the structure and loads are symmetric, one quarter of the antenna is idealized. Figure 11 is an Orthomat plot of the input data generated through the use of SAIL.

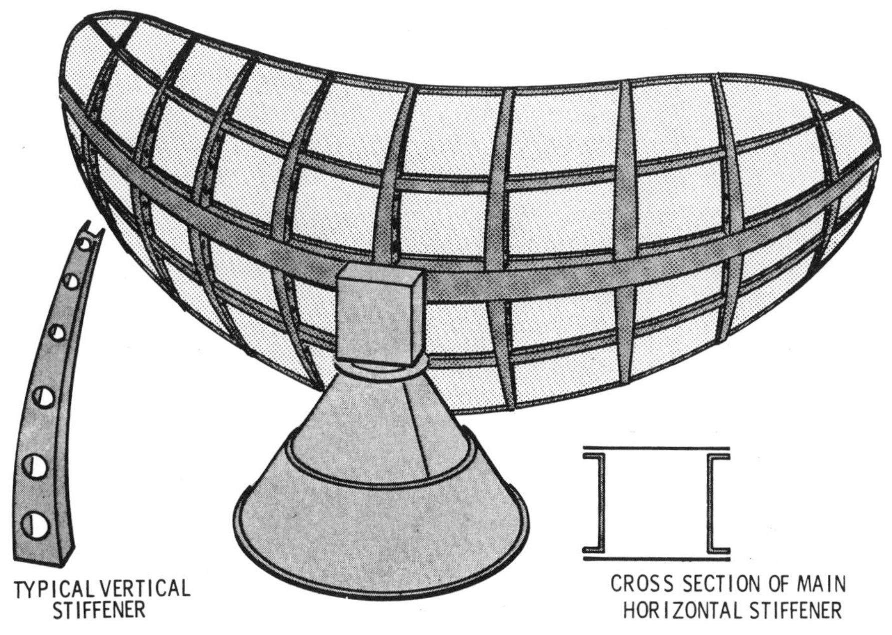

Figure 10. Typical antenna configuration.

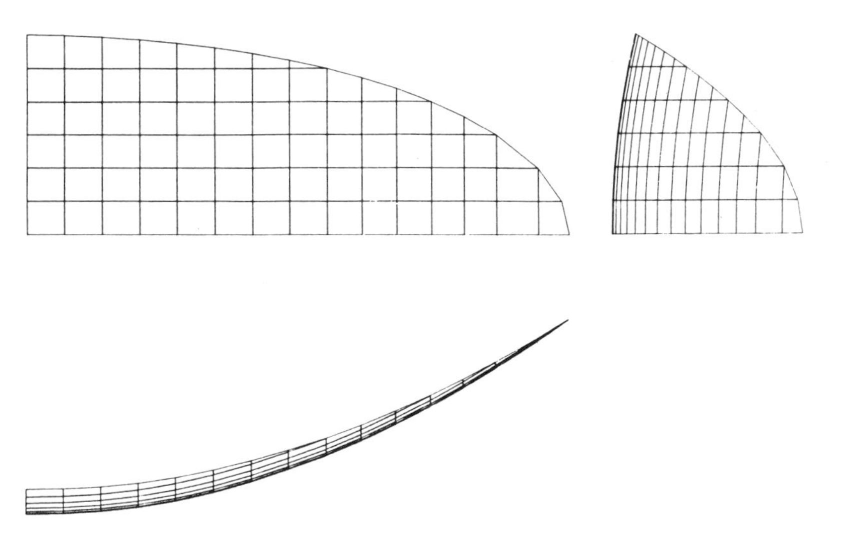

Figure 11. Orthomat plot of idealized structure–generated by SAIL.

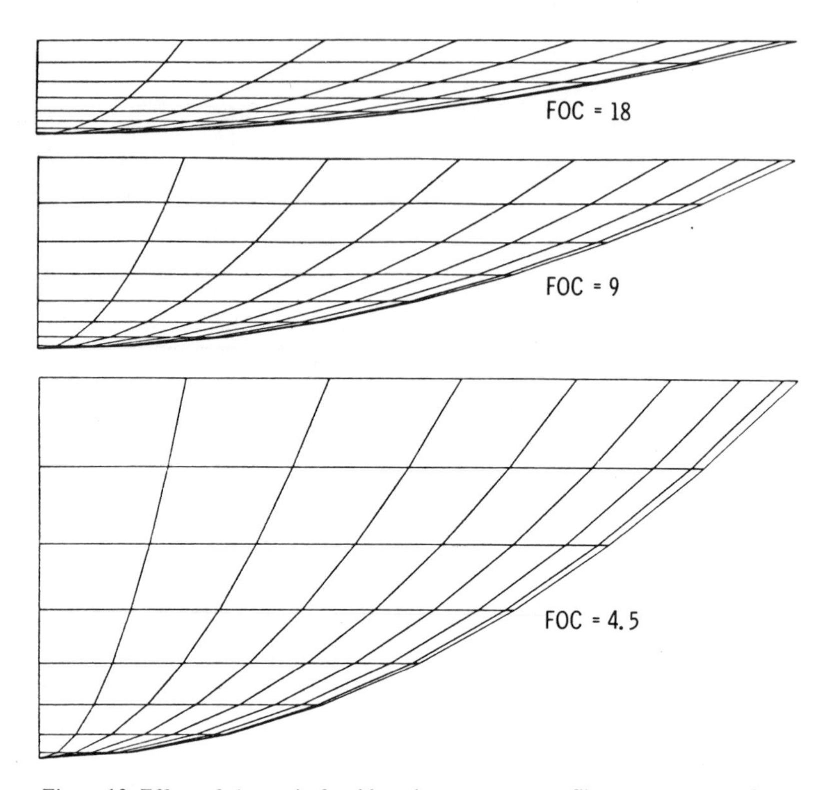

Figure 12. Effect of change in focal length on antenna profile geometry–x-z plane.

In this type of problem, where the geometry can be expressed in functional form, SAIL is particularly useful. The calculation of nodal coordinates from the equations of the paraboloid and the ellipse is expressed conveniently in loop statements. In essence, the engineer writes down the same equations he would use in performing slide rule calculations, but the computer performs the work. The choice of a convenient node numbering system allows the generation of structural elements within loop statements also.

SAIL input preparation required 8 hours and resulted in 153 data cards. Data preparation in the archaic tabular form would require an estimated 32 hours and 394 cards. Further advantages obtained through the use of SAIL in this particular application are: (1) nodal coordinates are calculated to an accuracy of seven significant figures, (2) fewer cards reduce the possibility of input and keypunch errors, and (3) much of the tedious hand calculations required to obtain the data for tabular input is eliminated.

Application of SAIL to Preliminary Design. External data generators are extremely useful in preliminary design analyses and trade studies where broad

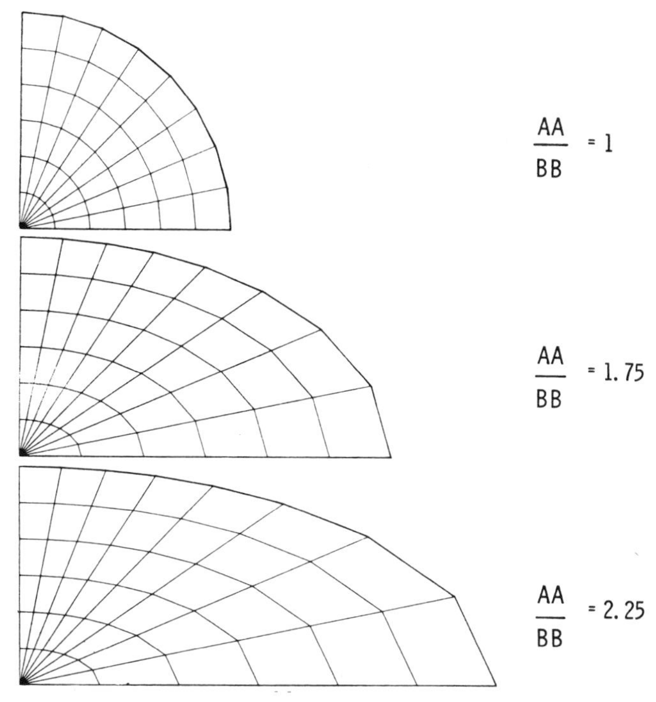

Figure 13. Effect of change in AA/BB ratio on antenna planform geometry–x-y plane.

guidelines may be known concerning a structure.

In the following example the geometry is assumed to be an arbitrary paraboloid of revolution with a projected contour of an arbitrary ellipse in the x-y plane. The structure may be idealized for an arbitrary number of nodes and elements. The shell that comprises the reflector surface has arbitrary material properties and an arbitrary thickness. One-quarter of the antenna is generated by the external data generator. This is to take advantage of the symmetry of the structure. For those cases where a complete shell is required, due to nonsymmetric loadings, the same external data generator may be used to generate each of the four quadrants.

This structure is idealized once in general form by an external data generator with twelve control parameters (FOC, AA, BB, N, M, NOID, TID, ELID, SX, SY, TMB, MAT). These twelve parameters are easily incorporated in the SAIL input. Each time the external data generator is used, explicit values are supplied for the control parameters and the desired configuration is generated. Figure 12 through 14 are Orthomat plots of the configurations generated by varying some of these control parameters.

The first three parameters (FOC, AA, BB) allow the user to change the geometry of the reflector surface. FOC is the parameter that defines the focal

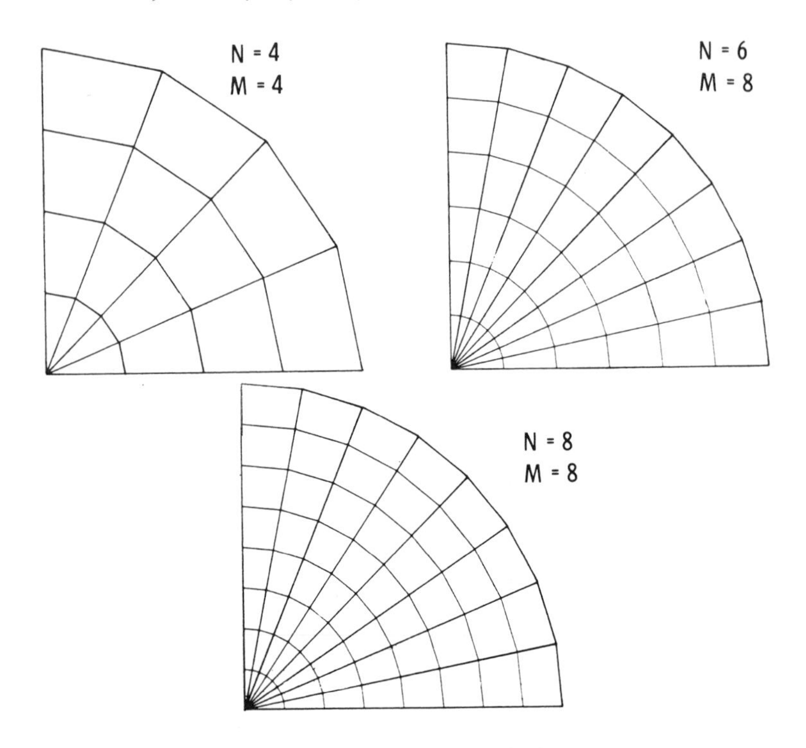

Figure 14. Effect of change in N and M on idealization mesh–x-y plane.

length of the paraboloid. Figure 12 illustrates the change in geometry when the focal length has the values 18, 9, and 4.5 while all other control parameters are held constant. The parameters AA and BB define semimajor and semiminor axes of the projected ellipse in the x-y plane. The configurations obtained when the ratio of AA/BB is 1.0, 1.75, and 2.25, while all other control parameters are held constant, are illustrated in Figure 13.

The fourth and fifth parameters (N, M) control the mesh size of the idealization. The number of radial segments being N and the number of circumferential segments, M. In preliminary work the user may wish to use a rather coarse mesh to reduce computer time. Structural sizing on final configurations may require a refined mesh which is obtained by simply increasing the values of N and M. Supplying values of (4,4), (6,8), and (8,8) to N and M while all other control parameters are held constant, produced the configurations shown in Figure 14.

The next five control parameters (NOID, TID, ELID, SX, SY) give the user the ability to generate the quarter shell in any of the four quadrants. This generality has been included so that the whole structure can be generated when loads or boundary conditions are not symmetric. There are a number of ways the user can do this using SAIL, therefore the starting values for node identification (NOID), table identification (TID), and element identification (ELID) have been left arbitrary. The parameters SX and SY represent the sign of the x and y coordinates, respectively.

The final two control parameters (TMB, MAT), the shell thickness and material table identification, respectively, allowed the user to analyze any combination of reflectors of constant thickness and composed of any material.

This example typifies the use of an external data generator. It can only begin to show the power given the analyst to solve his problems efficiently. The generality that can be incorporated in an external data generator is virtually unlimited. Engineering ability becomes the criterion in analyses of problems instead of the number of engineering manhours required.

A comparison of input data volume between SAIL and ordinary tabular input depends on the number of configurations to be analyzed when an external data generator is used. Some idea of how the advantage of SAIL increases with the number of configurations is given in the following table.

No. of	No. of input data cards	
configurations	Sail	Tabular
1	38	137
5	54	705
10	74	1370

Some Typical Analysis Results

To present some analytical results typical of those which ASTRA can produce, the antenna configuration shown in Figure 10, and described in the first part of this section, was analyzed. Some of the results obtained are shown in Figures 15, 16, and 17.

Figure 15 is an Orthomat plot of two views of the deformed structure with the undeformed structure superimposed. The scale of the deformation has been exaggerated 50 times. In the actual Orthomat plot two colors of ink are used to distinguish the deformed structure from the undeformed, but for clarity in the figure the undeformed structure has been shown as a dashed line. Also for the sake of clarity, the plotter was instructed to generate only the beam elements. Therefore, fewer lines are shown than in Figure 11 where the plate elements were generated.

Figure 16 is a plot of the bending moments and shears calculated for the main horizontal beam. The stepped shear diagram and the slightly sawtooth pattern of the moment diagram result from the assumptions of constant shear and linearly varying bending moment which were made in deriving the elemental beam properties. The discontinuities between elements in these diagrams represent the load transfer between the main beam and the plate and vertical

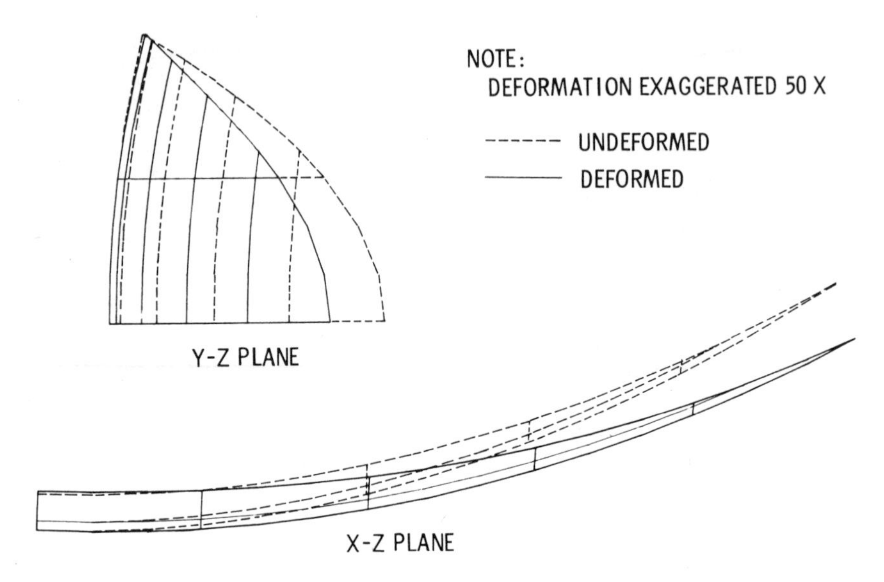

Figure 15. Deflection of a typical radar antenna–100 mph head-on wind.

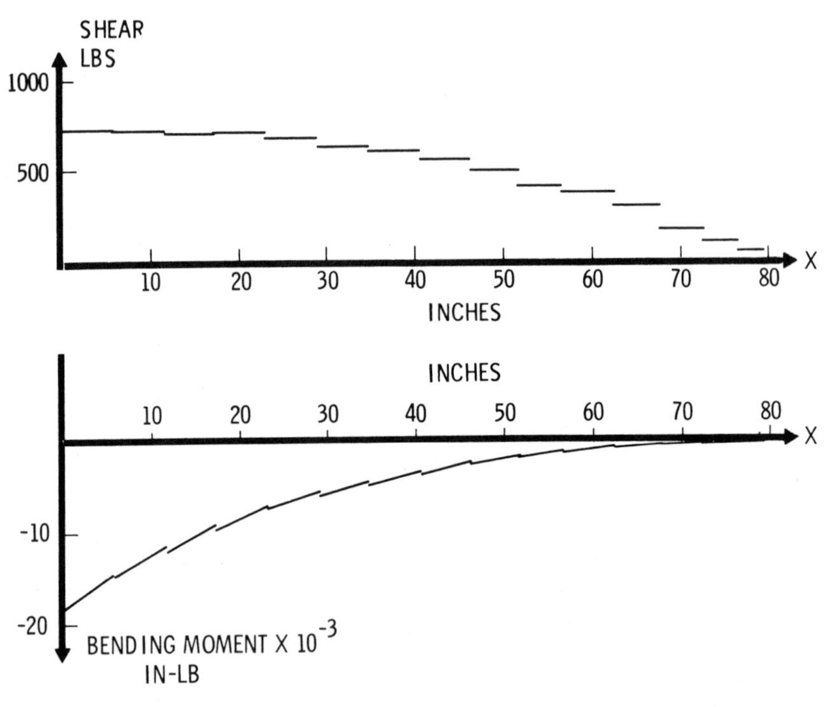

Figure 16. Shear and bending moment diagrams for main horizontal beam–100 mph head-on wind.

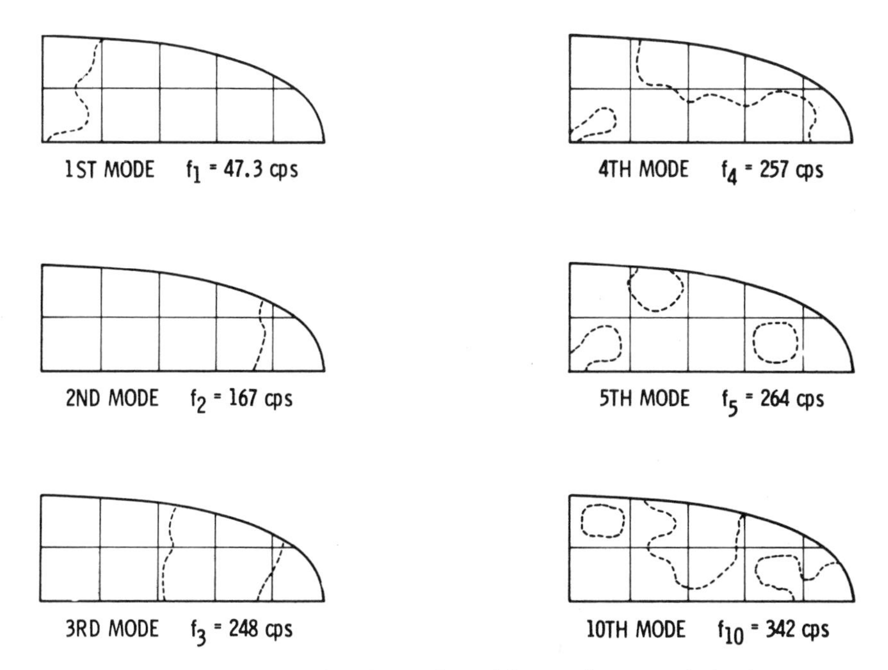

Figure 17. Symmetric natural vibration modes and frequencies of a typical radar antenna.

stiffeners. All of the beam elements used in this example contained the offset feature which allowed for proper coupling of bending and stretching due to the separation between the beam elastic axis and the surface of the reflector.

Some of the natural vibration modes frequencies obtained for this example are shown in Figure 17. Ten symmetric modes were calculated using a power iteration eigenvalue solution procedure. Since the antenna configuration chosen for this analysis was purely hypothetical, these modes should not necessarily be assumed typical of an actual antenna. In fact, a tendency toward local panel modes is apparent which might indicate a potential panel flutter problem requiring additional stiffening. The example does, however, show the ability of ASTRA to predict this type of behavior.

References

1. M. J. Turner, R. W. Clough, H. C. Martin, and L. J. Topp, "Stiffness and Deflection Analysis of Complex Structures," *J. Aeronaut. Sci.* **23** (No. 9), 805 (September 1956).

2. H. C. Martin, "Truss Analysis by Stiffness Considerations," presented at the American Society of Civil Engineers Meeting, San Diego, February 1955; published *J. Engr. Mech. ASCE Div.*, pp. 1182-1194 (October 1956).

3. M. J. Turner, "The Direct Stiffness Method of Structural Analysis," paper presented at a meeting of the Structures and Materials Panel, AGARD, Aachen, Germany (September

1959).
4. M. J. Turner, H. C. Martin, and R. C. Weikel, *Matrix Methods of Structural Analysis* (The Macmillan Company, New York, 1964), Agardograph 72, pp. 203-266.
5. R. J. Melosh and R. G. Merritt, "Evaluation of Spar Matrices for Stiffness Analyses," *J. Aeronaut. Sci.* **25** (No. 9), 537 (September 1958).
6. M. J. Turner, E. H. Dill, H. C. Martin, and R. J. Melosh, "Large Deflections of Structures Subjected to Heating and External Loads," *J. Aerospace Sci.* **27** (No. 2) 97 (February 1960).
7. R. C. Weikel, R. E. Jones, J. A. Seiler, H. C. Martin, and B. E. Greene, "Non-Linear and Thermal Effects on Elastic-Vibrations," ASD-TDR-62-156, W-PAFB (February 1962).
8. Boeing Document No. D6-29059, "SAMECS User's Manual (TS-172), Structural Analysis Method for Evaluation of Complex Structure" (May 1967).
9. P. E. Grafton and D. R. Strome, "Analysis of Axisymmetrical Shells by the Direct Stiffness Method," *AIAA J.* **1** (No. 10), 2342 (October 1963).
10. J. H. Percy, T. H. H. Pian, S. Klein, and D. R. Navaratna, "Application of Matrix Displacement Method to Linear Elastic Analysis of Shells of Revolution," *AIAA J.* **3** (No. 11), 2138 (November 1965).
11. R. E. Jones and D. R. Strome, "Direct Stiffness Method Analysis of Shells of Revolution Utilizing Curved Elements," *AIAA J.* **4** (No. 9), 1519 (September 1966).
12. R. W. Clough and Y. Rashid, "Finite Element Analysis of Axi-Symmetric Solids," *ASCE Eng. Mech. Div. J.* **91** (No. EM1), 71 (February 1965).
13. E. L. Wilson, "Structural Analysis of Axisymmetric Solids," AIAA Paper No. 65-143 (January 1965).
14. Boeing Document No. D2-24116-1, "Analysis of Axisymmetric Solids by the Direct Stiffness Method" (1966).
15. H. C. Martin, "Plane Elasticity Problems and the Direct Stiffness Method," *The Trend in Engineering,* University of Washington, Seattle, Washington (January 1961).
16. J. H. Argyris, "Continua and Discontinua, An Apercu of Recent Developments of the Matrix Displacement Method," AFFDL-TR-66-80, Proceedings of Conference on Matrix Methods in Structural Mechanics (November 1966).
17. Boeing Document D2-125170-1, "A Linear Strain Tetrahedron Element" (August 1966).
18. B. E. Greene, "Application of Generalized Constraints in the Stiffness Method of Structural Analysis," *AIAA J.* **4** (No. 9) (September 1966).
19. R. E. Jones, "A Generalization of the Direct-Stiffness Method of Structural Analyses," *AIAA J.* **2** (No. 5), 821 (May 1964).
20. J. S. Archer, "Consistent Matrix Formulations for Structural Analysis Using Influence-Coefficient Techniques," AIAA Paper No. 64-488 (June 1964).
21. J. S. Archer, "Consistent Mass Matrix for Distributed Mass Systems," *J. Struct. Div. ASCE,* pp. 161-178 (August 1963).
22. R. J. Guyan, "Reduction of Stiffness and Mass Matrices," *AIAA J.* **3** (No. 2), 380 (February 1965).
23. Boeing Document No. D2-125179-1, "Introduction to the Direct Stiffness Method" (November 1967).
24. Boeing Document No. D2-125179-2, "The Basic Theory of the Direct Stiffness Method" (November 1967).
25. Boeing Document No. D2-125179-3, "Derivation of Elements Contained in the ASTRA Structural Program" (November 1967).
26. Boeing Document No. D2-125179-4, "Structural Application and Idealization–The Direct Stiffness Method" (November 1967).
27. Boeing Document No. D2-125179-5, "User's Document–The ASTRA System" (November 1967).
28. Boeing Document No. D2-125179-6, "Programmers Document–The ASTRA System" (November 1967).
29. Boeing Document No. D2-139514-1, "Computer Language for Structural Dynamics (LSD)," Volume I–Engineer's Document (May 1966).

Caleb W. McCormick
Senior Scientist, The MacNeal-Schwendler Corporation
San Marino, California
Associate Professor of Civil Engineering
(Leave of Absence–California Institute of Technology,
Pasadena, California)

APPLICATION OF THE NASA GENERAL PURPOSE STRUCTURAL ANALYSIS PROGRAM TO LARGE RADIO TELESCOPES

Introduction

NASTRAN (NAsa STRuctural ANalysis), a new general purpose structural analysis program, is currently (October 1967) being developed under NASA sponsorship by a team consisting of Computer Sciences Corporation, The Martin Company (Baltimore) and The MacNeal-Schwendler Corporation. The program is designed to have the capability of solving static and dynamic structural problems by both the force and displacement methods, using a finite element modeling technique. The program has been specifically designed to treat large problems effectively and the modular organization of the program permits easy modifications and additions to the program.

NASTRAN is organized with complete separation of system functions, performed by an executive routine, from problem solution functions accomplished by modules separated strictly along functional lines. Each functional module is independent from all other modules in the sense that modification of a module will not, in general, require modification of other modules. This is accomplished by requiring that functional modules operate as independent subprograms that may not call or be called by other modules. Indirect communication between modules is accomplished through auxiliary storage files and parameter tables, both of which are maintained by the executive system.

The essential functions of the executive system are to control the sequence of module executions, to establish and communicate parameters, to control input-output to auxiliary storage devices, and maintain restart capability. The sequence of module calls is controlled either by rigid format tables stored in the executive system or by a chain of DMAP (Direct Matrix Abstraction Package) instructions prepared by the user. The executive system must allocate files for all data blocks generated during program execution and perform all input-output to auxiliary files requested by functional modules. Matrix packing routines that are used by the general input-output routines are designed so that both full and sparse matrices are efficiently stored. The executive system is open-ended in the sense that it can accommodate an essentially unlimited number of functional modules, files, and parameters.

In addition to basic static analysis by both the force and displacement methods, the program will include the solution of problems with nonlinear stress-strain relationship by piecewise linear analysis, a first approximation to the effects of large displacements (static analysis with differential stiffness), and the inclusion of inertia-relief effects for free bodies.

Four methods of real eigenvalue extraction are provided with the program for use in buckling and vibration problems. A relatively large number of methods is considered necessary in order to optimize efficiency over the range of structural analysis problems. In general, the most efficient methods have the smallest range of application. The version of the tridiagonalization method provided with the program has superior efficiency when many (or all) of the eigenvalues are required. The other three methods find eigenvalues one at a time and are, therefore, preferred when only a few eigenvalues are required. These methods are the standard power method, inverse power method with shifts, and the determinant method.

Dynamics problems are separated in NASTRAN into the following types:

1. Complex eigenvalue problems, such as flutter problems.
2. Frequency and random response problems.
3. Transient response problems.

Each of these problem types can be treated by either the force or the displacement method, and each can be solved using either a direct or a modal formulation.

NASTRAN is being developed in the FORTRAN language to the greatest extent possible in order to simplify the conversion from the development computer (IBM 7094-7040 DCS) to various third-generation computing systems. However, a limited amount of machine language is being used where its use achieves increased capability or substantial reductions in computing time. Current plans are to install NASTRAN on IBM 7094-7040 DCS, IBM System 360, Univac 1108 and CDC 6600 computers.

Structural Model

NASTRAN employs the finite element approach to structural analysis; i.e., the distributed physical properties of structures are represented by idealized lumped element models. The idealized models consist of "grid points" to which loads are applied and at which degrees of freedom are defined, and "elements" that are connected to or between the grid points and define the elastic, inertial, and damping properties of the structure.

The location of grid points and the orientation of the coordinate axes used to measure displacements may be specified by the user either in a basic rectangular system or in "local" coordinate systems that may be rectangular, cylindrical, or spherical. The user may specify a different local coordinate system at each grid point if he desires, and it is not necessary that the location of the grid point and the displacements of the grid point be specified in the same coordinate system. A special class of grid points, called "scalar points," is provided which do not have geometrical locations and for each of which only one degree of freedom is defined. These points are useful for the representation of control system variables and of abstract quantities such as the generalized coordinates of vibration modes.

The following types of structural elements are provided with both the force and the displacement methods:

1. Rod and tube elements.
2. Beam element.
3. Shear and twist panels.
4. Triangular panel with membrane and/or bending properties.
5. Quadrilateral panel with membrane and/or bending properties.
6. General element used for representation of part of a structure by its deflection influence coefficients.

The following types of elements are provided with the force method only:

7. Reaction element.
8. Beam element connecting three or more grid points.

The following types of elements are provided with the displacement method only:

9. Axisymmetric conical shell element.
10. Viscous damper element.
11. Scalar spring element.

The scalar spring element may be connected between any pair of degrees of freedom, including scalar points. Its stiffness is specified directly by the user rather than being calculated from geometry.

Material properties for elements are specified by reference to material properties tables, which include density, elastic coefficients, thermal expansion coefficients, and damping coefficients for each type of material. Material properties may be anisotropic and may be temperature dependent.

Mass properties of the structure are specified in three ways: by structural mass density obtained from material properties tables, by a nonstructural mass density specified for each element, and by mass concentrated at grid points.

Every effort has been made to make the input of the structural model simple and flexible in order to minimize the chances of errors in the model. The input is restricted to independent information, with all dependent quantities, such as side lengths and surface areas of elements, determined internally by the program. All identification numbers must be unique, but not necessarily sequential. No particular order is required for the data deck, since the deck is sorted by the program. The user may ask for a sorted and/or unsorted echo of the data deck.

The following four basic card types are used for the structural model input:

1. Grid cards – one per grid point.
2. Connection cards – one per element connected.
3. Property cards – one for each unique element.
4. Material cards – one for each material.

The grid cards contain the grid point number, grid point coordinates, identification of coordinate systems used for coordinate input and displacement calculations, and the identification of any degrees of freedom to be removed, such as all rotational displacements for a space truss. The connection cards contain the element identification number, a reference to a property card, the grid points connected, and the orientation of the element where appropriate. The property cards contain the property identification number, a reference to a material card, the nonstructural mass associated with the element, and the cross-sectional properties that are appropriate for the element under consideration. For simple elements, such as rods, the connection card and property card may be combined on a single card. The material card contains the elastic and mechanical properties of the material.

Problem Formulation

Problem formulation begins with the preparation of tables and the performance of auxiliary calculations associated with the geometry of the structure. All coordinate system transformation matrices that will be required in the solution are calculated and stored in a table. Another table is prepared which defines the several dependent and independent sets of displacement components.

The generation of the stiffness matrix begins with the preparation of a table, ordered by grid point number, collecting all of the required quantities for every element connected to the grid point under consideration. The stiffness matrix can then be assembled six columns at a time, one for each possible degree of freedom at the grid point, by reading one line at a time from the previously prepared table. If the problem requires a dynamic analysis, the mass and damping matrices are prepared from the same table. A similar table, sorted by element types and numbers, rather than by grid point number, is prepared for later use in stress recovery.

During the assembly of the stiffness matrix, the diagonal six-by-six matrices are checked for singularities at the grid point level. Singularities at the grid point level for a space truss structure, such as might be used for a large radio telescope, means that all rods intersecting at a grid point lie in a plane or are collinear. For structures in general, a singularity at the grid point level means that no elastic resistance has been provided for one or more of the displacement components. The list of singulatities generated during the assembly of the stiffness matrix is checked against the list of single point constraints provided by the user. If the lists do not match, a warning message only is given, since the remaining singularities may be removed by other constraints that are provided later in the problem and cannot be checked at this point in the solution. The lack of a warning message at this time does not necessarily mean that the complete stiffness matrix is nonsingular.

The assembled stiffness matrix contains six degrees of freedom for each grid point defined and one degree of freedom for each scalar point defined. The next step involves the removal of unwanted and dependent degrees of freedom. The first phase is a matrix operation associated with the removal of the dependent terms of the multipoint constraint equations. These equations permit the declaration of a linear relationship among any selected degrees of freedom. This procedure has a variety of uses, including the introduction of enforced displacements that are not parallel to the local coordinate system, simulation of very stiff members or rigid bodies, and the generation of non-standard structural elements.

The second phase is a partitioning operation associated with the removal of the single-point constraints. The single point constraints may be associated with unwanted degrees of freedom declared on the grid cards or enforced displacements along the local coordinate axes declared on a special card for this purpose. If the enforced displacement is not along one of the local co-ordinate axes, the removal of the dependent degree of freedom is not a simple partitioning operation and the multipoint constraint must be used.

The third phase involves the removal of selected coordinates, either to reduce the order of the problem or to partition the model into two or more segments. In order to accomplish this, the user specifies the degrees of freedom that will be omitted or lie along the boundary of a partition, and the stiffness matrix is automatically computed for the reduced set in such manner that the errors in elastic and kinetic energies are minimized. The reduction of the order of the problem is primarily for use in dynamic analysis, where a smaller number of degrees of freedom may be more appropriate than in a static analysis.

If the stiffness matrix, as constrained at this point in the formulation, still has rigid-body motions remaining, an additional set of statically determinate constraints must be supplied in order to remove the singularities associated with the rigid-body motions. In order to accomplish this, the user specifies a determinate set of reactions, and the program removes the singularities and calculates the rigid-body characteristics associated with the rigid-body motions.

Static Analysis

The first step in static analysis is the decomposition of the reduced stiffness matrix into upper and lower triangular forms. Matrix decomposition is the primary tool for solving static problems and frequency response problems, and it also has an important role for two of the program's methods of eigenvalue extraction (inverse power method with shifts and determinant method) and for the integration of transient problems. The method used senses the character of the matrix and minimizes solution time on the basis of bandwidth and general sparsity. A complete description of the triangular decomposition procedure will be given in a forthcoming NASA Contractor Report.

It is well known that band matrices significantly reduce the computing time for matrix operations. However, for many problems it is difficult to generate stiffness matrices having narrow bands, and the large radio telescope is an example of such a structure. It is easy to generate a band matrix for a dish with an opening in the center, but such is not the case when no opening is present and the supporting structure and quadripod are considered. However, it is still possible to generate a relatively narrow band matrix with a few scattered terms outside the band. Such matrices will be referred to as "partially banded" and will be described in terms of their "semiband" (number of columns from the diagonal term to the most remote column considered inside the band) and number of "active columns" (maximum number of nonzero columns existing outside the band at any stage of the decomposition).

In performing the decomposition, a preliminary pass is made over the matrix in order to select that combination of bandwidth and active columns that will give the minimum computing time to perform the decomposition. Since the terms inside the band can be handled far more efficiently than terms outside the band, and since stiffness matrices tend to have terms clustered about the diagonal with the density decreasing away from the diagonal, the program will extend the bandwidth out until the sparsity is such that it is more efficient to consider the remaining terms outside the band. Full matrices are handled automatically by considering the semiband equal to the order of the problem.

The times required to complete a triangular decomposition on an IBM 7094 DCS computer for various combinations of semibands and active columns are shown on Figure 1. The indicated times are based on the assumption that the

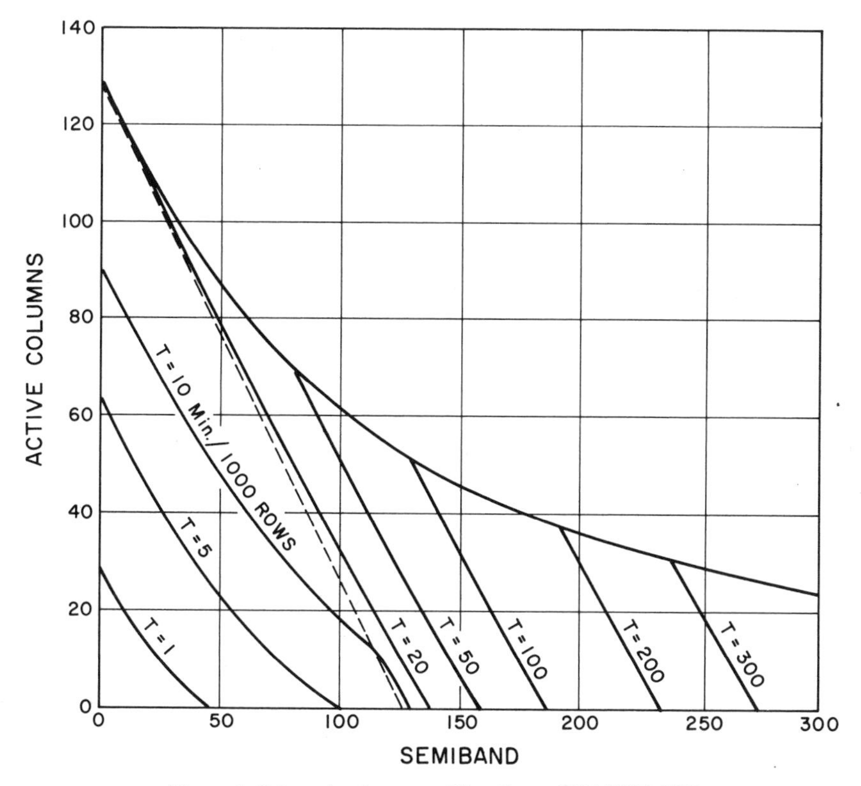

Figure 1. Triangular decomposition times–IBM 7094 CDS.

order of the problem is large compared to the semiband. For full, or nearly full, matrices where the semiband is equal, or nearly equal, to the order of the problem, the triangular decomposition times are about 30 per cent less than indicated on Figure 1. The upper boundary of acceptable combinations is determined by the requirement that all of the nonzero terms of a single column inside the band and all active column terms must be retained in core. The dotted line, coinciding with the line T = 20 near the top and crossing the line T = 10 near the bottom, divides the region into two parts. Combinations below the dotted line (no-spill region) do not require the use of auxiliary scratch files. Combinations above and to the right of this dotted line (spill region) require the use of auxiliary scratch files, and hence computing times are substantially greater in this region.

Larger core sizes increase the sizes of both the region of acceptable combinations and the no-spill region. For a 90,000-word working space, the dotted line would remain straight and intersect both axes at a value of 300. Computing times for combinations formerly located in the spill region would be reduced by almost an order of magnitude. Since the computing times for combinations located inside the no-spill region are proportional to the arithmetic speed of the computer, they will be substantially decreased for third-generation machines.

Since no effort is made by the program to reorder the terms, it is important that the user sequence the grid points in an effective pattern. Fortunately, this is not a difficult operation, and in any event, the computing time is not too sensitive as long as the pattern is reasonable. An attempt should always be made to completely band the matrix by numbering the grid points so that the difference between any connected points is a small number. If this does not appear to be possible, the structure should be broken into segments having small bandwidths, and the connecting grid points should be sequenced after all points in the segments connected.

An undesirable consequence of requiring a particular numbering scheme is that the user may wish to number the grid points in some other manner for convenience in design or interpretation of results. In order to allow this convenience for the user, as well as efficiency for the triangular decomposition, provision has been made for the use of any convenient external numbering scheme, while all calculations are made in terms of an internal sequence that is more suitable for matrix operations. All input is made up and all output is returned in terms of the external numbering scheme. The internal sequence is declared by the user with a paired list of external and internal numbers. All required conversions are made automatically by the program. This procedure also permits the selection of a new sequence for improved efficiency if the first sequence is not satisfactory.

The next step in static analysis is to introduce the static loads in one or more of the following ways:

1. Load components at grid points.
2. Pressure on panels.
3. Gravity loads.
4. Enforced displacements of grid points.
5. Enforced deformations of elements.
6. Temperatures at grid points.

Following the generation of the load vectors, the same constraints are applied that were applied to the stiffness matrix. The next operation is to solve for the displacement vectors by making a forward and backward pass on the load vectors with the previously determined triangular factors. A residual displacement vector is calculated in the usual manner by multiplying the solution vectors by the stiffness matrix and combining the result with the applied load vectors. A comparison of the inner product of the solution vector and the residual vector with the inner product of the applied load vector and the solution vector gives an indication of the magnitude of the accumulated roundoff error in the solution vector.

The final operation in static analysis is the recovery of the dependent displacements and forces in the elements of the structural model. All, or any subset of the displacements and forces or stresses, may be output on option by the user.

A preliminary investigation conducted by the design team indicated that double precision arithmetic (16 decimal digits) is both a necessary and sufficient requirement for the formulation and solution of large (100 - 10,000 degrees of freedom) problems by the displacement method. Accordingly, double precision arithmetic is used in most operations, including the formulation of the stiffness matrix. Exceptions where single precision arithmetic is used include the formation of load vectors, the recovery of stresses from displacements, and transformations involving previously computed modal matrices.

Problem size is practically unlimited, the only limitations being that the decomposition routine requires that the nonzero elements of a single column inside the band and all active column terms must be held in core, and some vector operations require that two complete vectors be held in core. The decomposition limitation is of no consequence, even on second generation computers, since the computing times will become excessive before the size limitation is reached. The vector operation limitation is of no consequence on third-generation computers, but does limit problem size on second-generation machines. With 32,000 words of core memory, statics problems are limited to about 8000 degrees of freedom.

Static Analysis of Radio Telescopes

As the first example, consider a typical small 30-ft antenna composed of 132 grid points and 364 rod elements. The unconstrained stiffness matrix contains 792 degrees of freedom, 6 degrees of freedom for each grid point. The removal of the rotational constraints and the application of the appropriate single point constraints around the boundary results in a constrained stiffness matrix having 301 degrees of freedom. This structure was analyzed under gravity loading with a 45° look. The total running time on an IBM 7094 DCS computer was 7 min, of which less than 1/2 min was spent on the triangular decomposition. Although the decomposition time is a small fraction of the computing time for small problems, it becomes a major part of the computing time for large radio telescopes.

An examination of Figure 2 indicates the combinations of semibands and active columns that are available for the triangular decomposition using two different grid point sequences. The dashed curve shows the result for a

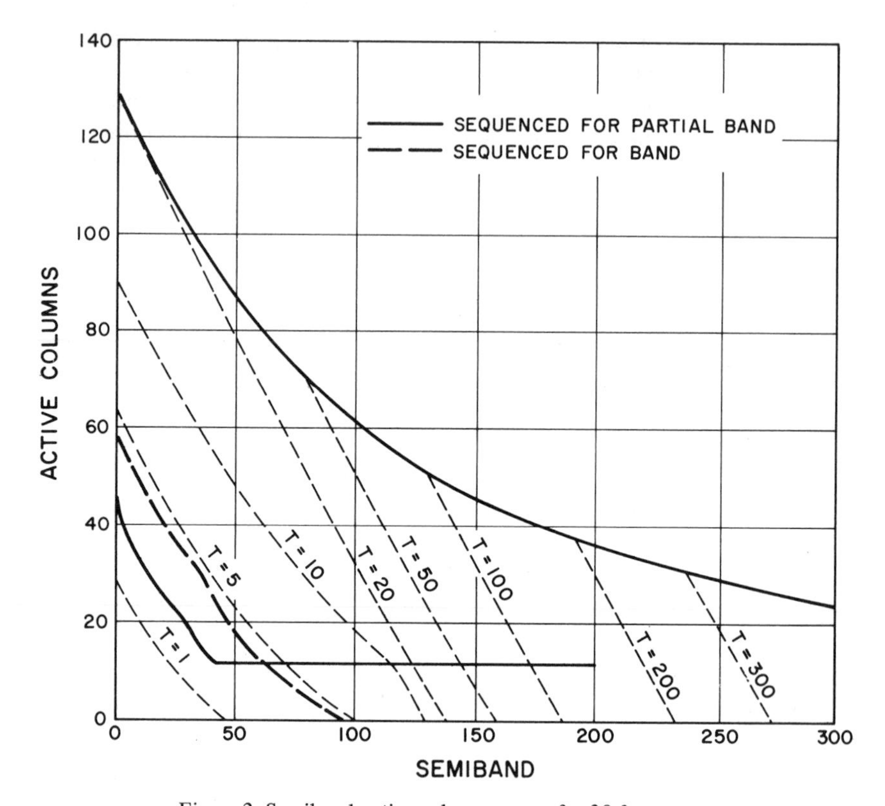

Figure 2. Semiband-active column curves for 30-ft antenna.

sequence that places all nonzero terms inside the narrowest possible band, resulting in a semiband of about 100 with no active columns. This is the sequence that would have to be used if the decomposition routine were restricted to band matrices. The partially banded routine detected a slightly more favorable situation for this sequence at a semiband of 55 with 15 active columns. However, since the semiband-active column curve closely parallels the constant time curves, the time saving for the partially banded combination is very slight.

The solid curve on Figure 2 indicates the result when the antenna dish and backup structure are banded separately, and then interconnected at the common grid points, The curve was arbitrarily terminated at a semiband of 200. This partially banded sequence results in 11 active columns from a semiband of 200 down to 40. From this point, the semiband-active column curve closely parallels the constant time curves. For semibands below 40, the program detected a slight advantage at a semiband of 18 with 24 active columns. The triangular decomposition time for this combination was about two-thirds of the time required for the alternate sequencing associated with the complete banding of the matrix.

As the second example consider the quarter section of a typical reflector structure shown in Figure 3. This structure is similar to the reflector used on

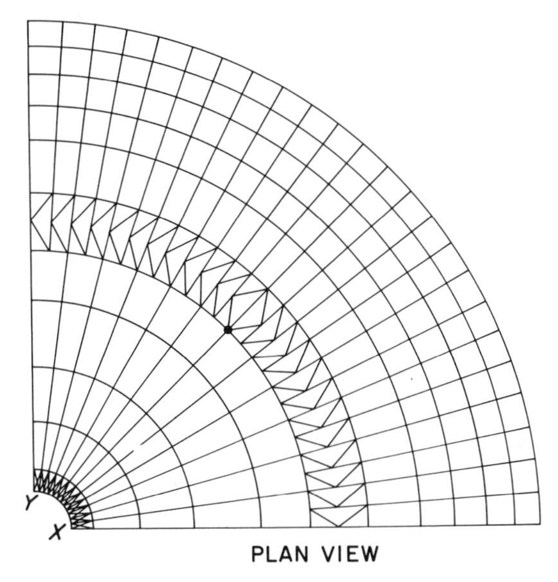

Figure 3. Quarter section of typical reflector.

the 210-ft antenna at Goldstone. The model is composed of 651 grid points and 2224 rod elements. The unconstrained stiffness matrix contains 3906 degrees of freedom, and the constrained stiffness matrix contains 1898 degrees of freedom.

This structure was analyzed under gravity loading, zenith look, with two planes of symmetry. The total running time on an IBM 7094 DCS computer was 68 min of which 25 min was spent on the triangular decomposition, 9 min on the backward read of the upper triangular factor, and the remaining 34 min on the rest of the problem. The portion of the running time associated with the backward read will disappear on third-generation computers where backward reading tapes will be available. On third generation machines the backward pass will be made immediately after the forward pass with no loss of time and at the same speed as the forward pass.

An examination of the dashed curve on Figure 4 indicates the combinations of semibands and active columns that were available for the triangular decomposition with the particular grid point sequencing used. Since only the reflector structure was used in the model, the stiffness matrix tends to be well banded. The flat portion of the curve between semibands of 60 and 100 is associated

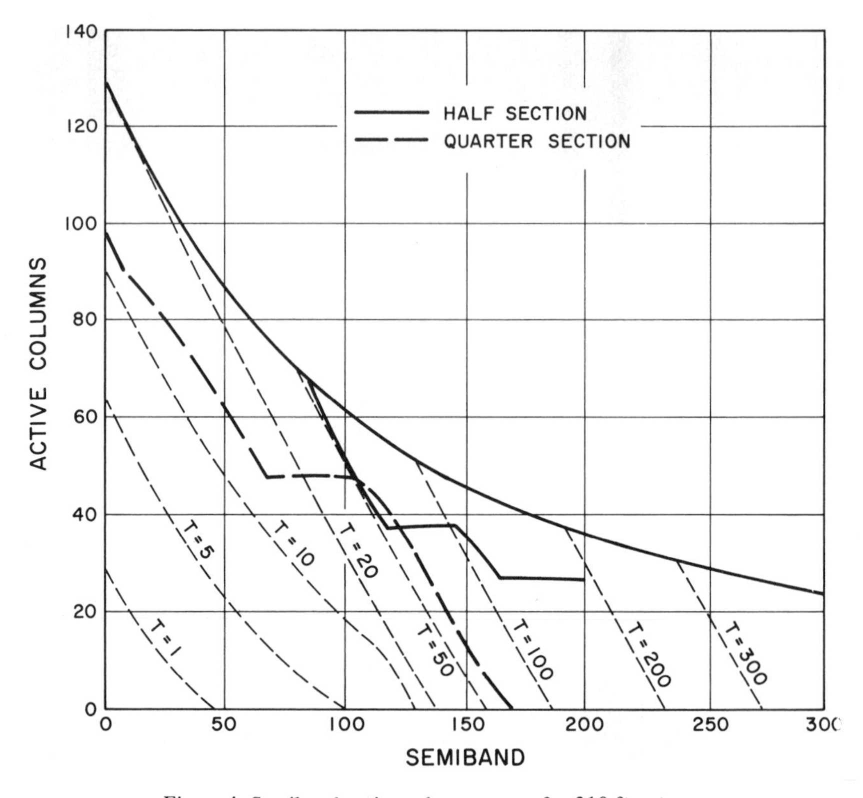

Figure 4. Semiband-active column curves for 210-ft antenna.

with the intermediate ribs in the outer portion of the reflector. The decomposition was actually performed at a semiband of 8 with 90 active columns, indicating that the sparsity of the matrix was such that most of the terms were considered outside the band. It can be seen that if a simple band matrix routine were used, the decomposition time would be about six times as long.

As the last example consider the 210-ft antenna at Goldstone shown in Figure 5. The model, consisting of the reflector, quadripod, and the backup structure, is composed of 952 grid points and 3596 rod elements. The unconstrained stiffness matrix contains 5712 degrees of freedom and the constrained stiffness matrix contains 2742 degrees of freedom. The reflector is shown in Figure 6 and the front and side elevations are shown in Figures 7 and 8, respectively.

The grid points were sequenced by considering the reflector as a banded segment, the backup structure as a second banded segment, and the quadripod as a third banded segment. The quadripod is connected to the reflector at two points as shown in Figure 8. The reflector is connected to the backup

Figure 5. Two hundred ten-foot antenna at Goldstone.

structure at nine points as shown in Figures 7 and 8.

This structure was analyzed under gravity loading for both zenith look and horizon look with one plane of symmetry. The total running time on an IBM 7094 DCS computer was 200 min, of which 135 min was spent on the triangular decomposition, 20 min on the backward read of the upper triangular factor, and the remaining 45 min on the rest of the problem.

An examination of the solid curve on Figure 4 indicates the combinations of semibands and active columns that were available for the triangular decomposition with the particular grid point sequencing used. The horizontal portion of the curve from a semiband of 160 to the arbitrary cutoff of 200

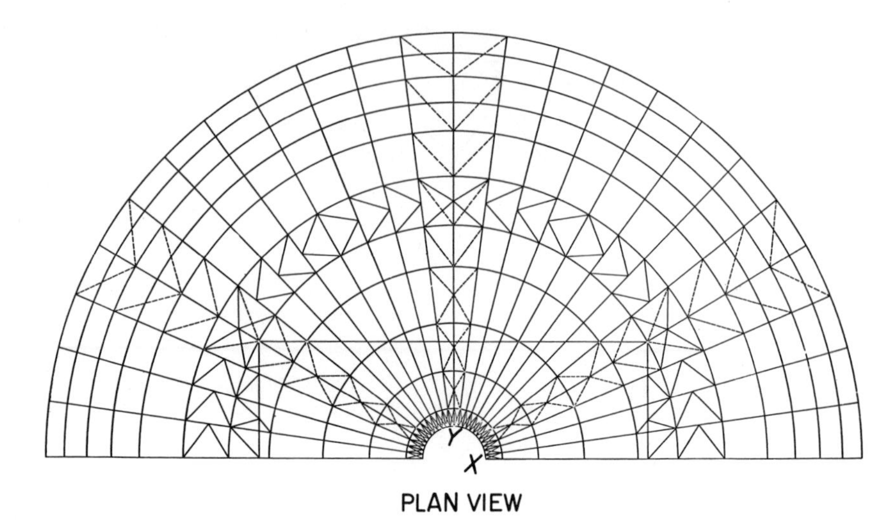

Figure 6. Half section of 210-ft reflector.

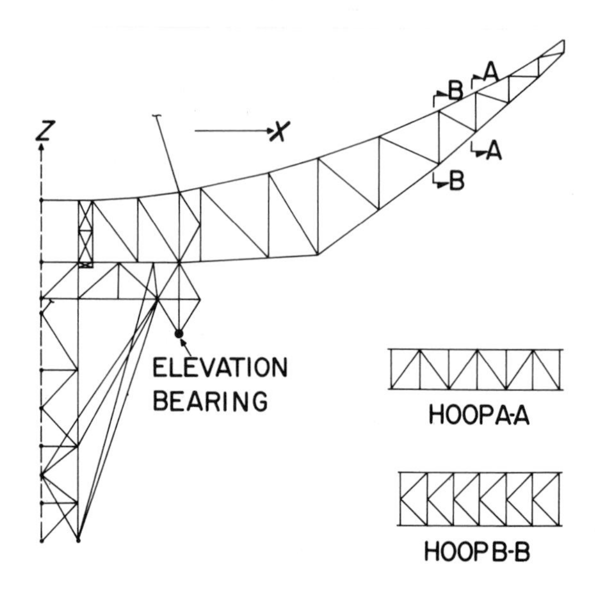

Figure 7. Front elevation of 210-ft antenna.

is associated with the nine points connecting the reflector and the backup structure. The horizontal portion of the curve below a semiband of 150 is associated with extra structure in some of the ribs at points of connection with the backup structure. From a semiband of 120 to about 80 the curve

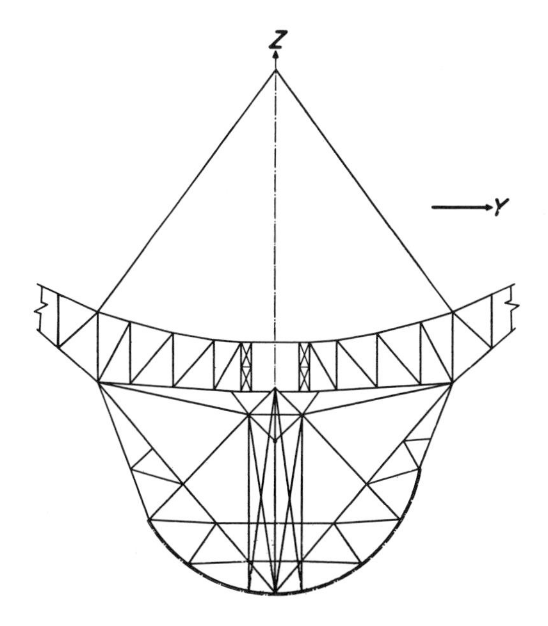

SECTIONAL VIEW THROUGH Y-Z PLANE

Figure 8. Side elevation of 210-ft antenna.

closely parallels the constant time curves in the spill region. The decomposition was actually performed at a semiband 110 with 44 active columns. Although there is insufficient core to perform the decomposition for semibands of less than 80, this is not a disadvantage for this particular problem as more favorable combinations do not exist in this range. However, if 8000 more core locations were available, the no-spill region would be increased to include the combination with a semiband of 120 and 38 active columns, and the decomposition could be performed about 2½ times faster.

Although the model used in this analysis consisted entirely of rods and did not include any structure below the elevation bearings, the computing time would not be substantially increased if the beams forming the alidade structure were included in the model. Also, the quadripod columns were considered as single rod elements in the model analyzed. Again, the addition of more elements in this part of the structure would not significantly increase the running time. The small increase in running time for these additions is explained by the fact that, although the order of the problem would increase, the semiband and number of active columns would not change, and hence only a linear increase in running time would result.

The computing times required for a complete static analysis of typical radio telescopes using the NASTRAN program on the IBM 7094 DCS computer can be summarized in the following manner. Models containing 500 grid points require about one hour, of which about half is spent on the triangular decomposition of the constrained stiffness matrix. Models containing 1000 grid points require about four hours, of which about three-quarters is spent on the triangular decomposition. Finally, models containing 2000 grid points require about forty hours, almost all of which is spent on the triangular decomposition Although the program can handle up to about 2500 grid points, it can be seen that the computing time becomes excessive before this limit is reached.

The outlook for the use of NASTRAN on third-generation computers for the analysis of large radio telescopes is very promising. Most problems will run about an order of magnitude faster due to increases in arithmetic speeds and improvements in auxiliary storage devices. Furthermore, the increase in core sizes will reduce the running times for large triangular decompositions by an additional order of magnitude. Hence the running times for static analysis of large radio telescopes can be expected to be reduced by between one and two orders of magnitude when using NASTRAN or third-generation computers.

Summary and Conclusions

1. NASTRAN is an effective program to use for the analysis of large radio telescopes.

2. Static analyses for radio telescopes containing several thousand grid points can be made in reasonable amounts of time on third-generation computers.

3. The use of double precision arithmetic will preserve eight decimal digits in the displacement vector for models containing several thousand grid points.

4. The capability to extract a declared number of eigenvalues from large matrices in an efficient manner will be useful in making vibration analyses of large radio telescopes.

5. The dynamics part of NASTRAN will be useful for determining the response of radio telescopes to dynamic loadings such as earthquakes and telescope drive systems.

Acknowledgments

In addition to the entire team associated with the development of NASTRAN, special thanks are due M. S. Katow of the Jet Propulsion Laboratory for furnishing the data and drawings for the sample problems, and to C. W. Hennrich of The MacNeal-Schwendler Corporation for preparation of the data and completion of the computer runs.

Robert D. Logcher
Associate Professor, Department of Civil Engineering, Massachusetts Institute of Technology, Cambridge, Massachusetts

ICES STRUDL—AN INTEGRATED APPROACH TO A STRUCTURAL COMPUTER SYSTEM

Introduction

Structural engineering is a field in which there are great differences in the characteristics of its many problems. The aerospace sector operates on complex problems with all the sophistication available to it. Payoffs for design refinement are high and, consequently, resources are relatively plentiful. In the building design field, however, payoffs in improved design are very modest. These designers operate under severe economic constraints with the typical structural design fee, each for a unique structure, less than one per cent of the construction costs. Designing for a relatively ill-defined criteria and using a crudely defined merit function, there exists little incentive beyond professionalism and competition for better design.

From their inception, computers have been widely used in structural engineering. Their use, however, has been limited primarily to analysis operations of various kinds. Attention to this area has been justified by the complexity of the problem and the infeasibility of a modern solution without a computer. The payoff in many problem areas is undeniable. But, in many sectors of building design, justification for sophisticated computer analyses rests primarily on the replacement it provides for unavailable manpower.

Although the computer has proved useful for structural analysis, the author believes that its full impact will arise only when the computer is integrated into the total design process, from conception to construction. With such integration structured so as not to constrain creative and varied design, the computer can create the incentive for better design by relieving the engineer of many noncreative functions while providing a vehicle for gaining an understanding of his problem. This integration requires the consideration of the computer not so much as a calculating device than as an information storage and retrieval system.

This concept is not foreign to developers of structural programs. They have always attempted to develop as broadly applicable programs as feasible within their technical and economic constraints. Any programmer developing a program that deals with more than a single well-defined process handling data only in core finds that the importance of the procedure in the program development diminishes with the breadth of his considerations. Data handling and bookkeeping functions gain in importance during program design and implementation. Programs such as STAIR[1] and FRAN[2] illustrate how data treatment even becomes a deciding factor in the development of the solution algorithm.

The program developers are rapidly increasing the breadth of programs into software systems. By system, we mean a computer program which has the following characteristics:

1. Contains one or more programming languages, which provide users with the facility to order procedures.
2. Allows for storage and retrieval of a wide variety of data which may be used selectively in procedures.
3. Contains a number of procedures for processing the information in the problem area.
4. Provides mechanisms and organization for continued expansion of the scope of the system.

Program developers have been hindered in the development of systems by programming techniques developed in the past and reinforced by the limitations of programming systems. The primary problem in computer usage was to get answers to specific problems. This process is shown in Figure 1, where the user supplies data to a specific program which, in turn, returns his results. The programs, then, are basically linear procedures, reading data, computing on the basis of an algorithm, and printing answers. The programs themselves are developed by individuals skilled in special programming languages. They are usually not the users of the program. In this way, the user often loses control over decisions embodied in the computer programs he used. And, as programs increase in complexity, the gap between programmers and users widens.

The approach we seek is random access and control of the engineering data and procedures by the engineer to solve his particular problems. Given random access and a mechanism for control, it no longer becomes necessary for program-

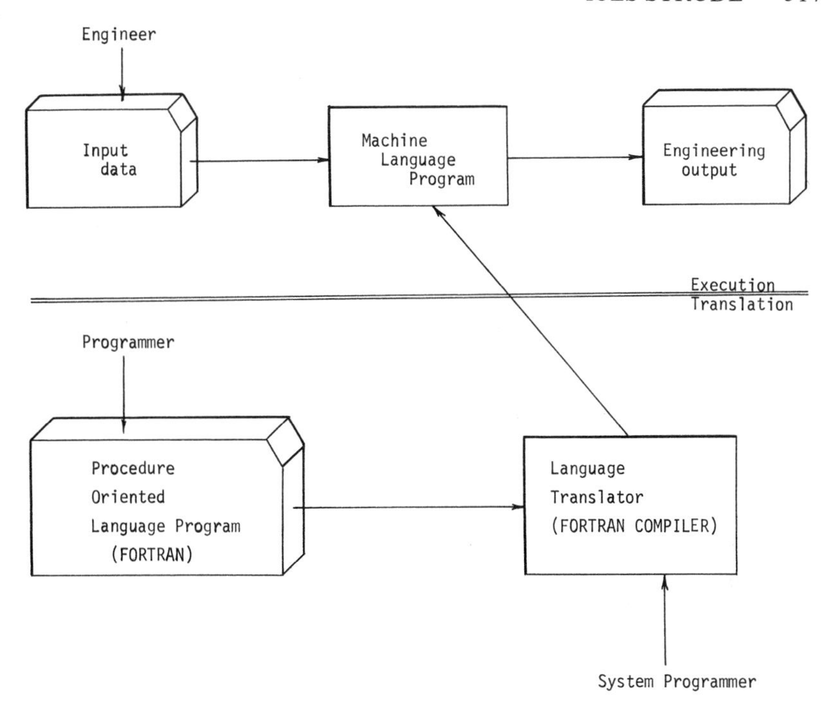

Figure 1. The computer process.

mers to develop highly complex programs. Rather, a system is developed which contains the basic operations in the problem area and the engineer, with his control mechanism, puts these together so as to solve his particular problem. This control mechanism is indeed programming, but in a very macro sense, sufficiently simple so that a program can be written for a problem and discarded.

This paper describes a computer system, the Integrated Civil Engineering System (ICES),[3] developed to support this outlook on engineering data processing. It consists of an operating and programming system with characteristics to support engineering, and of a number of applications subsystems in various engineering areas. The programming and operating characteristics of the system are first described. The Structural Design Language (STRUDL),[4] a subsystem of ICES, is described to illustrate the use of the system.

The ICES System

The ICES system[5,6] provides two integrated functions; an execution environment for problem solving, and modular programming facilities for the development of problem solving capabilities. The objectives of the system are to provide a fully dynamic facility, dynamic in the following senses:

1. Optimize the use of the computer resources as they are needed, at execution time.
2. Allow continuing development and easy modification of problem solving capabilities.
3. Structure operations and information at execution time on the basis of each particular problem and its data.
4. Allow random, but logically consistent access to procedures and information.

To describe the features of the ICES system, we must discuss five topics; the structure of the system, user communication or input languages, data structuring and handling, program structuring, and data management.

The ICES user communicates to the computer with a problem-oriented or command structured language. These commands state or imply the use of procedures developed for operation of the subsystem he is using and supply data to the procedures. Each command is read, translated, and executed sequentially. When the operations for one command are completed, the next command in the input stream is read. This execution process is controlled by the ICES Executive Program (ICEX). ICEX calls upon a translator, the Command Interpreter (COMINT), to translate each input command. COMINT is a universal translator in that the individual commands available to the user are not built within it. Rather, it uses dictionaries and tables (Command Data Blocks) which inform it how to translate each command. Commands, then, can be added at any time by a programmer by adding to the dictionaries and tables. A part of the language definition includes the names of programs that are to be executed for a command. When a command is translated, ICEX will call upon the programs which are required for that command. These programs operate and modify the problem data base, which is accessible to all programs, and may produce results for the user. Programs return to ICEX when they are finished. The lower left part of Figure 2 illustrates this user environment.

It is the subsystem programmers task to provide both the programs and command definitions for each of his subsystem commands. ICES provides facilities to assist him in these tasks. First, it provides the Command Definition Language (CDL) for the definition of how COMINT must translate a command. CDL is a command structured language and, as such, operates in the same manner as a user language. It uses ICEX, CDL command definitions, and CDL programs as shown at the top of Figure 2. Its primary output is disk files containing the subsystem dictionary and CDB's, which are used during subsystem execution. For program writing, the programmer is provided with ICETRAN (ICES FORTRAN). ICETRAN provides additional capabilities beyond FORTRAN which allow programs to be executed in a random sequence.

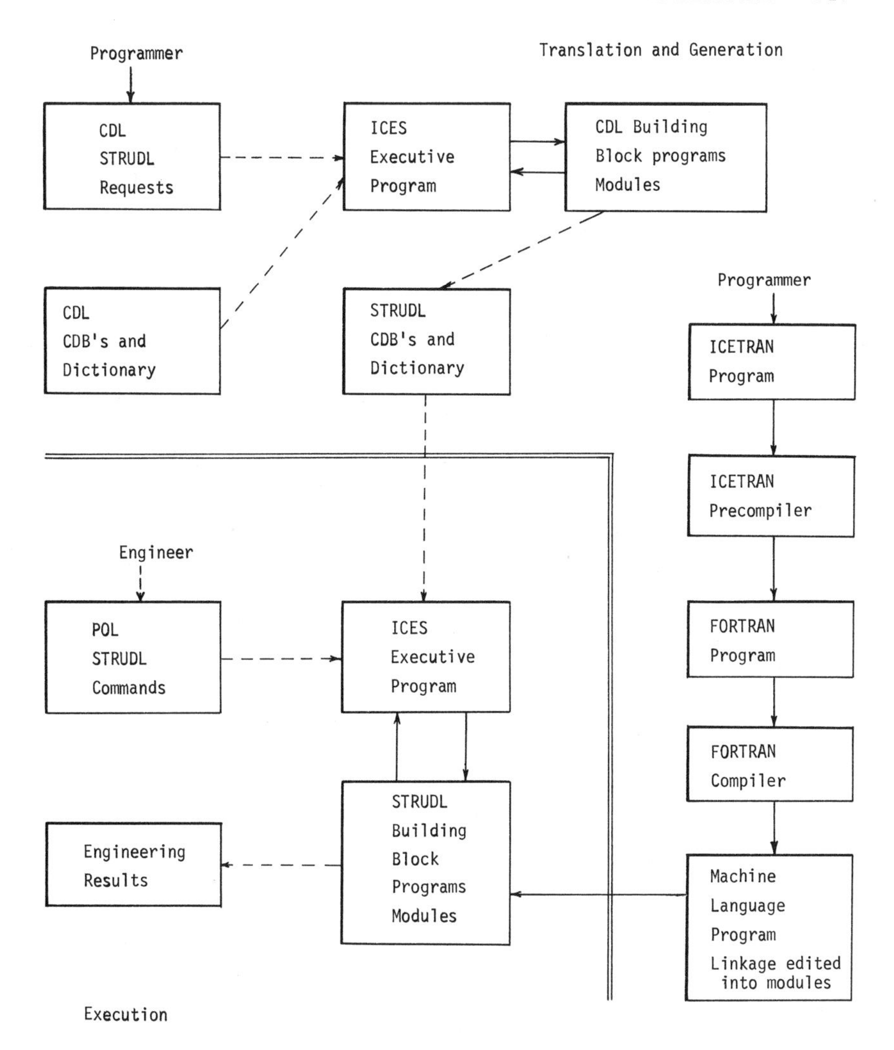

Figure 2. Subsystem generation and use.

The ICETRAN language is translated into legal FORTRAN and then compiled with a manufacture supplied compiler. A group of programs may then be linkage edited into a program load module which may be executed from ICEX. This aspect is shown on the right of Figure 2.

Figure 2, then, shows in very macro terms the integrated environment for subsystem generation and execution. For a subsystem, operations can take place at any time in any part of this figure. The only constraint is that for the execution of any command, its (and only its) definition must have been previously executed with CDL and the load modules required must have been formed and stored on the disk in one of the appropriate load module libraries. We will now discuss in more detail the features of these languages and how they relate to problem execution processing.

The command interpreter reads each command, determines its name to access its CDB. The CDB informs it what data are expected in the command and where to store them in a communications area, and what programs to call. Error and consistency checking are also performed. The basic premise in the command definition and translation is that commands are tree-structured. Figure 3 illustrates this structure for a STRUDLE command. Branching to

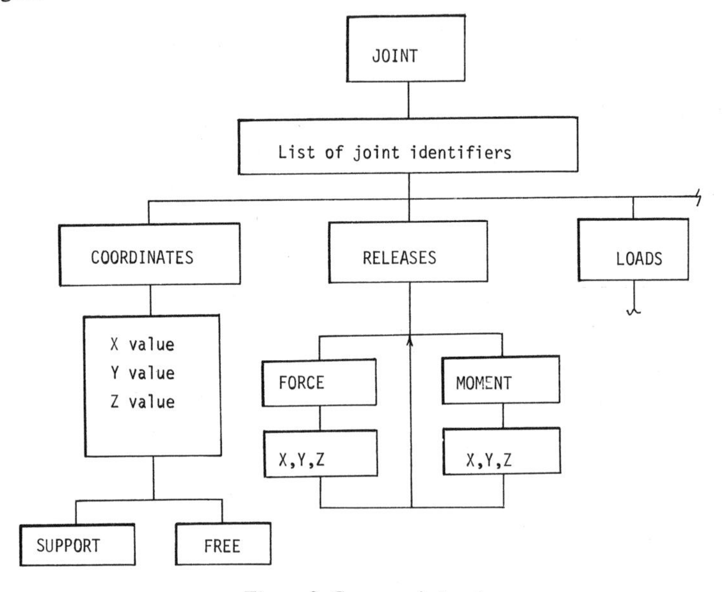

Figure 3. Command structure.

different parts of the command is specified in its CDL. COMINT, then, follows through the command translation, down the tree, by branching on the specified conditions. The most common branch is on a modifier word, such as COORDINATES or RELEASES.

CDL is a full programming language, although structured for a specific purpose. It includes such features as subroutines, looping on input, and output messages, and even allows recursive calling of its subroutines. Figure 4 provides an illustrative example of some of its features. T1, T2, T3, I1, and I10 are defined variables in the communications area. X, Y, and Z are data identifiers which *may* be included in the user input. If included, the data may be given in a random order. If any data item is not given, the standard value is stored (used in the example for checking by the program STCOOR). Data could also be specified as required. Alphameric information whose exis-

```
SYSTEM 'STRUDL' 'password'
ADD 'JOINT'
PRESET INTEGER 'I10' EQUAL 1
CALL 'GETID'
MODIFIER 'COOR'
     ID 'X' REAL 'T1' STANDARD -99999.
     ID 'Y' REAL 'T2' STANDARD -99999.
     ID 'Z' REAL 'T3' STANDARD -99999.
     EXISTENCE 'FREE' 'SUPPORT' SET 'I1' STANDARD 0
     EXECUTE 'STCOOR'
OR MODIFIER 'RELEASE'
     .
     .
     .
END MODIFIER
FILE
```

Figure 4. CDL example.

tence in the input stream implies data may also be treated. In the EXISTENCE statement, the variable is set equal to the position of the data word in the list if it is found in the input. Although not illustrated, CDL allows for the execution of any number of programs at different points within the translation of a command.

ICETRAN provides language facilities for dynamic data and program structuring and data management. ICEX provides the programs for handling the operations specified with the language. Its most interesting facility is the dynamic data structuring and dynamic memory allocation. Although a list processor is provided, we will discuss here only dynamic array data structures. In the system, all available core memory is managed by ICEX. It manages the positions of programs and dynamic data in a data pool. When memory becomes full, it is automatically reorganized, purging unneeded programs, moving data within core and to secondary storage. When data is accessed in a program, it is automatically retrieved from secondary storage if it had previously been moved there. Thus, a programmer need not be concerned with the location of data, but only with its logical structure. Herein lies a large measure of the ICES modularity. By not needing to know where data is stored, a program does not need to know what occurred before it or what will occur after it is finished. The ability to handle movable data rests on the use of a pointer scheme, where a single variable at a known location represents an array. When, and only when, this array contains data, this variable points to its location in the data pool or on the disk.

ICETRAN contains *executable* statements which may be used to manipulate arrays. These statements include DEFINE, DESTROY, RELEASE, and SWITCH. The data structures may be defined at execution time on the basis of the data for the particular problem, and may change during execution. By allowing a pointer to point to an array of pointers and by allowing each pointer to be uniquely defined, structuring flexibility is achieved. Figure 5

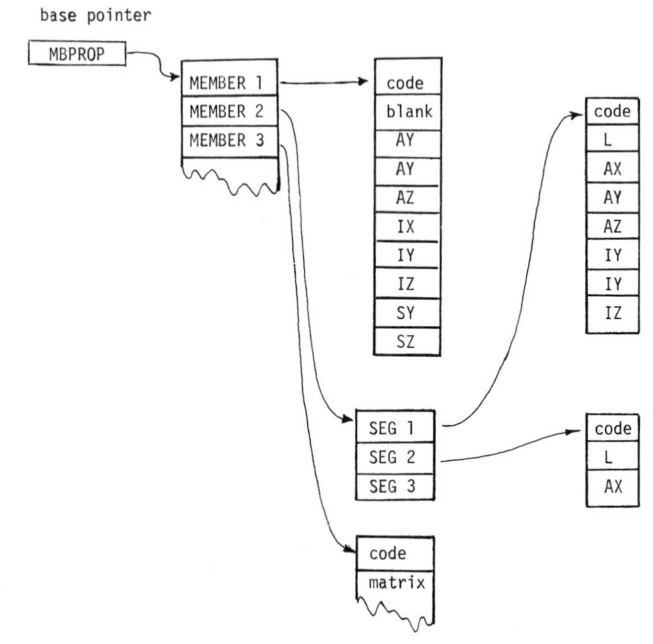

Figure 5. Example dynamic data structure.

illustrates a data structure for member properties which allows for both prismatic and segmented members, the mix of which is not known until execution time.

The ICETRAN language is basically FORTRAN with added statements. Dynamic array data referencing is similar to FORTRAN, with an array name and subscripts. Figure 6 provides a short programming example, performing the matrix operation A=A*B, where A is initially N1 by N2, B is N2 by N3, and the resultant A is N1 by N3. A significant programming feature is that such a program can operate on full arrays and on parts, regardless of the number of subscripts, as long as the parts passed contain the required structure. With reference to Figure 5, a program to calculate a segment flexibility matrix can be passed the data for member 1 (along with the length) or for segment 1 of member 2. This facility adds to program modularity.

The normal FORTRAN mechanism for program linkage is the CALL statement. This linkage requires all programs which are called to reside in one load module. In many cases programs, such as error processors, are needed only in pathological cases. Because of the large number of programs in a subsystem

```
      SUBROUTINE MATMPY(A,B,N1,N2,N3)
      DYNAMIC ARRAY A,B,TEMP
      DO  10  I=1, N1
      DESTROY TEMP
      DEFINE TEMP, N3,FULL
      DO  8  K=1,N3
      DO  8  J=1,N2
    8 TEMP(K)=TEMP(K)+A(I,J)*B(J,K)
      SWITCH(A(I),TEMP)
   10 RELEASE A(I)
      DESTROY TEMP
      RELEASE A
      RELEASE B
      RETURN
      END
```

Figure 6. ICETRAN example.

and its component procedures, because of the broad utility of some programs in many parts of a subsystem, and because of limited core memory, more flexibility is required. ICETRAN provides the following additional linkage mechanisms; LINK to a program in another module, which returns to the linking program, TRANSFER to a program in another module, LOAD a module, BRANCH to it, and DELETE a loaded mødule. In addition, ICEX maintains a pushdown linkage list which may be manipulated by an ICETRAN program. Whenever control returns to ICEX, it links to the program on top of the list. If the list is empty, ICEX links to COMINT. LINK and TRANSFER operations using the list may also be executed by a program.

The following are a few examples of how STRUDL utilizes dynamic program linkage. For finite element analysis, it is important to combine with general procedures the capability of treating a large and growing set of element formulations. Each formulation requires programs to perform the operations within a procedure which are dependent upon the formulation. Programmers are constantly adding new formulations, and users may use a number of different element types within the solution of a single problem. STRUDL has solved this dynamic problem by using very short control programs for element dependent phases of procedures and providing a command with which a programmer can define to STRUDL an element type and the program name of the program which may be used for that formulation with each phase control program. The control programs contain a loop on elements, retrieving for each the element type, searching the dictionary for the program name, and LINKing to the element program. A new element may be used as soon as the program names and programs in load module form have been provided.

Large multiphase procedures such as analyses require a number of load modules for their execution. The pushdown linkage list, or program stack, may then be used to control such procedures. For example, for both the stiffness and nonlinear analysis, ICEX links to a compilation phase. This program performs checking and setup operations and, depending upon procedures required for the particular analysis and on the data which needs to be treated, the stack is set up. For the nonlinear analysis, an iteration control program is placed on the bottom of the stack. This will be the last program executed and it will make decisions on what programs to reload onto the stack (depending on whether the process is iterating at a load level or incrementing the load), always putting itself on the bottom. If a fatal error is detected during any of this processing, the stack is emptied and control returned to ICEX.

ICES also provides a truly random access data management (file storage) mechanism. Although it can be used for sequential storage, its flexibility allows a programmer to create files with comparable complexity to any dynamic data structure previously discussed. The structure of a file and sizes of individual records within it may be determined and modified at execution time. Figure 7 illustrates the file structure used in STRUDL for the storage of tables of design section properties. The first record in the file contains "file information," including file record identifiers to six basic records which exist in pairs. The SECNAM-SECNUM pair store the name of each row and a record identifier for each row of property data. The PRPNAM-PRPNUM pair store the column name and record identifiers to the dimensional units of each column. Orders, used for sequencing design trials, are also shown. Note that with such a structure it is easy to expand and delete any of the types of information stored in the file. A new row may be added to the file by extending SECNAM, adding the name to the bottom, storing the row of properties in a new record, and adding that record identifier to the bottom of SECNUM.

The system described here is presently a reality. It is being used to develop and operate at least 20 different applications subsystems, both at MIT and elsewhere. In the first few months of its availability, the programs have been obtained from the IBM Program Library by over 300 installations. All indications point to enthusiastic acceptance by programmers and users.

The Structural Design Language

The use of a computer throughout the design process requires convenient access and a powerful applications software system or an interrelated group of systems. STRUDL is intended to be a research tool used in a variety of ways in the development of such systems. It is an easily modifiable and expandable system which can serve the following roles:

1. A test vehicle for new analysis and design procedures, data structures and languages.

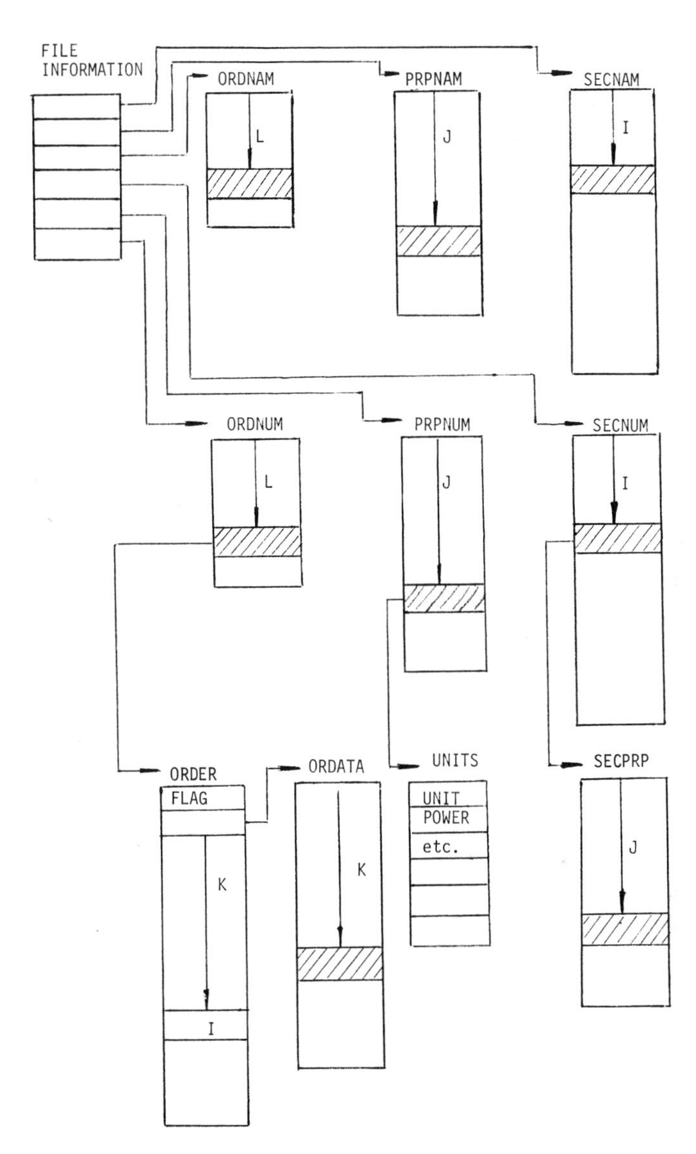

Figure 7. An example of ICES file structure.

2. A production tool that can provide real feedback from professional application to research.

3. A mechanism for teaching both students and practicing professionals more about modern design techniques.

4. A tool for studying the design process.

The first version of ICES STRUDL[7,8] in no way attempts to represent a complete structural design tool. In fact, it does not contain any actual design procedures. It is, however, a system structured for expansion and continuing development by many different groups. In addition to describing the present capabilities, this section will also describe the scope of expansion efforts currently under way.

Basically, STRUDL is designed as a structural information system. With it, a user may store his entire problem in the computer over a long period of time and make investigations on part or all of his problem. As a consequence, he may change, add, or delete any of the information in the machine, or make inquiries about any of the information stored in it. This approach allows the user to manipulate his problem in the machine, asking questions about his problem rather than, as with the special program approach, having to extract information from his manually kept data base to insert into the computer for each separate question.

Figure 8 provides an overview of ICES STRUDL. It shows the designer with a large bag of tricks at his finger tips, ready to be used in any sequence. These tricks include data input, analysis and design procedures, and output. Communication with other subsystems is accomplished with disk files and file access routines. The first version of STRUDL does not include the nonlinear or dynamic analyses or member selection and checking. The second version will.

As an information system, STRUDL has no way of knowing what operations are to be performed on its data. This requires that all information be maintained in its most general form, that is, for example, as a three-dimensional problem. Procedures operating on a subset of the general case, such as a plane frame, then extract their required information from the general form. This allows the user to change even the structural type. Conversely, the user need not specify more general information then needed for the particular conditions and operations he wishes to use for his solution. For a plane frame analysis, he need not provide a torsional constant for his members.

Consistent with this approach, each of the procedures have been developed in the most general sense feasible. For example, the stiffness analysis has been developed to operate on both planar and space structures, for trusses, frames, finite elements, or any combination of these. It was also developed with no topological or banding restrictions. Size limitations apply only for particular machine sizes and available computer time. The size limit inherent in the system itself is 16,383 members. Although efficiency improvements can and will be made using substructuring techniques and larger partitions for the stiffness matrix, the present version is already highly efficient. For a problem in which all information fits into core memory, it runs fast as a special-purpose analysis program. Providing the great flexibility does suggest overhead, and it

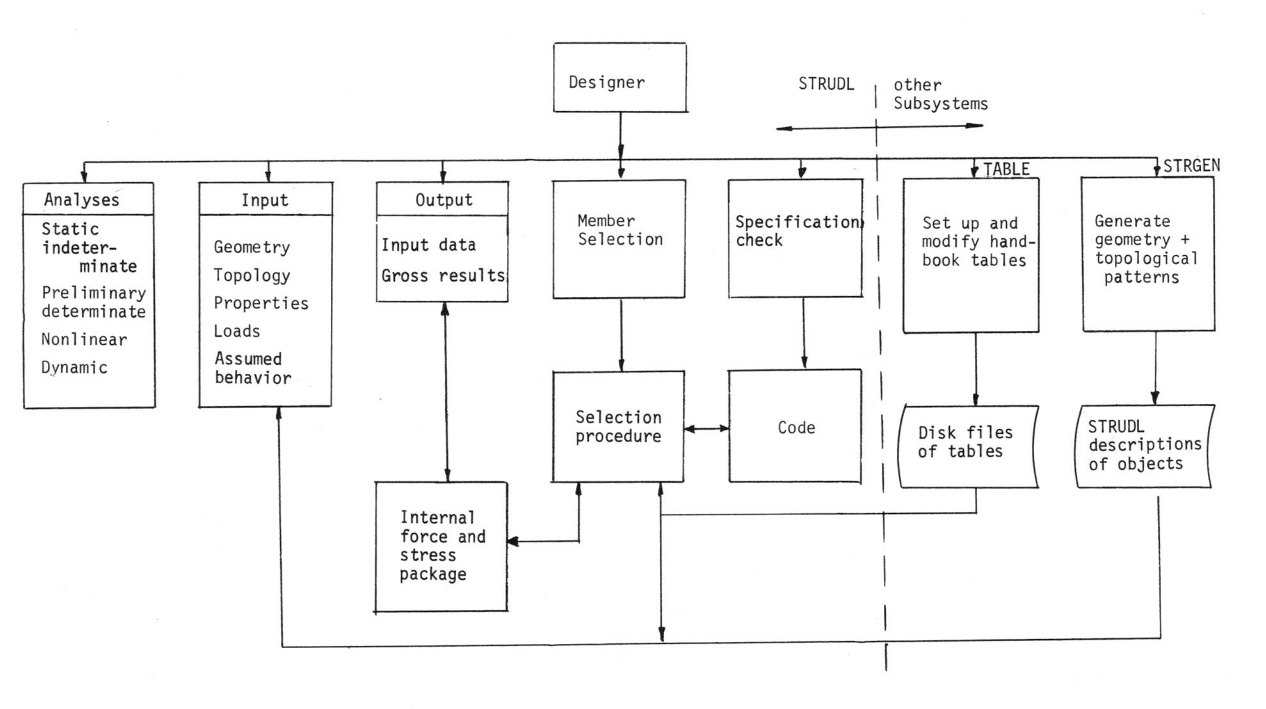

Figure 8. Overview of STRUDL.

is seen in some operations.

A companion subsystem, ICES TABLE, is a file management system with which a user can set up, maintain, and modify tabular files of data. Originally developed as a part of STRUDL for the storage of handbook tables, its capabilities were of sufficient interest to others to warrant its separation from STRUDL. Once a table has been setup, its data may be accessed by other subsystems through a general set of data retrieval functions provided to subsystem programmers. All references to tables are symbolic, using named rows and columns. Use of the table STEELWF may be noted in the example which follows.

The example shown in Figure 9 attempts to demonstrate concisely some of

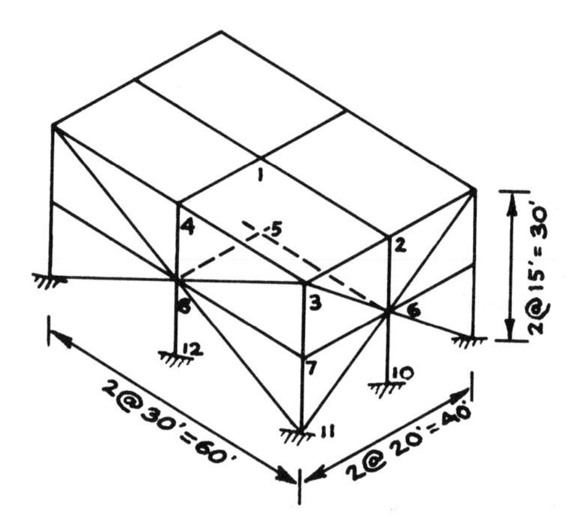

Figure 9. STRUDL example program.

the capabilities of ICES STRUDL-I. The building shown is to be designed for gravity loads, wind on the long face, and wind on the short face. Because of the behavior of the cross-braced exterior walls, the floor beams are not expected to contribute to the behavior under wind loads. For efficiency use may be made of symmetry and antisymmetry and the analysis performed on only one quarter of the building. The reader should note the data modification for different phases of the problem, the use of inactive status to subset the data for operations, the combination of results, even from different analyses, and selective result retrieval. The computer run is shown in Figure 10.

As early as a year before the release of STRUDL our group started adding capabilities. These were primarily a result of thesis work. As such, these new capabilities were valuable, but not sufficiently refined for general use. Our limited resources made us concentrate on the refinement of our committed capabilities instead of offering a larger but less effective system.

Almost at once we started working on a geometrically nonlinear frame analysis, an integration of finite elements and frame components in input, analysis, and output, and member selection from tables of section properties. The nonlinear analysis,[9] sponsored by Project CAMROC, was developed for the verification of an appropriate analytic procedure for the radome buckling problem with a large-scale model. The basic program development was completed by one research assistant in three months. Refinement and tuneup was done by Simpson, Gumpertz & Heger, Inc., using parts of the ICES system, outside STRUDL. The procedure was considered to be successful. An initial finite element system, using the approach described under dynamic program structures, and including four element types with the stiffness analysis, was developed at the same time. The refinement presently in progress will integrate the finite element capabilities with all analyses, static, nonlinear, and dynamic. Nonlinear plate and curved shell elements are being added.

The member selection capability allows the user to select from all available design procedures and codes those that he wishes. It would be a relatively simple task for any programmer to provide a designer with a new or revised procedure at any time. Again, this is accomplished through dynamic program linkage, with the user naming the program he wishes to apply. A large number of other additions are in development, including graphical output, reinforced concrete member checking, design, and detailing, a general dynamic analysis, including static condensation, and optimization studies.

Conclusions

We are living professionally in a very exciting and dynamic time. New dimensions of engineering tools are becoming available to us. The nature of the structural engineering profession will depend on how broadly we can make use of these tools.

This paper has attempted to describe one approach towards the use of computers which the author feels represents a broadening outlook on the use of this tool – as an information processing system, not a calculator. The system described is also sufficiently flexible so that larger numbers of professionals can contribute to large-scale integrated systems. Those developing these tools have the opportunity to help shape the outlook and direction of the profession, for it will surely be mirrored in these tools. This is a task in which we should all participate, and participate actively.

Acknowledgments

Support for the ICES Project has been provided to the Civil Engineering Systems Laboratory by a number of research sponsors. Special recognition is due the following: the IBM Corporation, the Massachusetts Department of Public Works in cooperation with the U.S. Bureau of Public Roads, the Mas-

Figure 10. Computer run for example problem.

```
*********************************************
*                                           *
*              ICES STRUDL-I                *
*     THE STRUCTURAL DESIGN LANGUAGE        *
*                                           *
*  CIVIL ENGINEERING SYSTEMS LABORATORY     *
* MASSACHUSETTS INSTITUTE OF TECHNOLOGY     *
*        CAMBRIDGE, MASSACHUSETTS           *
*           JUNE, 1968  MOD 2               *
*                                           *
*********************************************

$ ONE QUARTER OF STRUCTURE

TYPE SPACE FRAME

UNITS KIP FEET

JOINT COORDINATES

$ Y IS UP FROM CENTER OF BUILDING

1 Y 30. SUPPORT

2 30. 30. S

3 30. 30. 20.

4 Y 30. Z 20. S

5 Y 15. S

6 30. 15. S

7 30. 15. 20.

8 Y 15. Z 20. S

10 X 30. S

11 X 30. Z 20. S

12 Z 20. S

MEMBER INCIDENCES

$ BEAMS

1 1 2
```

```
2 2 3
3 3 4
4 1 4
5 5 6
6 6 7
7 7 8
8 5 8
$ COLUMNS
9 10 6
10 6 2
11 11 7
12 7 3
13 12 8
14 8 4
$ CROSS BRACES
15 3 6
16 11 6
17 3 8
18 11 8
UNITS INCHES
CONSTANTS E 30000. ALL
MEMBER 1 TO 18 PROPERTIES TABLE 'STEELWF' '16WF96'
UNITS FEET
LOADING 'GRAVITY' 'FOR EXAMPLE, UNIFORM ON ALL BEAMS'
MEMBER 1 TO 8 LOAD FORCE Y UNIFORM W -1.5
LOADING 'WINDLF' 'WIND ON LONG FACE'
JOINT 3, 4 LOAD FORCE Z -4.5
JOINT 7, 8 LOAD FORCE Z -9.0
```

Figure 10. (continued)

```
LOADING 'WINDSF' 'WIND ON SHORT FACE'

JOINT LOADS

2,3 FORCE X -3.0

6,7 FORCE X -6.0

$ BOUNDARY CONDITIONS FOR SYMMETRY

JOINT RELEASES

1, 5 FORCE Y

2, 6 FORCE X Y MOMENT Z

4, 8 FORCE Y Z MOMENT X

LOADING LIST 'GRAVITY'

STIFFNESS ANALYSIS
TIME FOR MODULE STCCMP FOR   18 MEMBERS=    0.55 SECONDS
TIME FOR MODULE STSTDP FOR   18 MEMBERS=    0.48 SECONDS
TIME FOR MODULE STMDDP FOR    8 LOADS=      4.03 SECONDS
TIME FOR MODULE STASDP FOR   18 MEMBERS=    1.21 SECONDS
TIME FOR MODULE STJPDP FOR   11 JOINTS=    1.52 SECONDS
TIME FOR MODULE STSLDP FOR NSOL=    8 IS    1.28 SECONDS
TIME FOR MODULE STD1BS FOR   11 JOINTS EQUALS    0.02 SECONDS
TIME FOR MODULE STD3BS FOR   11 JOINTS EQUALS    3.38 SECONDS
TIME FOR MODULE STBKSB FOR   18 MEMBERS=   10.89 SECONDS

CHANGES

JOINT RELEASES  $ BOUNDARY CONDITION FOR SYM/ANTIS ON LONG/SHORT SIDE

2, 6 FORCE Z MOMENT X Y

LOADING LIST 'WINDLF'

INACTIVE JOINTS 1,5        $ IGNORE INFILL

STIFFNESS ANALYSIS
**** STRUDL WARNING 2.13 - MEMBER 1        INCIDENT ON INACTIVE JOINT - MEMBER IGNORED IN THIS ANALYSIS
**** STRUDL WARNING 2.13 - MEMBER 4        INCIDENT ON INACTIVE JOINT - MEMBER IGNORED IN THIS ANALYSIS
**** STRUDL WARNING 2.13 - MEMBER 5        INCIDENT ON INACTIVE JOINT - MEMBER IGNORED IN THIS ANALYSIS
**** STRUDL WARNING 2.13 - MEMBER 8        INCIDENT ON INACTIVE JOINT - MEMBER IGNORED IN THIS ANALYSIS
TIME FOR MODULE STCCMP FOR   14 MEMBERS=    1.22 SECONDS
TIME FOR MODULE STSTDP FOR   14 MEMBERS=    0.39 SECONDS
TIME FOR MODULE STASDP FOR   14 MEMBERS=    0.92 SECONDS
TIME FOR MODULE STJPDP FOR    9 JOINTS=    0.53 SECONDS
TIME FOR MODULE STSLDP FOR NSOL=    6 IS    0.87 SECONDS
TIME FOR MODULE STD1BS FOR    9 JOINTS EQUALS    0.0  SECONDS
TIME FOR MODULE STD3BS FOR    9 JOINTS EQUALS    0.18 SECONDS
TIME FOR MODULE STBKSB FOR   14 MEMBERS=    4.68 SECONDS
```

```
CHANGES  $ BOUNDARY CONDITIONS F SYM/ANTIS ON SHORT/LONG SIDE

JOINT RELEASES

2, 6 FORCE X Y MOMENT Z

4, 8 FORCE  X MOMENT Y Z

LOADING LIST 'WINDSF'

STIFFNESS ANALYSIS
**** STRUDL WARNING 2.13 - MEMBER 1        INCIDENT ON INACTIVE JOINT - MEMBER IGNORED IN THIS ANALYSIS
**** STRUDL WARNING 2.13 - MEMBER 4        INCIDENT ON INACTIVE JOINT - MEMBER IGNORED IN THIS ANALYSIS
**** STRUDL WARNING 2.13 - MEMBER 5        INCIDENT ON INACTIVE JOINT - MEMBER IGNORED IN THIS ANALYSIS
**** STRUDL WARNING 2.13 - MEMBER 8        INCIDENT ON INACTIVE JOINT - MEMBER IGNORED IN THIS ANALYSIS
TIME FOR MODULE STCOMP FOR   14 MEMBERS=    1.20 SECONDS
TIME FOR MODULE STSTDP FOR   14 MEMBERS=    0.44 SECONDS
TIME FOR MODULE STASDP FOR   14 MEMBERS=    0.93 SECONDS
TIME FOR MODULE STJPDP FOR    9 JOINTS=    1.08 SECONDS
TIME FOR MODULE STSLDP FOR NSOL=    6 IS    0.96 SECONDS
TIME FOR MODULE STD1BS FOR    9 JOINTS EQUALS    0.0  SECONDS
TIME FOR MODULE STD3BS FOR    9 JOINTS EQUALS    0.11 SECONDS
TIME FOR MODULE STBKSB FOR   14 MEMBERS=    7.17 SECONDS

LOADING COMBINATION 1

COMBINE 1 'GRAVITY' 0.75 'WINDSF' 0.75 'WINDLF' 0.75
**** STRUDL WARNING 5.09 - RESULTS MISSING FOR MEMBER 1      , LOADING WINDSF   - MEMBER IGNORED
**** STRUDL WARNING 5.09 - RESULTS MISSING FOR MEMBER 4      , LOADING WINDSF   - MEMBER IGNORED
**** STRUDL WARNING 5.09 - RESULTS MISSING FOR MEMBER 5      , LOADING WINDSF   - MEMBER IGNORED
**** STRUDL WARNING 5.09 - RESULTS MISSING FOR MEMBER 8      , LOADING WINDSF   - MEMBER IGNORED
**** STRUDL WARNING 5.09 - RESULTS MISSING FOR MEMBER 1      , LOADING WINDLF   - MEMBER IGNORED
**** STRUDL WARNING 5.09 - RESULTS MISSING FOR MEMBER 4      , LOADING WINDLF   - MEMBER IGNORED
**** STRUDL WARNING 5.09 - RESULTS MISSING FOR MEMBER 5      , LOADING WINDLF   - MEMBER IGNORED
**** STRUDL WARNING 5.09 - RESULTS MISSING FOR MEMBER 8      , LOADING WINDLF   - MEMBER IGNORED

INACTIVE MEMBERS 1,4,5,8

LOADING COMBINATION 2

COMBINE 2 'GRAVITY' 0.75 'WINDSF' -0.75 'WINDLF' -0.75

LOADING COMBINATION 3

COMBINE 3 1 1.0 'WINDSF' -1.5

LOADING COMBINATION 4

COMBINE 4 1 1.0 'WINDLF' -1.5

LOADING LIST 'GRAVITY'

LIST FORCES ALL ACTIVE AND INACTIVE MEMBERS
```

Figure 10. (continued)

```
*****************************
*RESULTS OF LATEST ANALYSES*
*****************************

PROBLEM - EXAMPLE   TITLE -

ACTIVE UNITS  FEET KIP  RAD  DEGF SEC

ACTIVE STRUCTURE TYPE  SPACE   FRAME

ACTIVE COORDINATE AXES  X Y Z

LOADING - GRAVITY      FOR EXAMPLE, UNIFORM ON ALL BEAMS


MEMBER  FORCES
```

MEMBER	JOINT	FORCE			MOMENT		
		AXIAL	SHEAR Y	SHEAR Z	TORSIONAL	BENDING Y	BENDING Z
1	1	31.8206787	9.0794373	0.0	0.0	0.0	-120.5664978
1	2	-31.8206787	35.9205475	0.0	0.0	0.0	-282.0500488
2	2	0.2499619	16.4036865	0.1384247	0.0761486	-1.1081905	58.2319794
2	3	-0.2499619	13.5962925	-0.1384247	-0.0761486	-1.6603022	-30.1580048
3	3	7.5850544	21.4299774	0.0285590	-0.1798573	-0.6124219	90.1703186
3	4	-7.5850544	23.5700073	-0.0285590	0.1798573	-0.2443495	-122.2706604
4	1	18.4486847	-9.0794373	0.0	0.0	0.0	-332.0019531
4	4	-18.4486847	39.0794220	0.0	0.0	0.0	-149.5867462
5	5	-22.7742004	9.6214056	0.0	0.0	0.0	-91.2968292
5	6	22.7742004	35.3785706	0.0	0.0	0.0	-295.0605469
6	6	-0.4809099	17.4613800	-0.0263037	0.0181134	0.3154095	65.8229675
6	7	0.4809099	12.5386086	0.0263037	-0.0181134	0.2106639	-16.5952454
7	7	-4.4625893	21.7034454	0.0119150	-0.1025535	-0.2136403	95.9944916
7	8	4.4625893	23.2965240	-0.0119150	0.1025535	-0.1438096	-119.8905792
8	5	-11.3626518	-9.6214056	0.0	0.0	0.0	-282.2106934
8	8	11.3626518	39.6213989	0.0	0.0	0.0	-210.2173309
9	10	88.4280548	9.3044996	0.0	0.0	0.0	48.5476990
9	6	-88.4280548	-9.3044996	0.0	0.0	0.0	91.0197449
10	6	52.3242493	31.9591064	0.0	0.0	0.0	197.4128265
10	2	-52.3242493	-31.9591064	0.0	0.0	0.0	281.9738770
11	11	64.9849701	3.8046560	-0.7186272	0.0009100	3.5974970	19.4201202
11	7	-64.9849701	-3.8046560	0.7186272	-0.0009100	7.1819057	37.6497192
12	7	30.7429047	8.2409410	-1.1876221	0.0038864	9.5159006	58.3629608
12	3	-30.7429047	-8.2409410	1.1876221	-0.0038864	8.2984266	65.2511902
13	12	116.6839905	0.0	-7.0541162	0.0	35.3735962	0.0
13	8	-116.6839905	0.0	7.0541162	0.0	70.4381561	0.0
14	8	62.6494293	0.0	-18.4201202	0.0	126.8949890	0.0
14	4	-62.6494293	0.0	18.4201202	0.0	149.4069061	0.0
15	3	2.7879076	-1.1362190	0.4531975	0.0100254	-5.3640032	-18.8467102
15	6	-2.7879076	1.1362190	-0.4531975	-0.0100254	-5.9659319	-9.5587597
16	11	31.8340912	0.2717590	-0.3073012	0.0203098	2.6723213	3.3969898
16	6	-31.8340912	-0.2717590	0.3073012	-0.0203098	5.0102139	3.3969879
17	3	1.8768072	-0.9641008	-0.6394934	-0.0842565	7.3076296	-21.7849579
17	8	-1.8768072	0.9641008	0.6394934	0.0842565	14.1416245	-10.5519581
18	11	23.3367920	0.1660168	0.6309404	-0.0913845	-7.0747290	2.7841864
18	8	-23.3367920	-0.1660168	-0.6309404	0.0913845	-14.0876503	2.7841835

```
LOADING LIST 1 TO 4 'GRAVITY'

OUTPUT DECIMAL 3

LIST MAXIMUM STRESS MEMBER 2,3,6,7,9 TO 18 SECTIONS FR 3 0. 0.5 1.0



    ****************************
    *RESULTS OF LATEST ANALYSES*
    ****************************


     PROBLEM - EXAMPLE    TITLE -

     ACTIVE UNITS  FEET KIP  RAD  DEGF SEC

     ACTIVE STRUCTURE TYPE  SPACE   FRAME

     ACTIVE COORDINATE AXES  X Y Z


     INTERNAL MEMBER RESULTS
     -----------------------


      MEMBER MAXIMUM STRESS
            /---------------------------------------- ----  STRESS  --------------------------------------/
MEMBER                 MAX NORMAL    AT SECTION              MIN NORMAL    AT SECTION
2                       6167.020      0.0     FR             -6177.809      0.0     FR
3                      12717.051      0.0     FR            -12752.738      0.0     FR
6                       5556.574      0.0     FR             -5474.484      0.0     FR
7                      11785.625      0.0     FR            -11784.301      0.0     FR
9                        718.495      0.0     FR             -1399.595      0.0     FR
10                      2666.480      1.000   FR             -3200.477      1.000   FR
11                       405.772      1.000   FR             -1068.978      1.000   FR
12                       921.392      1.000   FR             -1235.140      1.000   FR
13                      3617.434      0.0     FR             -4514.172      0.0     FR
14                      6871.820      1.000   FR             -7511.191      1.000   FR
15                      1339.785      0.0     FR             -1450.818      0.0     FR
16                       484.096      1.000   FR              -531.427      1.000   FR
17                      2090.974      1.000   FR             -2153.574      1.000   FR
18                       836.200      0.0     FR              -922.475      0.0     FR
```

sachusetts Bay Transportation Authority, the National Science Foundation, the McDonnell Automation Company, the Wisconsin State Highway Commission, the Ford Foundation, the Union Pacific Railroad Foundation, M.I.T. Lincoln Laboratory, and Project CAMROC.

The author also wishes to recognize the large number (near 100) of people who have contributed to the development of this system and its applications subsystems. The success of this effort is due in larger measure to the sincere professionalism of all of these contributors.

Additional information on the availability of the publications and computer systems resulting from the ICES Project may be obtained from Professor Daniel Roos, Director, Civil Engineering Systems Laboratory, Room 1-163, Massachusetts Institute of Technology, Cambridge, Massachusetts 02139.

References

1. "STAIR (Structural Analysis Interpretive Routine) Instruction Manual," Lincoln Manual No. 48, M.I.T. Lincoln Laboratory, Lexington, Massachusetts (March 1962).

2. K. Eisemann, L. Woo, and S. Namyet, "Space Frame Analysis by Matrices and Computers," *J. Structural Div. ASCE,* 88 (No. ST6), Proc. Paper 3365 (December 1962), pp. 245-277.

3. D. Roos, *ICES System Design* (The M.I.T. Press, Cambridge, Massachusetts, 1966).

4. R. D. Logcher, *et al., ICES STRUDL-I Engineering User's Manual,* Report R67-56, Department of Civil Engineering, M.I.T. (September 1967).

5. D. Roos, *ICES System: General Description* Department of Civil Engineering, M.I.T. (September 1967).

6. J. Jordon, *ICES: Programmers Reference Manual,* Department of Civil Engineering, M.I.T. (September 1967).

7. J. M. Biggs and R. D. Logcher, *"ICES STRUDL-I, General Description,"* Department of Civil Engineering, M.I.T. (September 1967).

8. R. D. Logcher, B. B. Flachsbart, E. J. Hall, C. M. Power, and R. A. Wells, Jr., *"ICES STRUDL-I, Engineering User's Manual,"* Department Of Civil Engineering, M.I.T. (September 1967).

9. J. Connor, Jr., "Analysis of Geometrically Nonlinear Framed Structures," Chapter 23 of this volume.

INDEX